DIGITAL ENGINEERING

DIGITAL ENGINEERING

George K. Kostopoulos, Ph.D.

Honeywell, Inc.,

A WILEY-INTERSCIENCE PUBLICATION

JOHN WILEY & SONS New York · London · Sydney · Toronto

Library of Congress Cataloging in Publication Data

Kostopoulos, George K. 1939–
Digital engineering.

"A Wiley-Interscience publication."
Bibliography: p.
1. Digital electronics. I. Title.

TK7868.D5K69 621.3815 74–13427
ISBN 0–471–50460–2

Printed in the United States of America

10 9 8 7 6 5 4 3 2 1

This book is dedicated to the educators
who believe that successful teaching
is effective transfer of knowledge.

PREFACE

Digital circuits are extensively used in the design of scientific as well as industrial and commercial electronic equipment, and they will soon dominate practically all aspects of electronic design. The two major reasons for this extensive use and promising future are, first: the numerous advantages offered by the digital circuits and techniques, and second, the availability of these circuits in high-density packages.

Many books are available covering the subject of digital engineering as applied to digital computer design. Digital engineering, however, is rapidly expanding into almost all areas of equipment and system design, ranging from petroleum refineries and medical instruments to satellites. This demand for digital implementations has created a need for a book that treats the subject with a broader viewpoint.

This book has been written to fill that need. Its objective is to assist, as well as guide, the digital engineer and to allow him to concentrate on the development of digital concepts and systems rather than on the design of logic circuits.

New digital networks are available to the design engineers on, practically, a daily basis. To fully appreciate and benefit the most from this everincreasing advancement in digital techniques and digital circuit fabrication, the design engineer needs a solid background in digital engineering that will enable him to technologically progress along with advancements in digital engineering. It is the purpose of this book to provide the reader with this background.

To achieve these objectives the book is heavily illustrated, having well over 300 figures, is as descriptive as possible, and contains more than 150 carefully selected problems ranging from simple routine exercises to practical engineering applications. Numerous examples are also included illustrating the use of the presented concepts, methods, and techniques.

This book has also been written to be used in colleges and universities that offer digital courses, and in order to facilitate the use of individual parts, the chapters have been kept as independent as possible.

The presented material has been grouped into three parts. Part One serves as an Introduction to Digital Engineering, Part Two covers the Digital Implementation of Binary Mathematics, and Part Three presents General Topics in Digital Engineering. The book, although broad and deep, is descriptive enough to require a minimum of engineering and mathematics prerequisites, thus being of value to the beginner as well as to the knowledgeable digital engineer. By integrating the acquired digital knowledge with the rest of his engineering background, the reader will develop into a digital systems architect able to easily evaluate design trade-offs, based on his acquired familiarity with the digital implementation of functions and concepts.

GEORGE K. KOSTOPOULOS

West Corina, California

CONTENTS

PART TWO: DIGITAL IMPLEMENTATION OF BINARY MATHEMATICS

DIGITAL ENGINEERING

Introduction

The desire for the utilization of the never-ending advances in science has created the need for electronic equipment of ever-increasing capability and complexity. This equipment should utilize the advancements of science and convert them from mere scientific discoveries to useful applications.

This need for advanced equipment, in conjunction with advances in digital engineering, has resulted in extensive use of digital methods in almost all electronic system designs. The advantages of digital methods are so overwhelming that they have created a strong tendency toward total digitalization, especially in equipment where functions were previously implemented by electromechanical or analog devices. The major advantages that make digital networks preferred are:

1. *Stability.* Digital networks are considerably less subject to noise than analog networks.
2. *High reliability.* Digital networks can tolerate more component value change (drift.)
3. *Maintainability.* Digital networks need no adjustments or compensating circuits.
4. *Digital readout.* The readability of digital readouts is unquestionable, while that of an analog output is often debatable.

Digital networks are widely used in both commercial and industrial equipment and, as a result, a new field, that of digital engineering, has been created. When a sequence of operations needs to be controlled, digital techniques provide the answer; the function may be plastics processing, blood-circulation-monitoring, missile-launching, or any other simple or complex operation.

This extensive application of digital networks fully justifies the continuous efforts for advances in their fabrication. Inexpensive fabrication process of digital integrated circuits has been the key factor to the wide use of digital networks. In this process, all components of a digital circuit or network are simultaneously fabricated on an extremely small piece of semiconductor material. Thus, digital networks that previously required three or four printed circuit boards of discrete components are now housed in a transistor "can," or other package of similar size.

Integrated circuits have been the answer to many equipment design problems, such as space, weight, power consumption, reliability, maintainability, and logistics. The small size of integrated circuits has made possible the construction of valuable complex digital equipment, the implementation of which would have been otherwise impractical due to size and cost.

Integrated circuits have had a very significant impact on digital system development. In the past, the process of digital system design normally had the following five steps:

1. System design in block diagram form—one or more levels.
2. Logic design of each of the blocks.
3. Design of each of the logic components in terms of discrete components—resistors, capacitors, diodes, transistors, etc.
4. Selection of the discrete components to be used. This amounts to several types of resistors, capacitors, diodes, transistors, or other components.
5. Packaging of the discrete component circuits.

Today, the same goal is achieved with four steps. These are:

1. System design in block diagram form—one or more levels.
2. Logic design of each of the blocks.
3. Selection of logical components—integrated circuits.
4. Packaging of selected logical components.

Thus, use of integrated circuits has eliminated digital circuit design completely and has considerably reduced the amount of time needed in component selection, testing, procurement and packaging. Advances in medium-scale

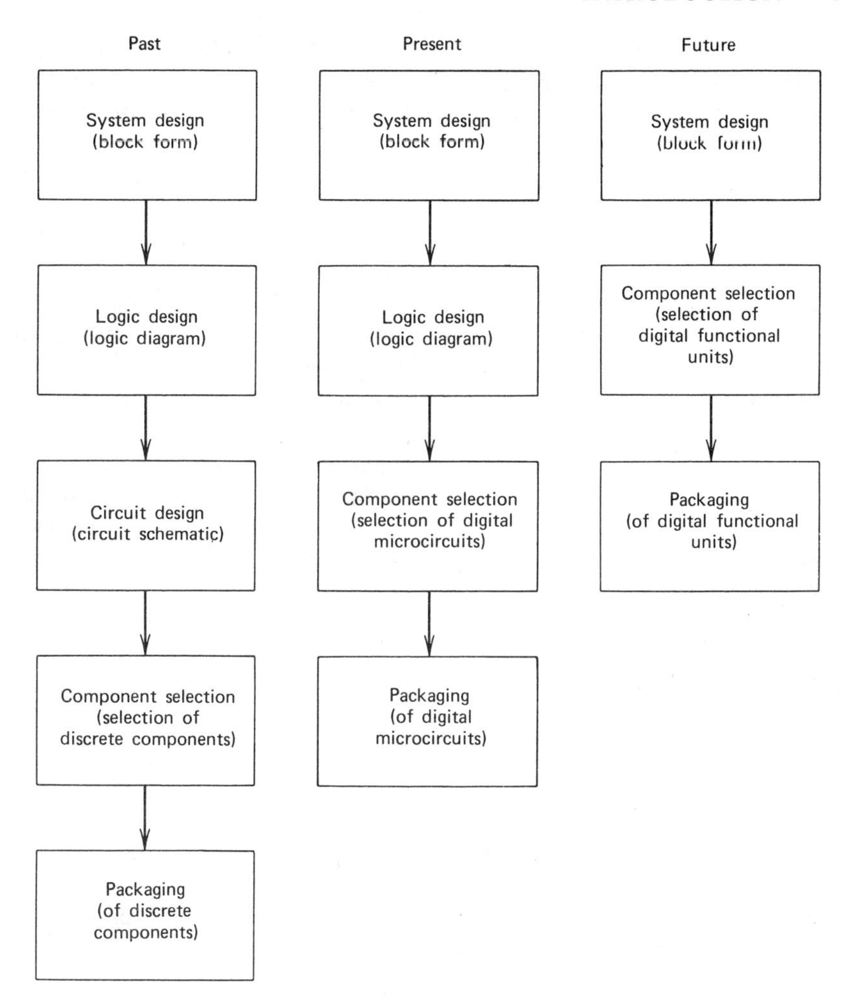

Figure I.1. Development of digital systems.

integration (MSI) as well as in large-scale integration (LSI) of digital networks often enable the digital systems designer to accomplish his task in three steps. These are:

1. System design in block diagram form—one or more levels.
2. Selection of functional components, MSI or LSI, that will perform the digital functions represented by the blocks.
3. Packaging of functional components.

Use of functional components, that is multibit adders, arithmetic units, memories, etc., always minimizes logic design making digital systems even more maintainable. Figure I.1 illustrates the evolution of digital system design.

For the best selection of logical networks available in integrated circuit form, the user must know the principles on which these networks operate. This knowledge will serve as a guide for optimal use of the selected logical networks and for simple interface between them.

This book presents the reader with a wealth of logical networks, digital techniques, and algorithms which can serve as the necessary background for thc understanding and design of digital systems.

In order for the digital engineer to best meet his design requirements, it is absolutely necessary that he be up-to-date on the availability of logical components, especially of logical networks made in IC form. This up-to-date knowledge, together with an extensive exposure to digital networks, methods, techniques, and algorithms, will enable him to invent and develop his own digital networks and techniques that best meet his digital system design requirements.

The material in the book consists of sixteen chapters and is grouped into three parts. Part One, entitled Introduction to Digital Engineering, covers material that will serve as the foundation on which the reader will build his digital engineering abilities, and consists of Chapters One through Six. In Part One, Binary System, Logical Analysis, Digital Integrated Circuits, Shift Registers, Sequential Networks, and Counting Networks are discussed.

Part Two is entitled Digital Implementation of Binary Mathematics and provides the reader with a thorough understanding on the various aspects of binary mathematical operations from the theoretical as well as the practical viewpoint. Included in this part are the binary-coded decimal system and the concepts and use of logical symmetry.

Part Three, called General Topics in Digital Engineering, further extends the reader's background by discussing Conversions, Miscellaneous Digital Operations, and Digital Memories.

Digital engineering is a developing discipline with concept and hardware advancements that need be followed by the digital engineer on a daily basis. To appreciate and benefit from these advances, a solid background in digital engineering is absolutely necessary. Making this available to the engineer is the aim of this book.

PART ONE

INTRODUCTION TO DIGITAL ENGINEERING

CHAPTER ONE

Binary System

1.1 INTRODUCTION

In a numerical system, the number of states each digit has is determined by the "base" of that system. In the decimal system, for example, where the base is 10, each digit may be in one of the following ten states:

0, 1, 2, 3, 4, 5, 6, 7, 8, and 9

In the octal numerical system, where the base is 8, each digit has eight states. These are

0, 1, 2, 3, 4, 5, 6, and 7

Similarly, in the binary numerical system where the base is 2, each binary digit, or "bit," has two states which are

0 and 1

Figure 1.1 shows a comparison of decimal and binary numbers.

Each number, regardless of the system it is in, can be considered as being an equation in abbreviated form. Decimal number 937.45, for example, is the abbreviation of

$$(9 \times 10^2) + (3 \times 10^1) + (7 \times 10^0) + (4 \times 10^{-1}) + (5 \times 10^{-2})$$

The above example follows the general equation for the expression of a number, which is

$$a_nR^n + \cdots + a_2R^2 + a_1R^1 + a_0R^0 + a_{-1}R^{-1} + a_{-2}R^{-2} + \cdots + a_{-m}R^{-m}$$

The number in abbreviated form is

$$(a_n \ldots a_2a_1a_0 . a_{-1}a_{-2} \ldots a_{-m})_R$$

where R is the base, or radix, of the numerical system used, and a is the coefficient of each power of the radix. The number of states of a is always R.

Decimal	Binary
0	0000
1	0001
2	0010
3	0011
4	0100
5	0101
6	0110
7	0111
8	1000
9	1001
10	1010
11	1011
12	1100
13	1101
14	1110
15	1111

Figure 1.1. Decimal and binary numbers from 0 to 15.

In the binary numerical system, the general equation is

$$a_n2^n + \cdots + a_32^3 + a_22^2 + a_12^1 + a_02^0 + a_{-1}2^{-1} + a_{-2}2^{-2} + a_{-3}3^{-3} + \cdots + a_{-m}2^{-m}$$

Performing the operations indicated by the exponents we obtain

$$a_n2^n + \cdots + a_38 + a_24 + a_12 + a_01 + a_{-1}0.5 + a_{-2}0.25 + a_{-3}0.125 + \cdots + a_{-m}2^{-m}$$

where coefficient a is either 1 or 0.

For example, binary number 1101.1 can be expressed as the following sum:

$$\begin{aligned} 1101.1 &= (1 \times 2^3) + (1 \times 2^2) + (0 \times 2^1) + (1 \times 2^0) + (1 \times 2^{-1}) \\ &= (1 \times 8) + (1 \times 4) + (0 \times 2) + (1 \times 1) + (1 \times 0.5) \\ &= 8 + 4 + 0 + 1 + 0.5 \\ &= 13.5 \text{ (in decimal form)} \end{aligned}$$

Similarly to decimal numbers, binary numbers are written with their most significant bit (MSB) occupying the extreme left place in the number, and their least significant bit (LSB) occupying the extreme right place in the number. The bits between the MSB and LSB are arranged so that their significance decreases from left to right. The significance of a bit is usually called "weight" and is expressed by some power of 2. Shown below are the weights of the bits in the order in which they appear in a binary number. Also shown is the binary point that separates the integral part of the number from the fractional part.

2^n	...	128	64	32	16	8	4	2	1	.	0.5	0.25	...	2^{-m}

For example, binary number 101101.01 will be 32 + 0 + 8 + 4 + 0 + 1 + 0 + 0.25, which is decimal 45.25.

To convert decimal numbers to binary, the decimal number is expressed as the sum of powers of 2 by continuously subtracting from that number the highest possible powers of 2 until the difference becomes 0. For example, decimal number 179 is converted to binary as follows:

179	51	51	19	3	3	3	1
− 128	− 64	− 32	− 16	− 8	− 4	− 2	− 1
51	*	19	3	*	*	1	0

* No subtraction is performed since the minuend is smaller than the subtrahend, and the coefficient in these cases is 0

Number 179 can be, then, expressed as the following sum:

$$(1 \times 128) + (0 \times 64) + (1 \times 32) + (1 \times 16) + (0 \times 8) + (0 \times 4) + (1 \times 2) + (1 \times 1)$$

Writing only the 1 and 0 coefficients, we obtain number 179 in binary, which is

10110011

Conversion of decimal numbers to binary can also be accomplished by means of the following simple algorithm:

Step 1. Divide the integral part of the decimal number by 2 until the quotient becomes 0. The remainder of the divisions represents bits of the integral binary equivalent of the decimal number. Bits are computed in ascending order—LSB first.

Example 1.1. Determine the binary equivalent of decimal number 193.

Conversion

Division	Quotient	Remainder
$193 \div 2 =$	96	1 LSB
$96 \div 2 =$	48	0
$48 \div 2 =$	24	0
$24 \div 2 =$	12	0
$12 \div 2 =$	6	0
$6 \div 2 =$	3	0
$3 \div 2 =$	1	1
$1 \div 2 =$	0	1 MSB

Therefore, the binary equivalent of decimal number 193 is 11000001. ▮

Step 2. Multiply the fractional part of the decimal number by 2 until the fractional part of the product becomes 0. The integral part of the products represents the bits of the fractional binary equivalent of the decimal number. Bits are computed in descending order—MSB first.

Example 1.2. Determine the binary equivalent of decimal number 0.7109375.

Conversion

	Product	
Multiplication	Integral part	Fractional part
0.7109375×2	1 MSB	0.421875
0.421875×2	0	0.84375
0.84375×2	1	0.6875
0.6875×2	1	0.375
0.375×2	0	0.75
0.75×2	1	0.5
0.5×2	1 LSB	0.0

Therefore, the binary equivalent of decimal number 0.7109375 is 0.1011011.

If a decimal number has integral and fractional parts, each one should be individually converted.

To convert binary numbers to decimal, individual bits are converted to decimal and their decimal equivalents are summed up. For example, binary number 1011.011 is converted to decimal as follows:

$$(1 \times 8) + (0 \times 4) + (1 \times 2) + (1 \times 1) + (0 \times 0.5) + (1 \times 0.25) + (1 \times 0.125)$$

which is 11.375 in decimal.

Conversion of binary numbers to decimal can also be accomplished by means of the following simple algorithm:

Step 3. Multiply the most significant integral bit by 2 and add the next bit in significance to the product. Multiply the formed sum by 2 and add the next bit to that product. Repeat the operation until the least significant integral bit has been added.

Example 1.3. Determine the decimal equivalent of binary number 10110.

1 0 1 1 0

$$((((1)2 + 0)2 + 1)2 + 1)2 + 0$$

Performing the indicated operations, we obtain 22. Indeed, binary 10110 is 22 in decimal. ∎

Step 4. Divide the least significant fractional bit by 2 and add the next bit in significance to the quotient. Divide the formed sum by 2 and add the next bit to that quotient. Repeat the operation until the most significant fractional bit has been divided.

Example 1.4. Determine the decimal equivalent of binary number 0.01101

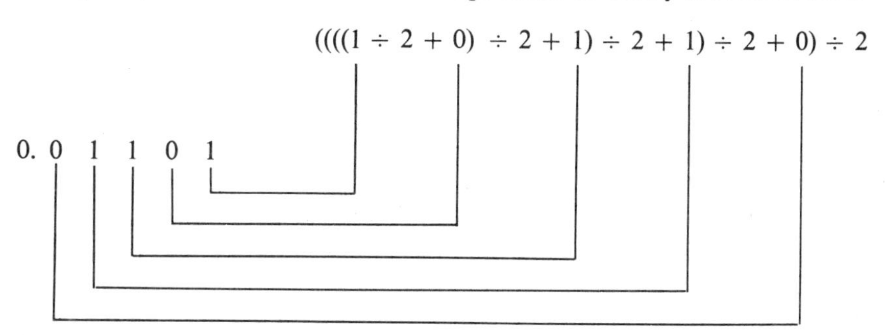

((((1 ÷ 2 + 0) ÷ 2 + 1) ÷ 2 + 1) ÷ 2 + 0) ÷ 2
0. 0 1 1 0 1

Performing the indicated operations, we obtain 0.40625. Indeed, binary 0.01101 is 0.40625 in decimal.

1.2 ADDITION OF BINARY NUMBERS

Addition of binary numbers is similar to that of decimal numbers. The only difference is that in the binary system we have a "carry" when the sum exceeds 1, while in the decimal system we have a "carry" when the sum exceeds 9.

For example, the addition of 1 and 1 in the two systems is as follows:

	Binary	Decimal
	1	1
	1	1
Sum	1 0	2
	↑ ↑	
Weight	2 1	

The binary sum reads "one, zero," not "ten." The bit on the right indicates the 1-s, and the bit on the left indicates the 2-s.

Addition of 3 and 3 is performed as follows:

	Binary	Decimal
	1 1	3
	1 1	3
Sum	1 1 0	6
	↑ ↑ ↑	
Weight	4 2 1	

In the binary sum, 110, the bit on the right indicates the 1-s, the bit at the center indicates the 2-s, and the bit on the left indicates the 4-s. This number reads "one, one, zero." In binary addition, the carry from the addition of the 1-s has a weight of 2, and is added to the 2-s, and the carry from addition of the 2-s has a weight of 4.

Similarly, the addition of the binary equivalents of decimal numbers 27 and 30 will be as follows:

	Binary	Decimal
	1 1 0 1 1	27
	1 1 1 1 0	30
Sum	1 1 1 0 0 1	57
Weight	32 16 8 4 2 1	

1.3 SUBTRACTION OF BINARY NUMBERS

Subtraction in the binary system is very similar to subtraction in the decimal system, with the difference being that in binary when the minuend is smaller than the subtrahend, the "borrow" is 2, rather than 10, which it is in decimal.

For example, the binary subtraction of numbers 101 (decimal number 5) and 11 (decimal number 3) is as follows:

	Binary	**Dccimal**
	101	5
	11	3
Difference	010	2

Similarly, subtraction of binary numbers 101101, (45), and 010111, (23), would be as follows:

	Binary	**Decimal**
	101101	45
	010111	23
Difference	010110	22

1.4 MULTIPLICATION OF BINARY NUMBERS

Multiplication of binary numbers is performed as follows: To multiply number 11011 (decimal 27) by 101 (decimal 5) we first align the two numbers as follows:

```
 11011      27
   101       5
 -----      --
```

Then we copy the multiplicand under these two numbers, once for each 1 of the multiplier. The multiplicand is copied with its LSB under the 1 of the multiplier to which it corresponds. For the above example it would be

```
   11011      27
     101       5
   -----      --
   11011      27
 11011       108
```

The next and final step is to add the numbers below the horizontal line. This will yield

```
   11011      27
     101       5
   -----      --
   11011      27
 11011       108
--------     ---
10000111     135
```

Similarly, multiplication of numbers 1011 and 1101 would be performed as follows:

```
    1011       11
    1101       13
    ----       --
    1011       11
   1011        44
  1011         88
--------      ---
10001111      143
```

1.5 DIVISION OF BINARY NUMBERS

In the binary system, division is performed by repeated subtraction of the divisor from the dividend. The divisor is first placed under the dividend so that the MSB of the divisor is under the MSB of the dividend. Next, the divisor is compared to the dividend portion which is directly above the divisor.

When that portion of the dividend is equal to or greater than the divisor, we write 1 in the quotient and we subtract the divisor from the dividend. When the dividend portion directly above the divisor is found to be less than the divisor, we do not perform the subtraction and we write 0 in the quotient place.

```
       1010.001                              Quotient
      _________
101 | 110011.
      101         subtraction                1 MSB
      ---------
      001011.
       101        no subtraction             0
      ---------
      001011.
        101       subtraction                1
      ---------
      000001.
         101      no subtraction             0.
      ---------
      000001.0
          10.1    no subtraction             0
      ---------
      000001.00
           1.01   no subtraction             0
      ---------
      000001.000
             101  subtraction                1 LSB
      ----------
      000000.011 Remainder
```

After the first step, the divisor is shifted one position to the right and the same procedure is repeated. We repeat the procedure for as many times as the desired number of bits for the quotient. When the unity bits line up, a point is placed to the right of the derived quotient bit, indicating separation between the integral and fractional part of the quotient.

A typical binary division where 51 is divided by 5 is shown on page 14.

1.6 COMPLEMENTS OF BINARY NUMBERS

The complement of a given binary number is the difference between that number and some other number that has been assigned as the reference number.

For example, if the given binary number is 1101 (13) and the reference number is 1000000 (64) the complement of the given number is the difference between 64 and 13, which is 51, or 110011 in binary. The reference number can be arbitrary, depending on the specific application. However, there are two types of complements most often used—the complement of 1 and the complement of 2. The 1's complement of a binary number is obtained by subtracting that number from a reference number, all bits of which are 1.

For example, when the reference number is 1111, the 1's complement of the binary number 1011 is 0100, and is obtained as shown here

1111	**Reference number**
− 1011	**Given number**
0100	**1's complement of given number**

If the reference number were 111111, the complement of 001011 would have been

111111	**Reference number**
− 001011	**Given number**
110100	**1's complement of given number**

The 1's complement of a given number can be also obtained by inverting each bit of that number. For example, by direct inversion of bits the 1's complement of 001011 is 110100, as was computed above.

The 2's complement of a given binary number is obtained by subtracting that number from a reference number, which is always greater than the given number, and all bits of which are 0 except the most significant one.

For example, the 2's complement of the binary number 1011, used in the preceding example, is

	10000	**Reference number**
−	1011	**Given number**
	0101	**2's complement of given number**

If the reference number were 1000000, the 2's complement of the given number would have been 110101.

$$\begin{array}{r} 1000000 \\ -\quad 1011 \\ \hline 110101 \end{array}$$

Therefore, before computing the complement of a given binary number, we need to know the exact number of bits of the reference number.

The 2's complement of a binary number can be also obtained by simply adding a 1 to the 1's complement of that number, that is,

$$\text{2's complement} = \text{1's complement} + 1 \text{ LSB}$$

This is possible because the reference numbers from which the complements are determined also differ by 1.

1.7 NEGATIVE BINARY NUMBERS

In the binary system there are two most frequently used methods of negative number representation. One is as a sign and magnitude, and the other is as a 2's complement of the positive number having the same magnitude.

The first method is a straightforward approach where the sign and the magnitude of the given negative number are explicitly expressed. For example, the numbers −5 and +5 are expressed as

$$-5 = 1 \quad 101 \qquad\qquad +5 = 0 \quad 101$$

where the single bit on the left indicates the sign of the number following it. The bit indicating the sign can be arbitrarily taken as 0 for positive and 1 for negative, or vice versa.

The 2's complement representation of negative binary numbers follows the assumption that when a negative number is added to its positive counterpart, the resulting sum should always be 0. For example, 5 − 5 should equal 0.

$$\begin{array}{rrr} & 101 & 5 \\ \text{(2's complement of 101)} & 011 & -\ 5 \\ \hline & (1)000 & 0 \end{array}$$

It should be noted that -5 in the 2's complement is 011 only in the 3 bit space. Three computations of -5 for 4, 5, and 6 bit spaces follow:

$$
\begin{array}{r} (1)\ 0000 \\ -\ 0101 \\ \hline 1011 \end{array}
\qquad
\begin{array}{r} (1)\ 00000 \\ -\ 00101 \\ \hline 11011 \end{array}
\qquad
\begin{array}{r} (1)\ 000000 \\ -\ 000101 \\ \hline 111011 \end{array}
\qquad
\begin{array}{r} 0 \\ -\ 5 \\ \hline -\ 5 \end{array}
$$

In all of the above cases, when 101 and its 2's complement are added together, all bits of the sum within the space are 0. As an example, in the 6 bit space $5 - 5$ is

$$
\begin{array}{r} 000101 \\ 111011 \\ \hline (1)\ 000000 \end{array}
$$

The extreme left bit of the sum is of no significance because it is outside the 6 bit space.

In most applications of digitally implemented binary arithmetic, the 2's complement representation of negative numbers is used because it requires less hardware than other methods.

Arithmetic operations can be performed involving positive and negative (2's complement) numbers, as if all numbers were positive.

Example 1.5. Perform the addition of binary numbers 1011 and -1101 employing the 2's complement for negative numbers. Since $-1011 = 0011$, we have

$$
\begin{array}{rr} 1011 & 11 \\ +\ 0011 & -\ 13 \\ \hline 1110 = 14 = & -\ \ 2 \end{array}
$$

∎

Example 1.6. Perform subtraction $(-0011) - (0111)$, employing 2's complement for negative numbers. Since $-0011 = 1101$, we have

$$
\begin{array}{rr} 1101 & -3 \\ -\ 0111 & -7 \\ \hline 0110 = +6 = & -10 \end{array}
$$

∎

Example 1.7. Perform multiplication $(-0101) \times (0010)$, employing 2's complement for negative numbers.

$$
\begin{array}{rr} 1011 & -5 \\ \times\ 0010 & \times\ \ 2 \\ \hline (1)0110 = +6 = & -10 \end{array}
$$

∎

Example 1.8. Perform multiplication $(-0101) \times (-1010)$, employing 2's complement for negative numbers.

$$
\begin{array}{rr}
1011 & -5 \\
0110 & -10 \\
\hline
10110 & \\
1011 & \\
\hline
(100)0010 = +2 = & +50 = 3 \times 16 + 2
\end{array}
$$

Since the numbers are in the 4 bit space or modulo 16, multiples of 16, indicated in the parentheses, are omitted. ∎

1.8 FORM OF BINARY NUMBERS

Binary numbers are available in either parallel or serial form. Transmission of binary numbers in parallel form requires a number of signal lines equal

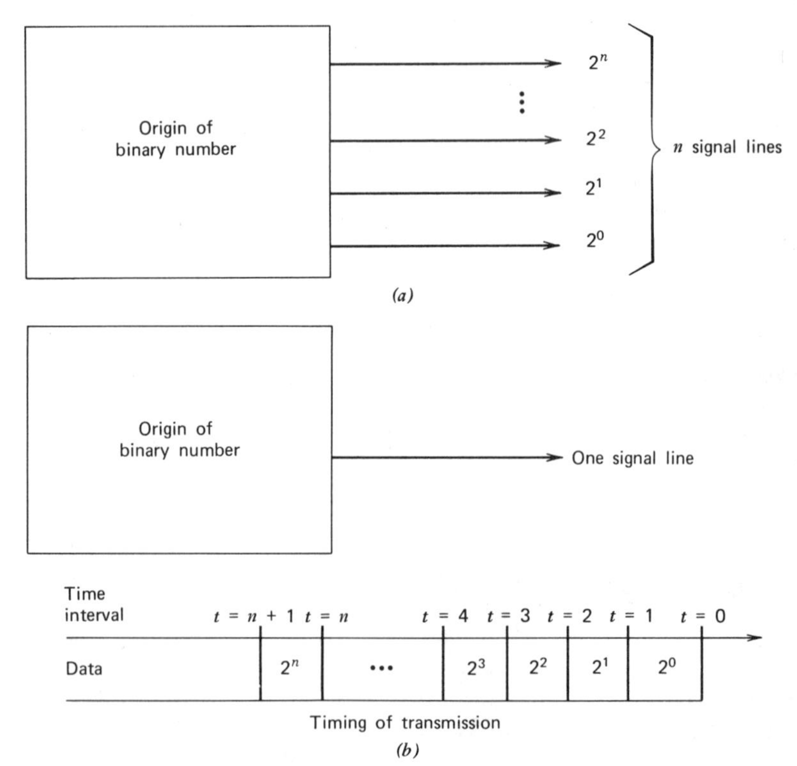

Figure 1.2. Forms of binary number transmission: (*a*) parallel; (*b*) serial.

to the number of bits of the number. As the size of the binary number increases, the number of signal lines required for the transfer of that number becomes impractical. At this point, the serial form is considered.

In the serial form all bits of the number are transmitted in sequence via a single signal line where each bit occupies the line only for a short time interval. Figure 1.2 illustrates the parallel and serial form of binary number transmission.

The advantage of parallel over serial data transmission is speed. In parallel transfer, there is practically no delay between information generation and information transfer. The disadvantage, on the other hand, is that for each bit of information to be transferred a separate signal line is needed.

In serial data transmission only one signal line is needed, which is shared by all bits of the transferred information. Consequently, the disadvantage of the serial transfer is a slow rate of transmission. In addition to that, conversion from parallel to serial will be needed in the event information is originally in parallel form, and conversion requires time and hardware. Selection between parallel and serial method of data transmission depends on the time and hardware available, as well as on the original form of the data.

1.9 BINARY-DERIVED RADICES

Numerical systems, the base or radix of which is a power of 2, are called binary-derived radices. Conversion from these systems to pure binary (base 2) or vice versa, is very simple, making their use desirable in many applications of the human–machine interface. In ascending order these systems are the base 4, base 8 or octal, base 16 or hexadecimal, base 32, etc.

To convert a binary number to any of these systems, the number is partitioned in m bit groups starting from the binary point, where m is the exponent of 2 that makes the power 2^m equal to the base of the numerical system to which the binary number is converted. Next, each such group is expressed in terms of 1 digit of the new system.

For example, to convert binary number 101101011100 to base 4, we partition the number as follows:

$$10 \mid 11 \mid 01 \mid 01 \mid 11 \mid 00 \qquad \text{(base 2)}$$

Each group is now expressed by 1 digit of the base 4 system. These digits are 0, 1, 2, and 3. The above pure binary number then becomes

$$231130 \qquad \text{(base 4)}$$

To convert the same binary number to the octal numerical system, we partition the number into groups of 3 bits.

$$101 \mid 101 \mid 011 \mid 100 \qquad \text{(base 2)}$$

Expressed in the octal radix, the above number becomes

$$5534 \qquad \text{(base 8)}$$

Conversion from a binary-derived system to binary is accomplished by converting each digit to binary, taking into consideration the number of bits each digit corresponds to. This number is m, where 2^m equals the base of the binary-derived system.

For example, to convert the base 4 number 20132 to binary, we replace each digit by the binary equivalent expressed in 2 bits. Since $2^m = 4 = 2^2$, then

2 0 1 3 2	(base 4)
10 00 01 11 10	(base 2)

or 1000011110.

If the given number 20132 were of base 8, then the conversion would have been

2 0 1 3 2	(base 8)
010 000 001 011 010	(base 2)

or 010000001011010.

Conversion from one binary-derived system to another can be best made via the pure binary system. For example, to convert the base 8 number 67524 to base 4, each digit of the given number is converted to base 2 and then to base 4, that is;

6 7 5 2 4	(base 8)
110 | 111 | 101 | 010 | 100	(base 2)
1 | 10 | 11 | 11 | 01 | 01 | 01 | 00	(base 2)
1 2 3 3 1 1 1 0	(base 4)

The octal system finds wide use in the machine-to-human interface where binary-to-decimal conversion is avoided. This is mainly in computer printouts of memory contents, as well as in memory-location-indexing.

PROBLEMS

1. Convert the following binary numbers to decimal:
   (*a*) 101011 (*b*) 000011
   (*c*) 11011 (*d*) 11001
   (*e*) 10.011 (*f*) 01.110
   (*g*) 0.0111 (*h*) 111.10

2. Convert the following decimal numbers to binary. Compute the binary fractional part to 7 bits.
   (*a*) 539 (*b*) 367
   (*c*) 101 (*d*) 11
   (*e*) 12.34 (*f*) 7.37
   (*g*) 0.0101 (*h*) 10.0101
3. Perform the following binary additions:
   (*a*) 1011 + 0101 (*b*) 110 + 101
   (*c*) 1110 + 1111 (*d*) 1.011 + 10.011
4. Perform the following binary subtractions:
   (*a*) 10101 − 01101 (*b*) 100 − 001
   (*c*) 1111 − 1011 (*d*) 11.001 − 1.11
5. Perform the following binary multiplications:
   (*a*) 101 × 11 (*b*) 11011 × 101
   (*c*) 10110 × 1100 (*d*) 11.01 × 10.11
6. Perform the following binary divisions. Compute the quotient to 7 bits.
   (*a*) 110110 ÷ 11 (*b*) 1011101 ÷ 10
   (*c*) 111111 ÷ 111 (*d*) 11011.11 ÷ 1.01
7. Compute the 1's and 2's complements of the following binary numbers in the 7 bit space:
   (*a*) 0101011 (*b*) 110111
   (*c*) 01000 (*d*) 111011
8. State the advantages and disadvantages of the parallel and serial forms of data transmission.
9. Convert the following binary numbers to octal:
   (*a*) 1010 (*b*) 11101
   (*c*) 101101 (*d*) 1010111
10. State the advantages of the octal over the decimal numerical system.

CHAPTER TWO

Logical Analysis

2.1 INTRODUCTION

The performance of logical networks is expressed in terms of equations that are governed by special theorems and rules. The mathematics that cover these equations is called boolean algebra.

By means of boolean expressions, the operation of logical networks can be explicitly defined, and the need for breadboarding is minimal. It must be noted that in the design or analysis of complex sequential logical networks, that is, networks where signals are time-dependent, the use of logical equations is impractical, and therefore timing diagrams, illustrating the waveforms of the various signals, are preferred.

2.2 LOGICAL QUANTITIES

In digital design a quantity may have either of two values. It can be 1 or 0. When it is 1, the quantity is considered to be in its TRUE state, indicating that all conditions leading to the generation of this quantity have been satisfied. When it is 0, the quantity is considered to be in its FALSE state, indicating that these conditions have not been satisfied.

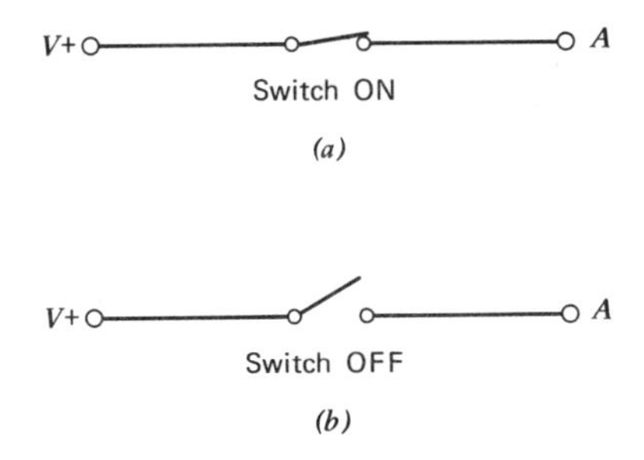

Figure 2.1. Single-pole single-throw switch.

Since a logical quantity, by definition, can have either of two states, it may be compared to an electric switch which can be either ON or OFF, where ON may be arbitrarily called 1, and OFF 0. Figure 2.1 shows a single-pole single-throw switch in its two states. When the switch is ON, output A is $+V$, and when OFF, output A has no voltage. Thus, the state of the switch can be remotely determined by sensing output A.

Logical quantities may be combined in a "multiplication" or "addition" fashion. Logical quantities A and B, for example, may be combined as follows:

$$A \times B \rightarrow C \qquad \text{or} \qquad A + B \rightarrow C$$

Since the two quantities A and B may have only two values each, they can produce a total of four combinations, which are

For $A \times B \rightarrow C$

A	B	C
0	0	0
0	1	0
1	0	0
1	1	1

For $A + B \rightarrow C$

A	B	C
0	0	0
0	1	1
1	0	1
1	1	1

In order for the $A \times B$ "product" to be 1, both quantities A and B must be 1. As a result, the logical multiplication $A \times B$ is called the AND function, and $A \times B$ is read "A AND B." Similarly, ABC is read "A AND B AND C." The multiplication sign $\times$ is eliminated and is sometimes replaced by a dot.

In order for the $A + B$ "sum" to be 1, either quantity A *or* B must be 1. As a result, the logical addition $A + B$ is called the OR function, and $A + B$ is read "A OR B." Similarly, $A + B + C$ is read "A OR B OR C."

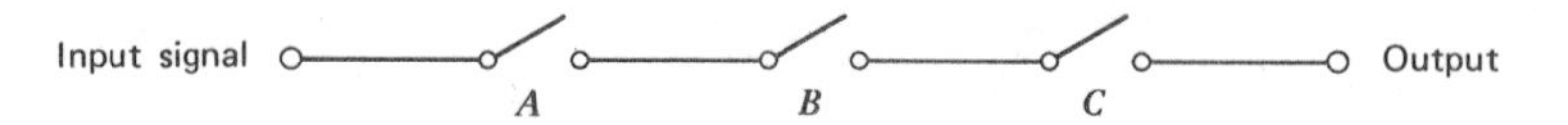

Figure 2.2. Three switches forming an AND circuit.

The OR function is indicated by the plus sign, which should always appear between the referenced quantities.

In the three-variable space, the logical expression ABC can be represented by three switches connected in series, as shown in Figure 2.2. Here, in order for information to flow from the input to the output, all switches must be ON. If ON is associated with 1 and OFF with 0, information will flow from the input to the output only when all the quantities A, B, and C, are 1, making product ABC a 1. The obtained ABC product can be found in any of the following eight combinations.

A ·	B ·	$C \rightarrow$	D
0	0	0	0
0	0	1	0
0	1	0	0
0	1	1	0
1	0	0	0
1	0	1	0
1	1	0	0
1	1	1	1

Of the above eight combinations only one gives a 1 output. The above operations are not binary multiplications but merely logic expressions where the dot is read "AND." Here, however, the result of the indicated logical operation is the same as that of binary multiplication.

The logical expression $A + B + C$ can be represented by three switches connected in parallel, as shown in Figure 2.3. In this illustration, in order

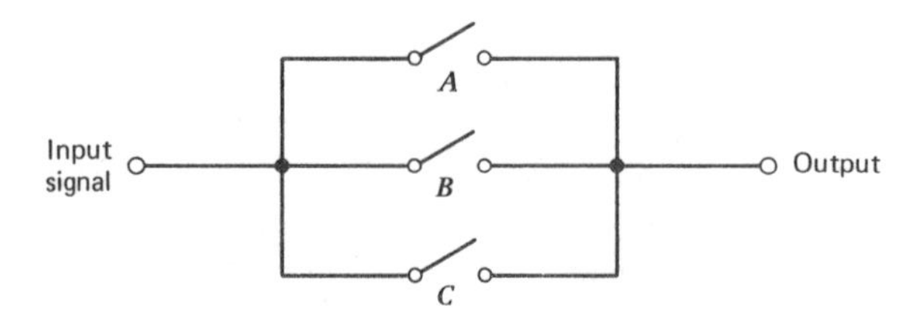

Figure 2.3. Three switches forming an OR circuit.

for information to flow from input to output, only one of the switches need be ON. If ON is associated with 1 and OFF with 0, information will flow from input to output only when one or more of the logical quantities A, B, and C is 1. The $A + B + C$ sum obtained by this method can be found in any of the following eight combinations:

A +	B +	$C \rightarrow$	D
0	0	0	0
0	0	1	1
0	1	0	1
0	1	1	1
1	0	0	1
1	0	1	1
1	1	0	1
1	1	1	1

Of the above combinations all but one give an output of 1. The above operations are not binary addition, but logical expressions where the plus sign is read as "OR."

The inverse of a boolean quantity A is written as $\bar{A}$, and it is read as "A bar." The value of $\bar{A}$ is, by definition, the logical opposite of that of A at all times, that is, when $A = 1$, $\bar{A} = 0$, and when $A = 0$, $\bar{A} = 1$.

Considering the above rules on logical expressions, expression $\bar{A}BC$, read as "A bar BC," will produce a 1 when

$$\bar{A} = 1$$
$$B = 1$$
and
$$C = 1$$

Similarly, logical expression $A + \bar{B} + \bar{C}$, read as "A OR B bar OR C bar," will produce a 1 when either

$$A = 1$$
$$\bar{B} = 1$$
or
$$\bar{C} = 1$$

which is the same as saying that $A + \bar{B} + \bar{C}$ is 1 when either

$$A = 1$$
$$B = 0$$
or
$$C = 0$$

2.3 LOGICAL EXPRESSIONS

Logical expressions are governed by a set of rules that constitute the basics of boolean algebra. The following rules have been derived from the philosophical logic and are expressed in a mathematical form, easy to be used in logical network design and analysis:

Rule 1 $A + 0 = A$

Rule 2 $A + 1 = 1$

Rule 3 $A + A = A$

Rule 4 $A + \overline{A} = 1$

Rule 5 $A \cdot 0 = 0$

Rule 6 $A \cdot 1 = A$

Rule 7 $A \cdot A = A$

Rule 8 $A \cdot \overline{A} = 0$

Rule 9 $A + B = B + A$

Rule 10 $A \cdot B = B \cdot A$

Rule 11 $(A + B) + C = A + (B + C)$

Rule 12 $(A \cdot B) \cdot C = A \cdot (B \cdot C)$

Rule 13 $\overline{ABC} = \overline{A} + \overline{B} + \overline{C}$

Rule 14 $\overline{A + B + C} = \overline{A} \cdot \overline{B} \cdot \overline{C}$

Rule 15 $A(B + C) = AB + AC$

Rule 16 $\overline{AB + CD + EF} = \overline{AB} \cdot \overline{CD} \cdot \overline{EF}$
$= (\overline{A} + \overline{B}) \cdot (\overline{C} + \overline{D}) \cdot (\overline{E} + \overline{F})$

Rule 17 $\overline{(A + B) \cdot (C + D) \cdot (E + F)} = \overline{(A + B)} + \overline{(C + D)} + \overline{(E + F)}$
$= \overline{A}\overline{B} + \overline{C}\overline{D} + \overline{E}\overline{F}$

The above rules find wide use in the simplification of logical expressions, where the implementation of simplified forms results in a considerable saving of hardware. The next three examples illustrate the use of the rules of logical expressions.

Example 2.1. Using the above listed rules simplify the logical expression

$$f(A, B, C) = A\overline{B}\overline{C} + A\overline{B}C + AB\overline{C} + ABC$$

First, we factor out quantity A, since it is a common term, and we obtain

$$f(A, B, C) = A(\overline{B}\overline{C} + \overline{B}C + B\overline{C} + BC) \qquad \text{(Rule 15)}$$

Next, we factor out common terms within the parentheses, which yields

$$f(A, B, C) = A[\overline{B}(\overline{C} + C) + B(\overline{C} + C)] \qquad \text{(Rule 15)}$$

Since $\bar{C} + C = 1$, it may be omitted and we now have

$$f(A, B, C) = A(\bar{B} + B) \qquad \text{(Rule 4)}$$

Similarly, $\bar{B} + B = 1$ and expression reduces to

$$f(A, B, C) = A \qquad \text{(Rule 7)}$$

The handling and the hardware implementation of the simplified form is considerably simpler than that of the original expression, which contains two variables, B and C, that do not in any way affect $f(A, B, C)$. ∎

Example 2.2. Using the above listed rules simplify the logical expression

$$f(A, B, C) = (A + B)C + \bar{A}C + AB + ABC + \bar{B}C$$

Simplification of this expression starts with the performance of the multiplication indicated in the first term.

$$f(A, B, C) = AC + BC + \bar{A}C + AB + ABC + \bar{B}C \qquad \text{(Rule 15)}$$

Next, we factor out common terms aiming at the elimination of expressions resulting in parentheses, and we obtain

$$f(A, B, C) = (A + \bar{A})C + BC(1 + A) + AB + \bar{B}C \qquad \text{(Rule 15)}$$

The expressions in the parentheses can be eliminated, since they equal 1, reducing the original expression to

$$f(A, B, C) = C + BC + AB + \bar{B}C \qquad \text{(Rules 4 and 2)}$$

Factoring out term C we obtain

$$f(A, B, C) = C(1 + B + \bar{B}) + AB \qquad \text{(Rule 15)}$$

which is the same as

$$f(A, B, C) = C + AB \qquad \text{(Rule 2)}$$

∎

Example 2.3. Using the above listed rules simplify the logical expression

$$f(A, B, C) = \overline{AC + \bar{A}BC + \bar{B}C} + AB\bar{C}$$

Here, we start by removing the main bar from the first group of terms. Removal of that bar results in the inversion of every logical symbol that previously existed under the bar, that is, each symbol as well as each term are logically inverted. The OR functions become AND functions, the AND become OR and a bar is placed over each term. Hence, we obtain

$$f(A, B, C) = \overline{AC} \cdot \overline{\bar{A}BC} \cdot \overline{\bar{B}C} + AB\bar{C} \qquad \text{(Rule 13)}$$

Applying the same rule on the first three groupings we have

$$f(A, B, C) = (\bar{A} + \bar{C})\cdot(\bar{\bar{A}} + \bar{B} + \bar{C})\cdot(\bar{\bar{B}} + \bar{C}) + AB\bar{C} \quad \text{(Rule 13)}$$

Since $\bar{\bar{A}}$ is the same as A inverted twice, and double inversion leaves a term unchanged, the two bars can be removed. Therefore,

$$f(A, B, C) = (\bar{A} + \bar{C})\cdot(A + \bar{B} + \bar{C})\cdot(B + \bar{C}) + AB\bar{C}$$

Multiplying out the first two groupings, we obtain

$$\begin{aligned} f(A, B, C) &= [\bar{A}(A + \bar{B} + \bar{C}) + \bar{C}(A + \bar{B} + \bar{C})]\cdot(B + \bar{C}) + AB\bar{C} \\ &= (\bar{A}A + \bar{A}\bar{B} + \bar{A}\bar{C} + A\bar{C} + \bar{B}\bar{C} + \bar{C}\bar{C})\cdot(B + \bar{C}) + AB\bar{C} \end{aligned}$$

(Rule 15)

Term $\bar{A}A$ becomes 0 and it is eliminated, in accordance with Rule 1, and term $\bar{C}\bar{C}$ becomes $\bar{C}$, reducing the expression to

$$f(A, B, C) = (\bar{A}\bar{B} + \bar{A}\bar{C} + A\bar{C} + \bar{B}\bar{C} + \bar{C})\cdot(B + \bar{C}) + AB\bar{C}$$

(Rules 1, 7, and 8)

We now factor out $\bar{C}$ obtaining

$$f(A, B, C) = [\bar{A}\bar{B} + \bar{C}(\bar{A} + A + \bar{B} + 1)]\cdot(B + \bar{C}) + AB\bar{C} \quad \text{(Rule 15)}$$

The expression in the parentheses is the same as 1, yielding

$$f(A, B, C) = (\bar{A}\bar{B} + \bar{C})\cdot(B + \bar{C}) + AB\bar{C} \quad \text{(Rule 2)}$$

Multiplying out the first two terms, we obtain

$$f(A, B, C) = \bar{A}\bar{B}B + B\bar{C} + \bar{A}\bar{B}\bar{C} + \bar{C}\bar{C} + AB\bar{C} \quad \text{(Rule 15)}$$

Term $\bar{A}\bar{B}B$ becomes 0, term $\bar{C}\bar{C}$ becomes $\bar{C}$ and, when $\bar{C}$ is factored out, it yields

$$f(A, B, C) = \bar{C}(B + \bar{A}\bar{B} + 1 + AB) \quad \text{(Rules 8 and 15)}$$

which reduces to

$$f(A, B, C) = \bar{C} \quad \text{(Rule 2)}$$

∎

Because of the considerable advantages simple expressions have in hardware implementation as well as in mathematical manipulations, proficiency in the simplification of logical expressions is a very useful tool that must be possessed by every logical designer.

2.4 LOGICAL DIAGRAMS

Another way by means of which logical expressions can be represented is diagrams. Figure 2.4 illustrates four such diagrams for the logical space of one, two, three, and four variables. These diagrams are often called "Veitch

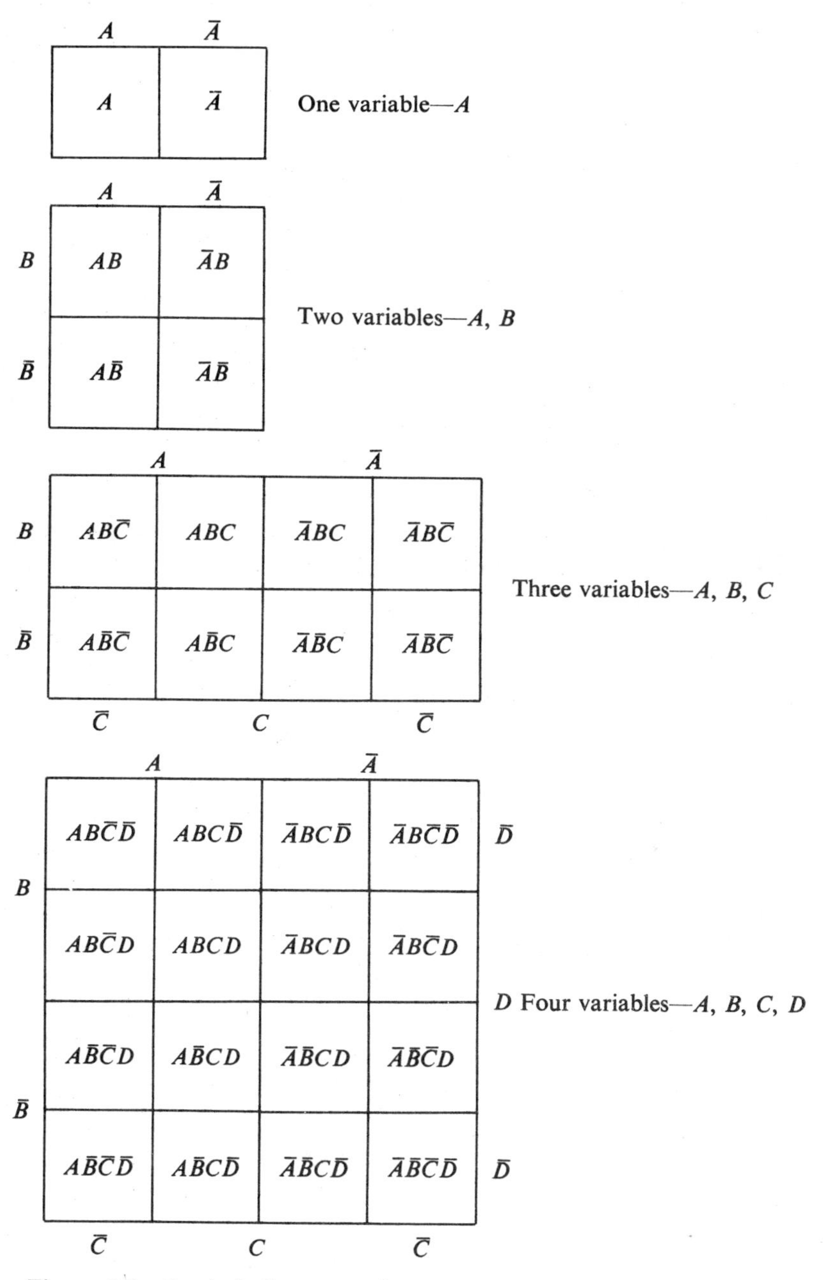

Figure 2.4. Logical diagrams of one-, two-, three-, and four-variable logical spaces.

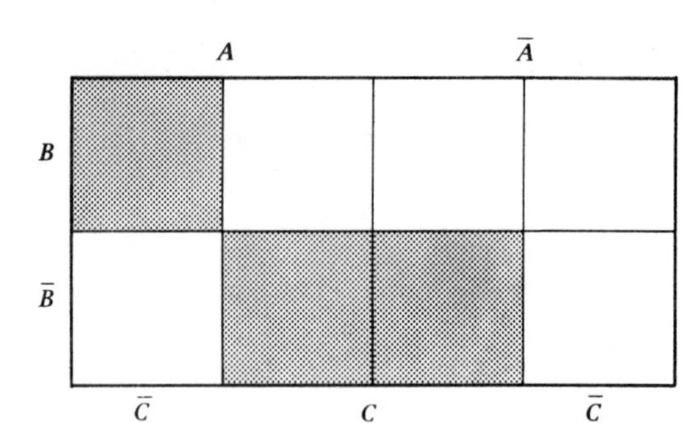

Figure 2.5. Mapping of the logical expression $f(A, B, C) = AB\bar{C} + \bar{B}C$.

diagrams." In these diagrams, each square corresponds to one logical term defined by as many variables as the space of the diagram has.

The number of squares corresponding to a logical term is 2^{n-m}, where n is the number of variables of the diagram space and m is the number of variables by means of which the term is defined. The number of variables of a term cannot be greater than the number of variables in the space. In that case, the term will correspond to a different space. Also, all variables defining a term must be included in the diagram where the logical term is to be mapped. For example, the logical expression

$$f(A, B, C) = AB\bar{C} + \bar{B}C$$

can be mapped onto a three-variable logical diagram, as shown on Figure 2.5. Similarly, the logical expression

$$f(A, B, C) = (A + B)C + \bar{A}C + AB + ABC + \bar{B}C$$

of Example 2.2, is mapped on a three-variable diagram appearing on Figure 2.6. From this diagram it can be seen that the covered area can be represented

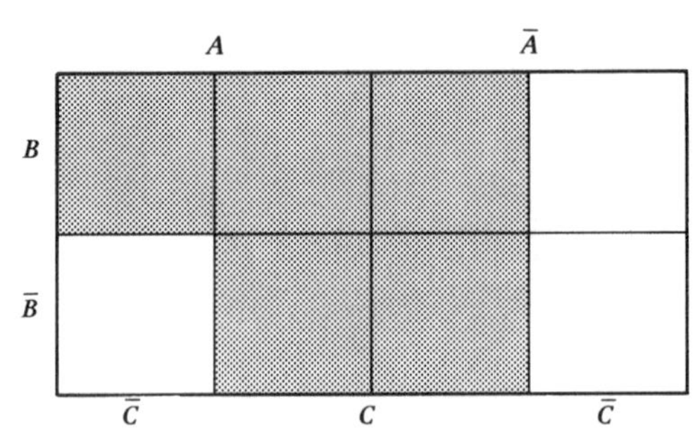

Figure 2.6. Mapping of the logical expression $f(A, B, C) = (A + B)C + \bar{A}C + AB + ABC + \bar{B}C$.

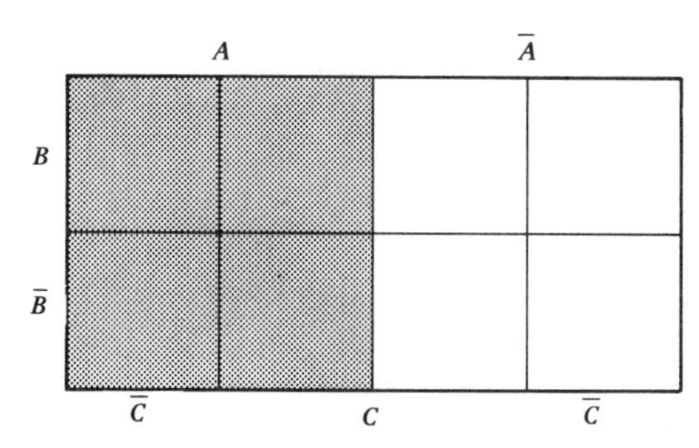

Figure 2.7. Mapping of the logical expression $f(A, B, C) = A\bar{B}\bar{C} + A\bar{B}C + AB\bar{C} + ABC$.

by $AB + C$. Thus, logical diagrams can be used for the simplification of logical expressions, as shown in the next two examples.

Example 2.4. Using logical diagrams, simplify the expression

$$f(A, B, C) = A\bar{B}\bar{C} + A\bar{B}C + AB\bar{C} + ABC$$

The first step is to map the expression onto a three-variable diagram with variables A, B, and C, as shown on Figure 2.7. From this figure it can be seen that the mapped function covers the entire A space and only that. Therefore, a simplified form of the expression will be $f(A, B, C) = A$. The same result was obtained in Example 2.1 where the same logical expression was simplified by means of boolean algebra. ∎

Example 2.5. Using logical diagrams, simplify the expression

$$f(A, B, C) = \overline{AC + \bar{A}BC + \bar{B}C} + AB\bar{C}$$

In this case we may either expand the expression by removing the main bar, or we may define the barred area and combine $AB\bar{C}$ to its complement.

The latter is easier and it is illustrated in Figure 2.8(b). Diagram (a) of the figure is an intermediate step which shows the area indicated by the expression under the main bar, $AC + \bar{A}BC + \bar{B}C$. The remainder of the total area in diagram (a) is by definition, $\overline{AC + \bar{A}BC + \bar{B}C}$, and it is shaded in diagram (b). It is obvious that the term $AB\bar{C}$ does not add anything to the shaded area, because the area of $AB\bar{C}$ is contained within $\overline{AC + \bar{A}BC + \bar{B}C}$. That is, the term $AB\bar{C}$ is redundant.

Veitch diagrams are also used for the synthesis of logical expressions. In this case, the expression is given in the form of a summation of individual terms, where the terms explicitly define the represented logical function. The following two examples illustrate synthesis by means of logical diagrams:

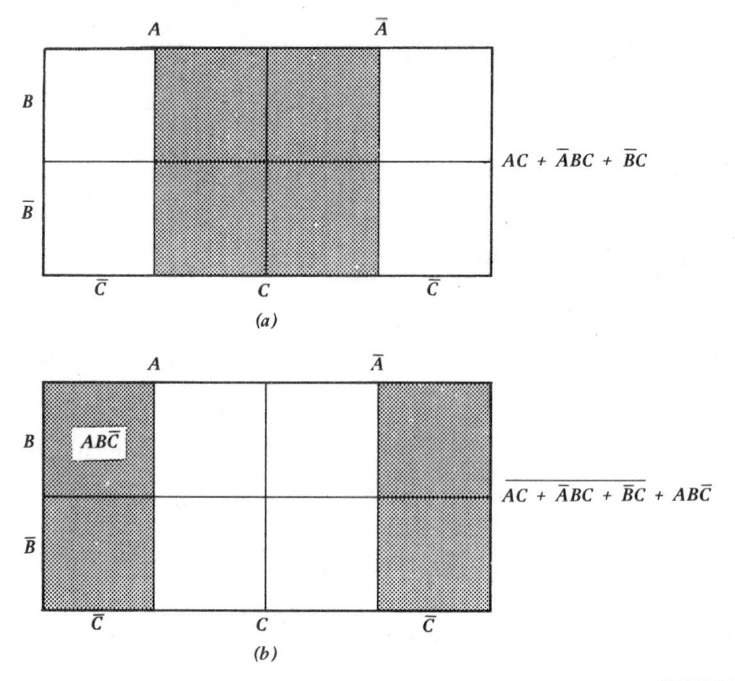

Figure 2.8. Mapping of the logical expression $f(A, B, C) = \overline{AC + \bar{A}BC + \bar{B}C} + AB\bar{C}$.

Example 2.6. With the aid of diagrams determine the logical expression for the function

$$f(A, B, C, D) = \sum(1, 3, 6, 7, 9, 11, 14, 15)$$

The first step is to map the given expression on a logical diagram; the number and type of variables of which are the same as those of the expression. Next, the terms are combined in logical groups as large as possible so that they can be expressed with as few variables as possible, as shown in Figure 2.9.

In this example, the terms 1, 3, 9, and 11 are combined together forming a group that can be defined as $\bar{B}D$. Similarly, terms 6, 7, 14, and 15 are defined as BC. ∎

Example 2.7. With the aid of diagrams determine the logical expression for the function

$$f(A, B, C, D) = \sum(0, 1, 5, 7, 8, 9, 14)$$

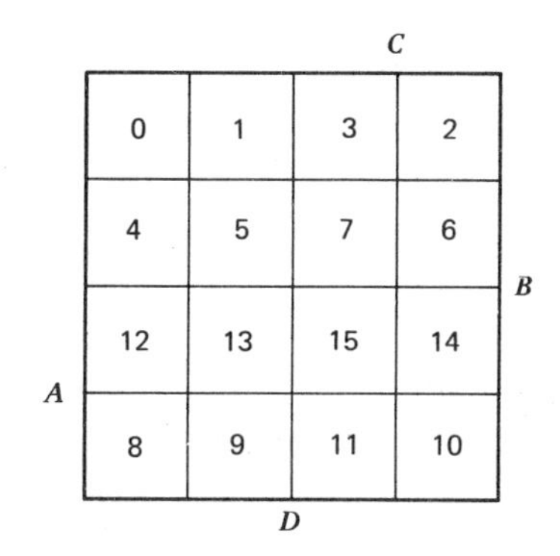

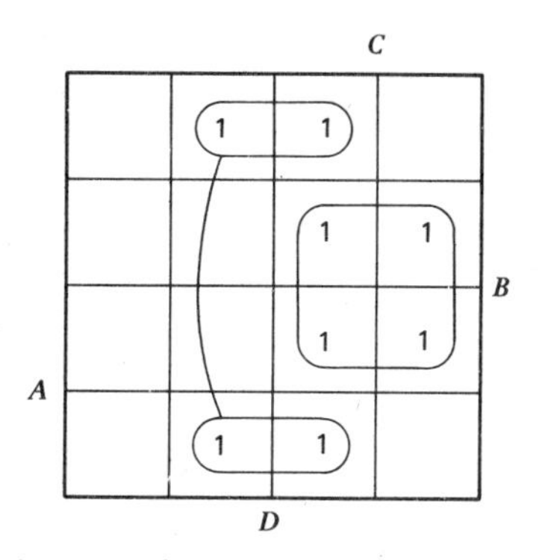

Figure 2.9. Mapping of the logical function $f(A, B, C, D) = \sum(1, 3, 6, 7, 9, 11, 14, 15) = \sum[(1, 3, 9, 11), (6, 7, 14, 15)] = \bar{B}D + BC$.
(*Note:* The numbers were allocated into the squares based on the following binary weight designation: $A = 8$, $B = 4$, $C = 2$, $D = 1$.)

After having mapped the terms of the function, groups are defined in order to represent the function via logical variables A, B, C, and D, and in a form as simple as possible. The mapping and grouping of terms are illustrated in Figure 2.10, where adjacent terms have been combined together to form three logical terms that define the given function.

From the combination of 0, 1, 8, and 9, the logical term $\bar{B}\bar{C}$ was derived, and from 5 and 7, the logical term $\bar{A}BD$. Term 14 has no adjacent terms and it will therefore, be explicitly defined by a logical term that contains all the variables. This is $ABC\bar{D}$. ∎

Very often in logical design one may encounter a logical function $f(A, B, C, D)$ which, by definition, will never be in the domain of another

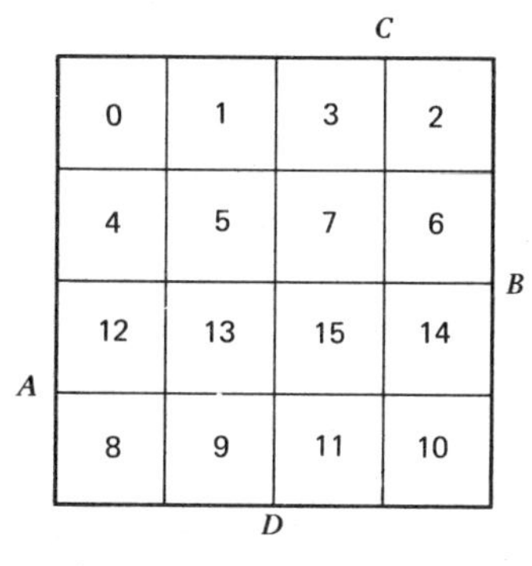

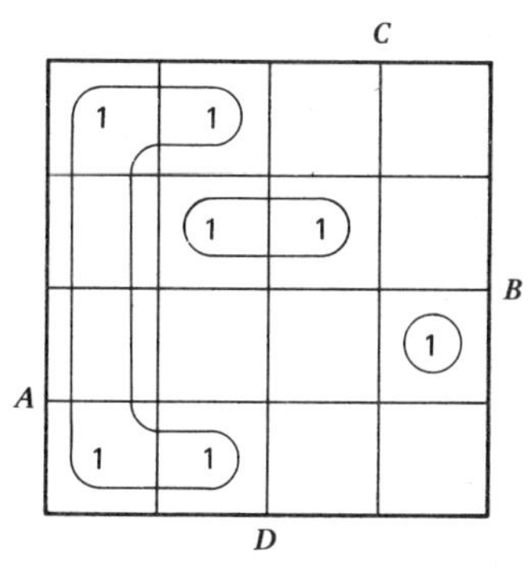

Figure 2.10. Mapping of the logical function $f(A, B, C, D) = \sum(0, 1, 5, 7, 8, 9, 14) = \sum[(0, 1, 8, 9), (5, 7), (14)] = \bar{B}\bar{C} + \bar{A}BD + ABC\bar{D}$.

function $g(A, B, C, D)$. If we wish to decode certain terms of the logical expression $f(A, B, C, D)$ that comprise function $h(A, B, C, D)$, we may combine function $h(A, B, C, D)$ with function $g(A, B, C, D)$ or with any part of $g(A, B, C, D)$. This may result in a function which is simpler to implement. Thus, since $f(A, B, C, D)$ will never be in the domain of $g(A, B, C, D)$

$$h(A, B, C, D) = h(A, B, C, D) + g(A, B, C, D)$$

Implementation of the right-hand side can be considerably simpler than that of the left-hand side since it contains terms that can only help in the simplification of expression $h(A, B, C, D)$. The terms of the function $g(A, B, C, D)$ are called "don't care terms."

Therefore, a "don't care term" is a term that represents a state in which a given function, by definition, is not expected to be, and consequently no harm can be done if that term be combined along with the actual terms of the function. Such operation may favorably affect the simplification of the expression that represents the given function.

For example, if $f(A, B, C, D)$ represents decimal numbers in binary code, the expression will be

$$f(A, B, C, D) = \sum(0, 1, 2, 3, 4, 5, 6, 7, 8, 9)$$

and, by definition, $f(A, B, C, D)$ will never be in the domain of logical function

$$g(A, B, C, D) = \sum(10, 11, 12, 13, 14, 15)$$

If we wish to decode the function

$$h(A, B, C, D) = \sum(8, 9)$$

we may combine that function with $g(A, B, C, D)$, or any part of it, in order to obtain an expression for $h(A, B, C, D)$ which is simpler than the direct implementation of $\sum(8, 9)$.

Figure 2.11 illustrates the Veitch diagrams of functions $h(A, B, C, D)$, $g(A, B, C, D)$, and that of the combined version $h(A, B, C, D) + g(A, B, C, D)$. It can be easily seen that the combined form yields a much simpler expression compared to that of the original function.

The simplicity offered by the use of the don't care terms is accompanied by a risk of satisfying the function with erroneous conditions. This will occur any time the function due to external error finds itself in the domain where, by definition, it is not supposed to be. Therefore, the convenience extended to the user by the don't care terms is paid with a decrease in the function's reliability. Consequently, convenience versus reliability must always be considered before using don't care terms.

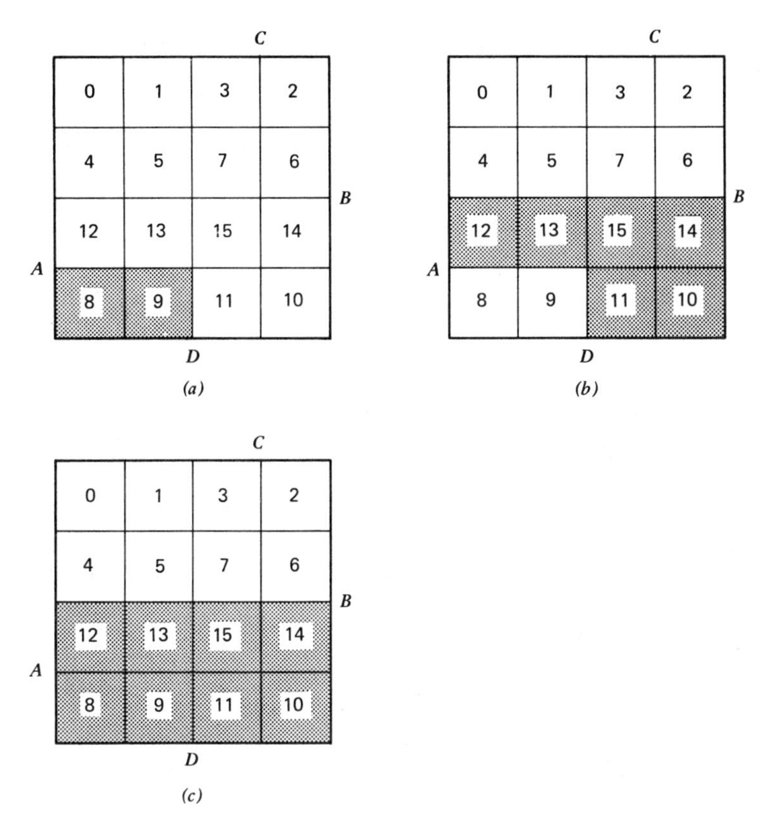

Figure 2.11. Use of don't care terms: (*a*) $h(A, B, C, D) = \sum(8, 9) = A\bar{B}\bar{C}$; (*b*) $f(A, B, C, D) = \sum(10, 11, 12, 13, 14, 15)$ = don't care terms; (*c*) $h(A, B, C, D) + g(A, B, C, D) = \sum(8, 9, 10, 11, 12, 13, 14, 15) = A$.

2.5 LOGICAL ELEMENTS

The simplest logical element is the GATE. The gate is a device that has two or more inputs and one output where the output depends on the state of the inputs. Gates are classified into two main categories: The AND gates and the OR gates.

The output of an AND gate is in its 1 or TRUE state when all of its inputs are in their 1 state—that is, if an AND gate has three inputs, A, B, and C, the gate output will be 1 only when $A = 1$, $B = 1$, and $C = 1$. For any other combination of input states, the output of the three-input AND gate will be in the 0 of FALSE state. Shown here are all possible input combinations of a three-input AND gate and the resulting output state. This presentation is called a truth table.

A	B	C	AND gate output
0	0	0	0
0	0	1	0
0	1	0	0
0	1	1	0
1	0	0	0
1	0	1	0
1	1	0	0
1	1	1	1

The output of an OR gate is in its 1 or TRUE state when either of its inputs is in the 1 state—that is, if an OR gate has three inputs, A, B, C, the gate output will be 1 when either $A = 1$, $B = 1$, or $C = 1$ and only then. Shown here are all possible input combinations of a three-input OR gate and the resulting output state.

A	B	C	OR gate output
0	0	0	0
0	0	1	1
0	1	0	1
1	1	1	1
1	0	0	1
1	0	1	1
1	1	0	1
1	1	1	1

Another logical element is the inverter. This is a single-input single-output device, the output of which is always the logical opposite of its input. That is, when the input is 1, the output is 0, and vice versa. When an AND gate is followed by an inverter, that is, when the output of the AND gate is connected to the input of the inverter, the inverter output is the opposite of that of the AND gate. This AND-gate-inverter arrangement is called NAND gate, NOT AND. When an OR gate is followed by an inverter, the inverter output is the opposite of that of the OR gate. This OR-gate–inverter arrangement is called NOR gate, NOT OR. Figure 2.12 illustrates the equivalent of a NAND and a NOR gate.

There are two variables associated with the input and output of a gate. One is the voltage level and the other is the significance assigned to the

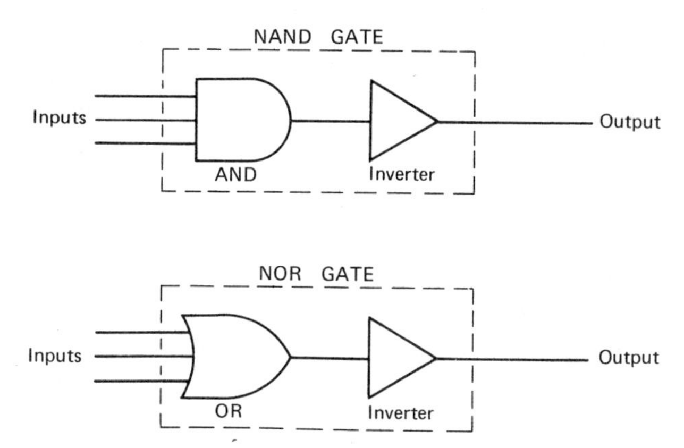

Figure 2.12. Equivalent form of NAND and NOR gates.

voltage level. A logical state, which may be the input or the output of a gate, is physically expressed by means of either of two voltage levels. The most positive level is called, by definition, HIGH, and the less positive, which may even be negative, is called LOW. In some digital circuits, HIGH is 0 V and LOW −3 V, while in others, HIGH is 5 V and LOW is 0 V—that is, the designation as to which voltage be called HIGH and which LOW depends on the voltages themselves, and there can be no arbitrary assignment.

The logical value of a logical signal is expressed in terms of 1 and 0, also defined as TRUE and FALSE, respectively. A voltage level, either HIGH or LOW, is called 1, when it has resulted from the satisfaction of a given requirement. The output of the AND gate performing the function $A \cdot B \cdot C$ is called 1, when all inputs A, B, and C are 1. The output of this gate will be called 1 regardless of its voltage level, which may be either HIGH or LOW. Similarly, the voltage output of an OR gate performing the function $A + B + C$ shall be called 1, anytime one of its inputs A, B, or C is 1, because that voltage indicates the satisfaction of the logic function $A + B + C$. Inversely, a voltage level is called 0 when it results from conditions that fail to satisfy a given requirement.

The two voltage levels HIGH and LOW, and the two logical values 1 and 0, produce the following two sets of logical rules:

1 = HIGH 0 = LOW
1 = LOW 0 = HIGH

The first set is defined as the Positive Logic Rule and the second set as the Negative Logic Rule. Figure 2.13 shows the symbols for a two-input AND gate, a two-input OR gate, and an inverter. Also shown are the respective tables of operation. By definition, the small circle at the input or output of the logic elements indicates that at this point the Negative Logic Rule is employed. Absence of a circle means that at that point Positive Logic Rule is

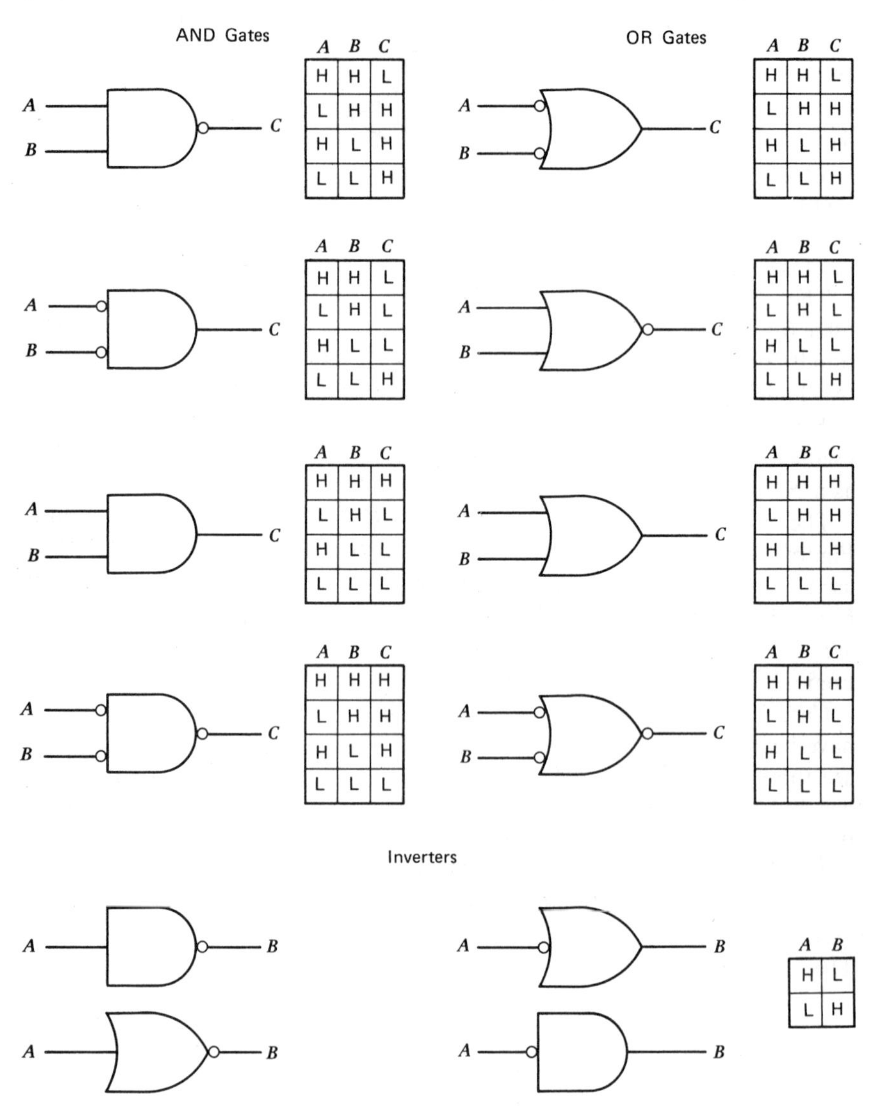

Figure 2. 13. Symbols and tables of operation for gates and inverter.

employed. The first OR gate symbol and the first AND gate symbol of Figure 2.13 are the most commonly used symbols, because they correspond to digital circuits extensively employed in logic design. The gates that are generally used in the industry are the NAND/NOR type, so described because the same component can be used to perform either a NAND or a NOR function. This can be achieved because the truth table of a NAND function is the same as that of a NOR function if the logic rules are reversed.

This can be easily seen when a NAND/NOR gate circuit is analyzed. Figure 2.14 illustrates such a circuit. This circuit can be defined as a logical component only after logic rules have been assigned to its inputs and output. This assignment will also define the circuit as a NAND or a NOR gate.

The basic property of the circuit of Figure 2.14 is that when both of its inputs are at 5 V, or open-circuited, the transistor is saturated and its output is close to 0 V (<0.5 V). Also, when one or both of its inputs are at 0 V, or close to it (<0.8 V), the output of the circuit is at 5 V.

The characteristics of this circuit are tabulated on Figure 2.15, where table (*a*) presents the operation of the circuit, and (*b*) illustrates the information on table (*a*) in terms of HIGHs and LOWs. Applying positive and negative logic rules onto table (*b*), tables (*c*) and (*d*) are respectively generated. Table (*c*) is the truth table of the circuit of Figure 2.14 when the circuit is used as a NAND gate and table (*d*) is the truth table when the circuit is used as a NOR gate.

The inverter, although a single circuit, has four symbols, illustrated in Figure 2.13, all having the same truth table. The two symbols on the left are used when the inverter transforms the logic information from the Positive Logic Rule to the Negative Logic Rule, and the two symbols on the right are used when the inverter transforms the logic information from the Negative Logic Rule to the Positive Logic Rule.

Certain arrangements of gates are so commonly used that separate symbols have been assigned for them. The two most often encountered arrangements are those of the exclusive-OR denoted by $\oplus$; and of the coincidence gates denoted by $\odot$. Figure 2.16 illustrates the symbols and logical diagrams of the exclusive-OR and coincidence gates.

The exclusive-OR gate accepts two inputs and provides an output of 1 when the two inputs are different. This is the same as saying that the exclusive-OR gate provides the LSB of the binary sum of two input bits. The logical expression for the exclusive-OR is

$$A \oplus B = A\bar{B} + \bar{A}B$$

$$A \oplus 0 = A \qquad \text{(if } B = 0\text{)}$$

$$A \oplus 1 = \bar{A} \qquad \text{(if } B = 1\text{)}$$

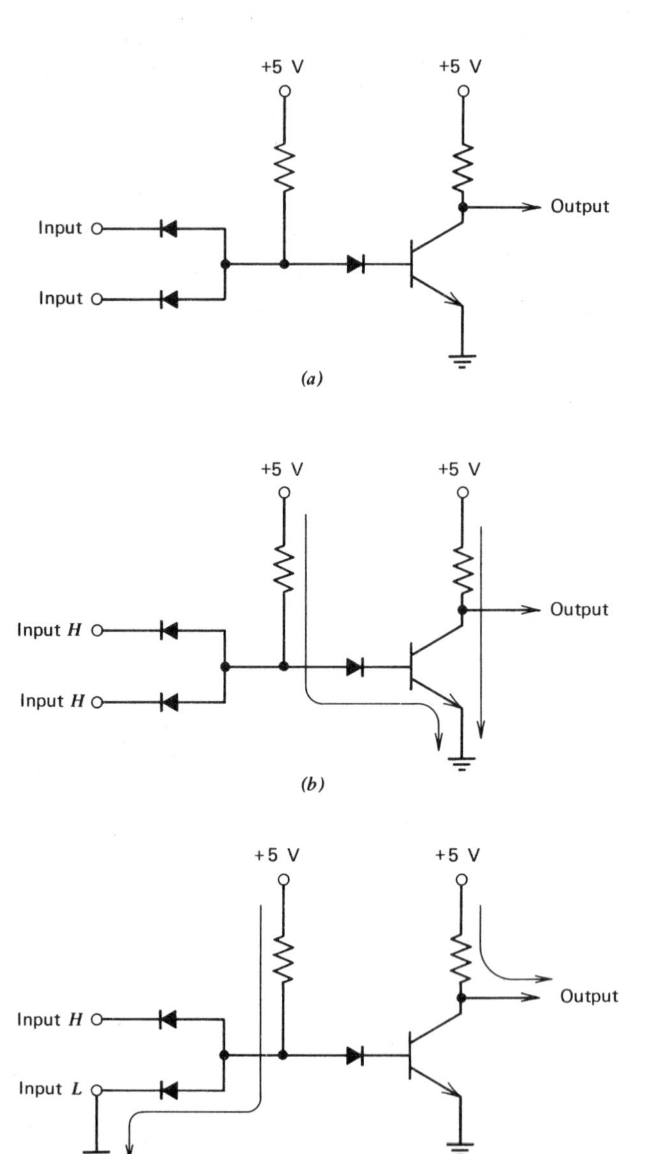

Figure 2.14. Schematic and operation of typical NAND/NOR gate: (*a*) simplified schematic of a NAND/NOR gate; (*b*) both inputs are open-circuited, or at +5 V; the transistor is saturated and the collector voltage is less than 0.5 V; (*c*) one input is open-circuited and the other is at ground; the transistor is cut-off and the collector voltage is 5 V.

Input		Output
A	*B*	*C*
Open circuit or 5 V	Open circuit or 5 V	0 V
Open circuit or 5 V	0 V	5 V
0 V	Open circuit or 5 V	5 V
0 V	0 V	5 V

(*a*)

Input		Output
A	*B*	*C*
HIGH	HIGH	LOW
HIGH	LOW	HIGH
LOW	HIGH	HIGH
LOW	LOW	HIGH

(*b*)

Input		Output
A	*B*	*C*
1	1	0
1	0	1
0	1	1
0	0	1

(*c*)

Input		Output
A	*B*	*C*
0	0	1
0	1	0
1	0	0
1	1	0

(*d*)

Figure 2.15. Characteristics of the NAND/NOR gate of Figure 2.14; (*a*) circuit operation; (*b*) definition; (*c*) application of Positive Logic Rule—circuit performs as NAND gate; (*d*) application of Negative Logic Rule—circuit performs as NOR gate.

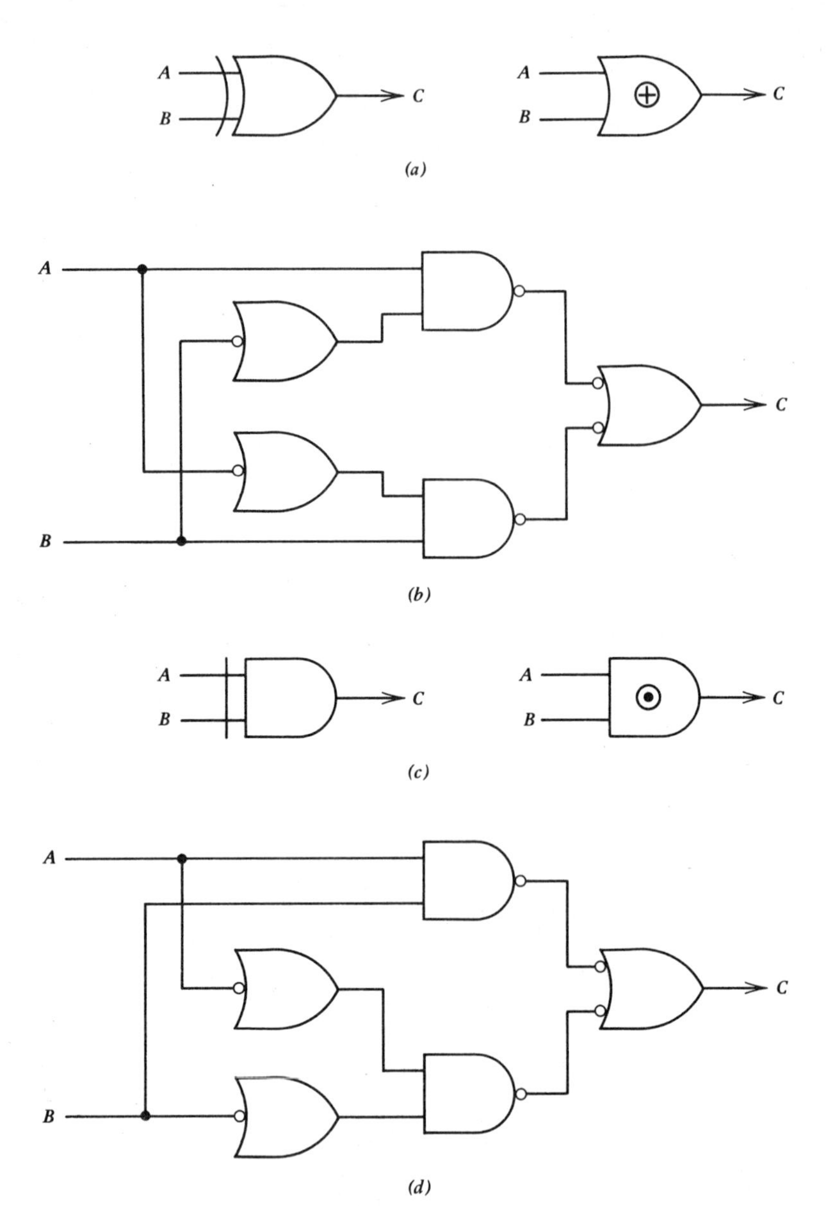

Figure 2.16. Symbols of (*a*) exclusive-OR gate and (*c*) coincidence gate. The corresponding logic diagrams are (*b*) and (*d*), respectively, where in (*b*) $C = \overline{A}B + A\overline{B} = A \oplus B$ and in (*d*) $C = \overline{A}\overline{B} + AB = A \odot B$.

When exclusive-OR operations can be linked together, a linear relationship exists which allows direct interchangeability of terms on either or both sides of the equation:

$$\begin{aligned} &\text{if} && A \oplus B = 1 \\ &\text{then} && A \oplus 1 = B \\ &\text{and} && B \oplus 1 = A \end{aligned}$$

Similarly,

$$\begin{aligned} &\text{if} && A \oplus B \oplus C \oplus D = 0 \\ &\text{then} && A \oplus C \oplus D \oplus B = 0 \\ &\text{and} && 0 \oplus B \oplus A \oplus D = C \end{aligned}$$

For example, if $A = 1$, $B = 0$, $C = 1$, and $D = 0$, we have

$$\begin{aligned} A \oplus B \oplus C \oplus D &= 1 \oplus 0 \oplus 1 \oplus 0 \\ &= 1 \oplus 1 \\ &= 0 \end{aligned}$$

In chains of exclusive-OR operations it is the number of 1s that counts. An even number of 1s produces a 0, while an odd number of 1s produces a 1. The number of 0s does not affect the outcome of an exclusive-OR operation. This is because the exclusive-OR of two 0s is also a 0. For example;

$$\begin{aligned} A \oplus B \oplus 0 \oplus 0 \oplus 0 \oplus 0 &= C \\ A \oplus B \oplus 0 \oplus 0 &= C \\ A \oplus B \oplus 0 &= C \\ A \oplus B &= C \end{aligned}$$

The coincidence gate accepts two inputs and provides an output of 1 when the two inputs are the same. The logical expression for the coincidence gate is

$$\begin{aligned} A \odot B &= AB + \overline{A}\overline{B} \\ A \odot 0 &= \overline{A} && \text{(if } B = 0\text{)} \\ A \odot 1 &= A && \text{(if } B = 1\text{)} \end{aligned}$$

Coincidence operations can be linked together, and terms can be interchanged in the same way as in the exclusive-OR operations; thus,

$$\begin{aligned} &\text{if} && A \odot B \odot C = 1 \\ &\text{then} && A \odot 1 = B \odot C \\ &\text{and} && B \odot 1 \odot C = A \end{aligned}$$

The coincidence gate is the counterpart of the exclusive-OR gate with the role of 1s and 0s reversed. Consequently, in chains of coincidence operations it is the number of 0s that counts. In this case, an even number of 0s produces a 1, while an odd number of 0s produces a 0. The number of 1s in a chain of coincidence operations does not in any way affect the result. This is because the coincidence of two 1s is also a 1. For example,

$$
\begin{aligned}
A \odot B \odot 1 \odot 1 \odot 1 \odot 1 &= C \\
A \odot B \odot \quad 1 \quad \odot \quad 1 \quad &= C \\
A \odot B \odot \qquad 1 \qquad &= C \\
A \odot B &= C
\end{aligned}
$$

Exclusive-OR and coincidence operations may be combined together and the terms of the formed chain can be freely interchanged. For example, the expression

$$f(A, B, C, D) = A \oplus B \odot C \oplus D \odot E$$

can take several forms, such as

$$
\begin{aligned}
f(A, B, C, D, E) &= A \odot B \odot C \oplus D \oplus E \\
&= A \oplus B \oplus C \odot D \odot E \\
&= C \odot E \oplus D \odot A \oplus B \\
&\text{etc.}
\end{aligned}
$$

A comparison of the logical expressions of the exclusive-OR and the coincidence operations, when one of the inputs is inverted shows the following:

$$
\begin{aligned}
A \oplus \bar{B} &= A\bar{\bar{B}} + \bar{A}\bar{B} \\
&= AB + \bar{A}\bar{B} \\
&= A \odot B
\end{aligned}
$$

Therefore, the exclusive-OR and the coincidence symbols can be interchanged if one of the inputs is inverted. For example,

$$
\begin{aligned}
f(A, B, C) &= A \odot B \odot C \\
&= A \oplus \bar{B} \odot C \\
&= A \oplus \bar{B} \oplus \bar{C}
\end{aligned}
$$

It should be noted that exclusive-OR and coincidence symbols cannot be indiscriminately mixed with OR and AND symbols. In these cases parentheses must be placed to explicitly indicate the terms of the OR and AND operations.

In addition to the exclusive-OR and coincidence gates, there is the AND-OR-Invert or AOI gate, which consists of AND gates, the outputs of which have been NORed. There is no standard logical configuration for the AOI gates, neither is there a symbol for them; AOI gates can be found with two, three, or four AND gates with each gate having two, three, or four inputs.

When AOI gates are defined, they are referred to by the number of inputs their AND gates have. For example, an AOI gate that has four AND gates—two of which have two inputs, one has three inputs, and one has four inputs—is referred to as a 2-2-3-4 AOI gate. Plain noninverting AND-OR gates are also available and they are similarly defined. Figure 2.17 illustrates typical AOI and AO gates as well as symbols for them recommended by the author.

All the logical elements and the arrangements of such elements described above process logical information without storing it. A logical arrangement that stores information is that of the "flip-flop."

The flip-flop is a logical element composed of gates which are connected in such a fashion that the overall arrangement gives a bistable operation—that is, the overall arrangement can rest in either of two states. The outputs of flip-flop are labeled Q and $\bar{Q}$, where Q is the nominal output and $\bar{Q}$ is the inverted output. Outputs Q and $\bar{Q}$ are always complementary; that is, when Q is 1, $\bar{Q}$ is 0, and vice versa.

The simplest form of a flip-flop, which consists of two gates that are cross-connected, is illustrated in Figure 2.18. This is called set-reset flip-flop, or *R-S* flip-flop. It is often called latch flip-flop. The *R-S* flip-flop has two inputs. The set input, which when activated makes $Q = 1$ and $\bar{Q} = 0$, and the reset, or clear input, which when activated makes $Q = 0$ and $\bar{Q} = 1$.

Set and reset inputs should never be activated at the same time, because they will cause improper operation of the flip-flop. The flip-flop of Figure 2.18 operates with negative logic (1 = LOW) at the input, and positive logic (1 = HIGH) at the output. In this circuit, when the set input is 1, LOW, the reset input must be 0, HIGH.

When the set input is LOW, the Q output becomes HIGH. This results in both inputs to gate G_2, Q and reset to be HIGH causing that gate's output to be LOW. Gate G_2 output, which is $\bar{Q}$, feeds gate G_1, thus maintaining the state $Q = 1$ and $\bar{Q} = 0$ initiated by the LOW signal at input set.

Similarly, when reset is LOW, 1, and set is HIGH, 0, Q is LOW, 0, and $\bar{Q}$ HIGH, 1. After the LOW input to reset is removed, the flip-flop remains in the above state ($Q = 0$, $\bar{Q} = 1$) until the set input becomes LOW, 1.

The next flip-flop in complexity is the "clocked" *R-S* flip-flop. The symbol of that flip-flop, and its logic diagram are shown in Figure 2.19. This flip-flop has two pairs of S-R inputs. One pair of inputs is called "direct" because it has direct control over the flip-flop's state and the other pair of

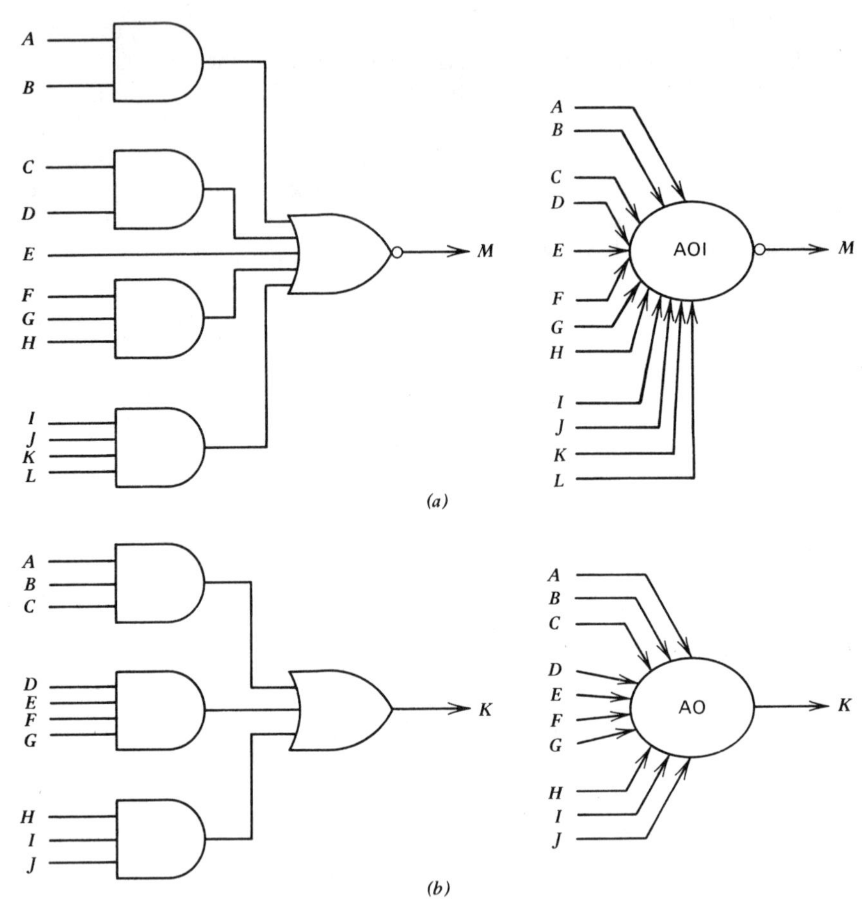

Figure 2.17. (*a*) Typical 2-2-3-4 AND-OR-INVERT (AOI) gate, its logic diagram, and a recommended symbol (logical equation. $M = \overline{AB + CD + E + FGH + IJKL}$). (*b*) Typical 3-4-3 AND-OR (AO) gate, its logic diagram, and a recommended symbol (logical equation: $K = ABC + DEFG + HIJ$).

inputs is called "clocked," because it can control the flip-flop state only when the clock input is in the 1 state. The operation of the direct set and direct reset inputs is identical to that of the set and reset inputs of the *R-S* flip-flop of Figure 2.18. The clocked *R-S* inputs of this flip-flop provide the advantage that information destined to enter the flip-flop will enter it not when it arrives at the input of the flip-flop, but when the clock input permits it. This way, entry of information to a flip-flop can be synchronized with other operations in the system.

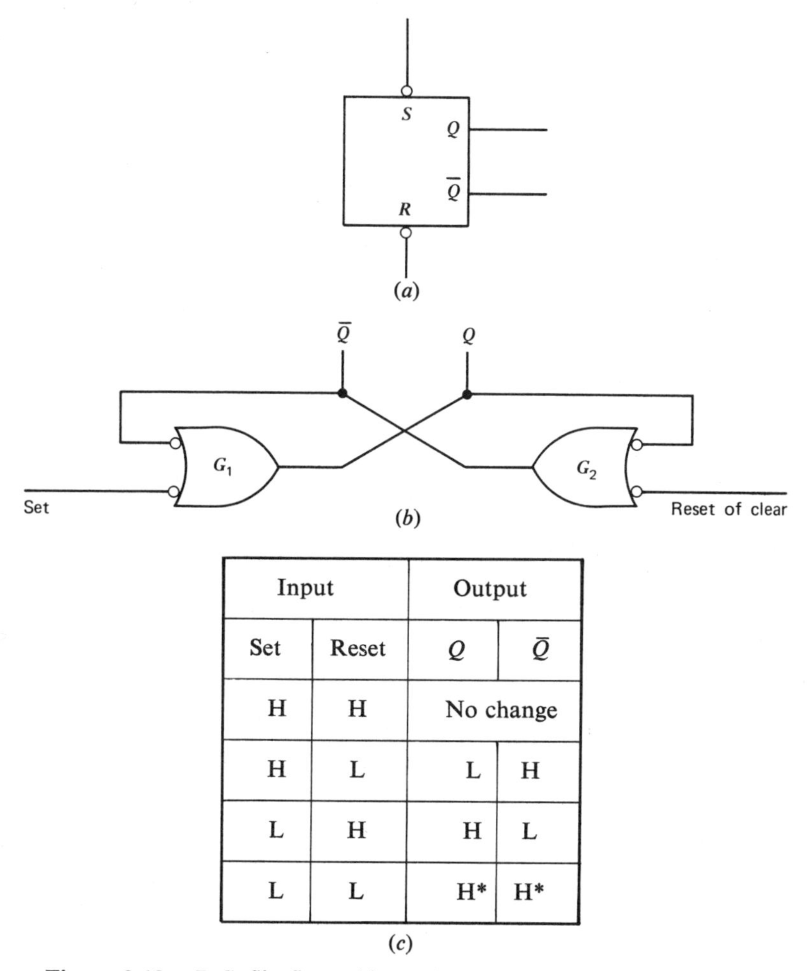

Input		Output	
Set	Reset	Q	$\bar{Q}$
H	H	No change	
H	L	L	H
L	H	H	L
L	L	H*	H*

(c)

Figure 2.18. R-S flip-flop: (*a*) symbol; (*b*) logic diagram; (*c*) table of operation. (*Note:* *Output is undetermined after simultaneous removal of inputs.)

The most commonly used flip-flop is the "master-slave" flip-flop. The master-slave flip-flop consists of two clocked *R-S* flip-flops connected in the configuration illustrated in Figure 2.20. The lower portion which accepts synchronous inputs is called "master" and the upper portion which provides the outputs is called "slave."

Information may enter the flip-flop either directly, by means of S_d and C_d, or under clock control by means of S_s and C_s. The direct entry is identical

(a)

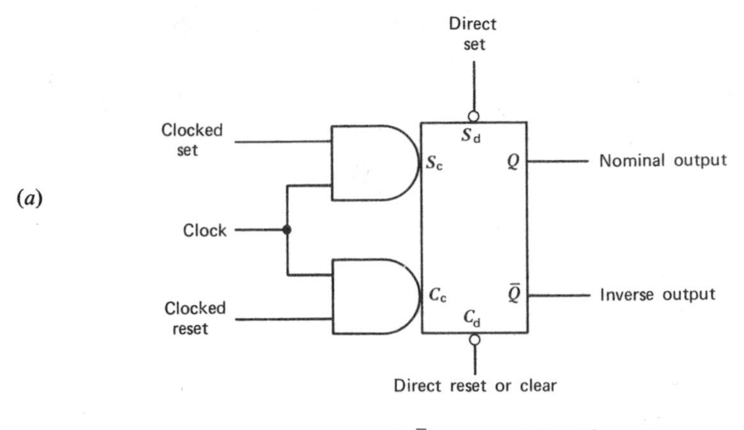

(b)

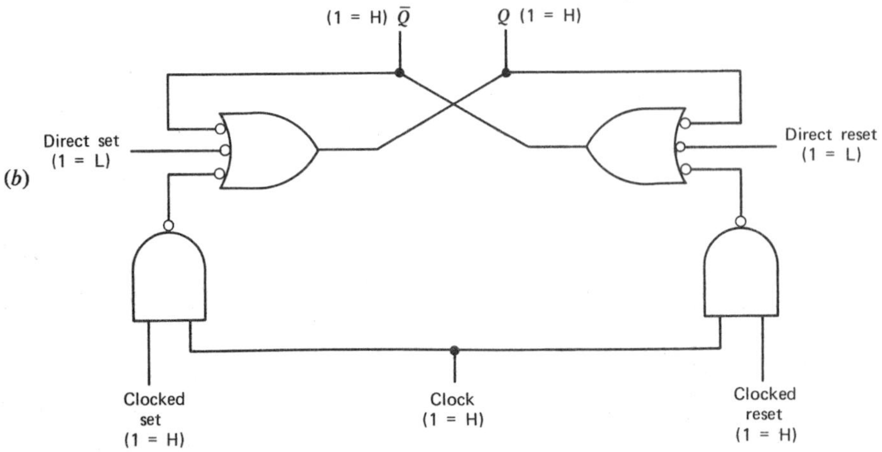

(c)

Direct Inputs*		Outputs	
Set	Reset	Q	$\bar{Q}$
H	*H*	No change	
H	*L*	*L*	*H*
L	*H*	*H*	*L*
L	*L*	*H***	*H***

Clocked Inputs***		Outputs	
Set	Reset	Q	$\bar{Q}$
H	*H*	*H***	*H***
H	*L*	*H*	*L*
L	*H*	*L*	*H*
L	*L*	No change	

Figure 2.19. Clocked R-S flip-flop: (*a*) symbol; (*b*) logic diagram; (*c*) tables of operation. (*Note:* *Clock input LOW. **Undetermined after simultaneous removal of the inputs. ***Direct inputs and clock inputs all HIGH.)

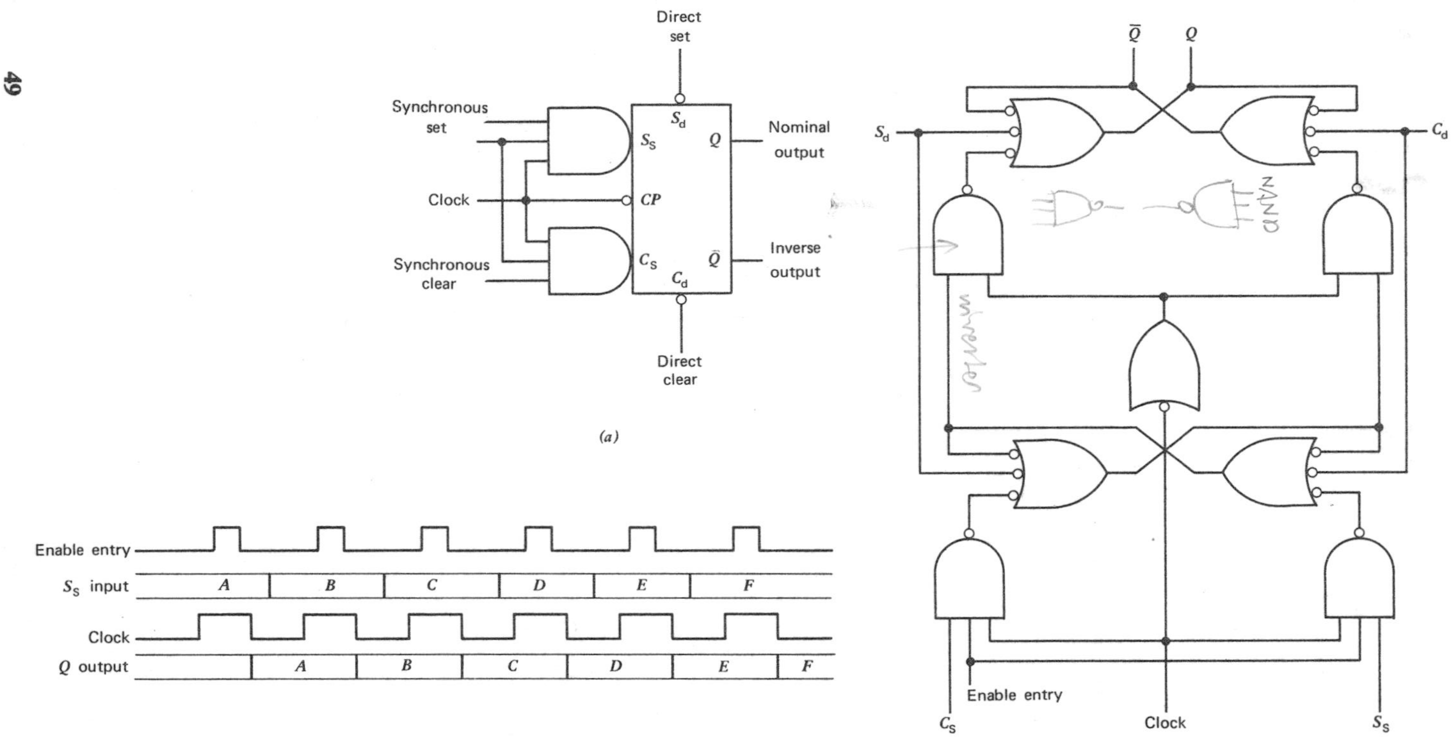

Figure 2.20. Master-slave flip-flop: (*a*) symbol; (*b*) logic diagram; (*c*) waveforms of operation. (*Note:* Information A–F is digital information and may be either HIGH or LOW.)

to that of an *R-S* flip-flop with the note that both upper and lower portions are activated simultaneously.

Synchronous data enter the flip-flop at the master portion with positive logic. When the clock and the "enable entry" inputs are both HIGH, information at S_s and C_s ($C_s = \bar{S}_s$) enters the master portion. If enable entry switches to LOW, the entered data will be protected against undesirable level changes at terminals S_s and C_s and will wait in the master portion until the next entry.

In the meantime, the entered data, now in the master portion, will be available to be copied into the slave portion. This will occur when the clock switches from HIGH to LOW. When the clock is LOW, the slave portion is directly affected by the output of the master portion, because the entry gates of the slave portion are enabled, while synchronous entry is disabled.

Thus, there is never a direct path between the synchronous inputs and the flip-flop output. Data either enter from outside into the master portion, or are copied from the master portion into the slave portion. For applications where the synchronous input is not expected to change during the time the clock is HIGH, the enable entry may be HIGH at all times.

The logical configuration of flip-flops labeled master-slave is not always the same. Therefore, before using such flip-flops, their logical diagrams must be defined in order to determine the exact conditions under which data are entered or altered.

Synchronous inputs S_s and C_s are also called J and K inputs, respectively. When this designation is used, the flip-flop is called *J-K*.

2.6 APPLICATION OF LOGICAL ELEMENTS

As an introduction to the use of logical components and to the relationship between logical expressions and logical components, the following three examples are presented:

Example 2.8. Design a logical network that implements the expression

$$G = (AB + CD)E + \bar{F}$$

This expression can be implemented in its present form or it can be expanded to an equivalent form and then implemented. Expanding the above expression we obtain

$$G = ABE + CDE + \bar{F}$$

Figure 2.21 illustrates the implementation of each of the above two expressions. A comparison of these two implementations shows that the expanded approach requires less logical components. ∎

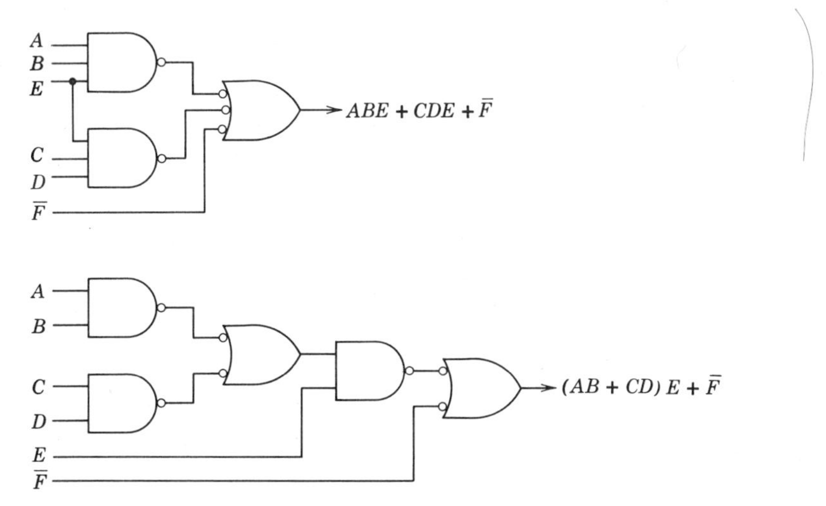

Figure 2.21. Implementation of the logic expression of Example 2.8.

Therefore, logical expressions should not be implemented directly from the form they are found in, but after an evaluation of all possible alternate approaches.

Example 2.9. Select the appropriate logical components that implement the expression

$$B^n = A^{n-1}$$

This expression indicates that B should have the value A had one clock period ago—that is, after each clock pulse, signal B must have the logical value signal A had immediately before that clock pulse. This value will be maintained by B until the next clock pulse arrives.

This operation can be achieved by means of a master-slave flip-flop as illustrated in Figure 2.22 where the appropriate logical network, as well as the waveforms of operation, are shown. ∎

Example 2.10. Select a logical component that implements the expression

$$A^{n+1} = \overline{A^n}$$

This expression indicates that the logical value of variable A at a reference time $n + 1$ is the logical inverse of what A was one period before, or at n. If A is 1 at n, it must become $\bar{1}$ at $n + 1$, or 0; and if A is 0 at n, it must become $\bar{0}$ at $n + 1$, or 1. In other words, A changes state at the beginning of each time period.

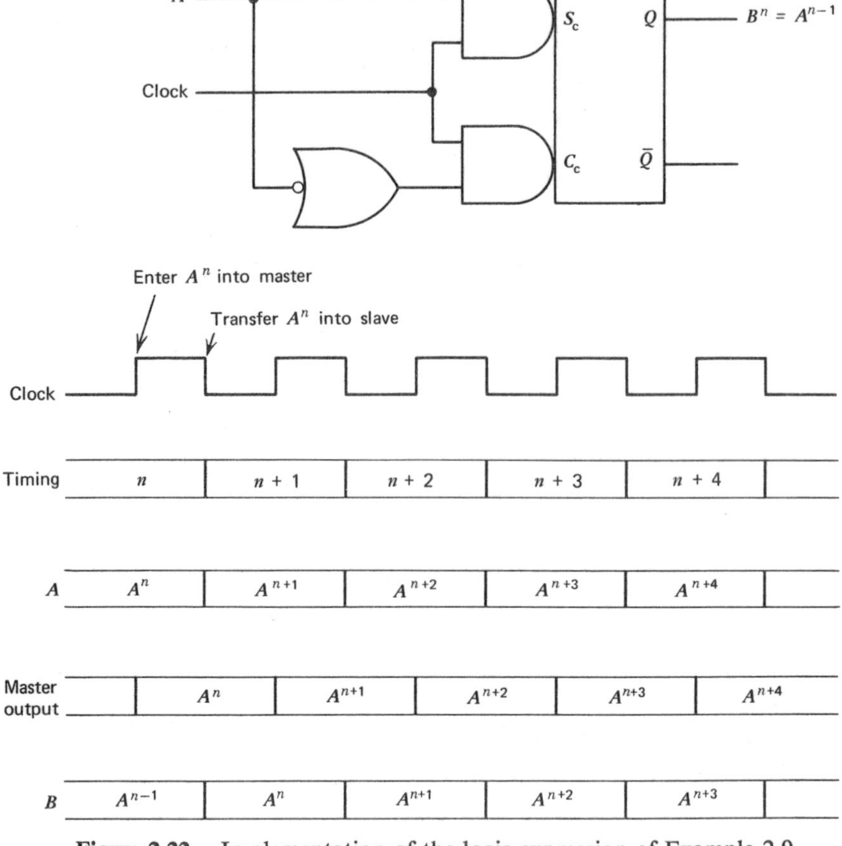

Figure 2.22. Implementation of the logic expression of Example 2.9.

The logical operation $A^{n+1} = \overline{A^n}$ can be implemented by means of a master-slave flip-flop connected in a cross-feedback fashion as shown in Figure 2.23, where the new value of A is always the inverse of its present value. ▮

PROBLEMS

1. Using boolean algebra, simplify the following logic expressions:
   (a) $AB(BC + A)$ (b) $(A + D)(A\bar{D})$
   (c) $\overline{AB + \bar{A}\bar{B} + \bar{A}B + A\bar{B}}$ (d) $\overline{\bar{A}BC}(B + \bar{C})$
   (e) $AB + B\bar{C} + ABC + AB\bar{C}$ (f) $(A + B + C)(\bar{A} + \bar{B} + \bar{C})$

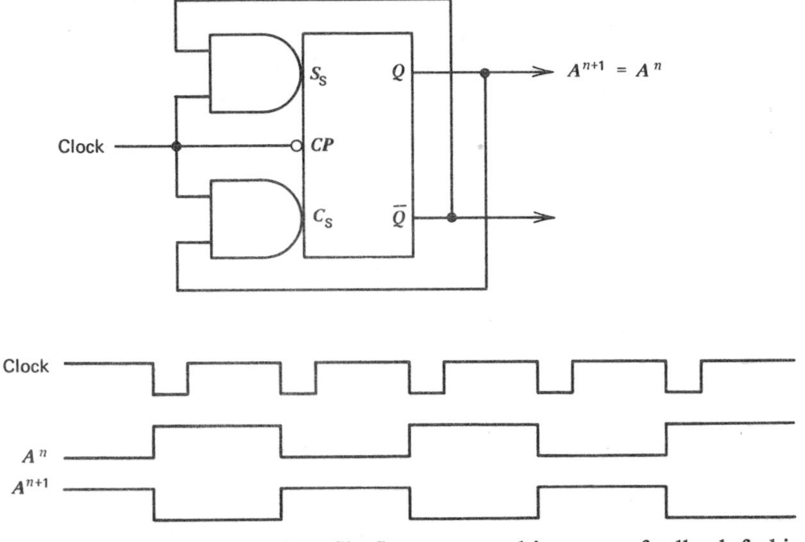

Figure 2.23. Master-slave flip-flop connected in a cross-feedback fashion.

2. By means of logical diagrams, simplify the following logical expressions (DC: Don't care terms)
   (*a*) $ABC + \bar{A}B\bar{C} + A\bar{B}C + \bar{A}\bar{B}\bar{C}$
   (*b*) $A\bar{B}\bar{C}D + AB\bar{C}\bar{D} + A\bar{B} + A\bar{D} + A\bar{B}C$
   (*c*) $AB + BC + CE + AD + \bar{A}BC\bar{D} + A\bar{B}\bar{C}\bar{D}$
   (*d*) $\sum(0, 1, 5, 7, 8, 11, 14)$, $DC(3, 9, 15)$
   (*e*) $\sum(3, 5, 9. 10, 12, 15)$, DC(0, 1, 13)
   (*f*) $\sum(2, 4, 5, 9, 10, 11, 14)$, DC(8, 12, 13, 15)
3. State the difference between the Positive and Negative Logic Rules.
4. Implement the following logical expressions using NAND/NOR gates. Input is available in positive logic rule and output should be in negative logic rule.

   (*a*) $(A + B)C$ (*b*) $(\overline{\bar{A} + B})C$
5. Implement the following logical expressions using the components discussed in this chapter:
   (*a*) $AB^n + CB^{n-1}$ (*b*) $(A^n + B^{n-1})\cdot(A^{n-1} + B^n)$
6. State a definition and an expression for the exclusive-OR and the coincidence gates.
7. State the advantages of a master-slave flip-flop over an *R-S* flip-flop.
8. State the advantages and disadvantages of don't care terms.

9. Write the truth table of the following logical expressions:
   (a) $(AB) \oplus (CD)$
   (b) $[A \odot B]C$
   (c) $\overline{A(B + CD)}$
10. Implement the following logical expressions:
   (a) $(A + B) \odot C$
   (b) $(A\bar{B} \oplus CD)B$
   (c) $(A\bar{B}C \odot \bar{D}) \oplus E$

CHAPTER THREE

Digital Integrated Circuits

3.1 INTRODUCTION

The extensive use of digital techniques in the design of electronic systems has created a need for miniature digital circuits. Miniaturization, which originally started as dense packaging of regular or miniature discrete components, has now developed into total component integration.

This integration is achieved by the simultaneous fabrication, on a single piece of material, of all components of a circuit. The result of this process has been the creation of integrated circuits which are so small that their volume is about one-thousandth of the volume of their housing. That is, the actual size of the integrated circuits is negligible, and the major consideration in their packaging is the circuits' interconnections with the outside world. Numerous items of advanced electronic equipment are presently being built using digital techniques implemented by digital integrated circuits. The construction of much of this equipment was impractical in the past because of size considerations.

The digital integrated circuits are classified into five major categories, often called "families." These are

Resistor-transistor logic	RTL
Diode-transistor logic	DTL
Transistor-transistor logic	TTL
Emitter-coupled logic	ECL
Metal-oxide semiconductors	MOS

Each of these categories is examined in this chapter, following the review of the general characteristics of digital integrated circuits.

3.2 CHARACTERISTICS OF DIGITAL CIRCUITS

To properly evaluate digital integrated circuits the engineer needs to carefully study the manufacturer's data sheets that accompany these circuits. The basic characteristics that specify the performance of digital integrated circuits are defined below.

Fan-in

Fan-in is the number of inputs that can control the operation of a digital circuit. Sometimes, extender circuits are used to increase the fan-in capability of digital circuits. Excessive increase of fan-in by means of extenders, however, increases the input capacitance of the circuit which, in turn, causes propagation delay to increase.

When inputs of a DTL or TTL circuit are unused, they should be connected to V_{CC}, the positive-supply voltage of the circuit. In the case of RTL and MOS, unused inputs should be connected to ground. When using ECL circuits, unused inputs should be connected to the negative supply voltage of the circuit.

Fan-out

Fan-out is a number indicating the ability of a digital circuit's output to drive other similar circuits. Each logic family has a "unit load" expressed in milliamperes. This unit load is the current required to activate a typical circuit of that logic family. The fan-out of a digital circuit is determined by normalizing its output current capability to the unit load current.

For example, let the unit load current of the TTL circuits of a given manufacturer be 1.3 mA. This is the current that flows from the input of the driven circuit into the output of the driving circuit. If the driving circuit has a current sink-in capability of 14.3 mA, the fan-out of that circuit is 11, 14.3 mA divided by 1.3 mA.

The number representing fan-out is valid only for loads to which it has been referenced. If the digital circuit, the fan-out of which is known, is to drive circuits of different unit load, its fan-out should be recomputed based on those unit loads.

The unit load of a logic family is usually the load provided by a gate, when that gate is driven by a circuit of that family. The load provided by the clock input of a flip-flop of the same family may not be the same as that of a gate. In most cases the load of the clock input of a flip-flop is twice that of a gate. Therefore, when evaluating the fan-out requirements of a digital circuit, the sum of the unit loads provided by the driven circuits should be computed. For example, if an output is to drive three single-unit load gates and four double-unit load flip-flop clock inputs, the fan-out capability of that output should be 11 ($3 \times 1 + 4 \times 2 = 11$).

Logic Levels

Under normal operation outputs of digital circuits are found to be in either of two states. These are HIGH and LOW. HIGH refers to the more positive of the two output states and LOW to the more negative one. The nomenclature for HIGH is V_H and for LOW is V_L.

For example, in the TTL family where the two states are nominally 0 V and 5 V, the former is called LOW and the latter HIGH. Similarly, in the ECL family where the two states are -1.55 V and -0.75 V, the reference is LOW and HIGH, respectively.

Some digital circuits are designed to have three states, with the third state being the equivalent of an open circuit. Mention of these circuits is made in Section 3.11.

Propagation Delay

Propagation delay is the time between the application of the input and the correct response of the output. There are two numbers that indicate propagation delay of digital circuits. These are tpd^+ and tpd^-. The first, tpd^+, is the time interval between the 50% point of the activating edge of the input waveform and the 50% point of the resulting positive-going edge of the output waveform. The second delay term, tpd^-, is the time interval between the 50% point of the activating edge and the 50% point of the resulting negative-going edge of the output waveform. In both cases, the activating edge may be either positive-going or negative-going. When the above two terms, tpd^+ and tpd^-, are added together and then divided by two, they produce a term known as "average propagation delay" designated by t_{av}.

The time interval between the 10% point and the 90% point of a positive-going edge is called "rise time" and it is designated by t_r. The time interval between the 90% point and the 10% point of a negative-going edge is called "fall time" and is designated by t_f.

Figure 3.1 shows the time relation between input and output waveforms of a digital circuit and illustrates the delay terms defined above.

Propagation delay of digital circuits varies with temperature, power-supply voltage, and fan-out. The graphs of Figure 3.2 show that the propagation delay varies inversely with supply voltage and directly with fan-out. The effect of temperature on propagation delay depends on the power-supply

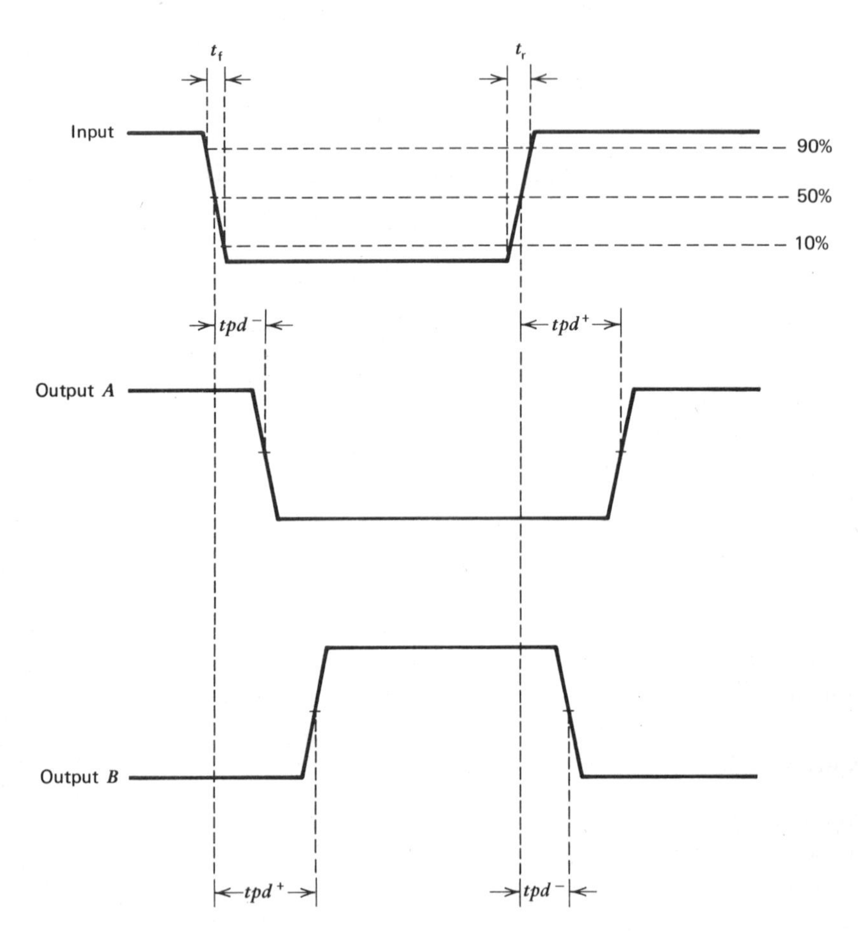

Figure 3.1. Definition of tpd^-, tpd^+, t_f, and t_r.

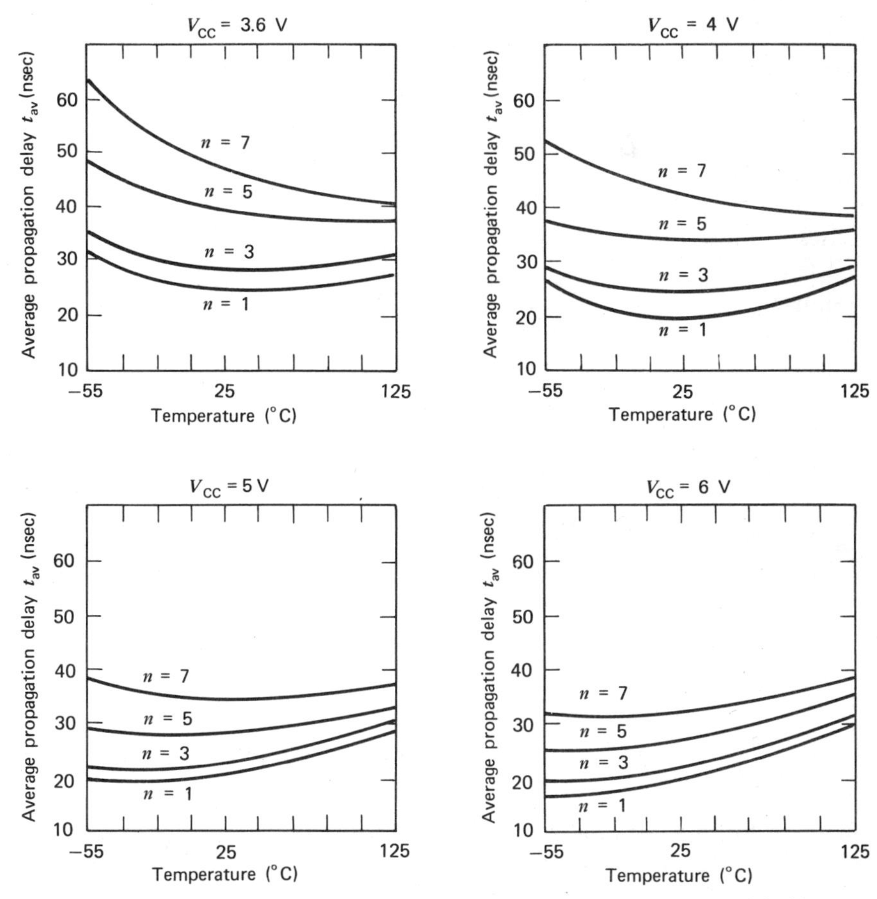

Figure 3.2. Average propagation delay versus temperature with fan-out as the parameter for various power-supply voltages.

voltage. With low-supply voltage the propagation delay varies inversely with temperature, while with high-supply voltage it varies directly.

Power Dissipation

Power dissipation, as in other electronic components, is a parameter expressed in watts or milliwatts and represents the actual power dissipated in the digital circuit. The number, representing this parameter, does not include the power being delivered by the digital circuit to other circuits. It does, however, include power delivered to the digital circuit from other circuits via the inputs or outputs of the digital circuit.

Noise Immunity

Noise immunity is the maximum amplitude of noise that does not cause erroneous output when it is added to the input signal of a digital circuit. There are two types of noise that should be considered in the design of digital systems. These are the dc noise and the ac noise. The dc noise is the steady drift in the voltage levels of the logical states, and the ac noise is the narrow pulses that are created, primarily, by switching transients.

The dc noise immunity of a digital circuit is the ability of that circuit to maintain a prescribed logic state in the presence of dc noise. The dc noise immunity is expressed by the following equations:

$$N_{L} = V_{IL\,max} - V_{OL\,max} \tag{3.1}$$

$$N_{H} = V_{OH\,min} - V_{IH\,min} \tag{3.2}$$

where

N_L = Noise immunity of the digital circuit input when the input signal is LOW

N_H = Noise immunity of the digital circuit input when the input signal is HIGH

$V_{IL\,max}$ = Maximum input voltage that can be read by the circuit as LOW

$V_{OL\,max}$ = Maximum output voltage that can represent LOW

$V_{OH\,min}$ = Minimum output voltage that can represent HIGH

$V_{IH\,min}$ = Minimum input voltage that can be read by the circuit as HIGH

Figure 3.3(*a*) is a pictorial presentation of dc noise immunity in digital circuits. The necessity for dc noise immunity will be clearer after the analysis of Figure 3.3(*b*), where a digital output of system *A* drives a digital input of system *B*. The output circuit of system *A* and the input circuit of system *B* are each powered by separate power supplies, nominally set to the same voltage level. It is possible, therefore, that the ground and V_{CC} levels of these two circuits differ, with the difference depending on power-supply regulation and voltage drops across power lines. As we shall see below, excessive difference in ground and V_{CC} levels, between system *A* and system *B*, causes transmission of erroneous signal.

Case 3.1 (Fig. 3.4(*a*)). The ground level of gate 1 has drifted upwards, relative to the ground level of gate 2, causing the LOW state of gate 1 output to be greater than $V_{IL\,max}$ of gate 2 at all times. The result is that the output of gate 2 is either LOW when gate 1 output is HIGH or it is undetermined when gate 1 output is LOW and has a value between $V_{IL\,max}$ and $V_{IH\,min}$ of gate 2. ∎

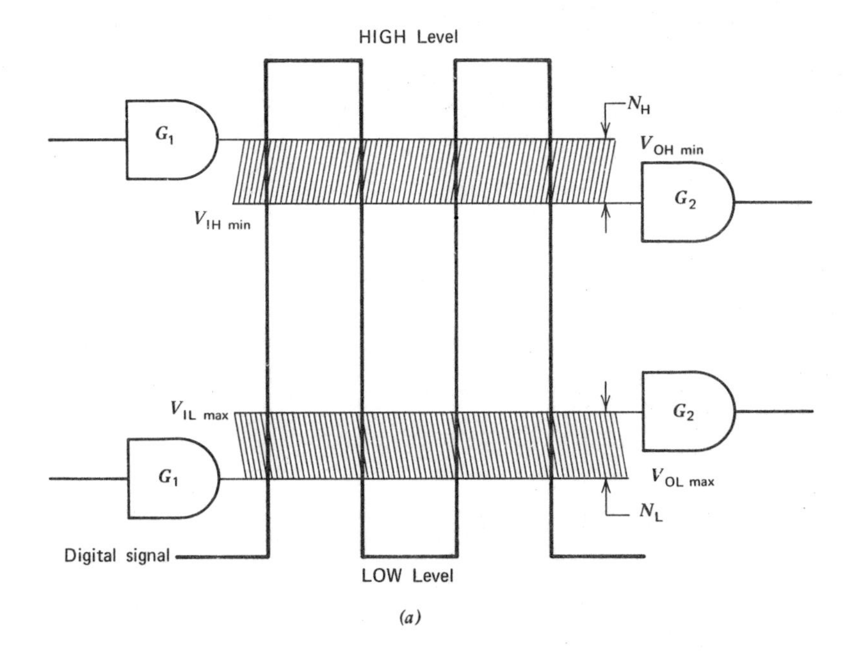

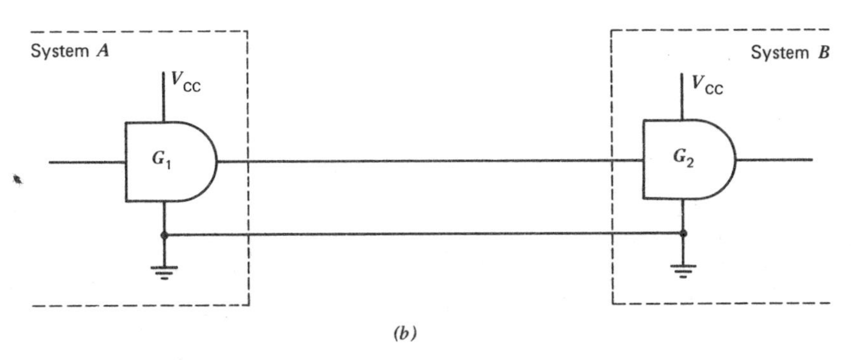

Figure 3.3. (*a*) Schematic illustrating dc noise immunity. (*b*) Actual connection of gates G_1 and G_2.

Case 3.2 (Fig. 3.4(*b*)). The ground level of gate 2 has drifted upwards, relative to the ground level of gate 1, causing the HIGH state of gate 1 output to be less than $V_{IH\,min}$ of gate 2 at all times. The result is that the output of gate 2 is either HIGH when gate 1 output is LOW, or it is undetermined when gate 1 output is HIGH and has a value between $V_{IL\,max}$ and $V_{IH\,min}$ of gate 2. ∎

Case 3.3 (Fig. 3.4(*c*)). The V_{CC} level of gate 1 has drifted downwards, relative to the V_{CC} level of gate 2, causing the HIGH state of gate 1 output to be less than $V_{IH\,min}$ of gate 2 at all times. The result is that the output of gate 2 is either HIGH when gate 1 output is LOW or it is undetermined when gate 1 output is HIGH and has a value between $V_{IL\,max}$ and $V_{IH\,min}$ of gate 2. ▮

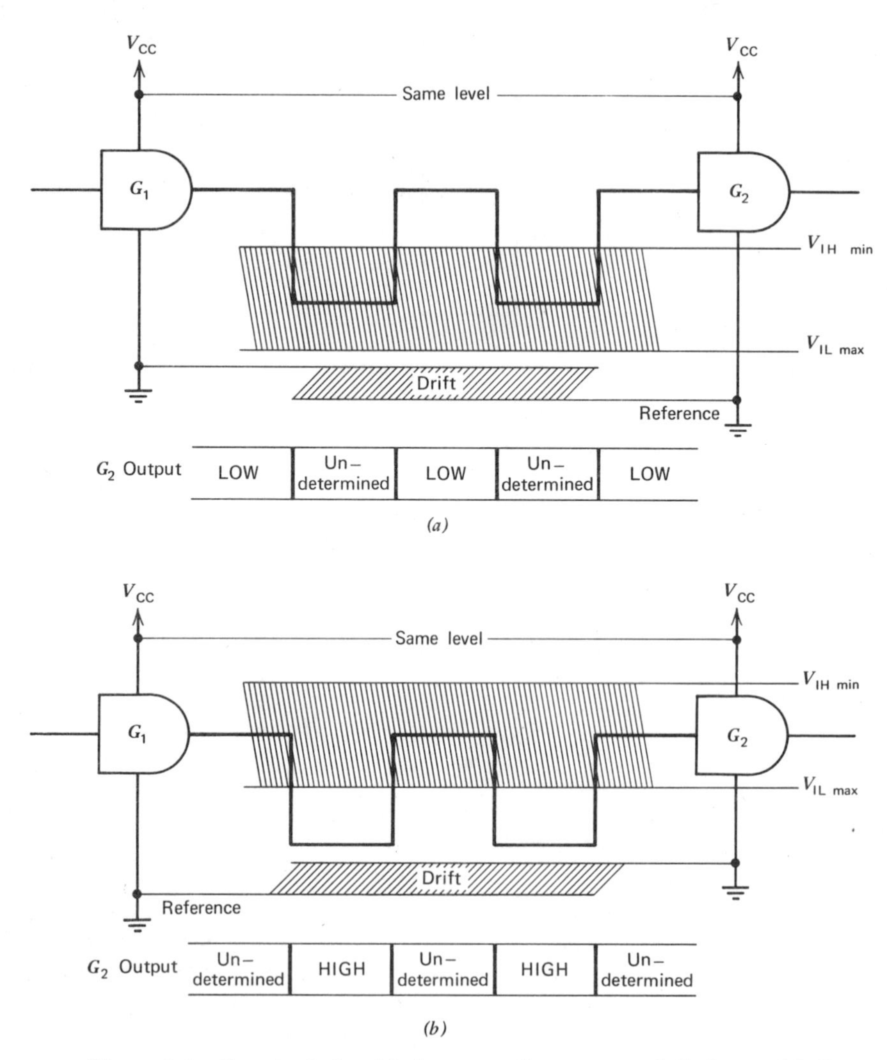

Figure 3.4. Level relationship between G_1 output and G_2 input: (*a*) Case 3.1, G_1 ground drifted upwards; (*b*) Case 3.2, G_2 ground drifted upwards;

Case 3.4 (Fig. 3.4(*d*)). The V_{CC} level of gate 2 has drifted downwards, relative to the V_{CC} level of gate 1, causing the LOW state of gate 1 output to be greater than $V_{IL\,max}$ of gate 2 at all times. The result is that the output of gate 2 is either LOW when gate 1 output is HIGH or it is undetermined when gate 1 output is LOW and has a value between $V_{IL\,max}$ and $V_{IH\,min}$ of gate 2. ∎

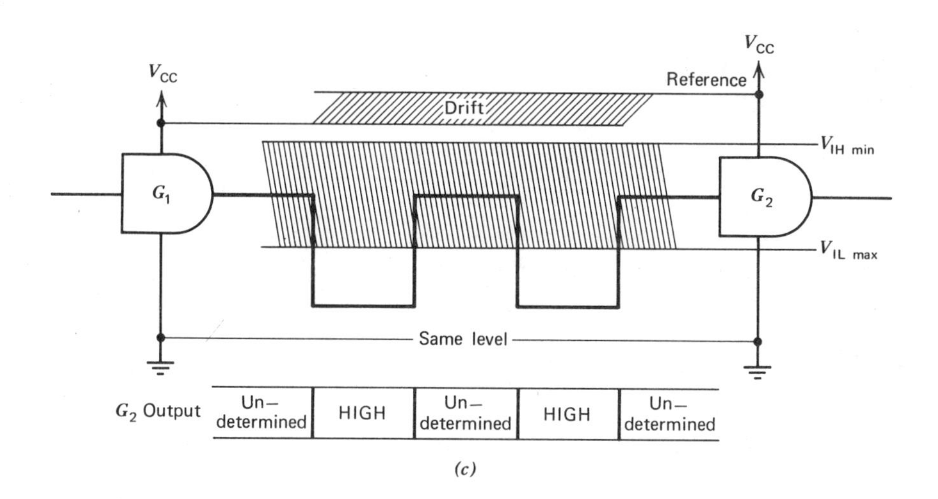

(*c*)

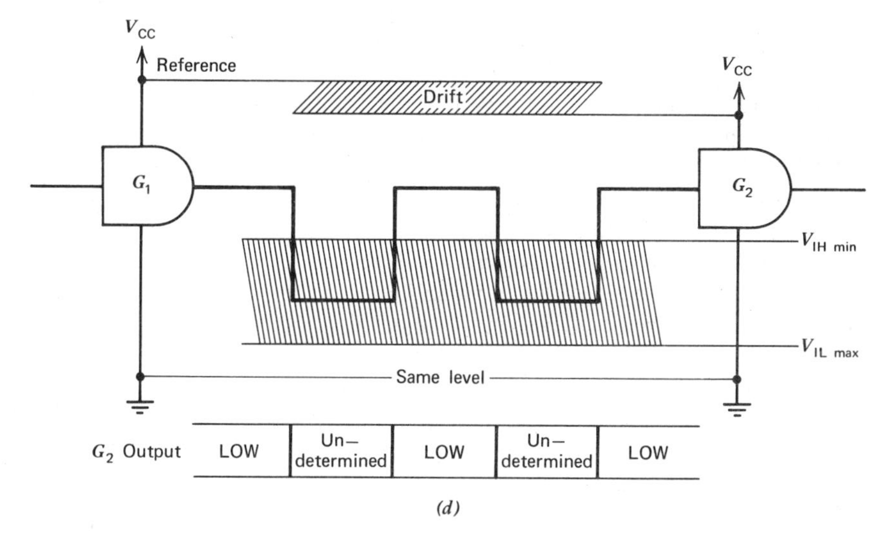

(*d*)

(*c*) Case 3.3, G_1 supply voltage drifted downwards; (*d*) Case 3.4, G_2 supply voltage drifted downwards.

Integrated circuit manufacturers do not always state the noise immunity of their digital circuits on the data sheets. They do, however, include the maximum and minimum input threshold levels as well as the limits of the output levels. From those values the dc noise immunity can be computed.

For example, a certain type of digital circuit has the following input and output level characteristics:

$$V_{\mathrm{IL\,max}} = 0.8\ \mathrm{V} \qquad V_{\mathrm{IH\,min}} = 2.1\ \mathrm{V}$$
$$V_{\mathrm{OL\,max}} = 0.4\ \mathrm{V} \qquad V_{\mathrm{OH\,min}} = 2.8\ \mathrm{V}$$

To determine the dc noise immunity of the above digital circuits, (3.1) and (3.2) are used

$$\begin{aligned} N_{\mathrm{L}} &= V_{\mathrm{IL\,max}} - V_{\mathrm{OL\,max}} \\ &= 0.8\ \mathrm{V} - 0.4\ \mathrm{V} \\ &= 0.4\ \mathrm{V} \end{aligned}$$

and

$$\begin{aligned} N_{\mathrm{H}} &= V_{\mathrm{OH\,min}} - V_{\mathrm{IH\,min}} \\ &= 2.8\ \mathrm{V} - 2.1\ \mathrm{V} \\ &= 0.7\ \mathrm{V} \end{aligned}$$

The noise immunity figures obtained above indicate that for reliable operation the dc noise on the signal lines should not exceed 0.4 V when the signal is LOW nor 0.7 V when the signal is HIGH.

It should be noted that the above values of noise immunity are valid only for the test conditions under which the four parameters, $V_{\mathrm{IL\,max}}$, $V_{\mathrm{OL\,max}}$, $V_{\mathrm{IH\,min}}$, and $V_{\mathrm{OH\,min}}$, were obtained. For the computation of the worst-case dc noise immunity, the worst-case values of the four parameters should be entered in the above calculations.

The noise immunity of digital circuits is a function of temperature, power-supply voltage, and fan-out. The graphs in Figure 3.5 show that noise immunity decreases (*a*) when the fan-out increases, (*b*) when the temperature deviates from 25°C in either direction, and (*c*) when the power-supply voltage deviates from the nominal level of 5 V in either direction.

The prime source of dc noise is voltage drops along power-supply lines. Proper attention to wire size and length exercised during system power harness design will minimize this dc noise.

The ac noise consists of the narrow pulses and jitter that are generated by high-frequency current transients. These transients may be in power lines, or in signal lines, and are caused by high-speed signal-switching. The noise generated from the transients is reactively coupled into the signal lines of the immediate vicinity of the noise source, and it is added to the digital

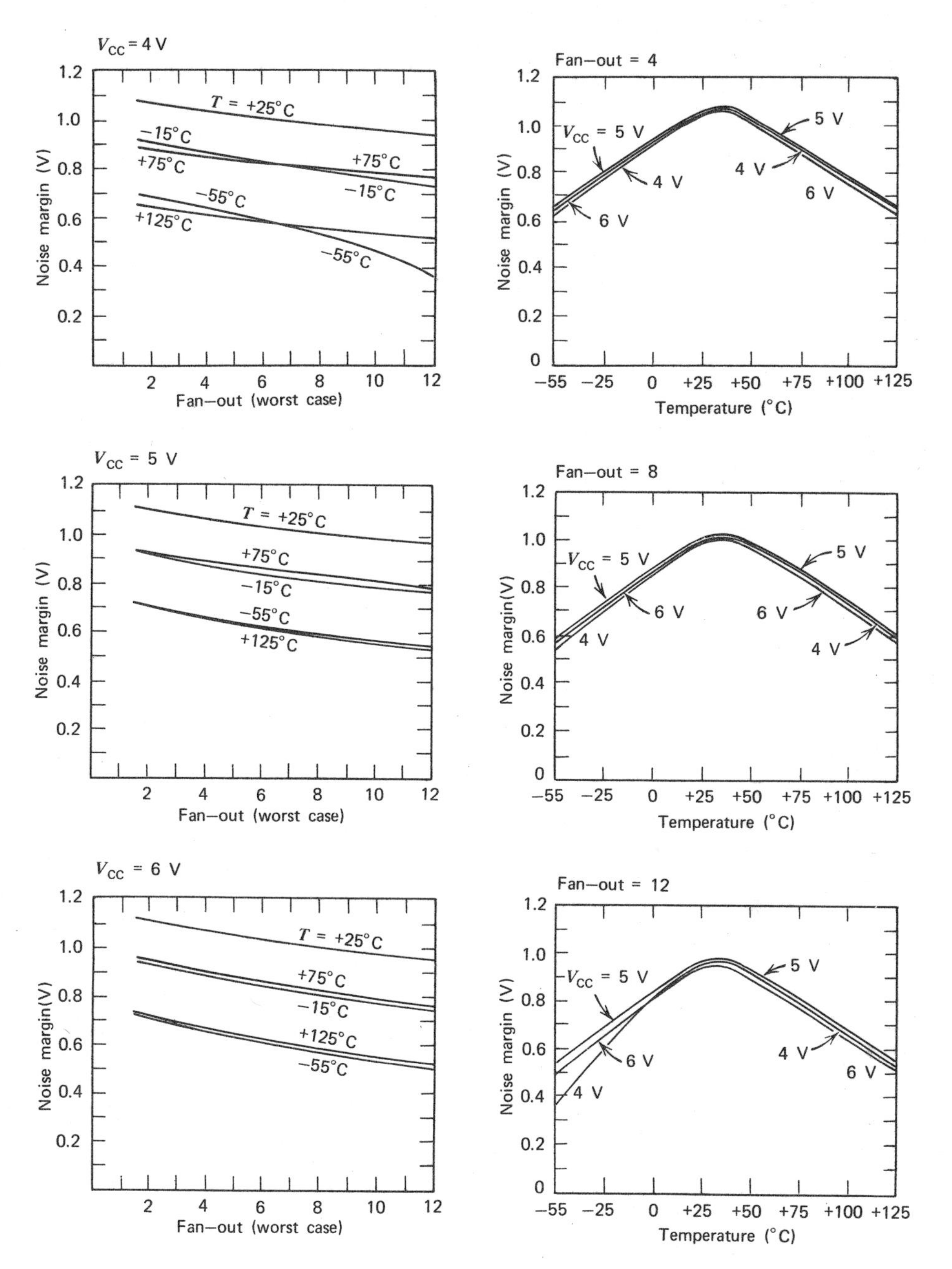

Figure 3.5. Noise immunity as a function of temperature, power-supply voltage, and fan-out.

signals carried by these lines. If the digital signals are driven by low-impedance outputs, the noise will not be able to alter the signals, and the digital circuits that are driven by these lines will not be affected by the noise.

The inherent input capacitance of digital circuits favorably contributes to the suppression of ac noise, especially that of small pulse width. This capacitance can be increased externally, if improvement in ac noise immunity may be obtained at the expense of high-frequency response.

The ac noise immunity is expressed in terms of amplitude and pulse width, because it is the amplitude-pulse-width product to which digital circuits respond. Figure 3.6 is a typical graphical presentation of noise immunity. The shaded area is the region where noise immunity exists. The figure indicates that as the pulse width of the noise increases, the noise amplitude to which the circuit is immune reaches the dc noise immunity of that circuit.

The prime source of ac noise is long power and signal wires that act as transmitting antennas during current transients. Special attention to wire

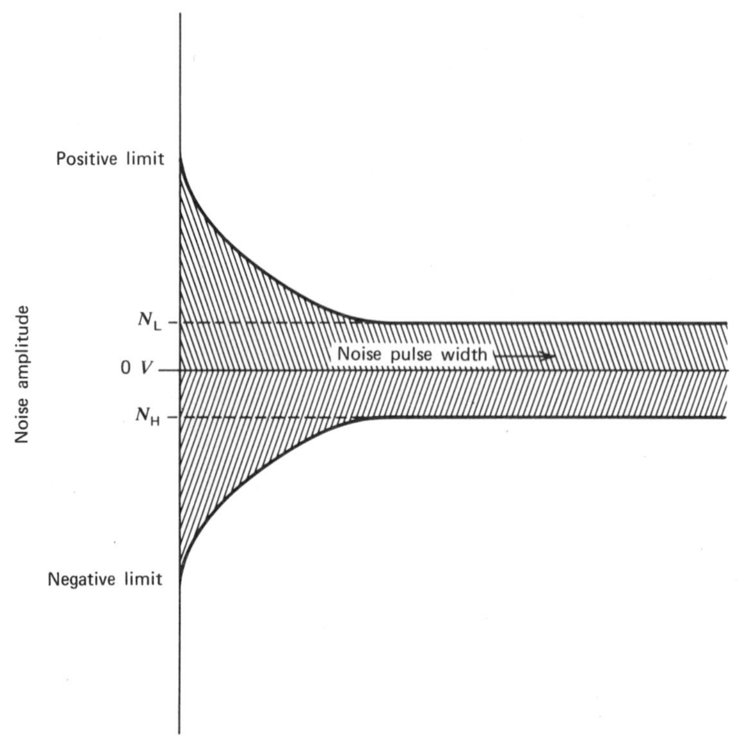

Figure 3.6. Amplitude versus pulse width of noise. Shaded area is region where noise immunity exists.

length and shielding exercised during component packaging will minimize ac noise. Equally beneficial is the use of low-output impedance circuits. In high-speed applications it is recommended that simple low-pass filters be placed on each printed circuit board across the power connections to serve as additional power-supply regulators.

Operation Temperature Range

This is the temperature range in which the output of the circuit is always within limits for the appropriate input.

Capacitance Drive Capability

Ability to drive capacitive loads is an important characteristic in digital integrated circuits, especially when long cables are driven by the output of these circuits. This ability is directly dependent on the output impedance of the circuit. The lower the impedance, the higher is the ability of the circuit to drive capacitive loads.

When a circuit drives a capacitive load, the rise and fall times of its output greatly depend on that load. A capacitive load increases the propagation delay of the circuit in a way proportional to the increase in the output time constant of the circuit caused by the capacitive load. Figure 3.7 is typical of the effect capacitive loads have on propagation delay.

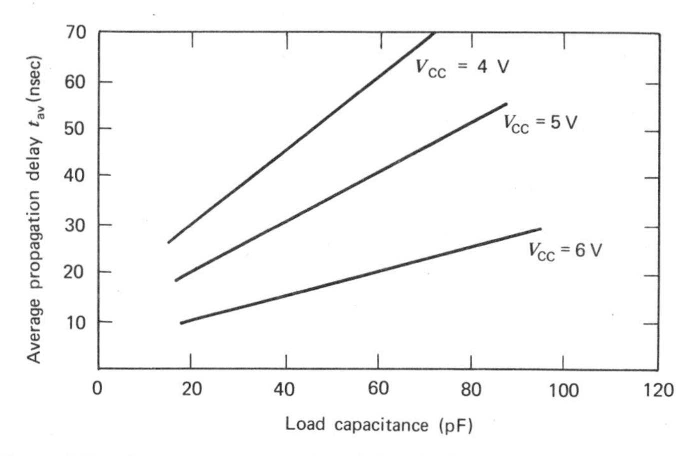

Figure 3.7. Average propagation delay (t_{av}) versus load capacitance with V_{CC} as the parameter for typical DTL circuits.

Short-Circuit Protection

This is protection of the circuit from being damaged, should one of its outputs be shorted to ground. The presence of this property can be determined from an inspection of the schematic of the circuit. This protection is provided by the resistor which is between the output pin and the power-supply pin of the circuit.

Power-Supply Tolerance

Knowledge of this characteristic is absolutely necessary in the evaluation of a digital circuit and its subsequent use in the system. Except for ECL circuits, digital circuits allow power-supply tolerance in the order of 10–20%.

Type of Logic Family

There are presently six major logic families a designer may choose from. These are

RTL	Resistor-transistor logic
DTL	Diode-transistor logic
TTL	Transistor-transistor logic
HTL	High-threshold logic
ECL	Emitter-coupled logic
MOS	Metal-oxide-semiconductor logic

Each of these logic families is examined in the following sections.

Type of Logic Circuit

Available in integrated circuit form are the basic logic elements, gates and flip-flops, and numerous logic networks of varying complexity consisting of gates, flip-flops, and special circuits. Depending on the number of logic elements they consist of, integrated circuits are classified into the following three categories:

SSI	Small-scale integration
MSI	Medium-scale integration
LSI	Large-scale integration

The SSI category covers the logic elements and simple logic networks, such as adders, counters, etc.; MSI covers larger logic networks like analog-to-digital converters, multiplexers, etc.; and LSI covers large and complex logic networks such as memories, elaborate arithmetic units, etc.

Type of Package

There are three main types of packages used for the housing of integrated circuits. These are the TO-5, the flat pack, and the dual in-line (Fig. 3.8).

The TO-5 package is the metal housing of typical transistors, and may have 8, 10, or 12 *pins*.

The flat pack comes in two standard sizes; one is 0.25 × 0.25 × 0.05 in., and the other is 0.25 × 0.15 × 0.05 in., and may have 10, 12, or 14 *pins*. The *pins* are usually flat flexible gold leads. Flat packs are also available with round copper leads.

The dual in-line package is larger than the other two and has hard leads numbering 14, 16, 24, or more. The height, width, and spacing between leads are standard, but the length of the dual in-line packages varies depending on the size of the circuit they house, especially in the case of MSI and LSI networks. Because of their ruggedness dual in-line packages are easily mounted on printed circuit boards and greatly simplify packaging.

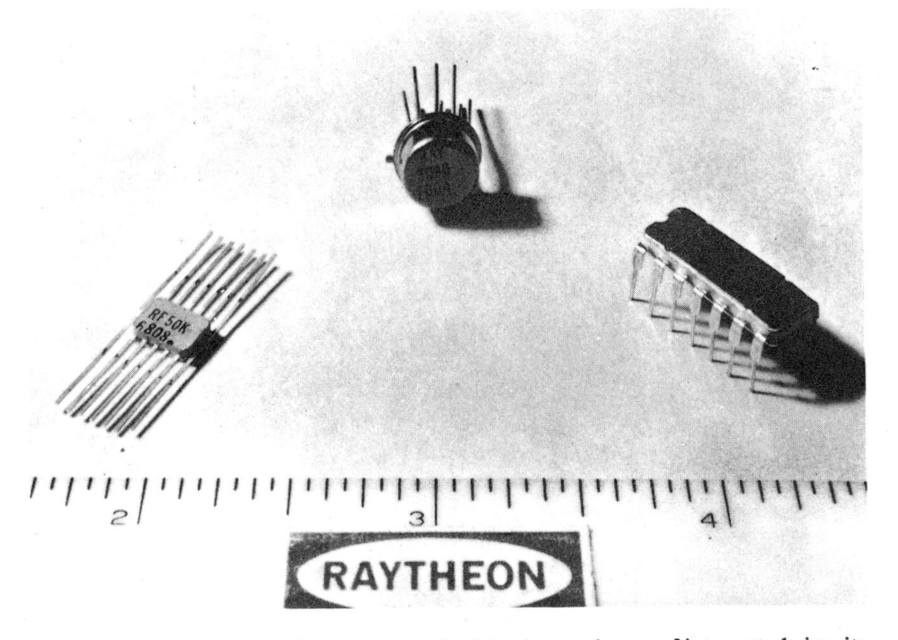

Figure 3.8. TO-5, flat pack, and dual-in-line packages of integrated circuits (courtesy of Raytheon).

3.3 RESISTOR-TRANSISTOR LOGIC CIRCUITS

The RTL circuits were the first digital integrated circuits to be developed, and their fabrication process is relatively simple. They use resistors and transistors only.

A typical RTL circuit appears in Figure 3.9. It is a dual two-input gate requiring a V_{CC} of 5 V and having an average power dissipation of 25 mW. In the LOW state the output voltage is the saturation voltage of the transistor, and in the HIGH state it is V_{CC} through the collector resistor if no load is applied. In RTL circuits the fan-out does not affect the voltage level of the LOW state. This is because, when the output of the driving circuit is LOW, there is no power transfer between the driving circuit and the driven ones.

A cascade of RTL circuits is shown in Figure 3.10. In this configuration the output of the driving circuit is assumed to be LOW, and its voltage V_L is the saturation voltage of transistor Q_1. There is no current flowing from the output of the driving circuit to the input resistors of the driven circuits because V_L, the saturation voltage of transistor Q_1, is not high enough to turn ON the transistors of the driven circuits. As a result, the driven circuits are not in any way affected by the output of Q_1 as long as it stays LOW.

A similar cascade of RTL circuits is shown on Figure 3.11. In this configuration the output of the driving circuit is assumed to be HIGH, and its voltage V_H is a function of the fan-out. The expression that relates to fan-out is

$$V_H = V_{CC} \frac{(R_1/n)}{(R_1/n) + R_2}$$

$$= V_{CC} \frac{R_1}{R_1 + nR_2} \tag{3.3}$$

$$V_H = \frac{V_{CC}}{1 + n} \qquad (\text{if } R_1 = R_2) \tag{3.4}$$

where

V_{CC} = Power-supply voltage
R_1 = Input resistor
R_2 = Collector resistor
n = Fan-out, i.e., the number of input resistors driven by the output

Equation (3.4) indicates that when the output of an RTL circuit is HIGH its voltage level is inversely proportional to the fan-out.

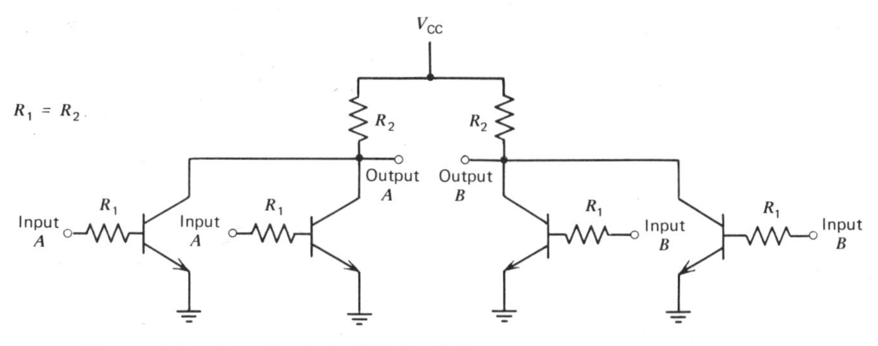

Figure 3.9. Dual NAND/NOR RTL gate.

Using (3.4) the voltage level at the collector of Q_1, in Figure 3.11, can be computed as follows:

$$V_H = \frac{V_{CC}}{1 + n} \qquad (n = 3)$$

$$= \frac{V_{CC}}{1 + 3}$$

$$= 0.25\ V_{CC}$$

The lower limit of V_H is

$$V_{H\,min} = R_1 I_{B\,min} + V_{BE} \tag{3.5}$$

where

R_1 = Input resistor of the driven circuit

$I_{B\,min}$ = Minimum current that can reliably saturate the transistor of the driven circuit

V_{BE} = Voltage drop across the base to emitter junction of the transistor of the driven circuit; this voltage must be first overcome in order for current to flow through the base of the transistor

Substituting (3.4) into (3.5) and solving for n, which in this case will be the maximum fan-out, we have

$$n_{max} = \frac{V_{CC}}{R_1 I_{B\,min} + V_{BE}} - 1 \tag{3.6}$$

Equation (3.6) indicates that in RTL circuits the fan-out is directly proportional to the power-supply voltage and inversely proportional to the voltage required to saturate the driven circuits. A fan-out of 5 is the typical RTL

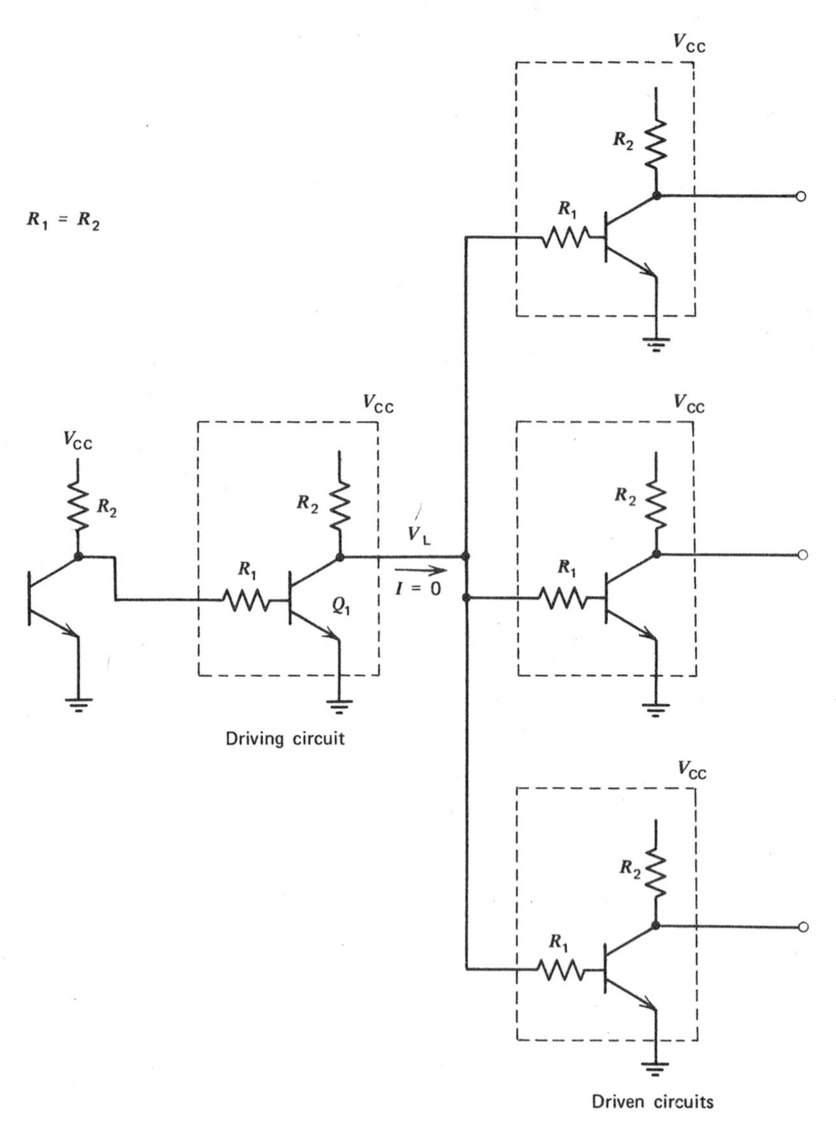

Figure 3.10. Cascade of RTL circuits. Output of driving circuit V_L.

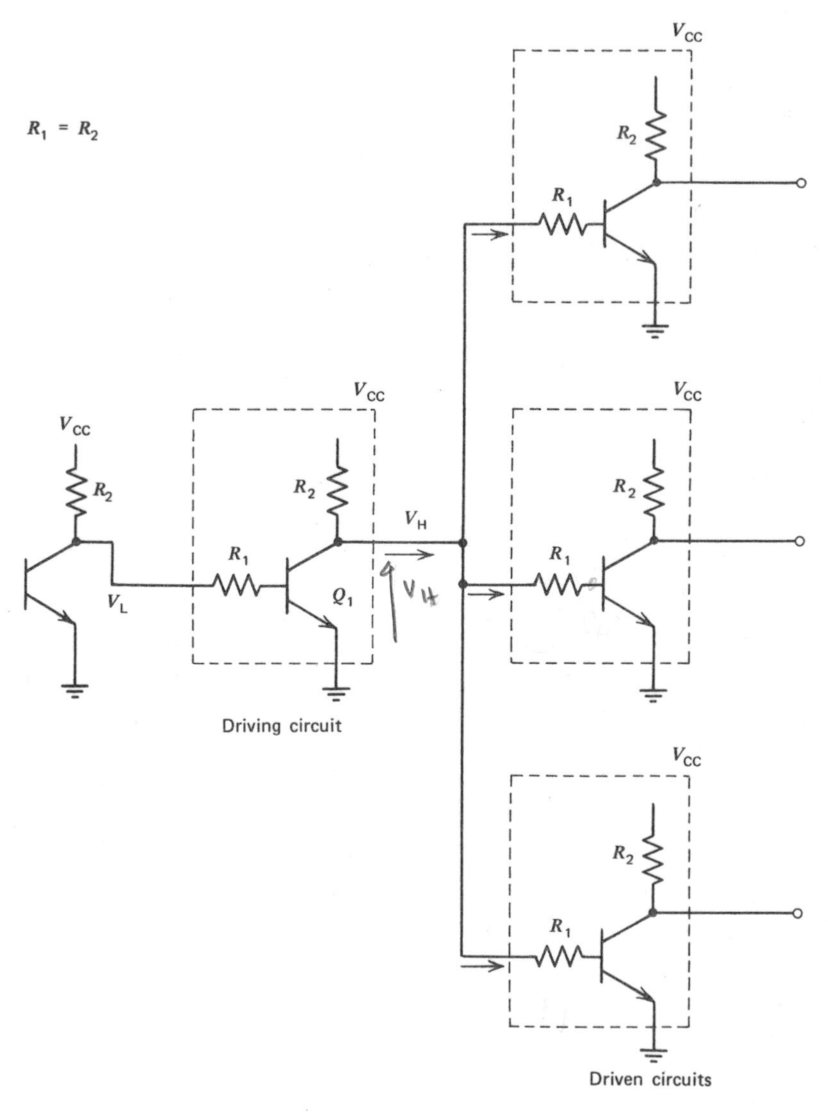

Figure 3.11. Cascade of RTL circuits. Output of driving circuit V_H.

fan-out. There are RTL circuits, however, with a fan-out of 30, but these are single gates mainly used as drivers rather than as actual logic elements.

Because of their circuit configuration, RTL circuits have a low noise immunity, typically 0.2 V. This means that, if the input to an RTL circuit deviates by more than 0.2 V from its nominal value, the circuit may give erroneous output. For example, if an RTL circuit has a fan-out of 5, its V_H output is

$$V_H = \frac{V_{CC}}{1 + n} \qquad \text{(assuming that } R_1 = R_2\text{)}$$

$$V_H = \frac{5\text{ V}}{1 + 5} \qquad \text{(if } V_{CC} = 5\text{ V)}$$

$$= \frac{5}{6}\text{ V}$$

$$= 0.833\text{ V}$$

If this voltage is lowered by some undesired dc noise of 0.25 V, it becomes 0.583 V. This voltage does not guarantee reliable operation, because it cannot supply enough current to saturate the transistors of the driven circuits.

Therefore, when using RTL circuits at their maximum fan-out capabilities, the engineer should be very cautious of undesired voltage drops (dc noise), and transients (ac noise), that may develop on the signal lines connecting the circuits. If the circuits are not used at their maximum fan-out capabilities, the problem of noise immunity is less critical.

The typical power dissipation of an RTL package is 10 mW times the number of gates contained in that package. In the case of a flip-flop, the power dissipation is equivalent to that of four gates. In digital design, this is considered medium power.

The propagation delay of the RTL circuits usually ranges from 10 to 30 nsec for gates and from 20 to 50 nsec for flip-flops. Compared to the speed of the other logic families this is considered low.

Several integrated circuit manufacturers produce RTL circuits. They come in complete lines—single, dual, triple, or quad gates, various types of flip-flops, inverters, etc. They find their prime use in low-speed applications in industrial and commercial equipment; RTL circuits are inexpensive when compared to the other families of digital circuits and their relatively simple fabrication process makes them very reliable.

3.4 DIODE-TRANSISTOR LOGIC CIRCUITS

The DTL circuits are similar to RTL circuits with the difference that DTL circuits have diode inputs instead of resistor inputs. Figure 3.12 shows typical DTL circuits. The power-supply voltage of DTL circuits is nominally 5 V and the nominal logic levels at which they operate are ground for LOW and 5 V for HIGH. The actual logic levels usually are

Circuit output capabilities $V_{OL} = 0.4–0.6$ V

$V_{OH} = 2.6–5.0$ V

Circuit input requirements $V_{IL} = 0.0–1.3$ V

$V_{IH} = 1.8–5.0$ V

These are typical values for a 25°C operation and are supplied by most DTL circuit manufacturers.

Using (3.1) and (3.2), the noise immunity of the circuit having the above characteristics can be computed as shown

$$\begin{aligned} N_L &= V_{IL\,max} - V_{OL\,max} \\ &= 1.3\text{–}0.6 \text{ V} \\ &= 0.7 \text{ V} \\ N_H &= V_{OH\,min} - V_{IH\,min} \\ &= 2.6\text{–}1.8 \text{ V} \\ &= 0.8 \text{ V} \end{aligned}$$

The noise immunity of DTL circuits having the above characteristics is, therefore, 0.7 V. In digital design this magnitude of noise immunity is considered relatively high and therefore desirable.

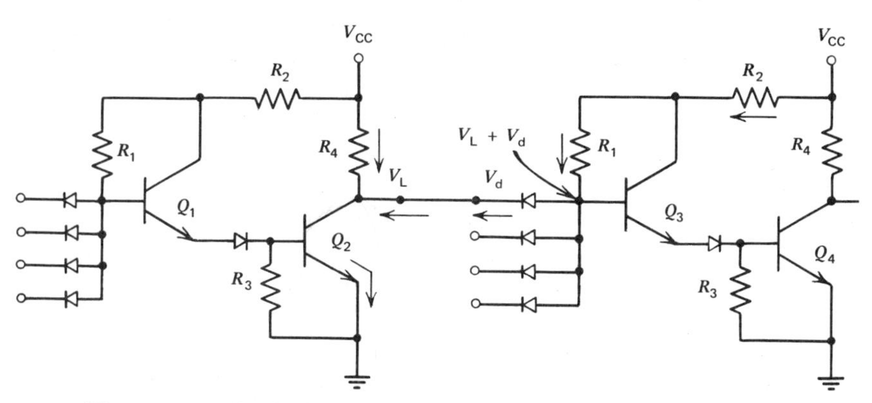

Figure 3.12. Typical DTL gates connected in cascade.

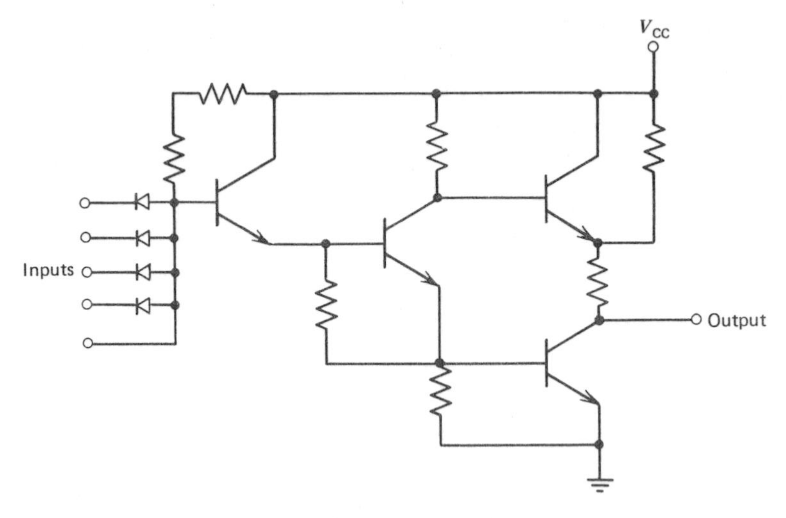

Figure 3.13. Typical high fan-out NAND/NOR DTL gate.

In a typical DTL circuit such as that shown on Figure 3.13, when the inputs are open-circuited or at a potential higher than that of the base of the first transistor, the output transistor of the gate is saturated. In this state, there is no current flowing through the input diodes, because the diodes are either open-circuited, or reverse-biased. Consequently, the gate does not draw any power from the circuit by which it is driven.

When one of the inputs is V_L (Fig. 3.12), the voltage at the base of the first transistor of the driven circuit is $V_L + V_d$, where V_d is the voltage drop across the input diode having V_L at its input. Voltage $V_L + V_d$ is low enough to keep the first transistor of the driven circuit OFF which, in turn, keeps the output transistor OFF. In this state, the output of the driven circuit is V_{CC} through R_4, while at the input of the driven circuit current flows from V_{CC} through R_2, R_1, and the input diode to the V_L input.

This current will flow into the collector of the output transistor of the driving circuit (Fig. 3.12). In order for the output transistor Q_2 of the driving circuit to sink-in this current without raising its collector voltage, the transistor must have a sufficient βI_B product. This product determines the amount of current the collector can sink-in, and consequently it determines the fan-out capabilities of the driving circuit.

Fan-out is an important circuit characteristic and should always be identified before using a circuit. The fan-out of the circuits of Figure 3.12 is typically 10, while that of the gate of Figure 3.13 may be as high as 30.

The per package power dissipation of DTL circuits varies from 5 to 50 mW,

Characteristics	Medium fan-out gate (Fig. 3.12)	High fan-out gate (Fig. 3.13)
Logic levels	Same	Same
Noise immunity	Same	Same
Fan-out	10	30
Power dissipation	5 mW	20 mW
Propagation delay	15 nsec	40 nsec

Figure 3.14. Comparison of medium and high fan-out DTL gates.

depending on the number of circuits contained in the package. The circuit of Figure 3.12, for example, has a power dissipation of about 5 mW per gate.

The propagation delay of DTL circuits usually varies from 10 to 100 nsec, depending mainly on the configuration of the circuit and the operating conditions. The propagation delay of the circuit of Figure 3.12 is given by most manufacturers as being 15 nsec per gate. In digital design, this is considered medium speed. Increase in power dissipation and propagation delay is usually the trade-off for higher fan-out. Figure 3.14 illustrates the basic characteristics of the DTL gates in Figures 3.12 and 3.13, which are functionally identical.

The DTL circuits are made by most integrated circuit manufacturers and come in complete lines—multiple gates, various types of flip-flops, binary adders, counters, etc.; DTL circuits are relatively inexpensive and find use in medium-speed applications. They will, however, be eventually replaced by TTL circuits whose cost because of high-volume production will soon compare to that of the DTL circuits.

3.5 TRANSISTOR-TRANSISTOR LOGIC CIRCUITS

The TTL circuits are very similar to DTL circuits with the only difference being that the inputs of the TTL circuits are transistor emitters instead of diode cathodes. A typical TTL circuit is shown in Figure 3.15. The power-supply voltage of TTL circuits and the logic level characteristics are the same as those of the DTL circuits.

Figure 3.16 illustrates the operation of a typical TTL two-input gate. When both inputs of this circuit are open-circuited or at a potential which is higher than that of the base of Q_1 (Fig. 3.16(*a*)), the emitter-to-base junction of Q_1 is reverse-biased, while the collector-to-base junction is forward-biased. This biasing allows current to flow through R_1 and turn ON transistors Q_2 and

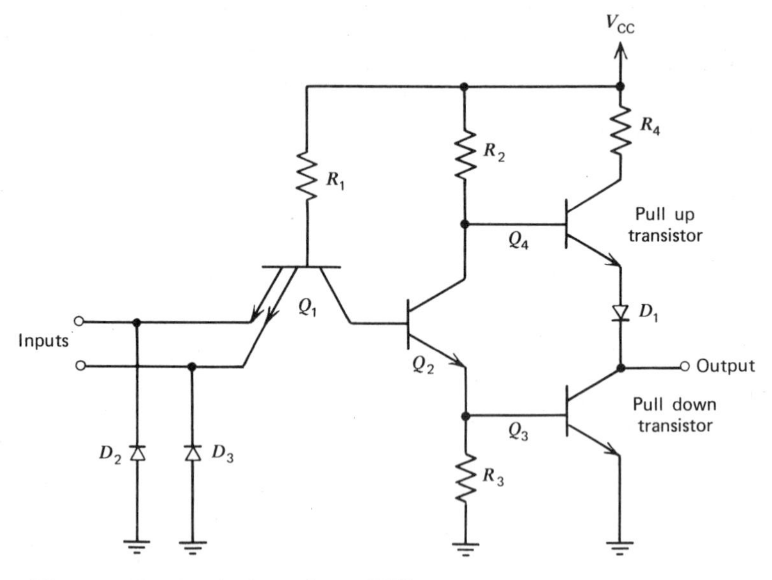

Figure 3.15. Typical two-input TTL gate.

Q_3. This condition is one of the two states of the circuit. In this state, the gate output is LOW, and its voltage level is the saturation voltage of Q_3.

When one of the inputs of a TTL gate is connected to a LOW (Fig. 3.16(*b*)), the output is HIGH. This is because the current through R_1 flows through the Q_1 base-emitter junction to the input voltage V_L, rather than through the Q_1 base–collector junction to the base of Q_2. With no current at its base Q_2 is OFF and, consequently, Q_3 is OFF also. With Q_2 OFF current flows from V_{CC} through R_2 to the base of Q_4, turning Q_4 ON and making the output of the gate V_H. The expression for V_H is

$$V_H = V_{CC} - (V_{Q_4}\text{sat} + V_{D_1} + I_\ell R_4)$$

where I_ℓ is the load current.

The noise immunity of TTL circuits ranges from 0.4 to 1.0 V with a typical value of 0.8 V depending on the circuit configuration and the manufacturer. Typical fan-out of TTL is 10. There are TTL circuits, however, with higher fan-out, going up to 50, used as drivers. The power dissipation of single TTL gates is 2 to 10 mW, and that of TTL flip-flops is 10 to 75 mW. The propagation delay of the TTL gates varies from 4 to 20 nsec for gates, and from 10 to 50 nsec for flip-flops. In digital design this magnitude of propagation delay is considered low and desirable.

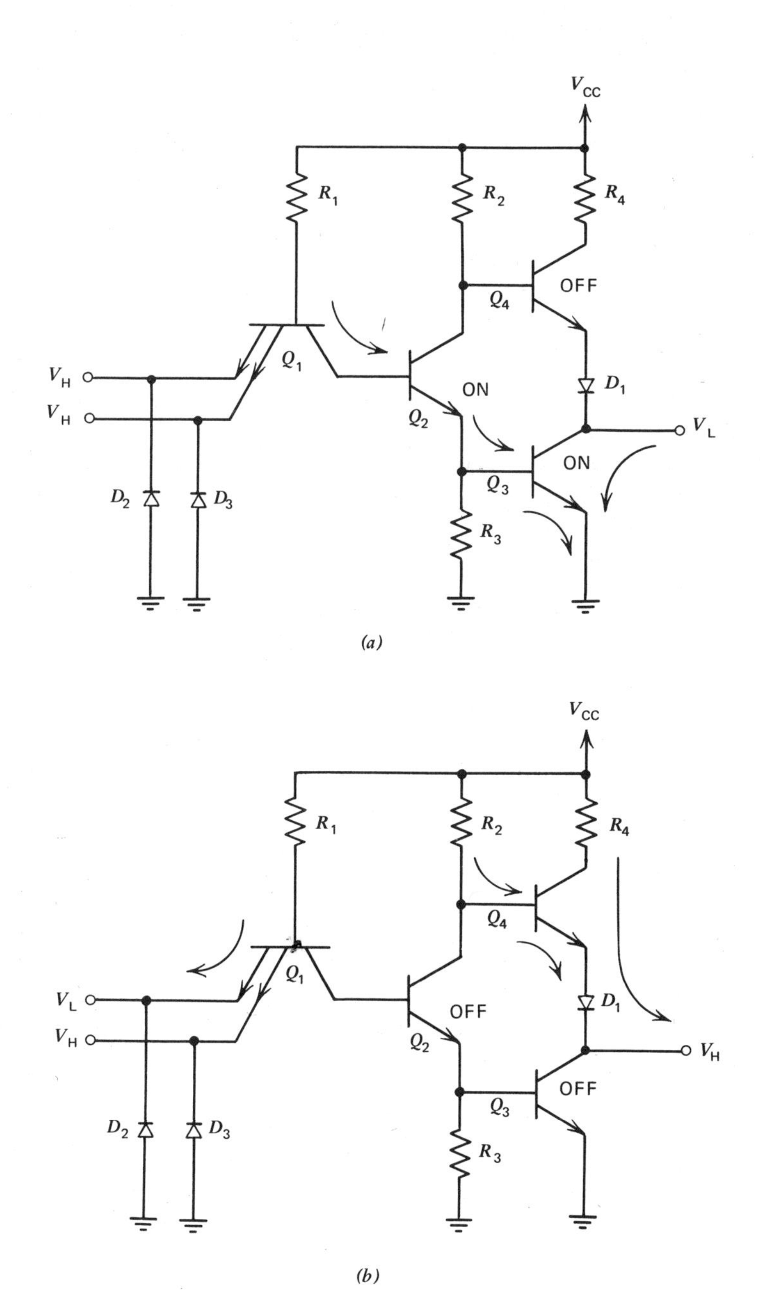

Figure 3.16. Operation of TTL gates: (*a*) output is LOW; (*b*) output is HIGH.

With advances in semiconductor fabrication, power dissipation and propagation delay figures have been steadily decreasing, and with time faster circuits with less power requirements and higher noise immunity will become available to the users.

3.6 HIGH-THRESHOLD LOGIC CIRCUITS

The HTL circuits are similar to the DTL circuits with the difference being that HTL circuits have a zener diode in series with the emitter of the input transistor. This diode raises the threshold of the circuit by a voltage equal to that of the zener-diode rating. Figure 3.17 shows a typical HTL circuit. The power-supply voltage of HTL circuits is nominally +15 V, and the nominal logic levels at which HTL circuits operate are ground for LOW and 15 V for HIGH. The actual logic levels usually are

Circuit output capabilities $V_{OL} = 0.7\text{–}1.5\ \text{V}$

$V_{OH} = 12.5\text{–}15\ \text{V}$

Circuit input requirements $V_{IL\,max} = 6.0\ \text{V}$

$V_{IH\,min} = 8.0\ \text{V}$

With the above typical input/output characteristics, the noise immunity can be computed using (3.1) and (3.2).

$$\begin{aligned} N_L &= V_{IL\,max} - V_{OL\,max} \\ &= 6.0 - 1.5\ \text{V} \\ &= 4.5\ \text{V} \\ N_H &= V_{OH\,min} - V_{IH\,min} \\ &= 12.5\text{–}8.0\ \text{V} \\ &= 4.5\ \text{V} \end{aligned}$$

Therefore, the noise immunity of HTL circuits, having the above characteristics, is 4.5 V. For their high-noise immunity, HTL circuits are used in high-noise applications, such as system interface.

Figure 3.18 illustrates the operation of a typical HTL four-input gate. Increase of its fan-in is accomplished by connecting additional diodes at the node input. When all inputs of the gate are open-circuited or at a potential higher than that of the base of Q_1 (Fig. 3.18(*a*)), transistor Q_1 is ON. This action causes transistor Q_3 to turn ON which, in turn, brings the base of transistor Q_2 at the saturation voltage of Q_3, thus preventing Q_2 from turning ON. The gate output is at approximately 1 V and the output

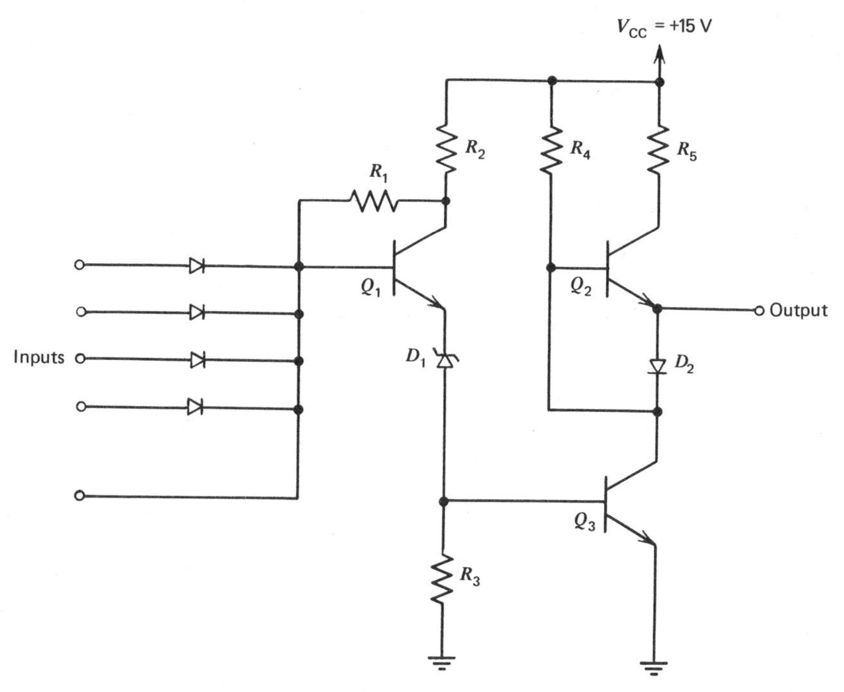

Figure 3.17. Typical four-input HTL gate with fan-in expanding node.

impedance is the path from the output terminal through diode D_2 and transistor Q_3 to ground.

When one of the inputs, either direct or through the node of the HTL gate, is connected to a LOW level (Fig. 3.18(*b*)), the output is HIGH. This is because transistor Q_1 is OFF, resulting in transistor Q_3 being OFF. Transistor Q_2, uninhibited by transistor Q_3 is now ON and the output of the HTL circuit is V_{CC} through *R5* and the saturated transistor Q_2. Typical fan-out of HTL circuits is 10 while their propagation delay and power dissipation are 60 nsec and 30 mW, respectively, per gate. Though HTL circuits are not as widely produced as TTL circuits, they are available from a small number of manufacturers.

3.7 EMITTER-COUPLED LOGIC CIRCUITS

The ECL circuits are unique in their configuration, and their outstanding characteristic is speed. The trade-off for speed is higher power dissipation and tighter power requirements; ECL gates are available with propagation

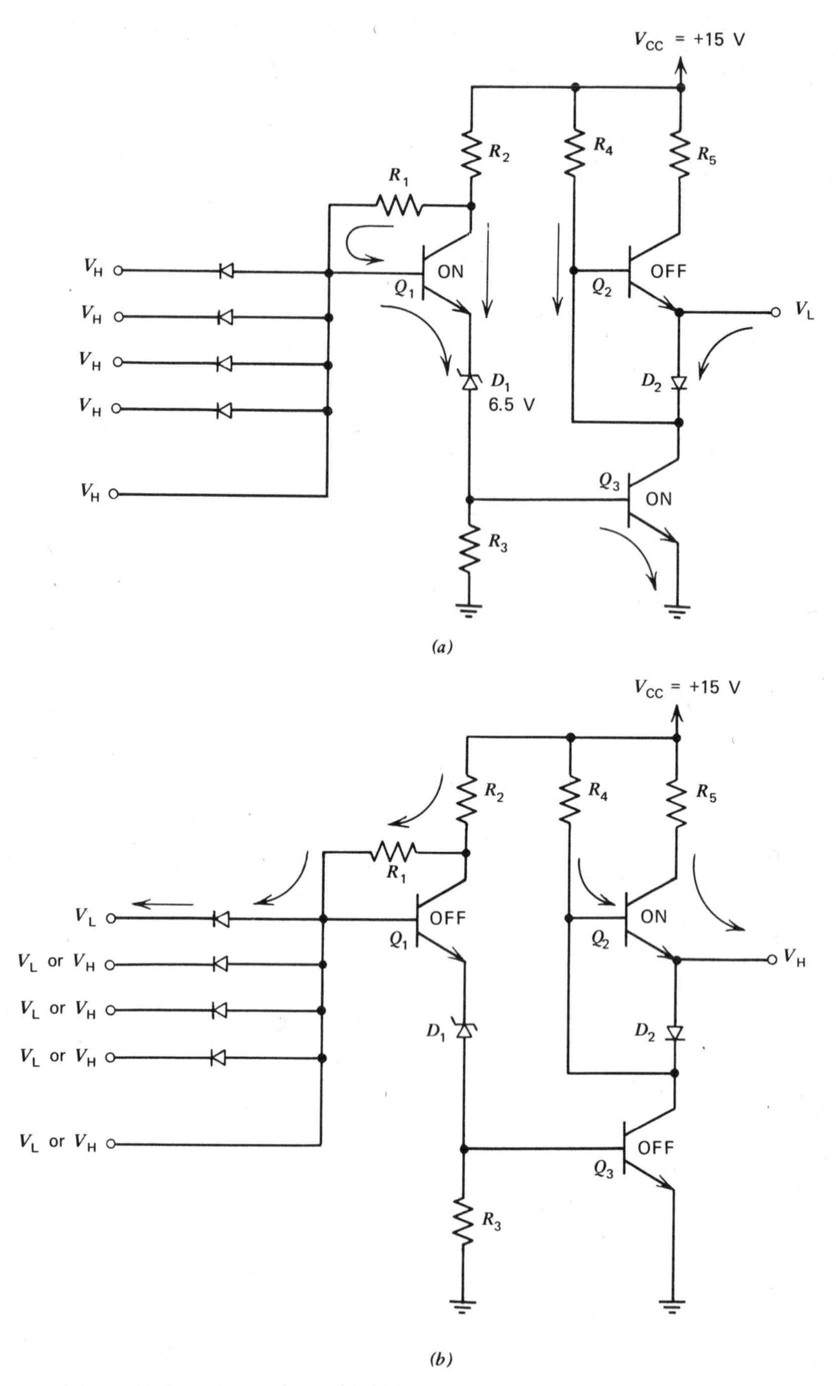

Figure 3.18. Operation of HTL gates.

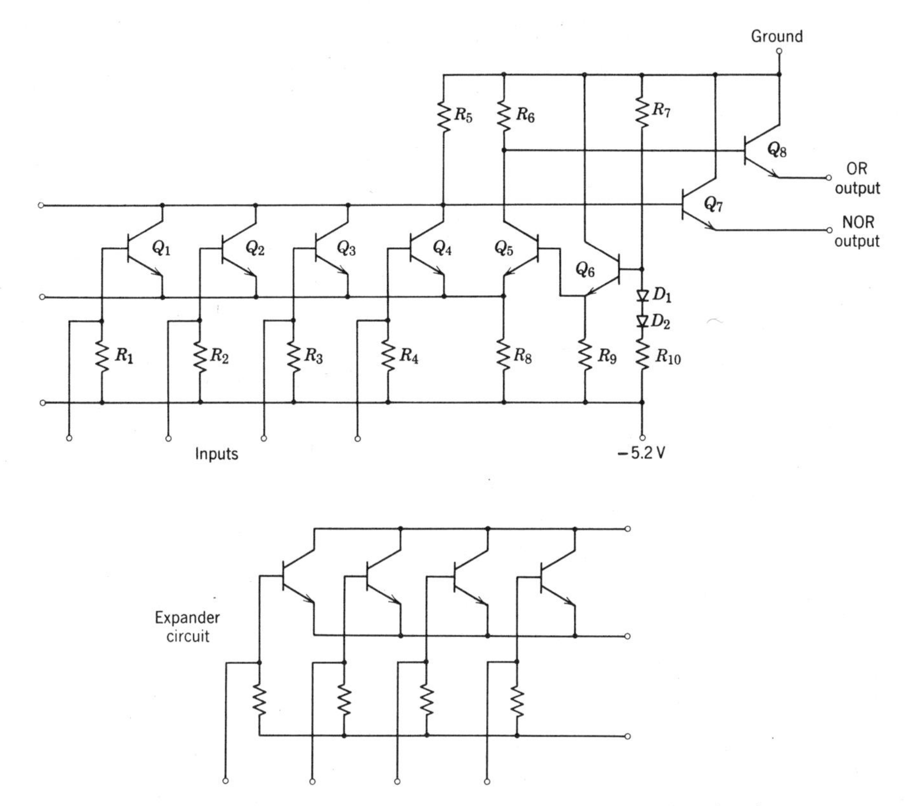

Figure 3.19. Typical four-input ECL gate and expander circuit.

delay ranging from 1 to 8 nsec and power dissipation ranging from 22 to 70 mW. The drawbacks of ECL circuits are the 10% maximum power-supply tolerance and the low-noise immunity, typically 0.125 V.

Figure 3.19 shows a typical ECL circuit. The power-supply voltage of ECL circuits is -5.2 V, and the nominal logic levels at which ECL circuits operate are -1.7 V for LOW and -0.9 V for HIGH. The actual logic levels usually are

Circuit output capabilities $V_{OL} = -1.85 - -1.60$ V

$V_{OH} = -0.98 - -0.75$ V

Circuit input requirements $V_{IL\,max} = -1.45$ V

$V_{IH\,min} = -1.10$ V

With the above typical input/output characteristics the noise immunity can be determined using (3.1) and (3.2)

$$
\begin{aligned}
N_{\mathrm{L}} &= V_{\mathrm{IL\,max}} - V_{\mathrm{OL\,max}} \\
&= -1.45 - (-1.60)\ \mathrm{V} \\
&= 0.15\ \mathrm{V} \\
N_{\mathrm{H}} &= V_{\mathrm{OH\,min}} - V_{\mathrm{IH\,min}} \\
&= -0.98 - (-1.10)\ \mathrm{V} \\
&= 0.12\ \mathrm{V}
\end{aligned}
$$

The noise immunity of ECL circuits is indeed low making it imperative that special precautions be taken to shield out ac noise and minimize dc noise wherever possible.

Figure 3.20 shows the operation of a typical ECL four-input gate. It has complementary outputs, thus providing the user with desirable flexibility, and the fan-out is 10 making it comparable to other families of digital circuits. When all inputs of the gate are open-circuited or at LOW, -1.7 V, the nominal output is LOW and the complementary output is HIGH (Fig. 3.20(*a*)). If one of the inputs is HIGH, -0.9 V, the nominal output is HIGH and the complementary output is LOW (Fig. 3.20(*b*)).

Since the two output transistors of ECL circuits always operate in the linear region, ECL circuits are referred to as nonsaturating. The ECL circuits find use in high-speed applications—mainly in advance communications equipment and in digital computers.

3.8 METAL-OXIDE SEMICONDUCTOR CIRCUITS

The MOS circuits are very different from the digital families described in the preceding sections. The digital MOS circuits consist of field-effect transistors only, which are voltage-sensitive rather than current-sensitive devices. The main advantages of MOS circuits are cost, resulting from the simplicity of their fabrication process, high-noise immunity, large power-supply tolerance, low-power quiescent operation, high density, and high fan-out capability. There are two types of MOS field-effect transistors. These are the *P*-channel and the *N*-channel devices.

In the *P*-channel device, when a negative voltage is applied to the gate with respect to the source, the drain-to-source resistance drops from 500 MΩ to about 750 Ω. When no voltage or a positive voltage is applied to the gate with respect to the source, the drain-to-source resistance is practically infinite. In the *N*-channel device a similar relationship exists with the only difference being that the polarities are reversed. Figure 3.21 shows the symbols, equivalent circuits, and transfer characteristics for MOS field-effect transistors.

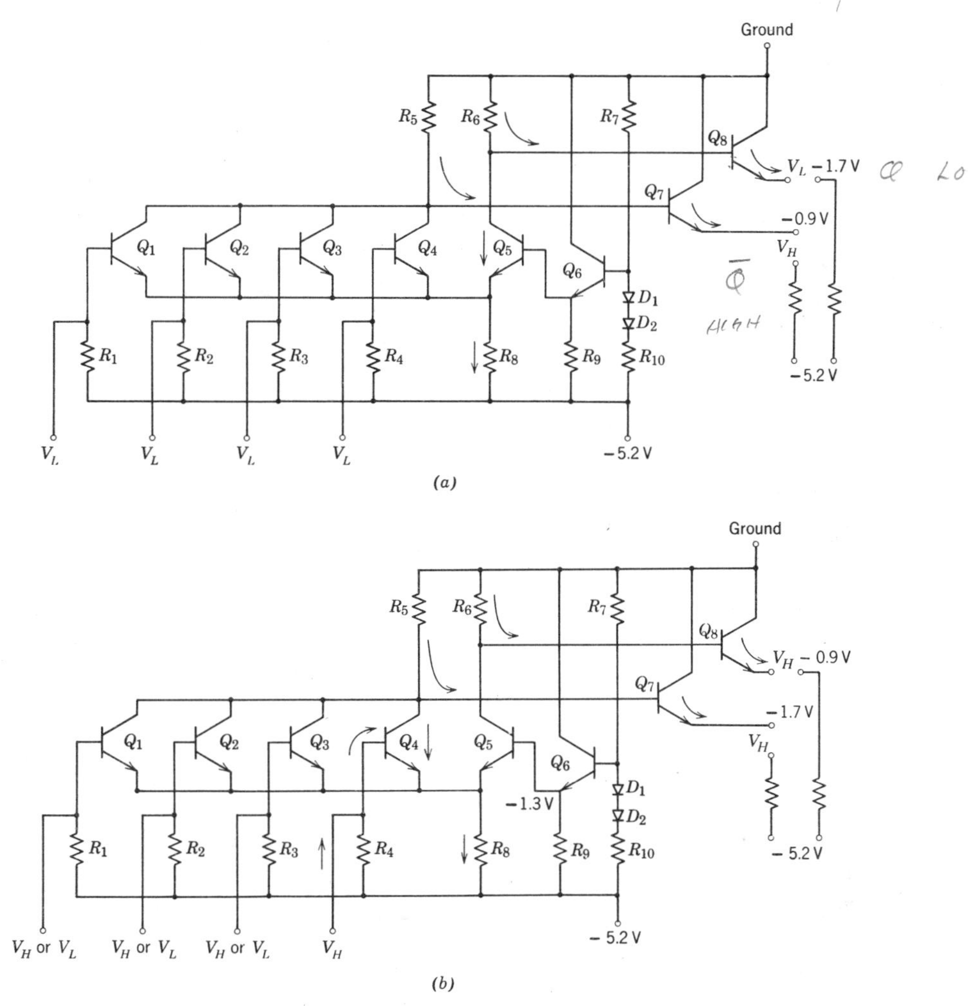

Figure 3.20. Operation of ECL gates.

The MOS circuits have found a very wide application in digital design, especially in medium-scale and large-scale integration designs. The MOS digital elements and networks are classified into "static," dc logic, and "dynamic," ac logic.

In static logic, MOS logic elements maintain their output states on a continuous basis the same way the logic families discussed in the preceding sections do. Figure 3.22 illustrates three MOS logical inverters, and their operating characteristics. In these circuits the output is at a HIGH or LOW

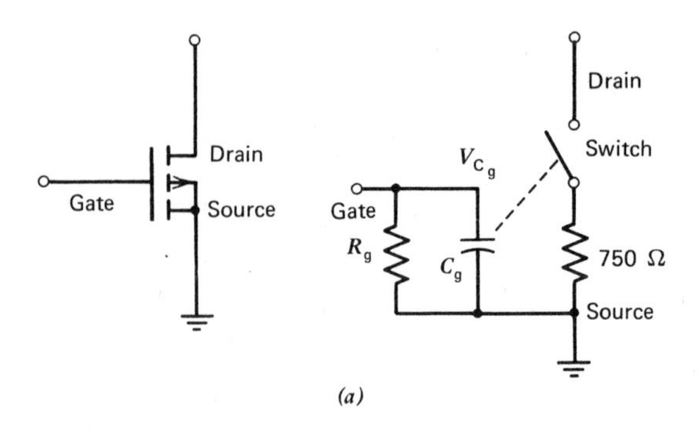

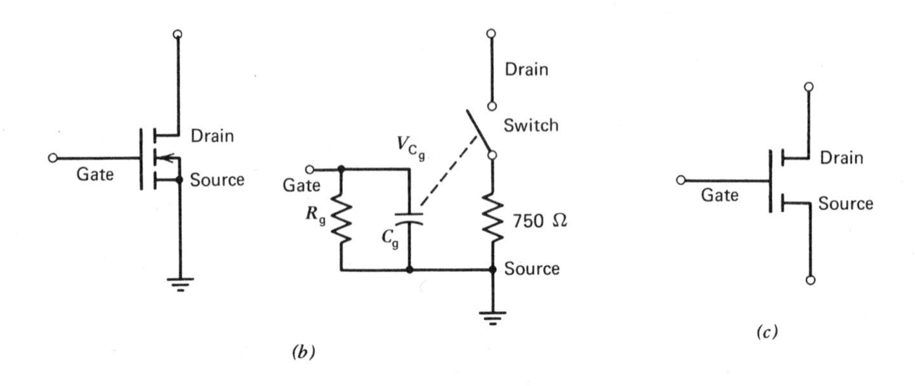

(*a*)

C_g voltage	Switch
Negative and greater than threshold	Closed
Positive or negative less than threshold	Open

(*b*)

C_g voltage	Switch
Positive and greater than threshold	Closed
Negative or positive less than threshold	Open

Figure 3.21. MOS field-effect transistors: (*a*) *P*-channel device; (*b*) *N*-channel device; (*c*) schematic symbol used when the type of channel is assumed.

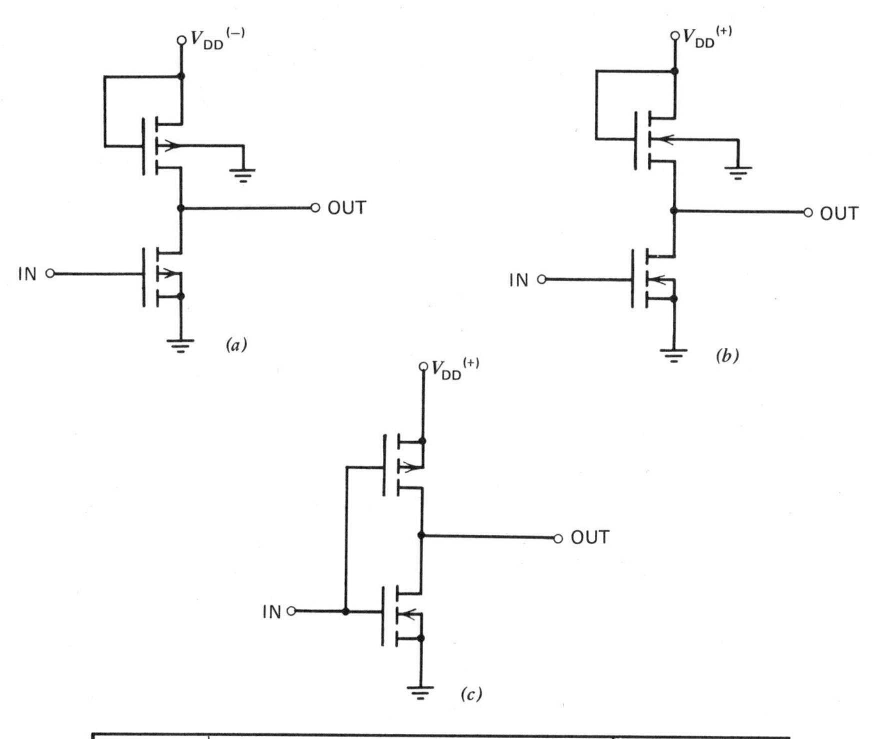

(a)

Input	Output	State of Lower transistor
HIGH	LOW = V_{DD} Z_{OUT} = 15 kΩ	OFF
LOW	HIGH = Ground Z_{OUT} = 750 Ω	ON

(b)

Input	Output	State of Lower transistor
LOW	HIGH = V_{DD} Z_{OUT} = 15 kΩ	OFF
HIGH	LOW = Ground Z_{OUT} = 750 Ω	ON

(c)

Input	Output	State of transistors	
		Upper	Lower
LOW	HIGH = V_{DD} Z_{OUT} = 750 Ω	ON	OFF
HIGH	LOW = Ground Z_{OUT} = 750 Ω	OFF	ON

Figure 3.22. MOS logical inverters: (*a*) *P*-channel inverter; (*b*) *N*-channel inverter; (*c*) complementary inverter.

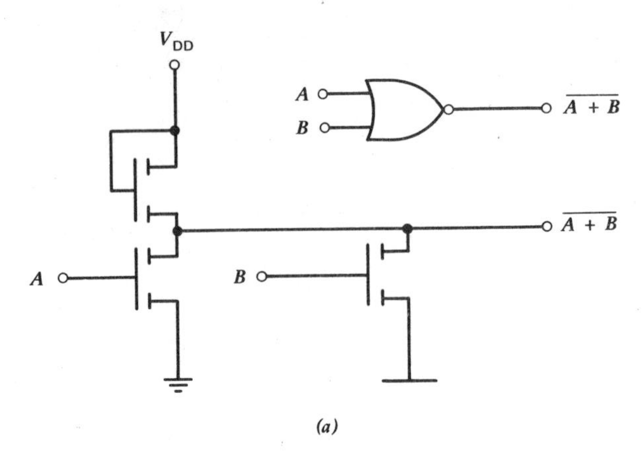

A	*B*	OUT
H	H	L
H	L	L
L	H	L
L	L	H

1 = H

(a)

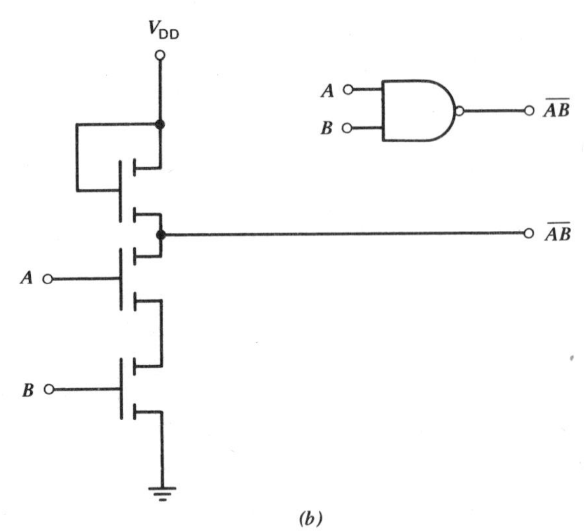

A	*B*	OUT
H	H	L
H	L	H
L	H	H
L	L	H

1 = H

(b)

$V_{DD}(+)$

A

B

$\overline{A + B}$

A	*B*	OUT
H	H	L
H	L	L
L	H	L
L	L	H

1 = H

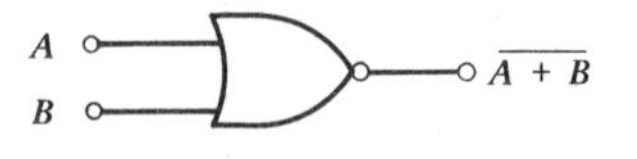

(c)

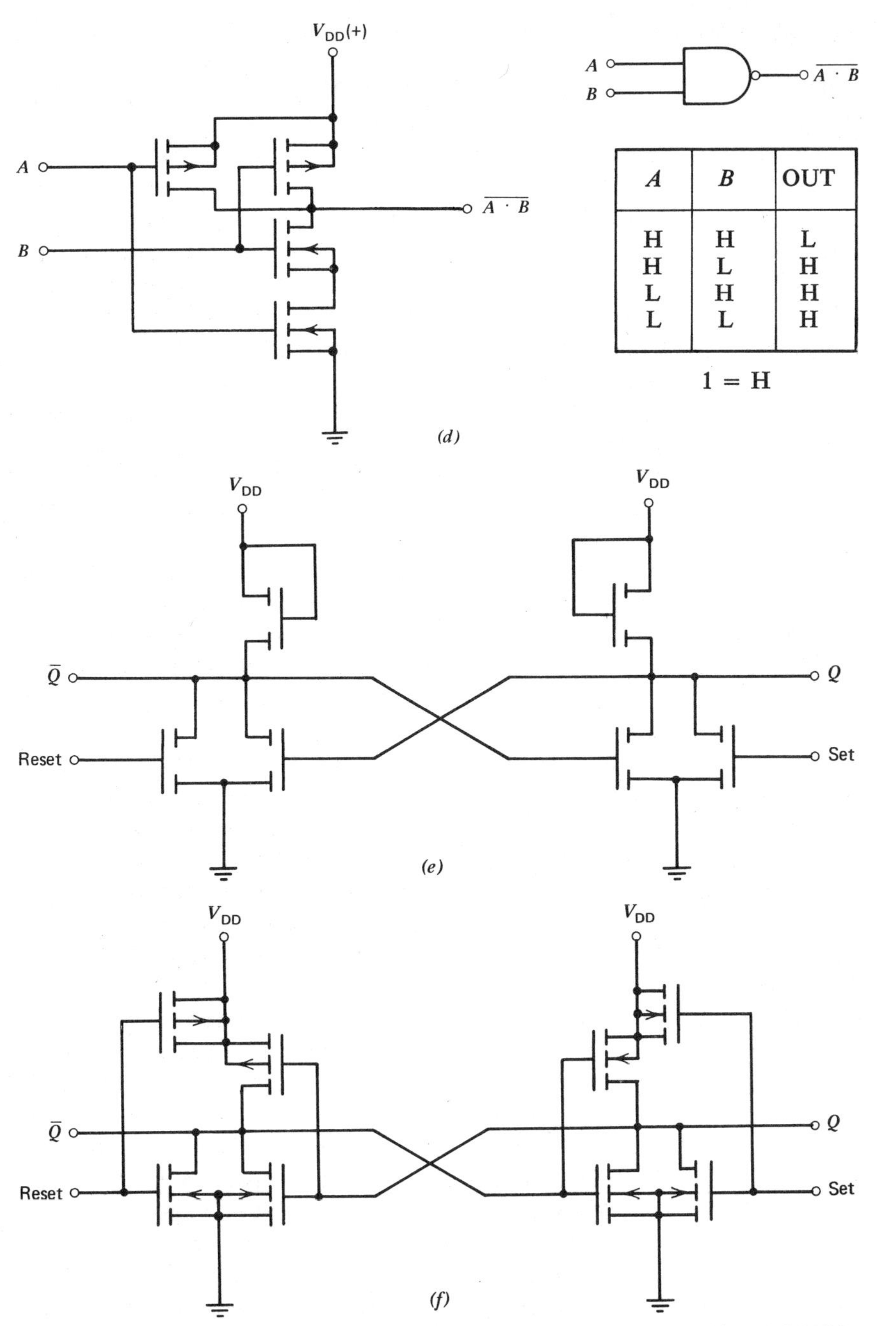

Figure 3.23. Typical MOS and CMOS gates and flip-flops: (*a*) MOS NOR gate; (*b*) MOS NAND gate; (*c*) CMOS NOR gate; (*d*) CMOS NAND gate; (*e*) MOS R-S flip-flop; (*f*) CMOS R-S flip-flop.

state as long as the input is at a LOW or HIGH state, respectively—that is, the circuits display no memory characteristics.

The circuit in Figure 3.22(*a*) consists of *P*-channel devices only where the upper transistor serves as a 15 kΩ resistor, and the switching transistor, the lower one, switches from 750 Ω to 500 MΩ. Thus, when the input is open-circuited or grounded, the lower transistor is OFF and the output is V_{DD} through 15Ω. When the input is at a negative potential, the lower transistor is ON and the output is at ground through 750 Ω. In this state power is dissipated across the upper transistor since it simply acts as a 15 kΩ resistor.

Figure 3.22(*b*) consists of *N*-channel devices only, and its operation is similar to that of Figure 3.22(*a*) with the roles of voltage polarity reversed.

Figure 3.22(*c*) consists of *P*-, as well as *N*-channel devices. The lower transistor is an *N*-channel while the upper transistor is a *P*-channel device. Because of this arrangement the circuit is called "complementary." When the input is grounded, the lower transistor is OFF while the upper one is ON. The output is V_{DD} through a 750 Ω resistance which is the conduction resistance of the upper transistor. This value is considerably less than that offered by the inverters that have all *P*-, or all *N*-channel devices. When the input is at a positive potential, the lower transistor is ON while the upper one is OFF, dissipating no power. The output is now at ground through the 750 Ω conduction resistance of the lower transistor.

Multi-input gates and flip-flops, as well as an infinite number of logical networks, can be built with MOS devices either in homogeneous MOS, or in complementary CMOS configurations. Figure 3.23 shows typical MOS and CMOS gates and flip-flops. The power-supply range of MOS and CMOS circuits varies from 3.0 to 18.0 V with the most recommended value being 15.0 V. Higher voltage increases noise immunity and speed at the expense of power, while lower voltage trades speed and noise immunity for low-power dissipation.

The advantages of CMOS logic over MOS logic are lower power requirements, higher speed of operation, and low-output impedance in both directions, HIGH and LOW, with the disadvantage being cost since its fabrication process requires more steps.

The dynamic MOS logic makes use of the high input and cut-off impedances of the field-effect transistors and treats their input capacitance as a charge storage element. Thus, if a given voltage from a low-output impedance driver is applied to the gate of an MOS transistor, the small (2.5 pF) input capacitance of the gate will be immediately charged. When this input voltage is disconnected from the gate, the only path through which the gate capacitance can discharge will be the high, 500 MΩ gate resistance. The charge may typically stay at a useful level for up to 2 msec. During that time, the MOS transistor will be in conduction or cut-off, depending on the amplitude of the applied voltage.

Figure 3.24 shows an inverter, the output of which need be sampled only during period ϕ_2 while the input is available only during ϕ_1. Since the inverter output is not considered at any other period of time, the inverter load transistor can be ON only during ϕ_2, thus saving power—that is, the power that

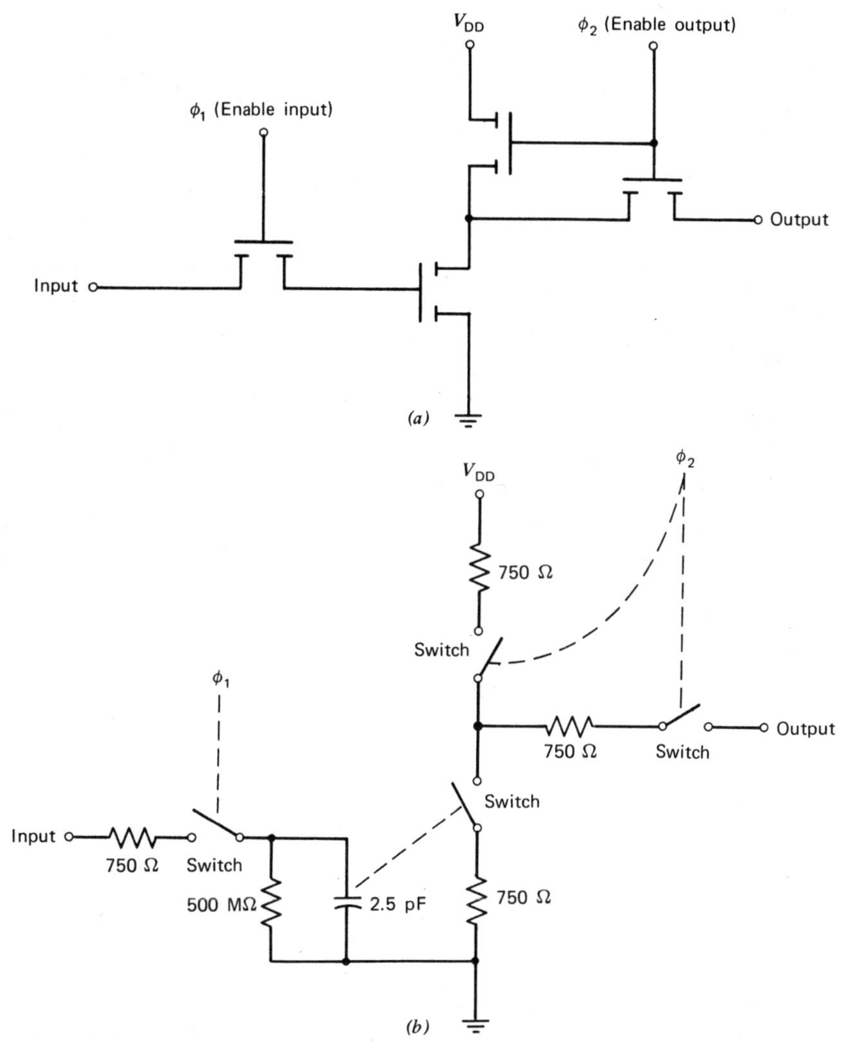

Figure 3.24. Clocked inverter using MOS dynamic logic: (*a*) schematic of clocked inverter; (*b*) equivalent circuit of clocked inverter.

would be normally dissipated across the transistor during time other than ϕ_2. No information loss will occur, because the charge that determines the output state of the inverter is stored in the gate capacitance.

If two clocked inverters are connected in cascade, as illustrated in Figure 3.25, the equivalent of a shift register is formed. The only restriction is that the entered information be exited before the gate capacitor, where it is held, discharges. That is, the time spacing between pulses ϕ_1 and ϕ_2 must be less than the time the gate capacitance voltage takes to drop below the threshold voltage of the gate.

In the circuit of Figure 3.25, the input information, which may be either at V_{DD} or ground, is applied to the gate of the first inverter at pulse ϕ_1. C_{g1} is charged if the input is at V_{DD}, or it is discharged if the input is at ground.

During ϕ_2 the load of the first inverter will be activated, switching from 500 MΩ to 750 Ω allowing the output of the inverter to be determined by the charge that has been stored on C_{g1}. Also, during ϕ_2, the path between the output of the first inverter and the gate of the second inverter is similarly switched from 500 MΩ to 750 Ω. Now, the logical complement of the information stored in G_{g_1} enters into C_{g_2}, the gate capacitance of the second inverter, where it will stay until the next ϕ_2 pulse.

At the arrival of the next ϕ_1 pulse, new information enters C_{g_1}, and the information now in C_{g_2} becomes available at the output. If many such stages are linked together, a digital delay line will be formed called "dynamic" shift register. It is called dynamic because the information stored in it must be continuously shifted in order for it to remain useful. Should motion stop, all gate capacitances will be discharged and the information stored in the shift register will be destroyed. Therefore, the minimum ϕ_1 and ϕ_2 frequency requirements, specified by the MOS manufacturer, should be carefully observed.

The advantages of static logic over dynamic are no need for clocks and simplicity in use, while the disadvantages are lower speed and higher power requirement.

The concept of switching power ON only when it is needed has been further developed, and now four-phase dynamic MOS logic is available. This logic provides its users with even higher speed, noise immunity, and power savings.

Figure 3.26 shows a one-stage dynamic shift register employing a phase concept, where the power lines have been replaced by clock lines. The operation of this circuit proceeds in the following manner: During clock ϕ_1 the gate capacitance of Q_3 is charged to the amplitude of ϕ_1 through transistor Q_2. During clock ϕ_2 and while clock ϕ_3 is ground, the charge in the gate capacitance of Q_3 will discharge through Q_1 into the input, should the input be LOW. If the input is HIGH, the charge will remain in the gate

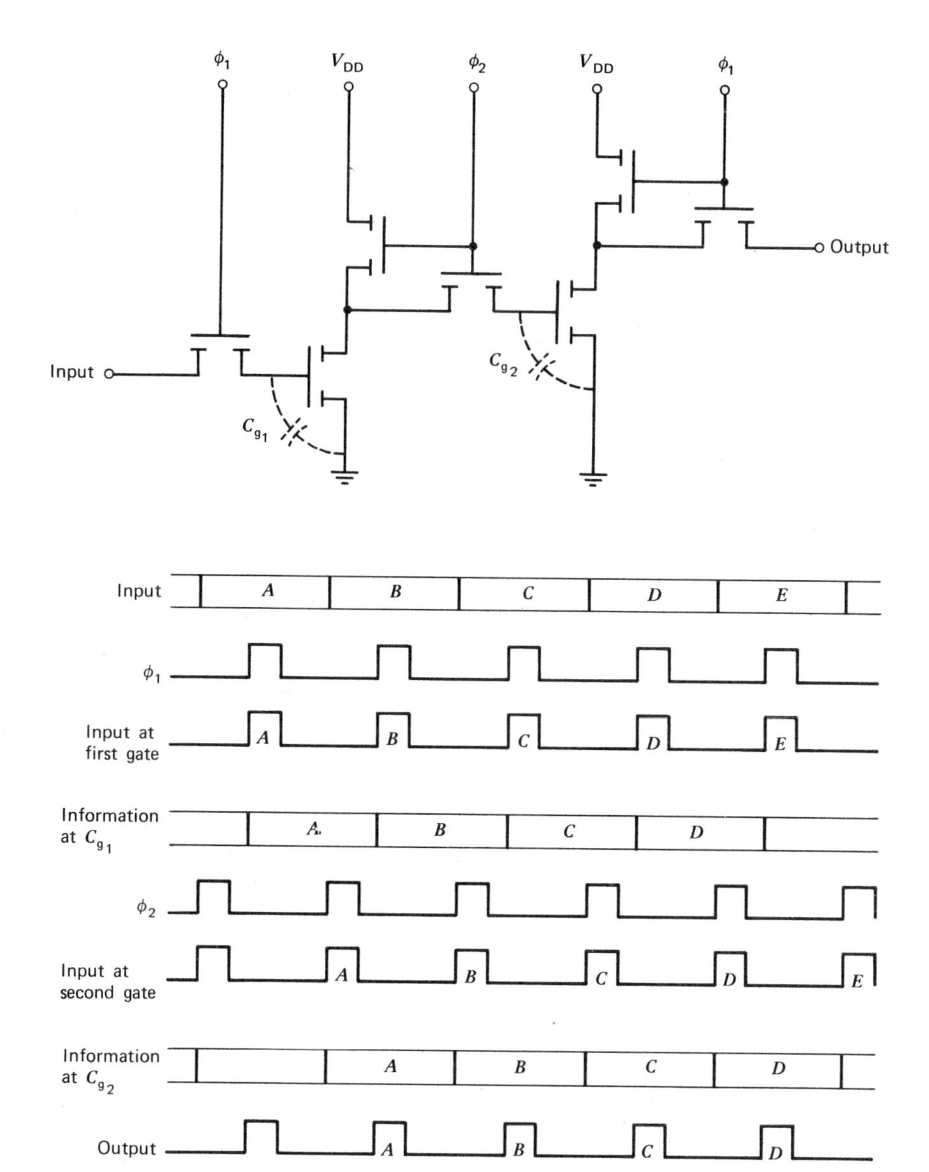

Figure 3.25. Cascade of two clocked inverters forming a dynamic shift register.

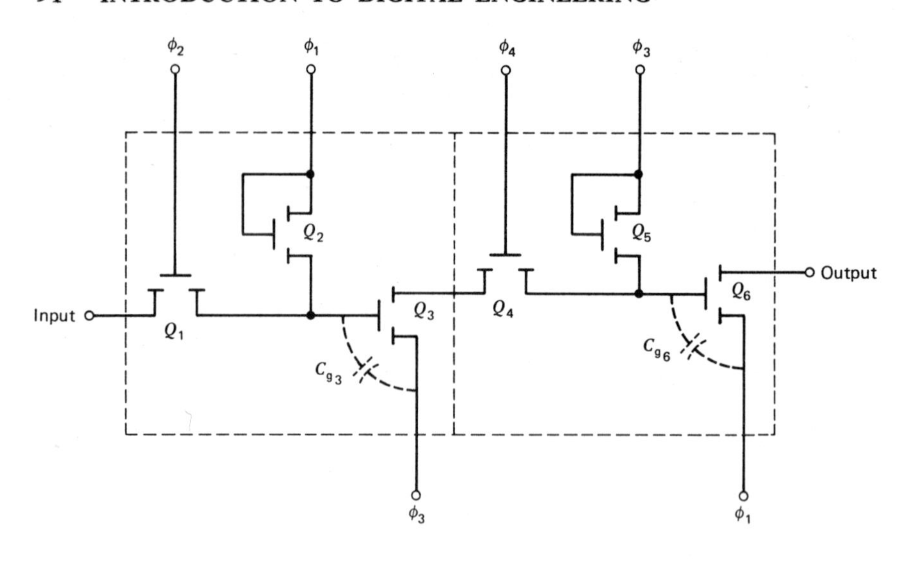

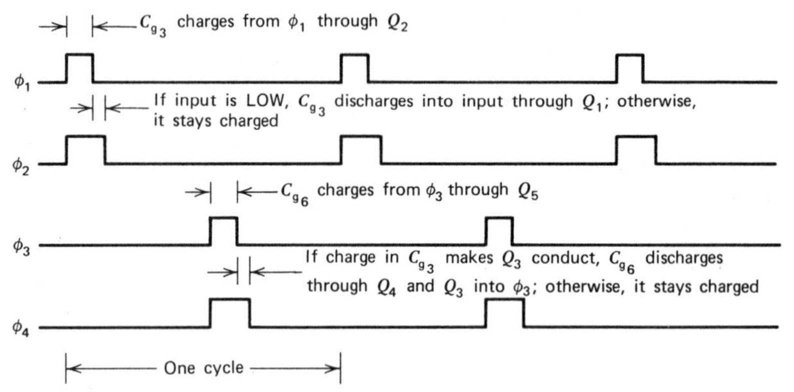

Figure 3.26. One-stage four-phase MOS dynamic shift register.

capacitance of Q_3. During clock ϕ_3 the gate capacitance of Q_6 is charged to the amplitude of ϕ_3, through transistor Q_5; although Q_4 is conducting, no current flows through it because the source of Q_3 is sitting HIGH at ϕ_3. During clock ϕ_4 and while clock ϕ_3 is ground, the gate capacitance of Q_6 will discharge through Q_4 into Q_3 if Q_3 is conducting. If Q_3 is not conducting, because its gate capacitance charge is LOW, the HIGH charge will remain in the gate capacitance of transistor Q_6.

That is, during clock ϕ_2 with the aid of clock ϕ_1, the logical information of the input is stored into the memory element—the gate capacitance—of the first half of the shift register. Similarly, during clock ϕ_4 with the aid of clock ϕ_3, the inverse of the logical information, stored in the first half of the register, is stored into the memory element—the gate capacitance—of the second half of the shift register. If many such shift register units are connected in cascade, a digital delay line will be formed.

Although the breakdown voltage of an MOS gate oxide normally exceeds 100 V, MOS sensitivity to externally applied charges makes it imperative to use protective circuits during testing and at interfaces. Thus, P-channel devices must be protected against extraneous positive voltage excursions, N-channel devices against negative, and CMOS circuits must be protected against both. Figure 3.27 illustrates typical protective circuits for the three MOS arrangements.

In the selection of MOS logical networks, in addition to the logical performance and general circuit characteristics, the following four items must be also considered:

1. *Compatibility with bipolar logic.* This is a necessary feature because without it special circuits will be required to interface the MOS inputs and outputs with outside TTL or DTL logic.
2. *Input protection against static charge.* Every input must internally have a diode protective network to prevent accidental static discharge into an MOS gate.
3. *Single-power supply.* It is always desirable to operate all logic circuits from one power supply and this feature should be given special consideration.
4. *Single-phase clock.* All MOS logical networks that provide storage normally require a two-phase clock—that is, two clock lines having waveforms one-half clock period apart. If the second phase is not generated internally in the MOS package, external generation will be needed.

3.9 A COMPARISON OF VARIOUS LOGIC FAMILIES

A comparison of the basic characteristics of the six major logic families is illustrated in Figure 3.28. Note that HTL logic cannot be really compared with the other logic families, because it is only used for system interfacing, and does not have the variety of logic elements and networks the other logic families have.

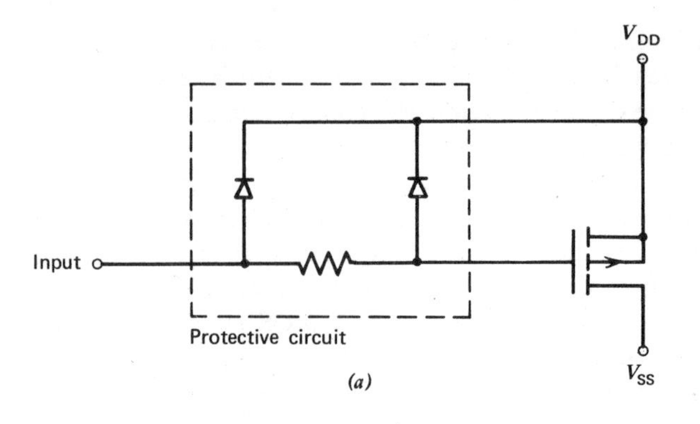

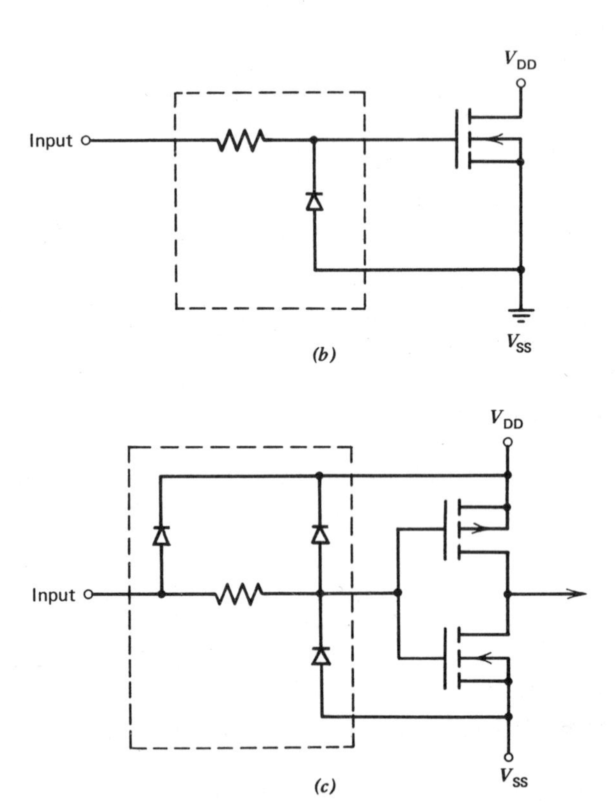

Figure 3.27. Protected MOS inverters (*a*) *P*-channel; (*b*) *N*-channel; and (*c*) CMOS.

Characteristics	Logic families					
	RTL	DTL	TTL	HTL	ECL	MOS
Propagation delay	Medium (12 nsec)	High (20 nsec)	Medium (6 nsec)	Medium (10 nsec)	Low (5 nsec)	High (30 nsec)
Maximum clock frequency	Low (7 MHz)	Medium (20 MHz)	High (50 MHz)	Medium (20 MHz)	High (150 MHz)	High (25 MHz)
Power dissipation	Medium (24 mW)	Low (5 mW)	Medium (20 mW)	High (40 mW)	High (35 mW)	Low (2 mW)
Noise immunity	Low (0.25 V)	Medium (1.0 V)	Medium (1.0 V)	High (4.0 V)	Low (0.12 V)	High (0.45 V_{DD})
Fan-out	Low (5)	Medium (10)	Medium (10)	Medium (10)	Medium (10)	High (50)
Power-supply voltage	3.6–5.0 V	5 V	5 V	15 V	5.2 V	3.0–18.0 V
Tolerance of power supply	Low (15%)	Medium (20%)	Medium (20%)	Medium (20%)	Low (10%)	High (50%)
Cost	Low	Low	Medium	High	High	Medium

Figure 3.28. Comparison of the basic characteristics of the six major logic families; values are based on the comparison of the equivalent four-input gates.

3.10 SELECTION OF MOST SUITABLE LOGIC FAMILY

In digital systems design selection of the most suitable logic family is of great importance. To arrive at this decision, the following four steps should be taken:

Step 1. Calculation of system digital circuit requirements (speed, power dissipation, noise immunity, etc).

Step 2. Comparison of the calculated system digital requirements to the characteristics of all logic families.

Step 3. Selection of logic family with characteristics that require a minimum of concessions in the system digital requirements established in Step 1.

Step 4. Modification of system digital requirements to match the characteristics of selected logic family.

The following example illustrates the implementation of the above procedure.

Example 3.1. Suppose that the system digital circuit requirements of a given application are

1.	Average fan-in	3
2.	Average fan-out	4
3.	Clock frequency	60 MHz
4.	Propagation delay	7 nsec
5.	Power dissipation	minimum possible
6.	Noise immunity	0.4 V
7.	Power-supply tolerance	15%
8.	Amount of space available	very limited

A comparison of the above digital circuit requirements is made to the characteristics of the logic families listed in Figure 3.28. Such a comparison shows that there is no logic family that satisfies all the digital requirements of the system.

The first two items in the list of requirements, which are average fan-in and average fan-out, pose no problem in the selection of a suitable logic family, since all logic families satisfy those two requirements. Items 3 and 4, clock frequency and propagation delay, favor ECL circuits, because they are the only ones that can operate at such high clock frequency. However, item 5, which is the power dissipation requirement, does not favor ECL circuits due to their high power needs. Item 6, the noise immunity requirement, rejects ECL circuits, favoring DTL, TTL, and MOS circuits. Item 7, the power-supply tolerance, also favors DTL, TTL, and MOS, over ECL which requires a 10% regulation. Finally, item 8, the space requirement, suggests that flat-packs be used rather than either TO-5 cans or dual-in-line packages.

The system digital circuit requirements are now reviewed and modified, so that they agree with the characteristics of one of the examined logic families. The major problem is the clock frequency requirement. One solution is to use ECL circuits since they meet that requirement and to improve the supply tolerance and noise immunity of the system to secure proper performance of ECL circuits. The alternative will be to lower the system clock-frequency requirements to that of either DTL, TTL, or MOS circuits. Such a selection would also relax the power supply requirements as well as the noise immunity requirements.

The decision, as to what logic family to select, of course, depends on the specific application. In this case, the high clock frequency requirements will have to be weighed against the need for tighter design criteria that ensure high noise immunity. After the selection is made, the designer modifies his system digital circuit requirements to match the characteristics of the selected logic family. ∎

In many applications the designer may consider using more than one logic family in the same system; for example, ECL for high-speed operations, MOS for memory or other MSI applications, and HTL for system interfacing. Therefore, extensive familiarity with all logic families and digital products is a necessary prerequisite for an optimum digital-component selection.

3.11 SPECIAL USE OF DIGITAL CIRCUITS

After the logic family is selected for a given application, all logic circuits that will be used should be analyzed from a circuit viewpoint as well as from a logic viewpoint. The circuit analysis will determine any special properties the circuits may have, while the logic analysis will familiarize the designer with the internal operation of the more complex logic circuits. Knowing the internal structure of digital circuits often leads to simpler designs that take advantage of circuit characteristics not revealed by the logic symbols that correspond to these circuits.

For example, the DTL circuit schematics show that the output of the low fan-out circuits are of high impedance. As a result, if the outputs of many such circuits are connected together, a gate is formed at the junction. If the HIGH level at the junction is considered as logical 1, the gate formed is AND. If the LOW level is considered as logical 1, the gate formed is OR. Such gates are called "wired," or "dotted."

Figure 3.29 illustrates typical examples of "wired-OR" gates. It is important to note that, if a gate output has pull-up transistor, that output should

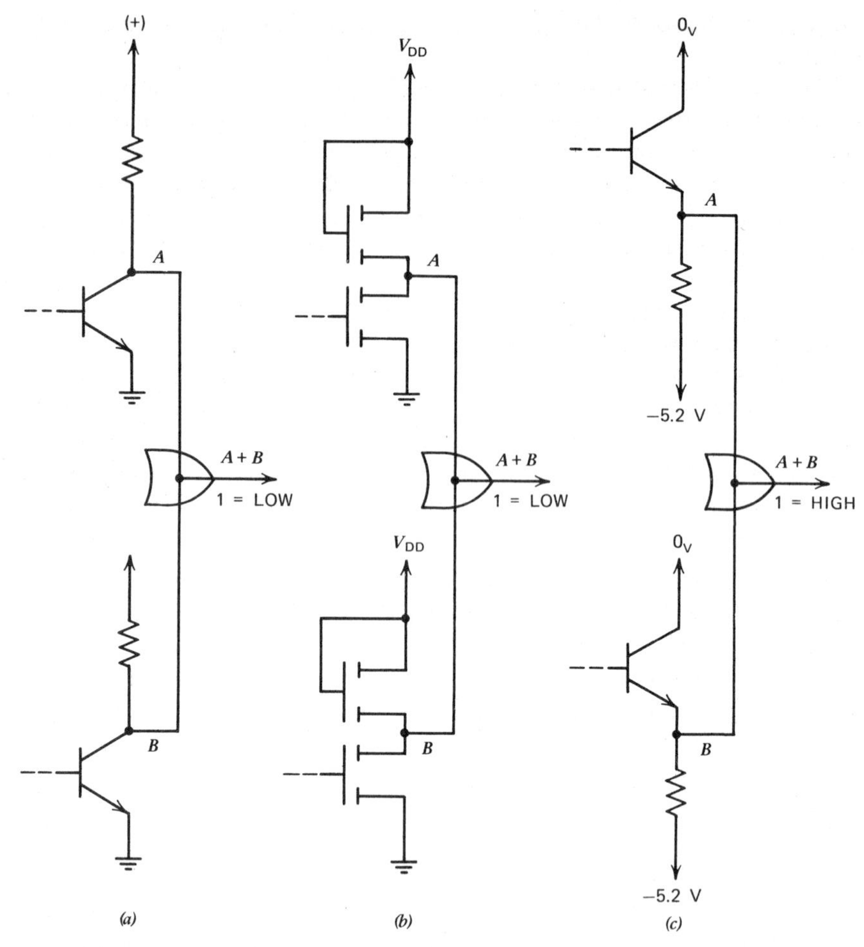

Figure 3.29. Wired-OR gates: (*a*) RTL or DTL circuit; (*b*) MOS circuit; (*c*) ECL circuit.

never be wired-OR. The reason is that the pull-up transistor provides the gate with a very low-output impedance, when the output is HIGH; a low-output impedance is also provided for the LOW state by the pull-down transistor. Therefore, if such low-impedance outputs are connected together and one of them is LOW, high currents from the HIGH outputs will rush into the one that is LOW, and definitely cause damage to the pull-down of the LOW output and to the pull-up transistors of the HIGH outputs.

There are, however, gates with output pull-up transistors that may have their outputs connected to outputs of similar gates but not for the purpose of forming a wired logical element. These gates, called "tri-state," are mainly used for "bussing," where they time-share a line for the transmission of data. The gates are provided with a special input that cuts off both of their output transistors, when other outputs are using the common data line. Figure 3.30 shows a tri-state TTL gate.

An advantage some DTL gates have is that their output fall time tpd^- can be increased by the addition of a small capacitor at their input node. Figure 3.31 shows a DTL gate the output tpd^- of which has been increased. Also shown are the waveforms of operation.

In the circuit of Figure 3.31, the external capacitor increases the input capacitance of the gate, thus increasing the time transistor *Q1* takes to turn ON. This is the time during which the circuit output switches from HIGH to LOW. Use of this circuit property may eliminate undesirable race conditions, resulting from simultaneous arrival of pulses, that often create hard-to-detect problems. Shortcuts, like forming wired gates and increasing delay by means of small capacitors, often simplify design and save hardware.

A logic analysis of the digital circuits that are to be used in a design, often reveals logical properties that could lead to design shortcuts, or to the discovery of weaknesses the circuit may have. In the case of gates logic

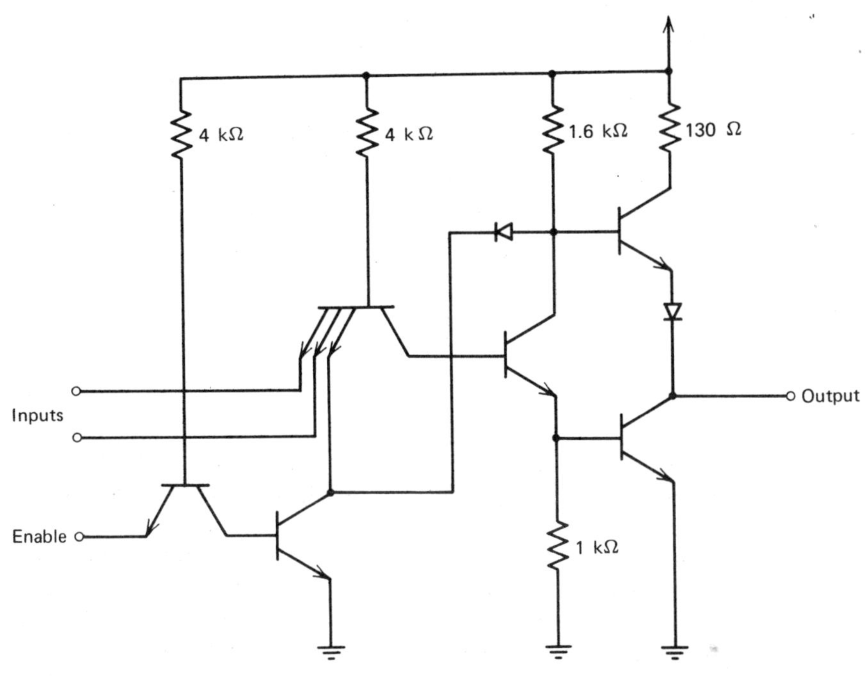

Figure 3.30. Tri-state TTL gate.

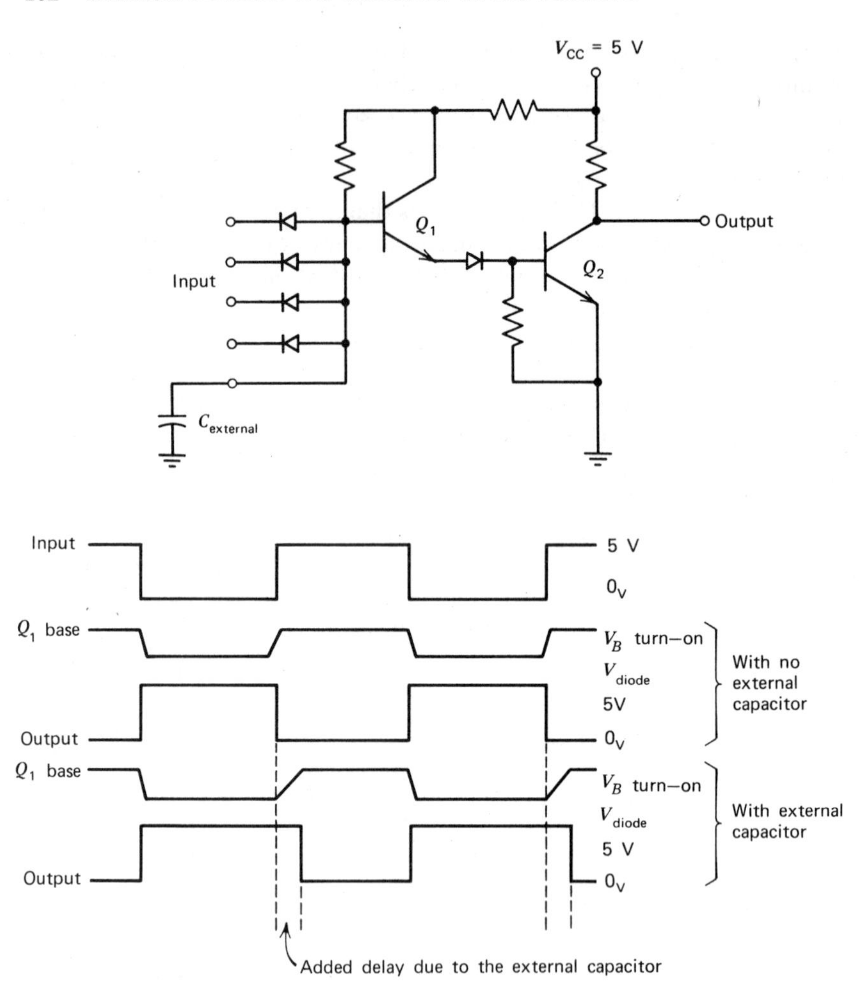

Figure 3.31. DTL gate with increased (tpd^-) delay.

analysis is an easy task, but with flip-flops the logic analysis can be elaborate since in many cases only a complicated discrete component schematic is provided by the manufacturer.

Figure 3.32 shows the logic diagram of the DTL 931 master-slave flip-flop. The logic diagram was not supplied by any of the integrated circuit manufacturers, but it was derived from the schematic of the circuit. A logic analysis of this diagram shows that new information can enter the master portion of the flip-flop via the C_s and S_s inputs, during the time the slave portion is being set or reset. This can be accomplished because the C_d and S_d inputs are not applied to the master portion of the flip-flop. Proper use of this property may result in significant design simplicity; while unawareness may cause logic operations that are difficult to explain.

It should be noted that information should not be transferred from the master portion to the slave portion while a set or reset is applied to the flip-flop. Such action may result in ambiguous outputs where both Q and $\bar{Q}$ output are forced to the HIGH state.

In the flip-flop of Figure 3.32, the principle of wire-ORing can be employed to provide alternate set and reset inputs. The alternate set input will be

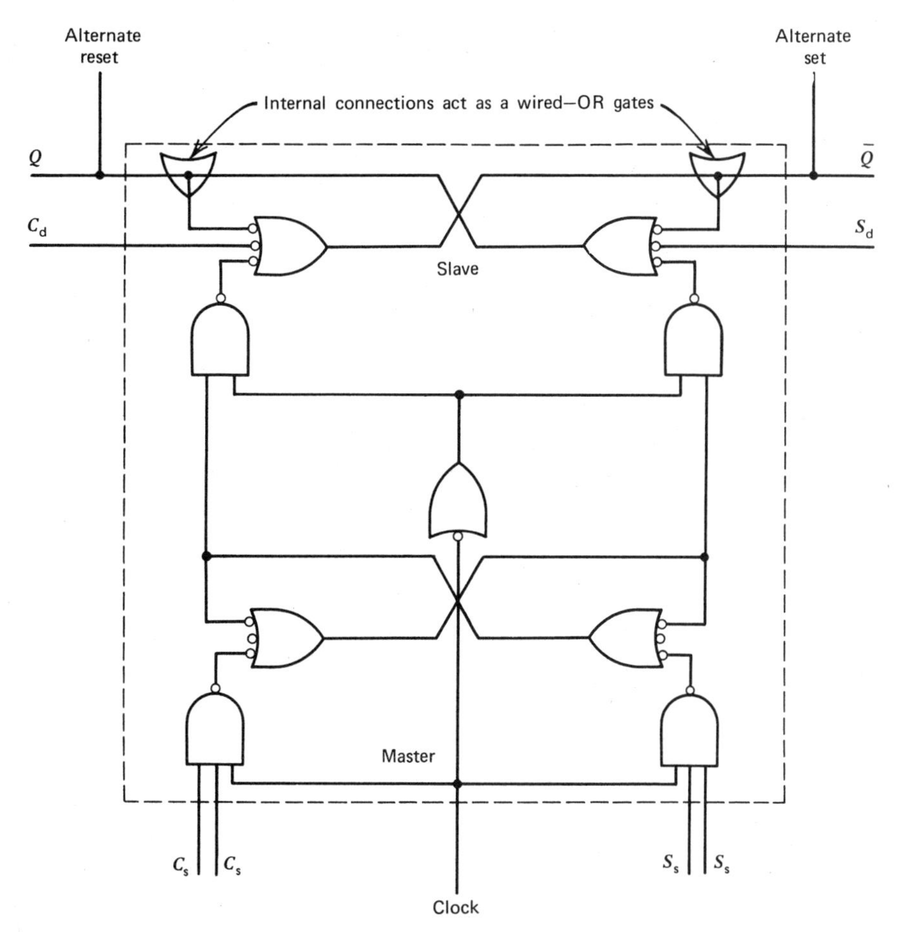

Figure 3.32. Logic diagram of a 931 DTL master-slave flip-flop.

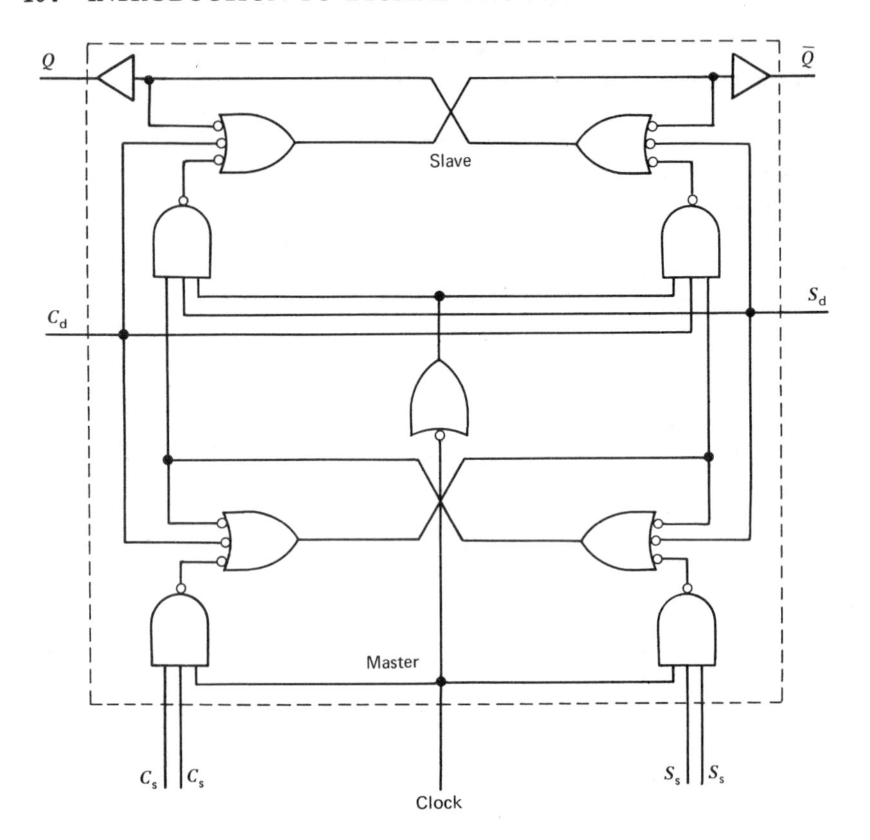

Figure 3.33. Logic diagram of a 945 DTL master-slave flip-flop.

directly connected to the $\bar{Q}$ output, and the alternate reset input to the Q output of the flip-flop.

The logic diagram of another digital circuit, that of the DTL 945 master-slave flip-flop, appears in Figure 3.33. The block diagrams given by the DTL manufacturers for the DTL 931 and DTL 945 flip-flops are identical; these two circuits however are not. In the DTL 945 flip-flop, unlike in the DTL 931, data cannot enter the master portion during set or reset, nor can alternate set or reset inputs be added by means of wire-ORing. Synchronous data may enter the flip-flop only when both set and reset inputs are HIGH.

The above circuit and logic analyses are examples of what a careful analysis may offer the digital designer. It may be concluded, therefore, that a thorough knowledge of the properties of the logic components used is necessary for optimum digital designs.

3.12 MEDIUM-SCALE AND LARGE-SCALE INTEGRATION NETWORKS

The extensive use of digital circuits has led to the development of integrated circuits that contain multielement logical configurations. Logical networks of high complexity, consisting of the equivalent of hundreds of gates and flip-flops, are now widely produced.

Digital integrated circuits are classified into three general categories depending on their complexity. Networks fabricated on a single chip replacing 20–100 logical elements, gates, or flip-flops are called medium-scale integration (MSI) circuits. Networks of less complexity are called small-scale integration (SSI) circuits, and networks of higher complexity are called large-scale integration (LSI) circuits.

The MSI and LSI circuits offer higher reliability and require less power, handling, testing, package care, and space than the SSI circuits they replace. They require less power because the circuit configuration of logical elements inside MSI or LSI packages is considerably simpler than that of similar SSI logical elements. The reason is that the interface requirements of the logical elements inside the MSI or LSI package are simpler. In an MSI package a gate, for example, may consist of three components, while the same gate in an SSI package will require ten components.

The MSI and LSI circuits employ either bipolar (TTL, ECL) or MOS fabrication technology with bipolar (TTL) and MOS dominating the high-complexity applications. The trade-off between the bipolar and MOS circuits has been speed versus power. Bipolars are faster at the expense of cost and power. However, MOS technology has been making considerable progress and it is expected that in the 1980s it will dominate the MSI/LSI market. The MSI/LSI circuits are presently offering a wide variety of logical networks in the areas of counters, shift registers, arithmetic units, decoders, and especially in memories.

The MSI/LSI technology is making one of the largest contributions to digital engineering. Consequently, it is to the benefit of digital designers to always keep up-to-date on MSI/LSI capabilities.

PROBLEMS

1. Compute the noise immunity of a logic family having the following input and output characteristics:

   $V_{OL} = 0.4 - 0.8$ V $\quad V_{OH} = 2.6 - 5.0$ V

   $V_{IL} = 0 - 0.9$ V $\quad V_{IH} = 2.1 - 6.0$ V

2. State the advantages and disadvantages of the ECL logic and compare it to TTL logic.
3. What kind of integrated circuit package is most suited for mass-produced equipment and why?
4. Select the most suitable logic family for the following digital-system requirements:

Average gate fan-in	3
Average fan-out	8
Clock frequency	25 MHz
Propagation delay (gates)	25 nsec
Power dissipation	medium
Noise immunity	0.4 V
Power-supply tolerance	10%

5. State the difference between *P*-channel and *N*-channel MOS transistors.
6. State the advantages and disadvantages of static and dynamic MOS logic.
7. State the principle of operation and the advantages of a multi-phased MOS dynamic operation.
8. Using the circuit of Figure 3.31, design a simple network that produces a pulse when its input signal changes from HIGH to LOW.
9. State the advantages and typical applications of tri-state gates.
10. State the advantages and disadvantages of bipolar and MOS MSI/LSI circuits.

CHAPTER FOUR

Shift Registers

4.1 INTRODUCTION

The shift register is a device in which information may enter either sequentially or in parallel and stay there until it is sequentially transferred out. During that period the entered information may be also accessed in parallel. However, large shift registers do not feature parallel entry and exit of data, due to *pin* limitation in the packages in which they are housed. The logical configuration of a shift register consists of a chain of master-slave flip-flops connected in cascade where all stages are triggered by the same clock.

Shift registers find wide application in the design of digital systems because they constitute the link between parallel and sequential format of digital data, and because data may be entered or accessed via single lines. Use of shift registers extends from pseudorandom generation of binary numbers and format conversion to counting and memory applications.

4.2 UNIDIRECTIONAL SHIFT REGISTERS

A typical unidirectional shift register and its waveforms of operation appear on Figure 4.1. In this circuit, when the input information is a parallel digital word, it enters by means of the direct set input of the flip-flops that comprise the shift register.

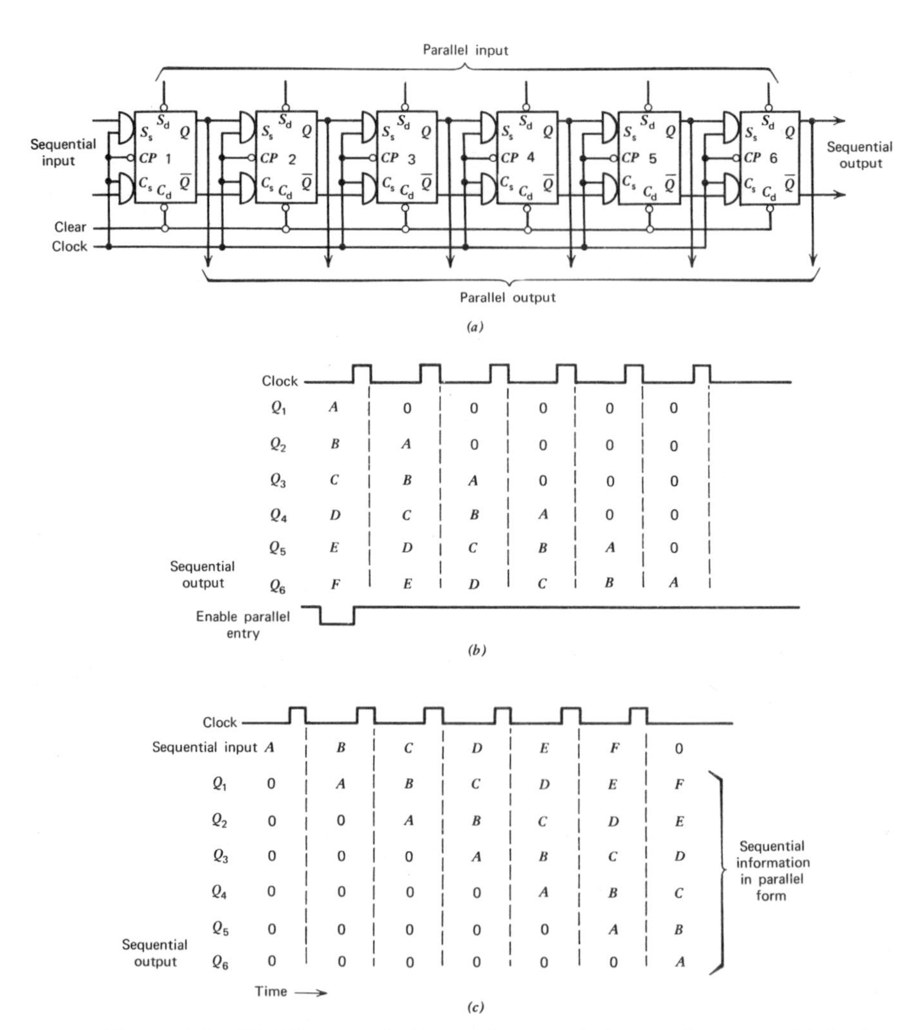

Figure 4.1. Unidirectional shift register and timing diagrams: (*a*) logic diagram; (*b*) timing diagram of a parallel entry (initial conditions: parallel inputs 0, shift register cleared, clear input 0); (*c*) timing diagram of a sequential entry (initial conditions: sequential input 0, clear input 0).

If the parallel word is *ABCDEF*, for example, the first stage of the register will be set to the value of *A*, the second to that of *B*, the third to that of *C*, the fourth to that of *D*, the fifth to that of *E*, and the sixth will be set to the value of *F*. Upon application of the clock to the shift register, digital word *ABCDEF* will start shifting to the right one stage per clock pulse. The parallel output of the shift register, before and after the application of the clock, will be

Stage	**1**	**2**	**3**	**4**	**5**	**6**
After parallel entry of word	*A*	*B*	*C*	*D*	*E*	*F*
After clock pulse 1	*X*	*A*	*B*	*C*	*D*	*E*
After clock pulse 2	*X*	*X*	*A*	*B*	*C*	*D*
After clock pulse 3	*X*	*X*	*X*	*A*	*B*	*C*
After clock pulse 4	*X*	*X*	*X*	*X*	*A*	*B*
After clock pulse 5	*X*	*X*	*X*	*X*	*X*	*A*
After clock pulse 6	*X*	*X*	*X*	*X*	*X*	*X*

Depending on the sequential input of the first stage, *X* will be either 0 or 1. The output of the sixth stage is also the sequential output of the shift register, through which digital word *ABCDEF* "parades" and is available for use, as shown in Figure 4.1(*b*).

The order of the bits appearing at the sequential output of the shift register is *FEDCBA*. The reverse order (i.e., *ABCDEF*) will be obtained should the parallel form of the entered word be applied in reverse order, which is *FEDCBA*.

For example, if the parallel word entered into the shift register is 101001, then the first flip-flop will be set, the second will remain cleared, the third will be set, the fourth and fifth will remain cleared, and the sixth will be set. When the clock starts triggering the shift register, the digital word 101001 will start moving to the right one stage per clock pulse and the parallel output of the shift register will change as follows:

Stage	**1**	**2**	**3**	**4**	**5**	**6**
After parallel entry of word	1	0	1	0	0	1
After clock pulse 1	0	1	0	1	0	0
After clock pulse 2	0	0	1	0	1	0
After clock pulse 3	0	0	0	1	0	1
After clock pulse 4	0	0	0	0	1	0
After clock pulse 5	0	0	0	0	0	1
After clock pulse 6	0	0	0	0	0	0

The result of the above operation is the parallel to sequential conversion of input digital word 101001.

Parallel-to-sequential conversion is often required in the interface of digital systems when the number of interfacing lines is to be minimized. Sequential transfer of digital information is necessary when multibit words are to be transmitted via slip-rings or via a limited number of signal lines such as telephone and radio data links.

Information may also enter a shift register sequentially by means of inputs S_s and C_s of the first flip-flop. In this case the entering information must be in synchronization with the clock—that is, its rate of entry into the shift register must be that of the clock, as shown in Figure 4.1(*c*). This is accomplished by making available the first bit of the entering information at input S_s and its complement at input C_s of the first flip-flop during the first clock pulse, the second bit during the second clock pulse, the third bit during the third clock pulse, and so on.

For a sequential input *ABCDEF*, where *A* is the first bit of information entering the shift register, the parallel output of the shift register is

Stage	**1**	**2**	**3**	**4**	**5**	**6**
Before clock pulse 1	0	0	0	0	0	0
After clock pulse 1	*A*	0	0	0	0	0
After clock pulse 2	*B*	*A*	0	0	0	0
After clock pulse 3	*C*	*B*	*A*	0	0	0
After clock pulse 4	*D*	*C*	*B*	*A*	0	0
After clock pulse 5	*E*	*D*	*C*	*B*	*A*	0
After clock pulse 6	*F*	*E*	*D*	*C*	*B*	*A*

Thus, after the sixth clock pulse, the 6 bit sequential input is available in parallel form and accessible at the outputs of the six flip-flops of the shift register. The result of this operation has been the sequential-to-parallel conversion of a given digital word *ABCDEF* which is the counterpart of parallel-to-sequential conversion and finds wide use in digital system interfacing.

4.3 BIDIRECTIONAL SHIFT REGISTERS

Sequential devices and systems are always sensitive to the order in which information enters them. A sequential binary adder, for example, requires that the LSB of the numbers it adds enter first. A sequential binary number comparator on the other hand requires that the MSB of the numbers it compares enter first.

The bidirectional shift register is a device that accepts sequential information in a given order and provides that information in the same or in the reverse order. Thus, this device may accept a binary word as *ABCDEF*, for example, and then provide it as *FEDCBA*. Information may also enter a bidirectional shift register by means of its parallel input.

A bidirectional shift register and is waveforms of operation are shown in Figure 4.2. This circuit is a chain of master-slave flip-flops where each flip-flop is linked with its preceding as well as with its following flip-flop. To shift information from left to right, the path that connects the output of a flip-flop to the input of the next right flip-flop is enabled. Similarly, to shift information from right to left, the path that connects the output of a flip-flop to the input of the next left flip-flop is enabled.

When a mode control input labeled RIGHT in the bidirectional shift register of Figure 4.2 is HIGH, all R gates are enabled and data flow from left to right. On the other hand, when control input LEFT is HIGH, all L gates are enabled and data flow from right to left. It should be noted that the mode control inputs should never both be HIGH at the same time.

Once information has entered the bidirectional shift register either sequentially or in parallel, it may be shifted in either direction by means of the triggering clock and of the mode control inputs. Therefore, if information *ABCDEF* enters the bidirectional shift register in parallel form via the direct-set inputs of the flip-flops, it may be sequentially transmitted as *ABCDEF* when the L gates are enabled or as *FEDCBA* when the R gates are enabled. In the first case the data will be available at the output of the first flip-flop and in the second case at the output of the sixth flip-flop.

With such unique properties the bidirectional shift register is a very valuable network for sequential data manipulations so often encountered in data transmission and reception.

4.4 MEDIUM-SCALE AND LARGE-SCALE INTEGRATION SHIFT REGISTERS

There is a wide variety of shift registers available in MSI and LSI form that meet almost every design requirement. Figure 4.3 lists the various categories of shift registers available in MSI or LSI form basically divided into MOS and bipolar, depending on their fabrication procedure.

The MOS branch which covers unidirectional multibit shift registers, 50 bits or more, splits into static and dynamic shift registers. The static shift registers consist of MOS flip-flops, similar to that of Figure 3.23, which are interconnected in the chain configuration of a conventional shift register, like the one of Figure 4.1.

In an MOS static shift register information stays for as long as desired and is shifted by means of a clock signal. The required clock signal must be a

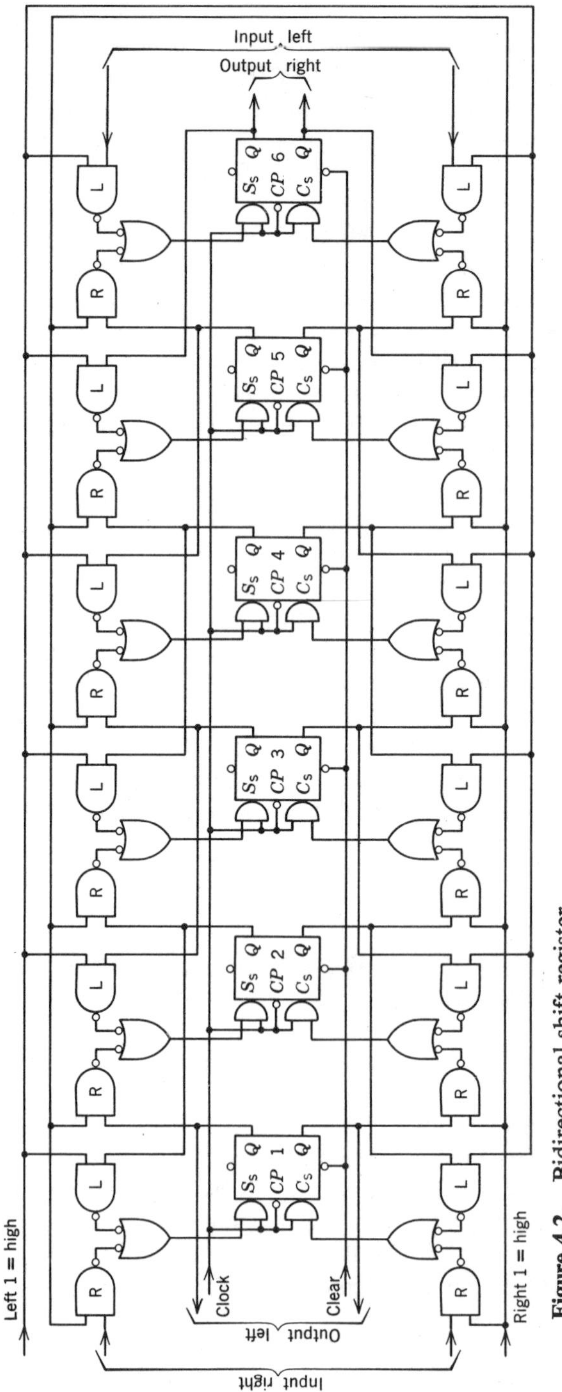

Figure 4.2. Bidirectional shift register.

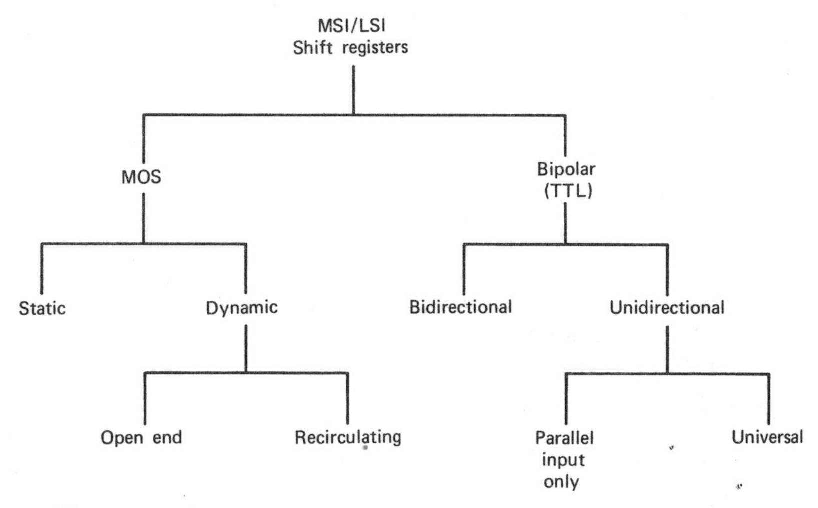

Figure 4.3. MSI/LSI shift registers.

two-phased one with the second phase being one-half period apart from the first. The second phase is sometimes generated internally in the shift register package thus minimizing external requirements as well as package *pins*.

The clock frequency of an MOS static shift register is not restricted and may range from dc to the maximum frequency of operation of the register—a value often found at 2 MHz. Figure 4.4 illustrates a static MOS shift register.

The MOS dynamic shift registers do not consist of flip-flops, but of inverters. Figure 3.25 illustrates a typical stage of a dynamic shift register. In this configuration data are stored in the form of electron charges and held in the MOS gate capacitances. The capacitance slowly discharges through the gate resistance and the represented data are lost, unless they are used or copied elsewhere before the capacitance charge goes below a reliable limit.

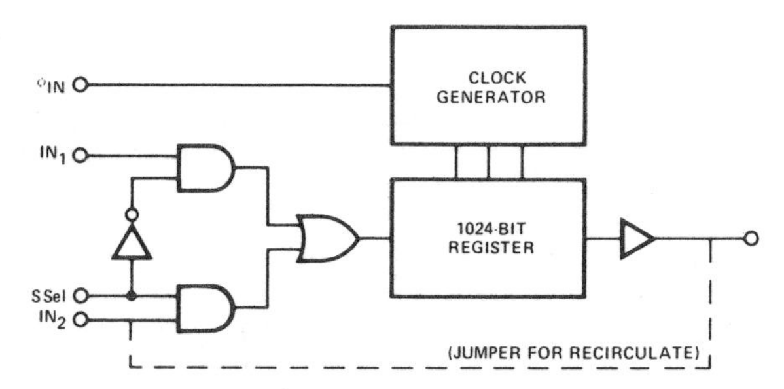

Figure 4.4. Static MOS shift register (courtesy of Signetics).

In an MOS dynamic shift register the above problem is solved by keeping data in continuous motion. This is accomplished by either using the data as soon as they arrive at the exit point, or by recirculating the data in the same shift register. Figure 4.5 illustrates a 1024 bit dynamic recirculating shift register. Recirculating types of shift registers find wide use in memory applications.

The advantages static shift registers offer over dynamic registers are no need for continuous data motion, and easier data accessibility because the location of the data in register is directly controlled. The disadvantages, on the other hand, are higher cost, lower speed, and higher power requirements.

Bipolar shift registers (the word bipolar indicates fabrication process) are of small bit length, usually 4–16 bits, and are classified as unidirectional and bidirectional. When a bidirectional shift register has parallel input as well as parallel output, it is called universal. Unidirectional shift registers can be found with parallel inputs and outputs or with parallel inputs only.

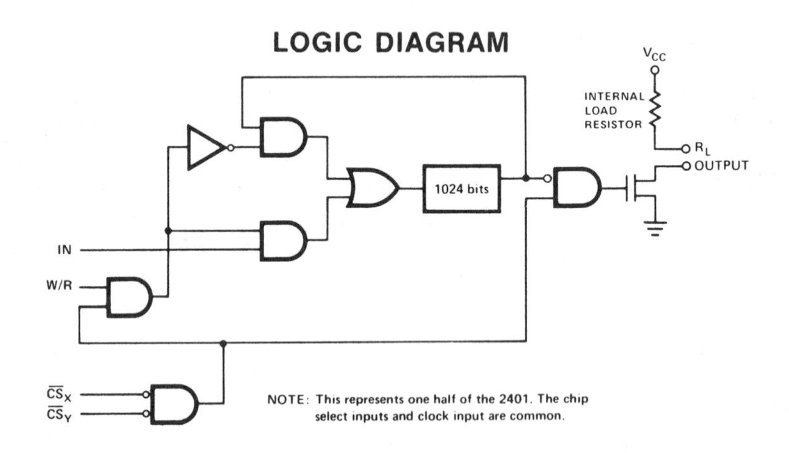

TRUTH TABLE

FUNCTION	PIN SYMBOL		
	W/R	$\overline{CS}_X$	$\overline{CS}_Y$
WRITE MODE	H	L	L
RECIRCULATE	L	X	X
	X	H	X
	X	X	H
READ MODE	X	L	L

H = Logic High Level L = Logic Low Level
X = Don't Care Condition

Figure 4.5. Recirculating dynamic MOS shift register (courtesy of Intel).

Shift registers with parallel inputs and outputs can also be used for parallel storage. In this case, the network is called storage register or simply register. Figure 4.6 illustrates two bipolar shift-register configurations.

4.5 APPLICATIONS OF SHIFT REGISTERS

An example of typical shift register applications is presented here. More applications are illustrated throughout this book where shift registers along with other logical networks are used in the design of various digital configurations. A very interesting one appears in Chapter Sixteen where the cyclomemory is described.

Example 4.1 (System Interfacing). The binary outputs of two systems, P and Q, are to be added by a sequential adder in a third system S. The output of system P is a 6 bit parallel binary number, and the output of system Q is a 6 bit sequential binary number available with its MSB first. A requirement of the sequential adder of system S is that both inputs be sequential, synchronized, and enter with their LSB first. Design the digital circuit that will interface systems P and Q to system S.

The first step is to draw a preliminary block diagram showing what should be done. In this case, there are three things to be done. First, to convert the P output from parallel to sequential making its LSB first; second, to reverse the order of the sequential Q output making its LSB first also; and third, to synchronize the two sequential binary numbers P and Q before they enter the adder of system S. Figure 4.7 shows the preliminary block diagram for the desired interfacing circuit.

The parallel-to-sequential converter of Figure 4.7 can be a unidirectional shift register similar to that illustrated in Figure 4.1 with the only difference being that the parallel inputs should be gated, so that they will not continuously keep the register set to the parallel P output. The parallel input P should enter the unidirectional shift register when the Q input enters the reversing circuit, and it should be disabled immediately before the sequential addition in system S starts.

The order-reversing circuit can be a bidirectional shift register similar to that of Figure 4.2 which will accept the Q output and switch direction after the number has entered—that is, six clock pulses after the start of the operation.

The synchronizer can be a 6 bit shift register that will delay the start signal for six clock periods. The delayed-start signal will be used to perform the following three functions:

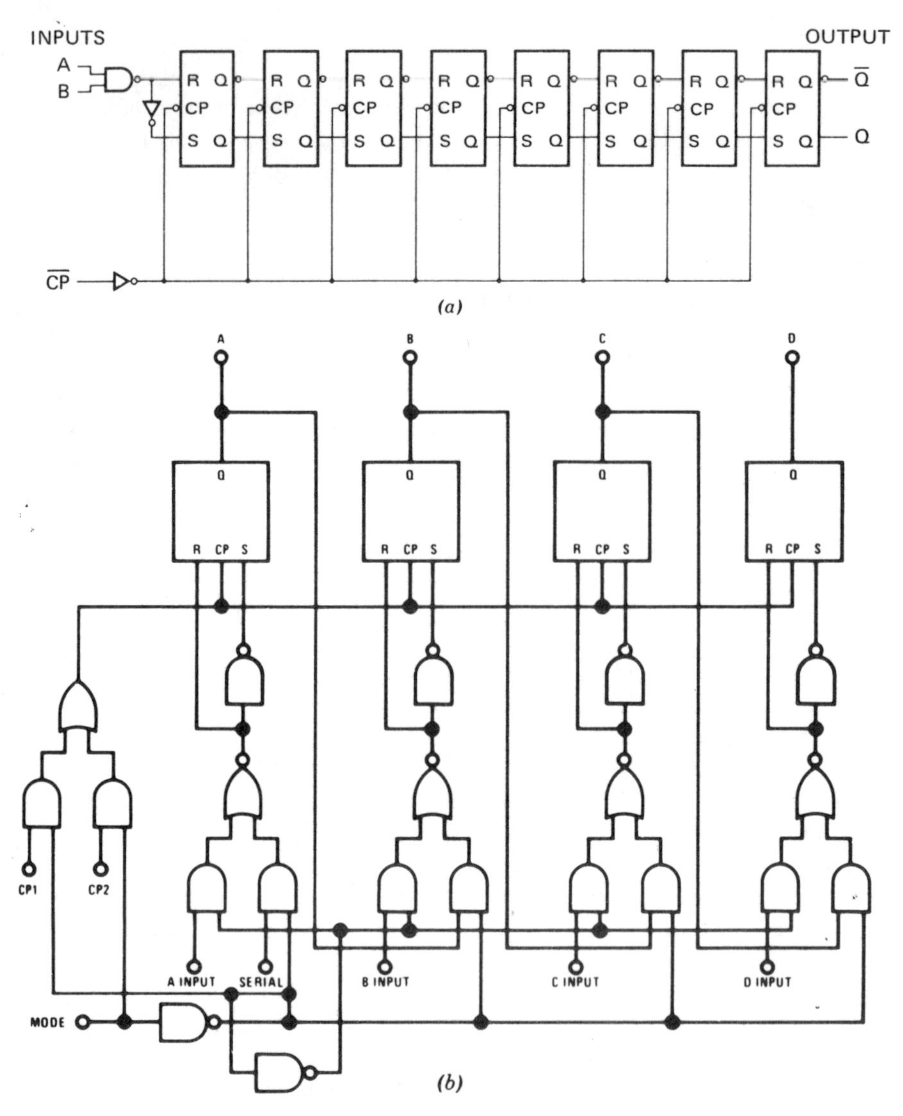

TRUTH TABLE

Mode	Ser_n	Inp A_n	Inp B_n	Inp C_n	Inp D_n	A_{n+1}	B_{n+1}	C_{n+1}	D_{n+1}
1		1	1	1	1	1	1	1	1
1		0	0	0	0	0	0	0	0
0	1					1	A_n	B_n	C_n
0	0					0	A_n	B_n	C_n

DM54L95/DM74L95

Figure 4.6. Bipolar shift register configurations: (*a*) 8 bit unidirectional shift register (courtesy of Fairchild Semiconductor); (*b*) 4 bit universal shift register (courtesy of National Semiconductor).

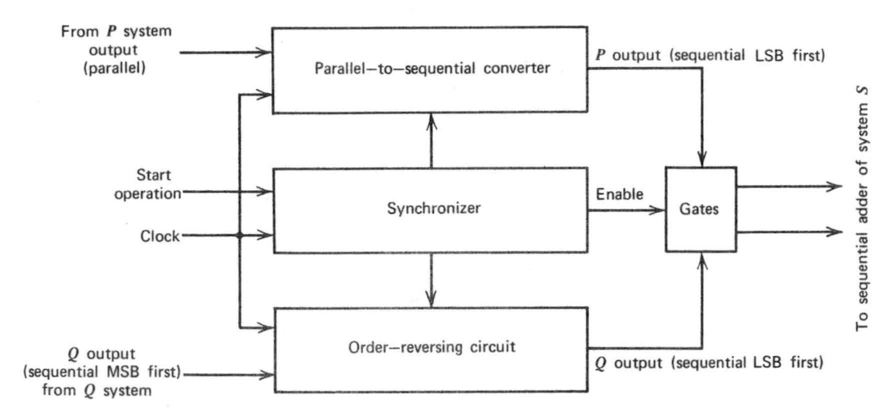

Figure 4.7. Preliminary block diagram of interfacing circuit of example 4.1.

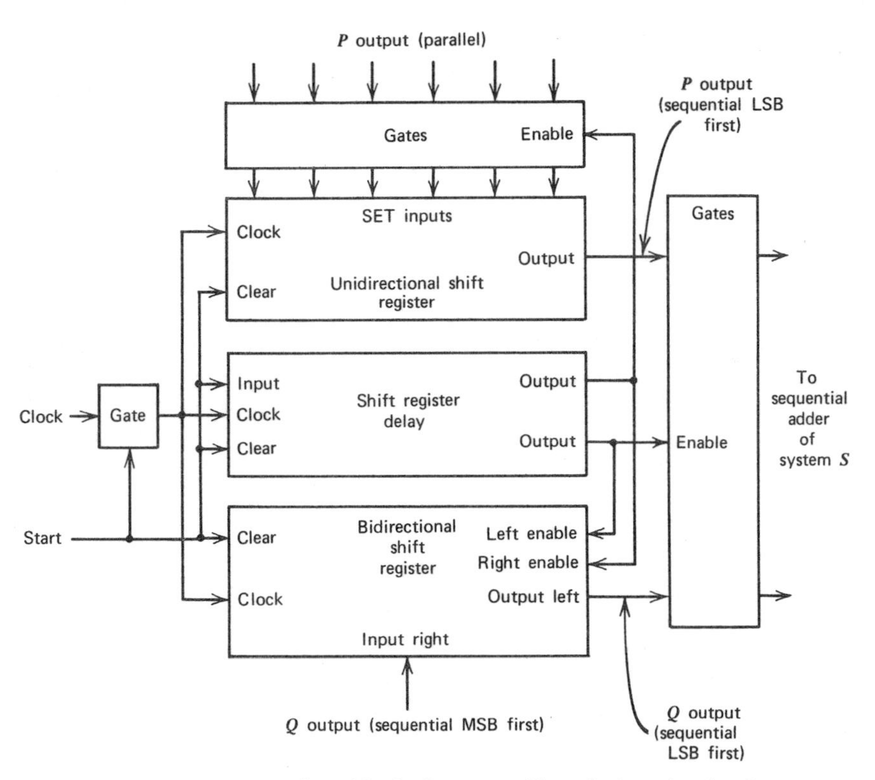

Figure 4.8. Intermediate block diagram of interfacing circuit of Example 4.1.

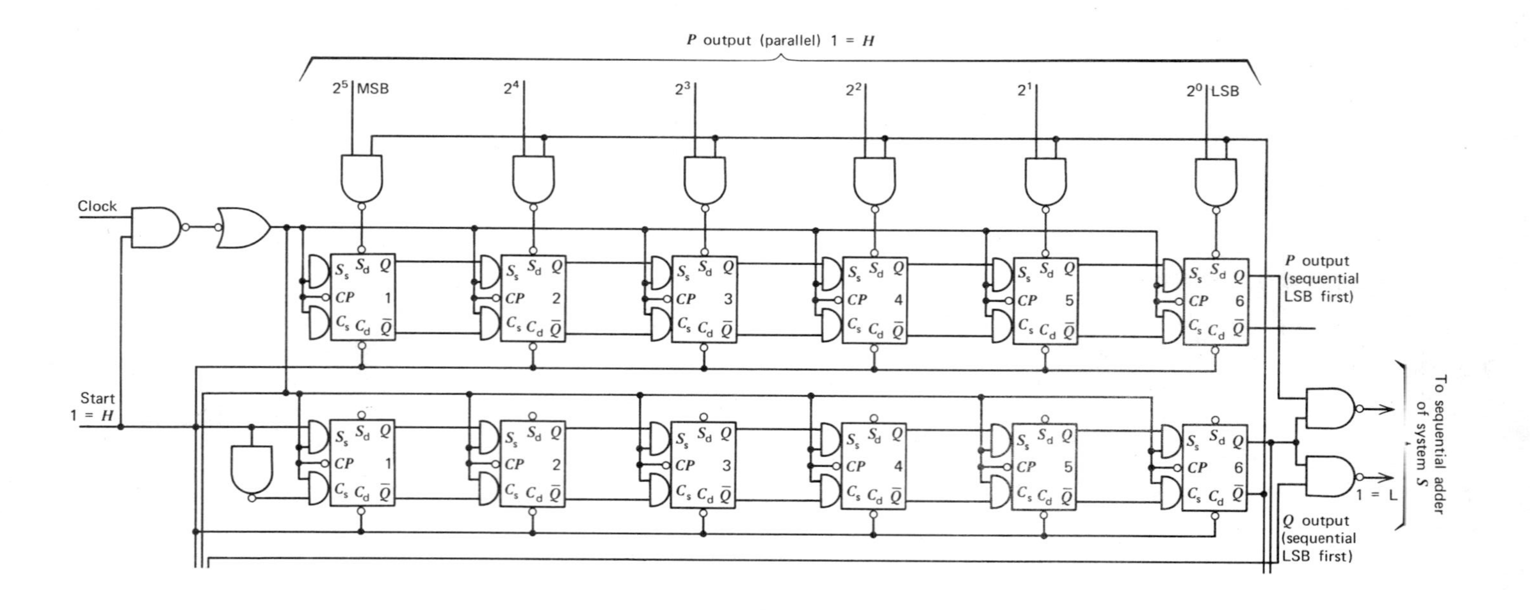

P output (parallel) 1 = H
2⁵ MSB
2⁴
2³
2²
2¹
2⁰ LSB
Clock
Start 1 = H
P output (sequential LSB first)
Q output (sequential LSB first)
To sequential adder of system S
1 = L

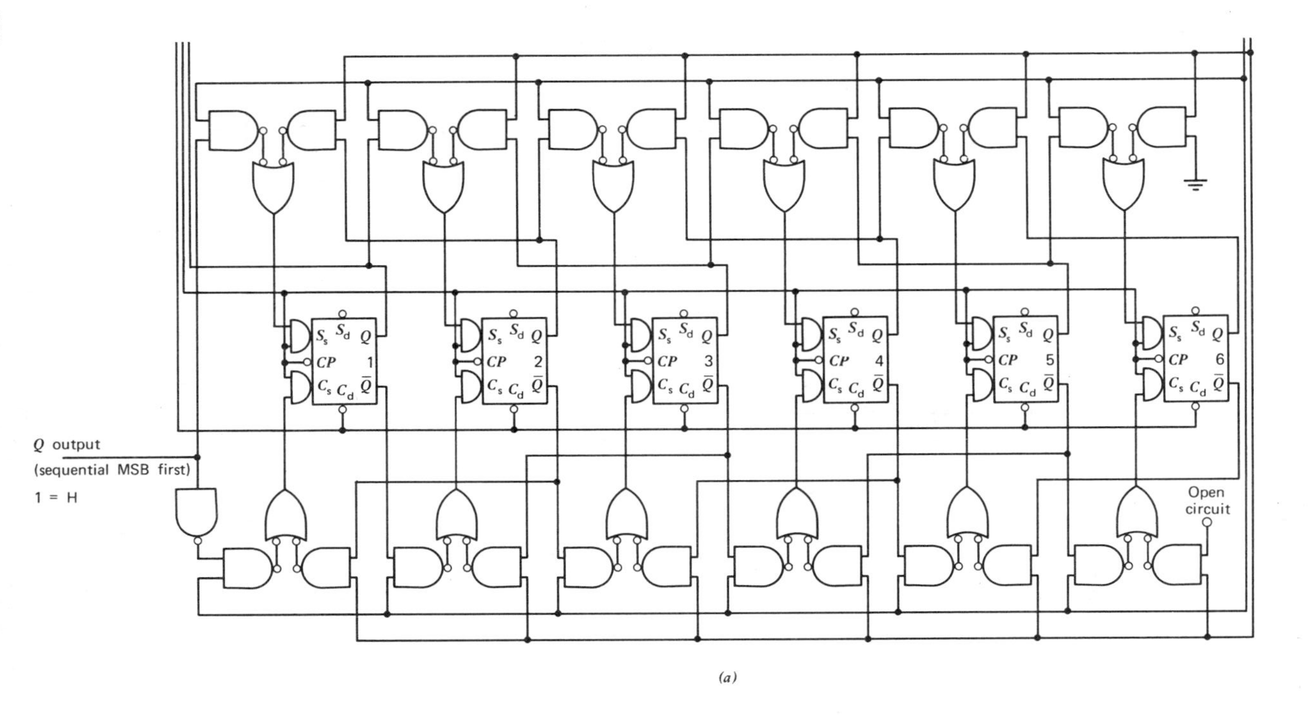

S_s S_d Q CP 1 C_s C_d $\bar{Q}$
S_s S_d Q CP 2 C_s C_d $\bar{Q}$
S_s S_d Q CP 3 C_s C_d $\bar{Q}$
S_s S_d Q CP 4 C_s C_d $\bar{Q}$
S_s S_d Q CP 5 C_s C_d $\bar{Q}$
S_s S_d Q CP 6 C_s C_d $\bar{Q}$
Q output
(sequential MSB first)
1 = H
Open circuit

(a)

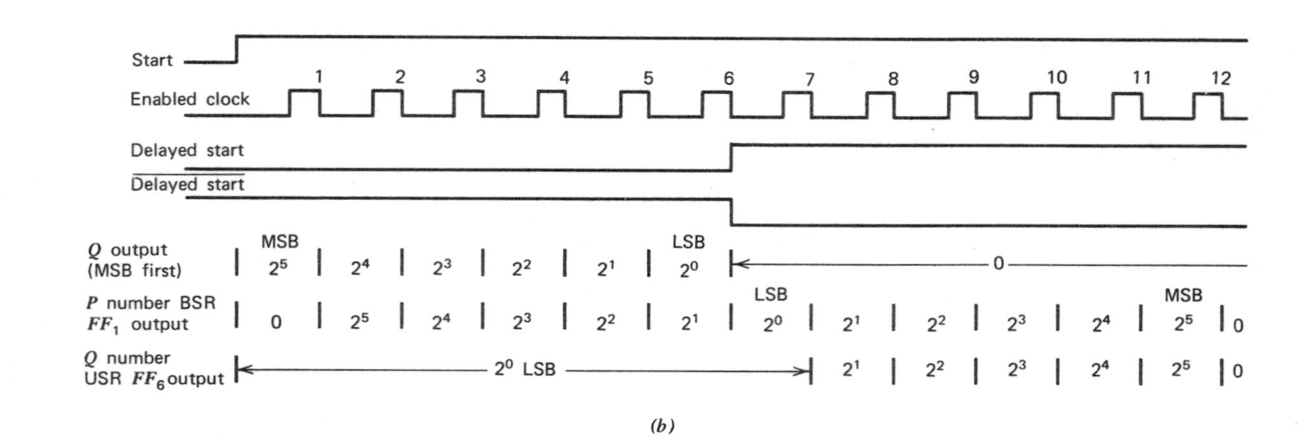

(*b*)

Figure 4.9. (*a*) Logic diagram and (*b*) timing diagram of interfacing circuit of Example 4.1.

1. Reverse the direction of the bidirectional shift register which contains the Q number.
2. Disable the input gates of the parallel-to-sequential converter, thus enabling the unidirectional shift register to start converting the P number from parallel to sequential.
3. Enable the two now-synchronized numbers to enter system S.

The above analysis of the preliminary block diagram of Figure 4.7 leads to the sketch of a more detailed diagram which is shown in Figure 4.8. From this diagram the appropriate design approach can be further defined. The next and final step will be to draw the logic diagram for each of the blocks and show the exact interconnections. This is shown in Figure 4.9, where the logical as well as timing diagrams are illustrated. ▮

The solution to this example was obtained by going through three stages of diagrams: The preliminary block diagram, showing the basic concept; the intermediate block diagram, showing the implementation of the concept; and the final logic schematic from which the circuit can be actually built. Some designs may require more than three stages, depending on the complexity, the size of the system or circuit, and on the availability of appropriate standard digital hardware.

PROBLEMS

1. Design a unidirectional shift register capable of handling 7 bits and of converting them from sequential to parallel form in less than 1 msec. Specify the minimum frequency of operation. Relate the design to currently available hardware.
2. Design a unidirectional shift register that will provide a 25 μ delay to a 3 bit serial digital word. The system clock frequency is 1 MHz.
3. Design a bidirectional shift register to be used for the reversing of a 12 bit serial binary number. Show the circuit timing and relate the design to currently available hardware.
4. Design a logical network that reverses the order of digital word $ABCDE$ and subsequently delays it for three clock periods. The system clock frequency is 5 MHz and the word is accompanied by a pulse occurring at bit A. Use this pulse as a "start operation" signal.
5. Describe a "universal" shift register.
6. State the limitations in the application of unidirectional shift registers that do not provide parallel output.

7. State the difference between a static and dynamic shift register, and identify two typical applications for each.
8. State the advantages and disadvantages of bipolar shift registers by comparing them to the MOS-type shift registers.
9. Design an interfacing network that accepts 4096 bits of information every 1 sec at a 5 MHz clock rate from one system and delivers that information to another system at a 250 kHz clock rate.
10. Using MOS dynamic shift registers, design a cathode ray tube (CRT) refresh circuit for a numerical display system. The system employs a raster scan with horizontal, vertical, and brightness resolutions of 8, 7, and 1 bit, respectively.

CHAPTER FIVE

Sequential Networks

5.1 INTRODUCTION

The function of sequential networks is to generate bit sequences or to detect bit sequences. The shift register, which stores time history, is extensively used in sequential designs because bit sequences are digital signals varying with time. A sequential network basically consists of a shift register and of a decoding or encoding logical circuit, as illustrated in Figure 5.1.

Figure 5.1(*a*) shows a network that decodes a given bit sequence. Its input is a bit sequence S that enters an n bit shift register where it is converted from sequential to parallel form. The parallel form of the input sequence is decoded by means of a logical decoding circuit consisting of gates appropriately interconnected to implement a given function Q. The input bit sequence S may have more than n bits, as long as any group of n consecutive bits of the sequence satisfies function Q. For each clock period during which function Q is satisfied, the output function f is 1. The arrrangement of Figure 5.1(*a*) is often called a "sequential filter" because its output indicates presence of a preselected code.

Figure 5.1(*b*) shows a network that encodes a desired bit sequence S in accordance with the function Q, the input function f and the current content

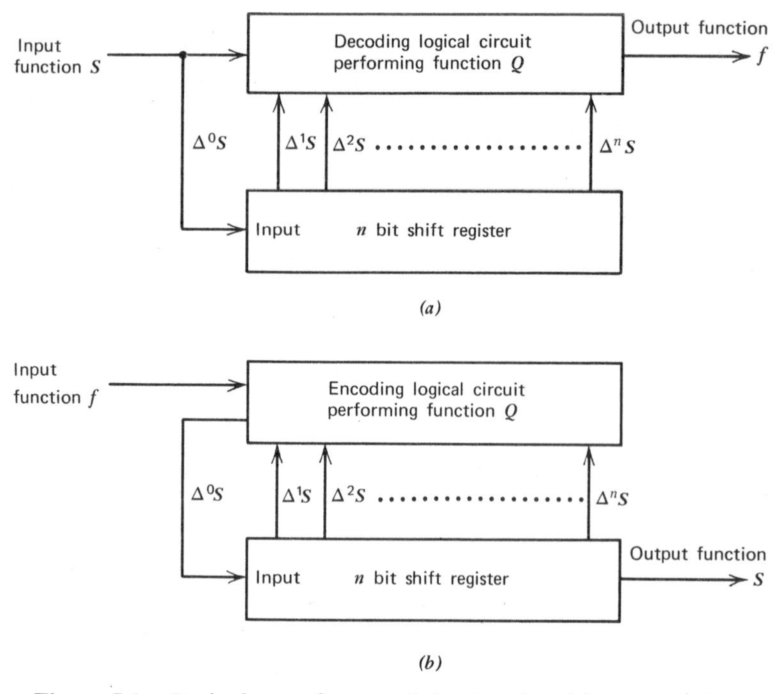

Figure 5.1. Basic form of sequential networks: (*a*) sequential network decoding input function S in accordance with function Q, resulting in output function f; (*b*) sequential network encoding input function f in accordance with function Q generating function S.

of the shift register. The initial content of the shift register must satisfy function Q. The network of Figure 5.1(*b*) is referred to as a "sequence generator."

The term Δ in Figure 5.1 indicates delay: Δ^0 is no delay, Δ^1 is one clock period delay, Δ^2 is two clock period delays, and so on, with Δ^n being n clock period delays. Thus, $\Delta^2 S$ indicates that at that point of the network the signal has the value input S had two clock periods ago. This delay notation has been adopted because it considerably facilitates manipulations of expressions that define the performance of sequential networks.

5.2 BASIC CONFIGURATIONS

A bit sequence can be continuous or discrete. In a continuous bit sequence, if the two ends are connected, every group of n successive bits satisfies a given logical relationship. That is, the sequence has neither a beginning nor an end but it is a closed loop. The term n, the number of decoded successive bits, is called "basic length" of the bit sequence. For example, S: ...1010101010... is a continuous bit sequence the basic length of which is 2. The sequence S can be defined as a bit sequence where no 2 successive bits are the same. The periods at the sides of the sequence indicate that the sequence repeats.

A discrete bit sequence has a beginning and an end and satisfaction of its logical relationship requires the entire sequence. For example, S: 1101110 is a discrete bit sequence. A discrete m bit sequence, however, can be also formed where its logical relationship needs only n successive bits to be satisfied, where n is less than m. For example, S: 1100110 is a discrete bit sequence where the first and the last bit in every 3 successive bits are always different. Thus, only 3 bits need be examined.

Figure 5.2 shows a sequential network that decodes bit sequences having a basic length of 4 bits. In the case of discrete 4 bit sequences there are 2^4 ways to design the decoding circuit, since there are 2^4 states in which the 4 bit sequence can be found. The decoding circuit takes the form of a 4 bit

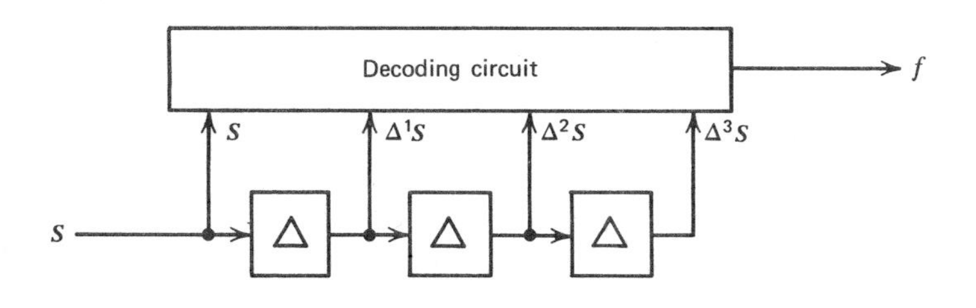

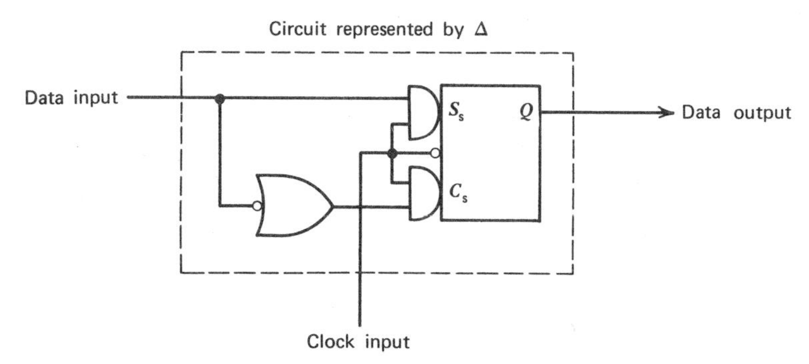

Figure 5.2. Sequential network with a 4 bit basic length; Δ represents one clock period delay. Clock connection is not shown.

AND gate that provides an output of 1 anytime the sought-after bit sequence appears across its input.

For example, to decode the presence of bit sequence 1011011 in a sequential string of bits the network of Figure 5.3 can be used. In this network the output f is 1 only when the sequential input S is 1011011. Depending on the definition of the sequential input S, the form of the output f can be any logical expression consisting of the ΔS terms. The sequential input S is usually defined as a fixed bit sequence, but it is possible to find it expressed in terms of events. Such an expression can be S: Two 1s followed by three 0s or one 1 followed by two 0s.

In the case of continuous bit sequences the decoding circuit must be able to identify each of the subsequences of the basic length that comprise the continuous bit sequence. Continuous bit sequences that are the easiest to implement are those defined by binary arithmetic relations, rather than boolean logical expressions not related to binary arithmetic. Such a sequence is S: . . .011010011010. . . where the binary sum of any 6 consecutive bits is always 3.

Of considerable simplicity in design, analysis, and implementation are the continuous bit sequences where the module-2 sum of selected bits of the basic length is either 0 or 1. Such as sequence

$$S: \ldots 101001110100 \ldots \tag{5.1}$$

where the basic length is 4 and the property is that the LSB of the binary sum of the first, third, and fourth bits is 0.

The LSB of a binary sum is also called a parity bit. Thus, the above sequence has a zero parity for bits one, three, and four of the basic length. Figure 5.4 illustrates the general form of the parity-oriented sequential networks. Network (*a*) is a sequence filter for continuous sequences, and network (*b*) is a continuous sequence generator. In network (*a*) the parity of the entering bit string is computed to indicate presence of a sequence, while in network (*b*) the parity of the sequence in the shift register is computed to determine the next bit of the sequence. Synthesis and analysis of these networks is described in the following two sections.

5.3 SYNTHESIS OF SEQUENCE FILTERS

Bit sequences can be generally classified into parity-oriented and logic-oriented sequences by the type of the equations they satisfy.

In the parity-oriented sequences, each bit is the parity bit of selected bits of the basic length that precedes that bit. For example, the arithmetic equation for sequence S: . . .01010101. . . is

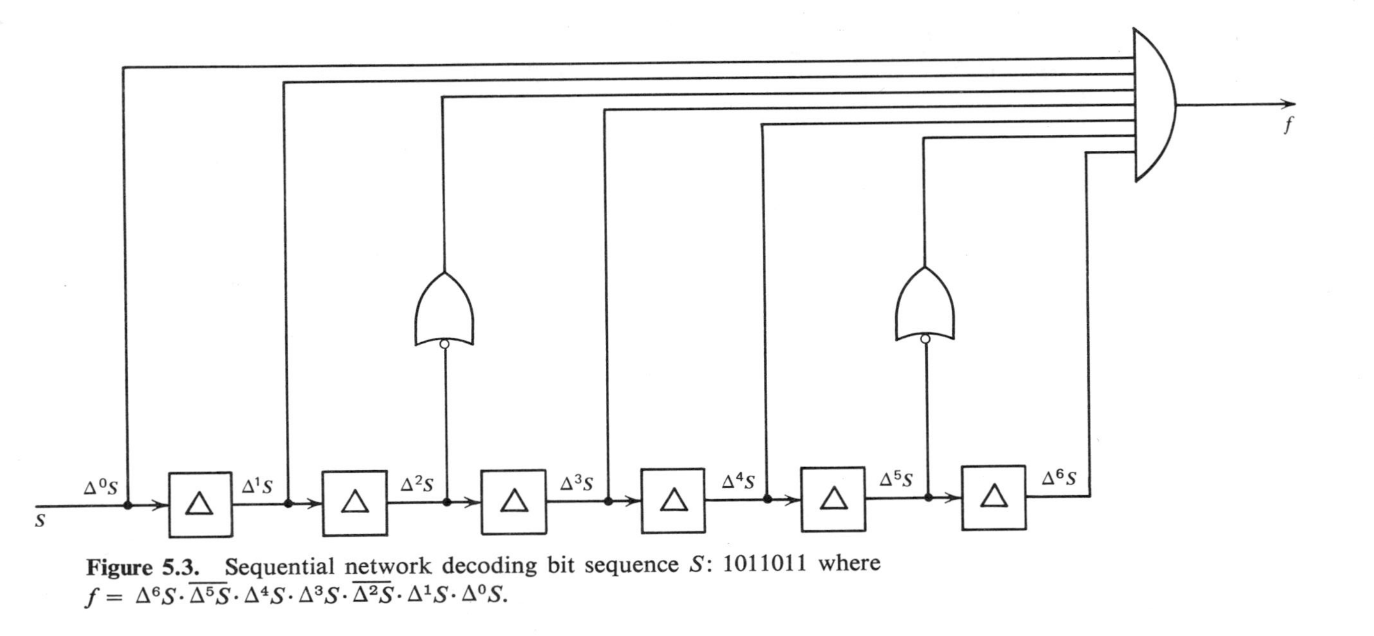

Figure 5.3. Sequential network decoding bit sequence S: 1011011 where $f = \Delta^6 S \cdot \overline{\Delta^5 S} \cdot \Delta^4 S \cdot \Delta^3 S \cdot \overline{\Delta^2 S} \cdot \Delta^1 S \cdot \Delta^0 S$.

$$S: \ldots 01010101 \ldots \tag{5.2}$$

$$f = S \oplus \Delta S \tag{5.3}$$

$$= \Delta^0 S \oplus \Delta^1 S \tag{5.4}$$

$$= S(\Delta^0 \oplus \Delta^1) \tag{5.5}$$

Similarly, the arithmetic equation for sequence

$$S: \ldots 101001110100 \tag{5.6}$$

where there is 0 parity for bits 1, 3, and 4, is

$$f = \Delta^1 S \oplus \Delta^3 S \oplus \Delta^4 S \tag{5.7}$$

$$= \Delta(S \oplus \Delta^2 S \oplus \Delta^3 S) \tag{5.8}$$

$$= \Delta[S(1 \oplus \Delta^2 \oplus \Delta^3)] \tag{5.9}$$

$$\begin{aligned} &= S(\Delta^0 \oplus \Delta^2 \oplus \Delta^3) \\ &= 0 \end{aligned} \tag{5.10}$$

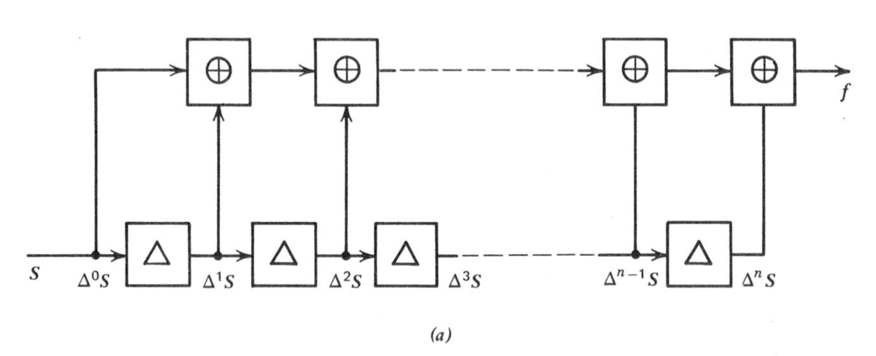

(a)

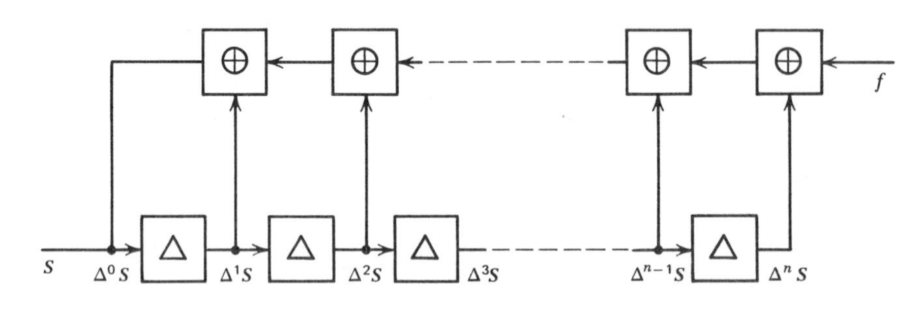

(b)

Figure 5.4. General form of parity-oriented sequential networks: (*a*) sequence filter; (*b*) sequence generator.

The term Δ can be eliminated, since removal of a delay does not change the composition of the sequence. The f expressions explicitly define the respective sequences and they suffice as the basis for the design of the necessary sequential network for the detection or generation of parity-oriented sequences.

The general form of the f expression that defines a parity-oriented sequence is

$$\begin{aligned} f &= S(\Delta^0 \ldots \Delta^{n-1}) \\ &= 0 \text{ or } 1 \end{aligned} \tag{5.11}$$

where n is the basic length of the sequence.

The sequence filter for the detection of parity-oriented sequences consists of an n-stage shift register the outputs of which feed exclusive-OR gates that generate the necessary parity bit, 0 or 1, indicated by the equation that defines the sequence. The bit sequence S that satisfies expression f is called a "characteristic" sequence and the resulting sequence of 1s or 0s of expression f is called "resonant" sequence.

Example 5.1. Design a sequential filter that detects the sequence

$$S: \ldots 101001110100 \ldots \tag{5.12}$$

defined by expression

$$\begin{aligned} f &= S(\Delta^0 \oplus \Delta^2 \oplus \Delta^3) \\ &= 0 \end{aligned} \tag{5.13}$$

The necessary sequential filter consists of a three-stage shift register and of a decoding circuit of two exclusive-OR gates. Figure 5.5 illustrates the sequential network that detects the sequence S: ...101001110100.... ∎

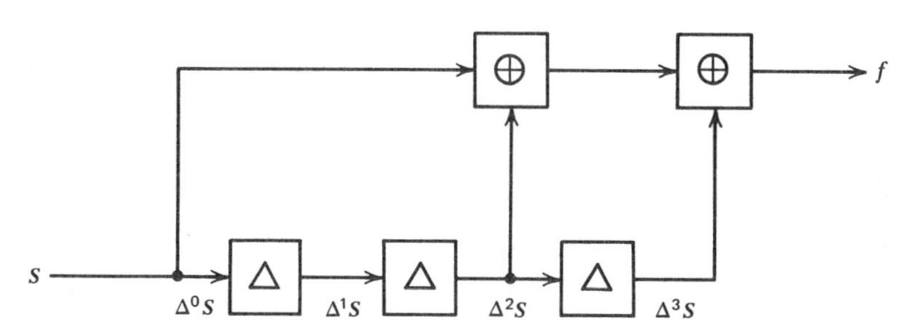

Figure 5.5. Sequential network detecting the continuous sequence S: ...101001110100... where $f = S(\Delta^0 \oplus \Delta^2 \oplus \Delta^3) = 0$. Example 5.1.

Example 5.2. Design a sequential network that detects the continuous sequence defined by the expression

$$f = S(\Delta^0 \oplus \Delta^1 \oplus \Delta^3) \qquad (5.14)$$
$$= 0$$

and determine the characteristic sequence of the designed network.

The sequential network implementing this expression consists of a three-stage shift register and of two exclusive-OR gates, as illustrated in Figure 5.6. To determine the continuous sequence that will keep the network's output at 0, a 4 bit sequence that satisfies expression f is developed. Let it be 0010 with the last bit arriving first. With this 4 bit sequence as a starting point, the entire continuous sequence is determined as follows:

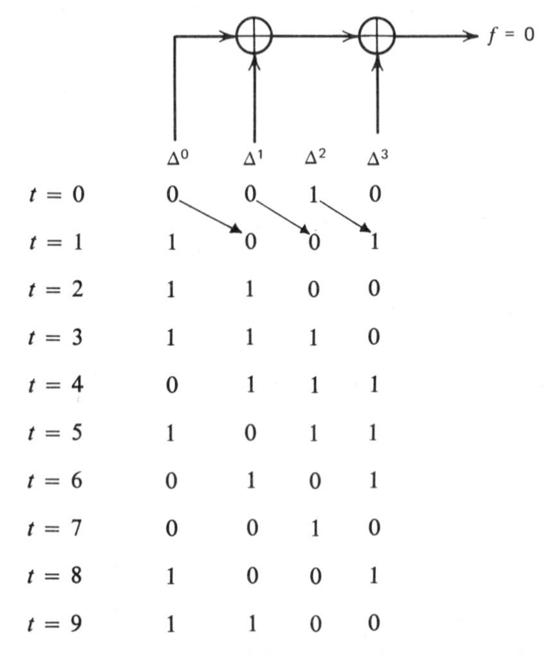

	Δ^0	Δ^1	Δ^2	Δ^3
$t = 0$	0	0	1	0
$t = 1$	1	0	0	1
$t = 2$	1	1	0	0
$t = 3$	1	1	1	0
$t = 4$	0	1	1	1
$t = 5$	1	0	1	1
$t = 6$	0	1	0	1
$t = 7$	0	0	1	0
$t = 8$	1	0	0	1
$t = 9$	1	1	0	0

The value on the second row of the first column was selected to be 1 so that the sequence of that row gives a 0 when entered into the expression

$$f = S(\Delta^0 \oplus \Delta^1 \oplus \Delta^3) \qquad (5.15)$$

giving

$$f = S(1 \oplus 0 \oplus 1) \qquad (5.16)$$
$$= 0$$

The value on the third row of the first column was selected on the same basis, so that the expression of the network be 0. In this case the expression takes the form

$$f = S(\Delta^0 \oplus \Delta^1 \oplus \Delta^3) \tag{5.17}$$

$$= S(1 \oplus 1 \oplus 0) \tag{5.18}$$

$$= 0$$

∎

Example 5.3. Design a sequential network that detects the continuous sequence defined by the expression

$$f = S(\Delta^0 \oplus \overline{\Delta^2} \oplus \Delta^3) \tag{5.19}$$

$$= 1$$

and determine the characteristic sequence of the designed network.

The sequential network that implements this expression consists of a three-stage shift register and of two exclusive-OR gates, as illustrated in Figure 5.7. To determine the continuous sequence that keeps output f at 1, a 4 bit sequence is developed that satisfies the expression $f = S(\Delta^0 \oplus \overline{\Delta^2} \oplus \Delta^3) = 1$. Bits 0011 satisfy this expression and they may serve as the starting content of the shift register. The continuous sequence detected by the given expression is computed as follows:

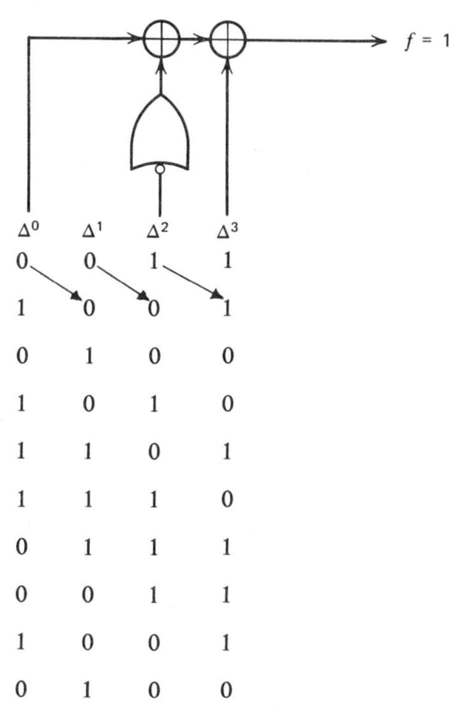

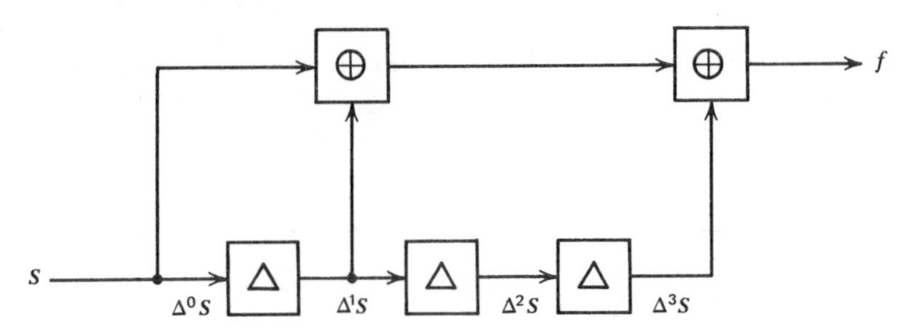

Figure 5.6. Sequential network detecting the continuous sequence S: ...0100111... where $f = S(\Delta^0 \oplus \Delta^1 \oplus \Delta^3) = 0$. Example 5.2.

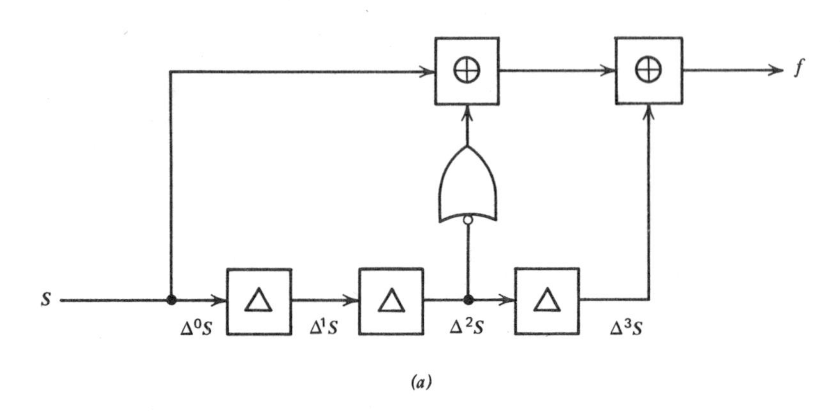

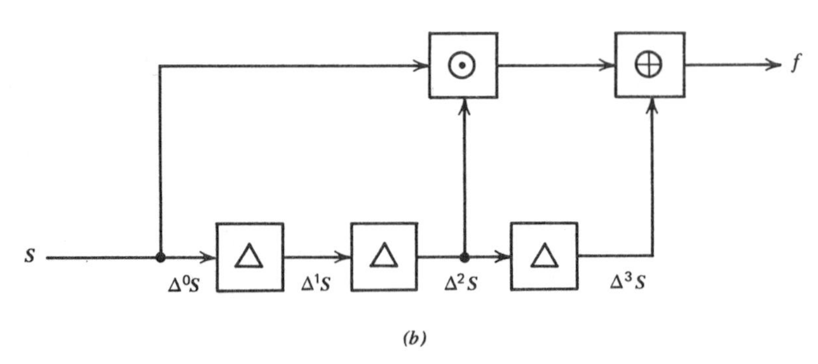

Figure 5.7. Sequential network detecting the continuous sequence S: ...1100101... where (*a*) $f = S(\Delta^0 \oplus \overline{\Delta^2} \oplus \Delta^3) = 1$ and (*b*) $f = S(\Delta^0 \odot \Delta^2 \oplus \Delta^3) = 1$. Example 5.3.

The values of the first column are determined by first entering the values of the other three columns into expression f and then selecting a value, 0 or 1, for the first column that makes f equal to 1. ▮

In the logic-oriented bit sequences the bits of the basic length have a place in a logical expression f, the result of which indicates presence of the sequence. As in the parity-oriented sequences, it is not necessary that all bits appear in the logic expression. For example, the logic expression for sequence

$$S: \ldots 1011110 \ldots \tag{5.20}$$

is

$$f = S(\Delta^0 + \Delta^1 \cdot \Delta^3) \tag{5.21}$$

Similarly, the logical expression that identifies discrete sequence

$$S: 101100 \tag{5.22}$$

is

$$f = S(\Delta^0\, \overline{\Delta^1}\, \Delta^2\, \Delta^3\, \overline{\Delta^4}\, \overline{\Delta^5}) \tag{5.23}$$

The sequential networks that implement the above two logic-oriented sequences are illustrated in Figure 5.8. Logic-oriented sequences are mainly discrete representing codes of fixed length.

5.4 SYNTHESIS OF SEQUENCE GENERATORS

Sequence generators can be generally classified into parity-oriented and logic-oriented generators in the same way sequence filters are classified.

Parity-oriented sequence generators consist of a shift register and of an encoding circuit. The generation of parity-oriented continuous sequences is accomplished by entering a starting sequence of basic length in the shift register, computing the parity of that sequence and entering the appropriate parity bit serially into the shift register. In the parity computation an input function also enters to determine whether 0 or 1 parity is being used.

Example 5.4. Design a sequence generator generating a bit sequence, where the parity of every 5 successive bits is 1 considering bits one, two, and five.

From the description of the design requirements, the equation of the encoding circuit can be identified as follows:

$$f = S(\Delta^1 \oplus \Delta^2 \oplus \Delta^5) \tag{5.24}$$

$$= S(\Delta^0 \oplus \Delta^1 \oplus \Delta^4) \tag{5.25}$$

$$= 1$$

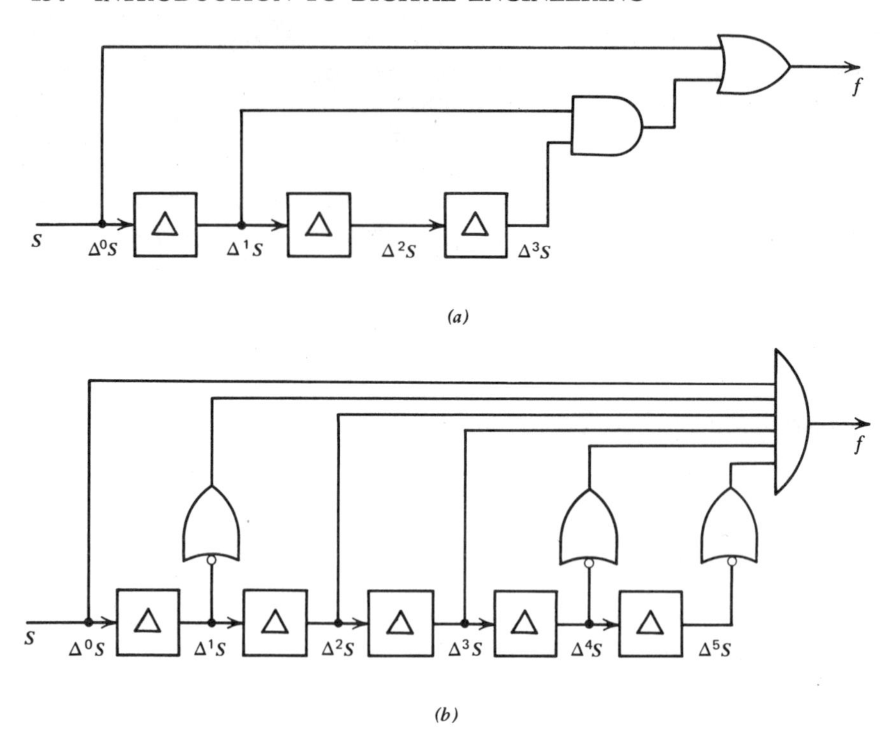

Figure 5.8. Sequential networks decoding logic-oriented sequences: (*a*) S: ...1011110... where $f = S(\Delta^0 + \Delta^1 \cdot \Delta^3) = 1$; (*b*) S: 101100 where $f = S(\Delta^0 \overline{\Delta^1} \Delta^2 \Delta^3 \overline{\Delta^4} \overline{\Delta^5}) = 1$.

The generated continuous sequence is

$$S\text{: } \ldots 010000111011001 \ldots \tag{5.26}$$

where the parity of any 5 successive bits, considering the first, second, and fifth bit, is 1. This sequence was determined by entering a 5 bit sequence that met the set requirements into the expression; the bits following that sequence were computed one by one, so that every new bit together with the preceding four satisfied the expression f.

Figure 5.9 illustrates the sequential network that generates the continuous sequence S: ...010000111011001.... The encoding circuit of this network was determined from the characteristic equation in the following manner:

$$\begin{aligned} f &= \Delta^0 S \oplus \Delta^1 S \oplus \Delta^4 S \\ &= 1 \end{aligned} \tag{5.27}$$

$$1 = \Delta^0 S \oplus \Delta^1 S \oplus \Delta^4 S \tag{5.28}$$

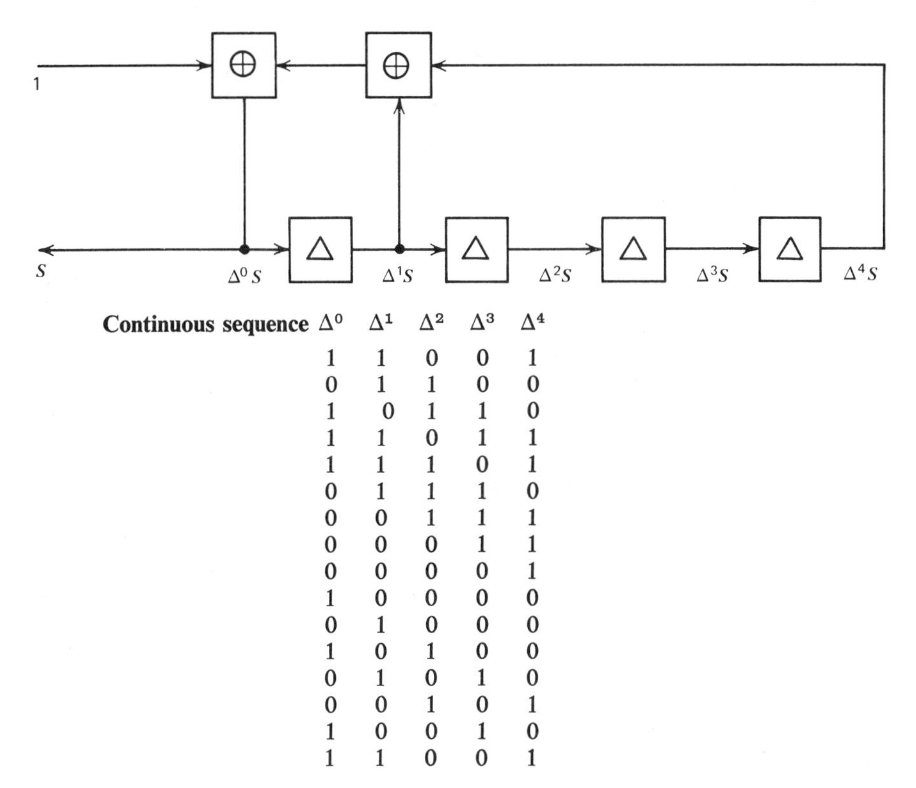

Continuous sequence Δ^0	Δ^1	Δ^2	Δ^3	Δ^4
1	1	0	0	1
0	1	1	0	0
1	0	1	1	0
1	1	0	1	1
1	1	1	0	1
0	1	1	1	0
0	0	1	1	1
0	0	0	1	1
0	0	0	0	1
1	0	0	0	0
0	1	0	0	0
1	0	1	0	0
0	1	0	1	0
0	0	1	0	1
1	0	0	1	0
1	1	0	0	1

Figure 5.9. Sequence generator producing the continuous sequence S: ...010000111011001... where $f = S(\Delta^0 \oplus \Delta^1 \oplus \Delta^4) = 1$ or $\Delta^0 S = 1 \oplus \Delta^1 S \oplus \Delta^4 S$.

Solving for $\Delta^0 S$, we obtain

$$\Delta^0 S = 1 \oplus \Delta^1 S \oplus \Delta^4 S \tag{5.29}$$

∎

Example 5.5. Design a sequential network that generates the following three continuous sequences:

S_1: ...1111000011110000...
S_2: ...1100000011000000...
S_3: ...1000000010000000...

An examination of sequence S_1 reveals that it repeats every 8 bits, having the unique sequence of ...00001111.... Out of this sequence an attempt is now made to determine the basic length of the sequence; 5 successive bits are taken as a starting sequence, so that 0s and 1s are contained. Let this be 01111. With this as a starting sequence the following order of sequences exists:

Starting sequence	0	1	1	1	1
	0	0	1	1	1
	0	0	0	1	1
	0	0	0	0	1
	1	0	0	0	0
	1	1	0	0	0
	1	1	1	0	0
	1	1	1	1	0
	0	1	1	1	1
$\Delta^0 S =$	$\uparrow$				

It can be observed that the first and fifth bit of each 5 bit sequence are always different. Or, that the parity of these 2 bits is always 1. In this case this characteristic is enough to define the requirements of the encoding circuit, which can be expressed by

$$f = S(\Delta^1 \oplus \Delta^5) \tag{5.30}$$

$$= S(\Delta^0 \oplus \Delta^4) \tag{5.31}$$

$$= 1$$

Solving for $\Delta^0 S$, we obtain

$$\Delta^0 S = 1 \oplus \Delta^4 S \tag{5.32}$$

This expression can be generalized and written as

$$\Delta^0 S = 1 \oplus \Delta^n S \tag{5.33}$$

which defines the encoding circuit of a sequential network generating a continuous sequence, where n 1s are followed by n 0s. Thus, the characteristic expression for the encoding circuit of a sequential network generating a sequence of two 1s followed by two 0s will be

$$\Delta^0 S = 1 \oplus \Delta^2 S \qquad (n = 2) \tag{5.34}$$

Similarly, for a continuous sequence of one 1 followed by one 0, the characteristic expression will be

$$\Delta^0 S = 1 \oplus \Delta^1 S \qquad (n = 1) \tag{5.35}$$

Therefore, for $n = 1$, 2, and 4, the following sequences are generated

$$S(n = 1)\text{: } \ldots 1010101010101010 \ldots \tag{5.36}$$

$$S(n = 2)\text{: } \ldots 1100110011001100 \ldots \tag{5.37}$$

$$S(n = 4)\text{: } \ldots 1111000011110000 \ldots \tag{5.38}$$

ANDing these sequences, we obtain

$$[S(n = 4)] \cdot [S(n = 2)]: \ldots 1100000011000000\ldots \qquad (5.39)$$

and

$$[S(n = 4)] \cdot [S(n = 2)] \cdot [S(n = 1)]: \ldots 1000000010000000\ldots \qquad (5.40)$$

Figure 5.10 illustrates a sequential network generating continuous sequences

$$S_1: \ldots 1111000011110000\ldots \qquad (5.41)$$

$$S_2: \ldots 1100000011000000\ldots \qquad (5.42)$$

$$S_3: \ldots 1000000010000000\ldots \qquad (5.43)$$

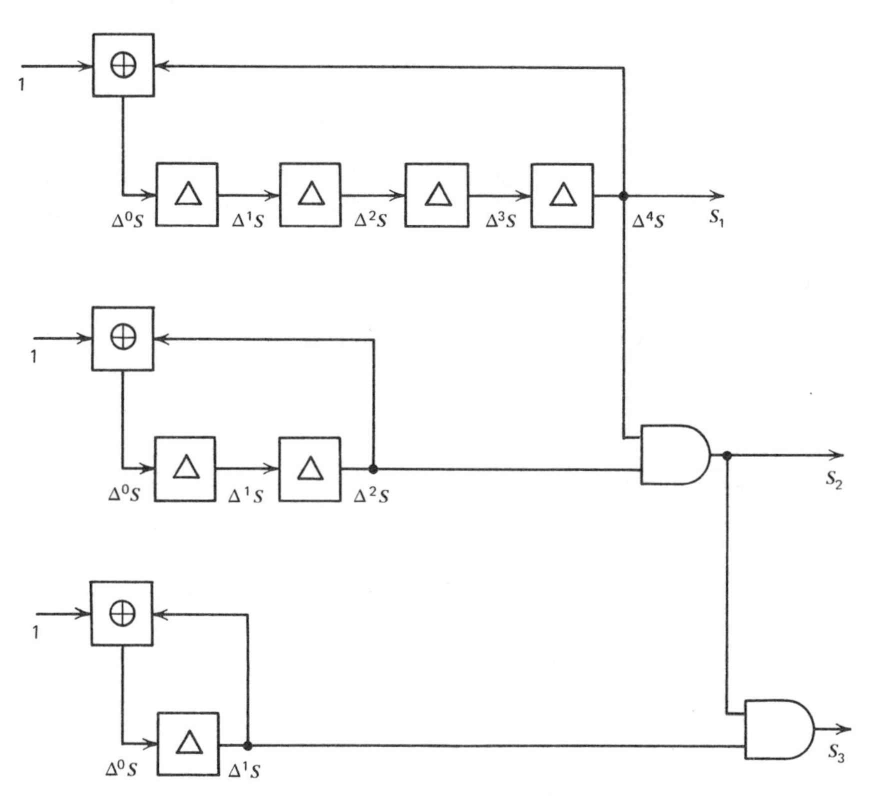

Figure 5.10. Sequence generator producing the continuous sequences S_1, S_2, and S_3. With all the delay units initially cleared, the generated sequences are S_1: $\ldots 1111000011110000\ldots$; S_2: $\ldots 1100000011000000\ldots$; S_3: $\ldots 1000000010000000\ldots$.

Logic-oriented sequence generators are mainly used for the generation of sequences that do not possess any parity characteristics, such as arbitrary sequences. Shown below is a typical example of a logic-oriented sequence generator.

Example 5.6. Design a sequence generator generating the following continuous sequence.

$$S: \ldots 1, 3, 9, 4, 11, 2, 5, \ldots$$

First, the sequence is converted to binary. Since the largest number in the sequence is expressed by 4 bits, the sequence generator must also have 4 bits. The binary conversion takes the following form:

	1	3	9	4	11	2	5
2^3 bit	0	0	1	0	1	0	0
2^2 bit	0	0	0	1	0	0	1
2^1 bit	0	1	0	0	1	1	0
2^0 bit	1	1	1	0	1	0	1

Next, the transitions from 0 to 1 and from 1 to 0 for each of the 4 bits are identified by means of the number that precedes them. If the bits are defined as W, X, Y, and Z in descending order of binary weight, the expressions will be as follows:

$$Z = (9, 4, 11, 2) + \mathrm{DC}(0, 6, 7, 8, 10, 12, 13, 14, 15) \tag{5.44}$$

$$Y = (1, 3, 4, 2) + \mathrm{DC}(0, 6, 7, 8, 10, 12, 13, 14, 15) \tag{5.45}$$

$$X = (9, 4, 2, 5) + \mathrm{DC}(0, 6, 7, 8, 10, 12, 13, 14, 15) \tag{5.46}$$

$$W = (3, 9, 4, 11) + \mathrm{DC}(0, 6, 7, 8, 10, 12, 13, 14, 15) \tag{5.47}$$

All 4 bit numbers that do not appear in the sequence can be considered as don't care terms. Using Veitch diagrams, the above four equations are simplified as shown on Figure 5.11, resulting in the following expressions:

$$W = A + CD + \overline{C}\overline{D} \tag{5.48}$$

$$X = B + \overline{D} + A\overline{C} \tag{5.49}$$

$$Y = \overline{D} + \overline{A}\overline{B} \tag{5.50}$$

$$Z = A + \overline{D} \tag{5.51}$$

These equations describe the conditions under which each of the 4 bits must change state. Figure 5.12 illustrates the sequence generator that produces the continuous sequence . . . 1, 3, 9, 4, 11, 2, 5, ... ∎

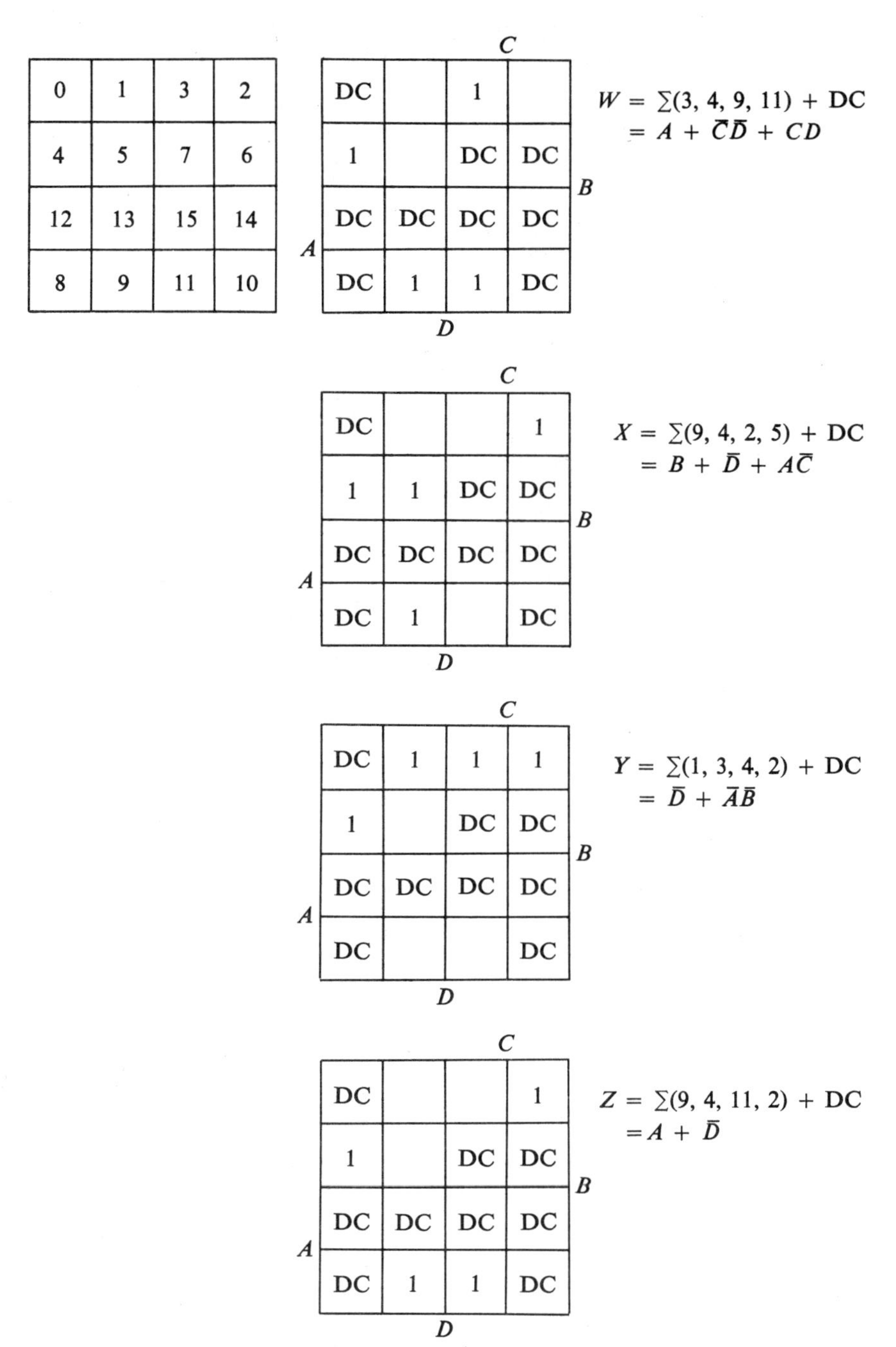

Figure 5.11. Simplication of W, X, Y, and Z expressions using Veitch diagrams where DC = don't care terms.

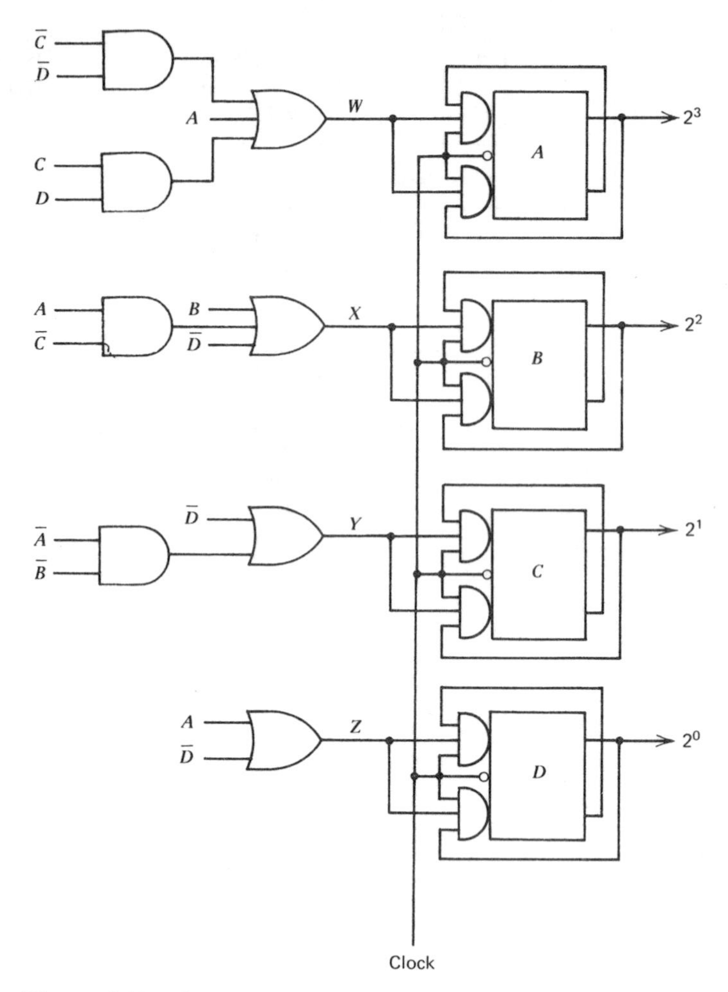

Figure 5.12. Sequence generator producing the continuous sequence . . ., 1, 3, 9, 4, 11, 2, 5, ... (*Note:* Flip-flops must be initially preset to one of the states of the sequence.)

5.5 ANALYSIS OF SEQUENCE FILTERS

Sequence filters of either parity-oriented or logic-oriented design can be analyzed by examining the decoding circuit of the filter and its relationship to the outputs of the delaying circuit. Two examples illustrate the approaches to analysis of sequence filters.

Example 5.7. Determine the continuous bit sequence that makes the output of the network of Figure 5.13(*a*) a 0.

The first step of the analysis is to express output f in terms of the sequence S by progressively identifying the signals at the various points of the network, as illustrated in Figure 5.13(*a*) where the output has been identified as

$$f = S(\Delta^0 \oplus \Delta^1 \oplus \Delta^3) \qquad (5.52)$$
$$= 0$$

From this expression a 4 bit sequence can be determined that makes output f a 0. Let this sequence be 1001, which is part of the continuous sequence that makes the output f a 0. With this sequence as a starting point the continuous sequence can be determined by computing one of its bits at a time. The bit to the left of sequence 1001 along with the three bits to its right, 001, must form a 4 bit sequence, $\Delta^0 100$, that satisfies expression

$$\Delta^0 \oplus \Delta^1 \oplus \Delta^3 = 0$$

Solving for Δ^0, the bit value of which is to be determined, we obtain

$$\Delta^0 = 0 \oplus \Delta^1 \oplus \Delta^3 \qquad (5.53)$$

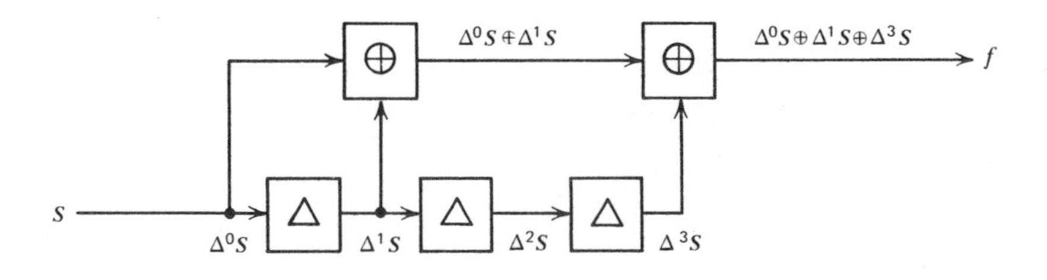

(*a*)

Δ^0	Δ^1	Δ^2	Δ^3
1	0	0	1
1	1	0	0
1	1	1	0
0	1	1	1
1	0	1	1
0	1	0	1
0	0	1	0
1	0	0	1

Figure 5.13. Sequential network decoding the continuous sequence S: ...10010111001...: (*a*) sequential network, $f = S(\Delta^0 \oplus \Delta^1 \oplus \Delta^3) = 0$; (*b*) detected sequence.

Substituting the Δs with their values, we have

$$\begin{aligned}\Delta^0 &= 0 \oplus 1 \oplus 0 \\ &= 1\end{aligned} \tag{5.54}$$

Now, the known bits of the continuous sequence are 11001. For the computation of the next bit sequence, $\Delta^0 110$ is considered. This process continues until the starting sequence 1001 is reached. Figure 5.13(*b*) lists the 4 bit sequences, each one having 3 bits in common with the sequence that precedes it as well as with the sequence that follows it. ∎

It should be noted that the number of different sequences, *Sc*, (other than all 0s or all 1s) of basic length that satisfy a sequential network of the parity-oriented configuration is

$$Sc = 2^{n-a} \sum_{i}^{i \leq a} \binom{i}{a} - K \tag{5.55}$$

where n is the size of the basic length, and a is the number of S terms that enter the f expression. When a is odd, $i = 1, 3, 5, \ldots$ and $K = 1$, and when a is even, $i = 0, 2, 4, \ldots$ and $K = 0$ for $f = 1$ and $K = 2$ for $f = 0$. If, for a given network, a continuous sequence is determined with less than Sc bits, then there is one or more additional continuous sequences that satisfy that network. The following example illustrates such a case.

Example 5.8. Determine the continuous bit sequence or sequences that make the output of the network of Figure 5.14(*a*) a 0, and those that make it 1.

The first step is to express the output f of the given network in terms of the input sequence S. This is easily accomplished by identifying in a progressive manner each of the points of the network, as shown on Figure 5.14(*a*). The resulting expression for the output of the network is

$$f = S(\Delta^0 \oplus \Delta^1 \oplus \Delta^3 \oplus \Delta^4) \tag{5.56}$$

Before we proceed with the computation of the sequences we first determine the number of basic length sequences that satisfy the expression of the given

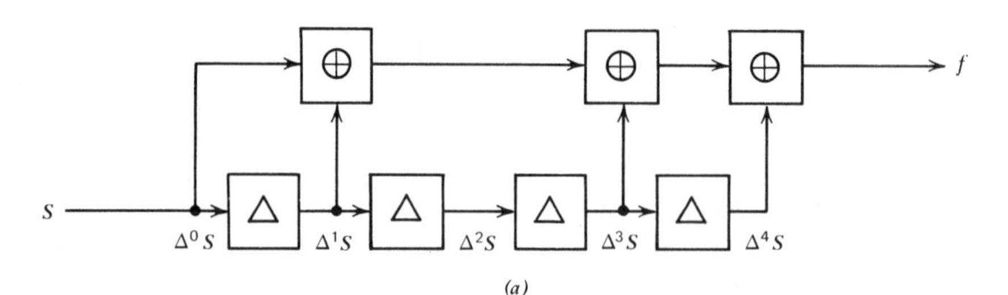

(*a*)

Δ^0	Δ^1	Δ^2	Δ^3	Δ^4	
1	1	0	1	1	
0	1	1	0	1	S_1:...1011011...
1	0	1	1	0	
1	1	0	1	1	
0	0	0	1	1	
1	0	0	0	1	
1	1	0	0	0	
1	1	1	0	0	S_2: ...0011100011...
0	1	1	1	0	
0	0	1	1	1	
0	0	0	1	1	
0	0	1	0	0	
1	0	0	1	0	S^3: ...0100100...
0	1	0	0	1	
0	0	1	0	0	
0	1	0	1	0	
1	0	1	0	1	S^4: ...101010...
0	1	0	1	0	

(*b*)

Δ^0	Δ^1	Δ^2	Δ^3	Δ^4	
0	0	0	0	1	
1	0	0	0	0	
0	1	0	0	0	
1	0	1	0	0	
1	1	0	1	0	
1	1	1	0	1	
1	1	1	1	0	S_5: ...0001011110100001...
0	1	1	1	1	
1	0	1	1	1	
0	1	0	1	1	
0	0	1	0	1	
0	0	0	1	0	
0	0	0	0	1	
0	0	1	1	0	
1	0	0	1	1	
1	1	0	0	1	S_6: ...01100110...
0	1	1	0	0	
0	0	1	1	0	

(*c*)

Figure 5.14. Sequential network decoding the continuous sequences S_1, S_2, S_3, S_4, S_5, and S_6: (*a*) sequential network where $f = \Delta^0 S \oplus \Delta^1 S \oplus \Delta^3 S \oplus \Delta^4 S$; (*b*) sequence that makes $f = 0$; (*c*) sequence that makes $f = 1$.

network. This is determined by means of (5.55), where $n = 5$ and $a = 4$, as shown below.

$$Sc = 2^{5-4} \sum_{i=0,2,4}^{4} \binom{i}{n} - 2 \qquad (\text{for } f = 0) \quad (5.57)$$

$$= 2(1 + 6 + 1) - 2$$

$$= 14 \qquad (5.58)$$

$$Sc = 2^{5-4} \sum_{i=0,2,4}^{4} \binom{i}{n} \qquad (\text{for } f = 1) \quad (5.59)$$

$$= 16 \qquad (5.60)$$

We now compute all 5 bit sequences that satisfy expressions

$$f = S(\Delta^0 \oplus \Delta^1 \oplus \Delta^3 \oplus \Delta^4) = 0 \qquad (5.61)$$

and

$$f = S(\Delta^0 \oplus \Delta^1 \oplus \Delta^3 \oplus \Delta^4) = 1 \qquad (5.62)$$

In each of the above cases, (5.61) and (5.62), a 5 bit sequence that satisfies the respective f expression is selected and the continuous sequence is determined by computing 1 bit of that sequence at a time.

The expressions for the computation of each bit following a known 5 bit sequence are as follows:

$$\text{New bit} = 0 \oplus \Delta^0 \oplus \Delta^2 \oplus \Delta^3 \qquad (\text{for } f = 0) \quad (5.63)$$

$$\text{New bit} = 1 \oplus \Delta^0 \oplus \Delta^2 \oplus \Delta^3 \qquad (\text{for } f = 1) \quad (5.64)$$

For the computation of the sequence that satisfies (5.61), the sequence 11011 has been selected. Using (5.63) the bits to the left of that sequence are computed:

Starting sequence: 1 1 0 1 1 $= \Delta^0\Delta^1\Delta^2\Delta^3\Delta^4$

Next Δ^0 bit $= f \oplus \Delta^0 \oplus \Delta^2 \oplus \Delta^4$
$= 0 \oplus 1 \oplus 0 \oplus 1 = 0$

Next sequence: 0 1 1 0 1

Next Δ^0 bit $= 0 \oplus 0 \oplus 1 \oplus 0 = 1$

Next sequence: 1 0 1 1 0

Next Δ^0 bit $= 0 \oplus 1 \oplus 1 \oplus 1 = 1$

Next sequence: 1 1 0 1 1

We have returned to the starting sequence which indicates that a continuous sequence has been determined. This is

$$S_1 :\ldots \underbrace{10}_{\text{Computed bits}}\underbrace{11011}_{\text{Starting sequence}} \tag{5.65}$$

Since only 3 of the 14 different 5 bit sequences that satisfy (5.61) have been determined, a new starting sequence is selected and the continuous sequence of which it is a part is computed. Figure 5.14(*b*) shows the four continuous sequences that satisfy expression

$$f = S(\Delta^0 \oplus \Delta^1 \oplus \Delta^3 \oplus \Delta^4) = 0 \tag{5.66}$$

In a similar manner using (5.64) the sequences that satisfy expression

$$f = S(\Delta^0 \oplus \Delta^1 \oplus \Delta^3 \oplus \Delta^4) = 1 \tag{5.67}$$

are computed. These sequences are listed in Figure 5.14(*c*). ∎

5.6 ANALYSIS OF SEQUENCE GENERATORS

Sequence generators are analyzed in the general way sequence filters are analyzed. The following example illustrates the analysis of a sequential network generating continuous bit sequences.

Example 5.9. Determine the continuous sequences that can be generated by the sequential network of Figure 5.15(*a*).

The first step toward determining these sequences is the definition of the network in terms of an expression that relates the output sequence S to the network configuration and to the input signal f. This is obtained by defining each point of the network (Fig. 5.15(*a*)). The expression that defines that output of the network is

$$S = f \oplus \Delta^2 S \oplus \Delta^4 S \tag{5.68}$$

The number of 5 bit sequences that satisfy this expression, for input f either 0 or 1, is computed using (5.55),

$$\begin{aligned} Sc &= 2^{5-3} \sum_{i=1,3}^{3} \binom{i}{3} - 1 \\ &= 2^2(3 + 1) - 1 \\ &= 15 \end{aligned}$$

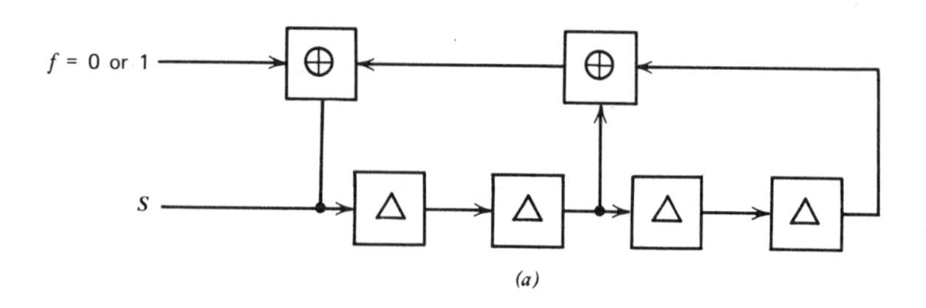

f = 0 or 1
S

(a)

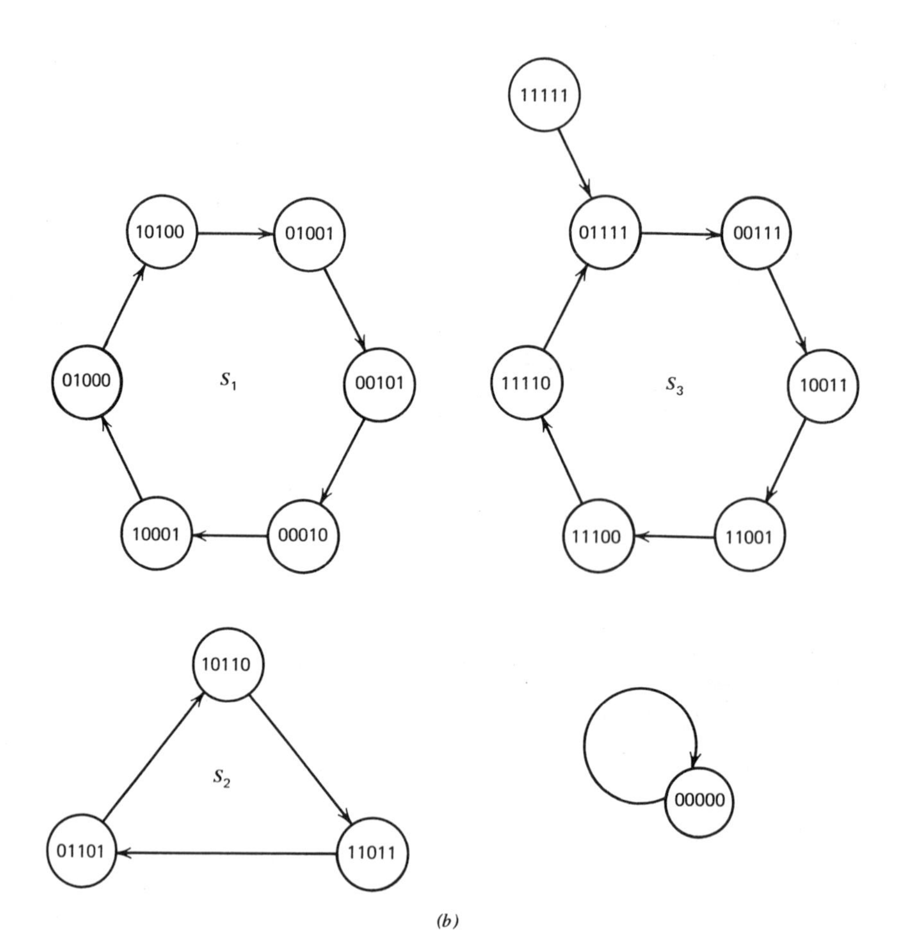

11111
10100
01001
01111
00111
01000
S_1
00101
11110
S_3
10011
10001
00010
11100
11001
10110
S_2
00000
01101
11011

(b)

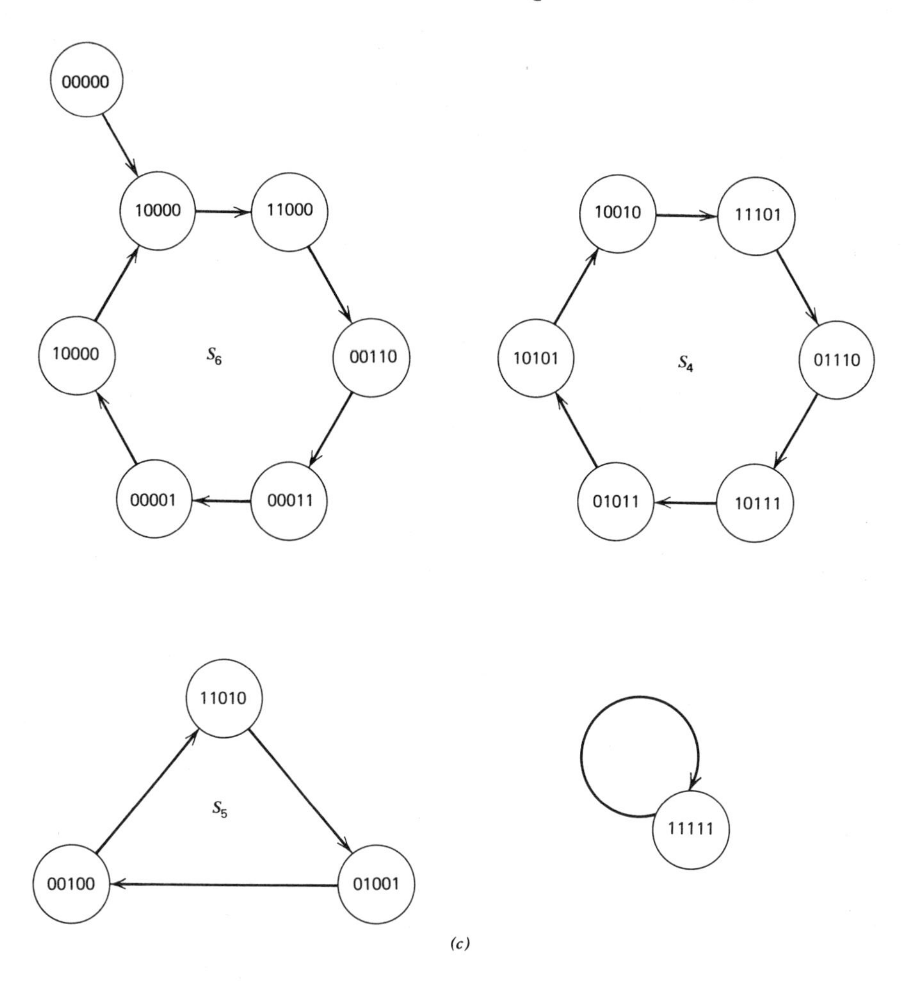

Figure 5.15. Implementation and operation of function $f = \Delta^0 S \oplus \Delta^2 S \oplus \Delta^4 S$: (*a*) sequential network; (*b*) continuous sequences when $f = 0$; (*c*) continuous sequences when $f = 1$.

The sequences that satisfy this expression for each of the two possible values of f are computed. For $f = 0$, (5.68) becomes

$$S = \Delta^0 S \tag{5.69}$$

$$= 0 \oplus \Delta^2 S \oplus \Delta^4 S \tag{5.70}$$

$$= \Delta^2 S \oplus \Delta^4 S \tag{5.71}$$

Now, a starting sequence is selected and the generated continuous sequence is determined the same way the continuous sequences of the sequence filters were determined. In this case the criterion is

$$\Delta_0 S = \Delta^2 S \oplus \Delta^4 S \tag{5.72}$$

Starting with 5 bit sequence 10100, we have

	Δ^0	Δ^1	Δ^2	Δ^3	Δ^4	
Starting sequence	1	0	1	0	0	
	0	1	0	1	0	
	0	0	1	0	1	
	0	0	0	1	0	
	1	0	0	0	1	
	0	1	0	0	0	
	1	0	1	0	0	**Return to starting sequence**

Thus, S_1: . . .0100010100. . . .

The analyzed sequence generator must have more than one continuous sequence, since the above list contains only 6 of the 15 states in which the generator may be found. Therefore, a new starting sequence is selected and the continuous sequence of which it is a part is computed. Starting with 01101, we have

	Δ^0	Δ^1	Δ^2	Δ^3	Δ^4	
Starting sequence	0	1	1	0	1	
	1	0	1	1	0	
	1	1	0	1	1	
	0	1	1	0	1	**Return to starting sequence**

Thus, S_2: . . .1101101. . . . Starting again, this time with 11110, we have

	Δ^0	Δ^1	Δ^2	Δ^3	Δ^4	
Starting sequence	1	1	1	1	0	
	0	1	1	1	1	
	0	0	1	1	1	
	1	0	0	1	1	
	1	1	0	0	1	
	1	1	1	0	0	
	1	1	1	1	0	**Return to starting sequence**

Thus, S_3: . . .1110011110. . . . Therefore, the given sequence generator for $f = 0$ has three distinct sequences, which it may generate depending on the sequence that has been preset into its shift register. Should an all 0s sequence be set into the shift register, the generator will endlessly remain in that sequence. Figure 5.15(*b*) shows the three continuous sequences that can be generated by the analyzed network when $f = 0$.

For $f = 1$ (5.68) becomes

$$\begin{aligned} S &= \Delta^0 S \qquad (5.73) \\ &= 1 \oplus \Delta^2 S \oplus \Delta^4 S \\ &= 1 \odot \overline{\Delta^2 S} \oplus \Delta^4 S \\ &= \overline{\Delta^2 S} \oplus \Delta^4 S \\ &= \Delta^2 S \odot \Delta^4 S \qquad (5.74) \end{aligned}$$

Starting with the 5 bit sequence 10101, the following continuous sequence is obtained

	Δ^0	Δ^1	Δ^2	Δ^3	Δ^4	
Starting sequence	1	0	1	0	1	
	1	1	0	1	0	
	1	1	1	0	1	
	0	1	1	1	0	
	1	0	1	1	1	
	0	1	0	1	1	
	1	0	1	0	1	**Return to starting sequence**

Thus, S_4: ...0101110101.... Starting again with 00100, we have

	Δ^0	Δ^1	Δ^2	Δ^3	Δ^4	
Starting sequence	0	0	1	0	0	
	1	0	0	1	0	
	0	1	0	0	1	
	0	0	1	0	0	**Return to starting sequence**

Thus, S_5: ...0100100.... Starting again, this time with 10000, we have

	Δ^0	Δ^1	Δ^2	Δ^3	Δ^4	
Starting sequence	1	0	0	0	0	
	1	1	0	0	0	
	0	1	1	0	0	
	0	0	1	1	0	
	0	0	0	1	1	
	0	0	0	0	1	
	1	0	0	0	0	**Return to starting sequence**

Thus, S_6: ...000011.... Therefore, for $f = 1$ the sequence generator has three distinct sequences also, the generation of which depends on the starting sequence. In the event an all 1s sequence enters into the generator, this sequence will stay indefinitely until the generator is forced out of it by means of the parallel inputs of the shift register. Figure 5.15(*c*) shows the three continuous sequences that can be generated by the examined network when $f = 1$. ∎

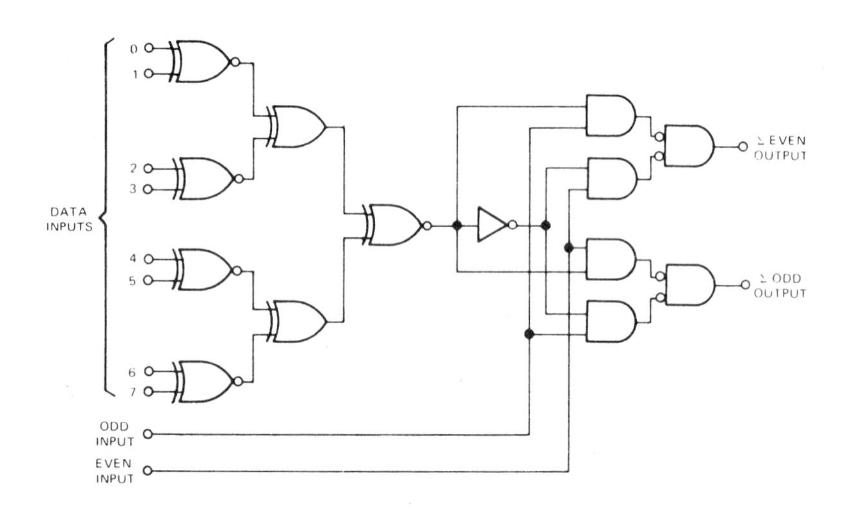

TRUTH TABLE

INPUTS			OUTPUTS	
Σ OF 1's AT 0 THRU 7	EVEN	ODD	Σ EVEN	Σ ODD
EVEN	H	L	H	L
ODD	H	L	L	H
EVEN	L	H	L	H
ODD	L	H	H	L
X	H	H	L	L
X	L	L	H	H

X = irrelevant

(*a*)

5.7 MEDIUM-SCALE INTEGRATION IN SEQUENTIAL DESIGNS

Sequential networks, except for the parity-oriented ones, are of a unique design meeting unique requirements. As a result, no convenient MSI networks can be developed that can satisfy such diverse needs. Design of parity-oriented networks, however, is facilitated with the availability of parity generators such as those of Figure 5.16.

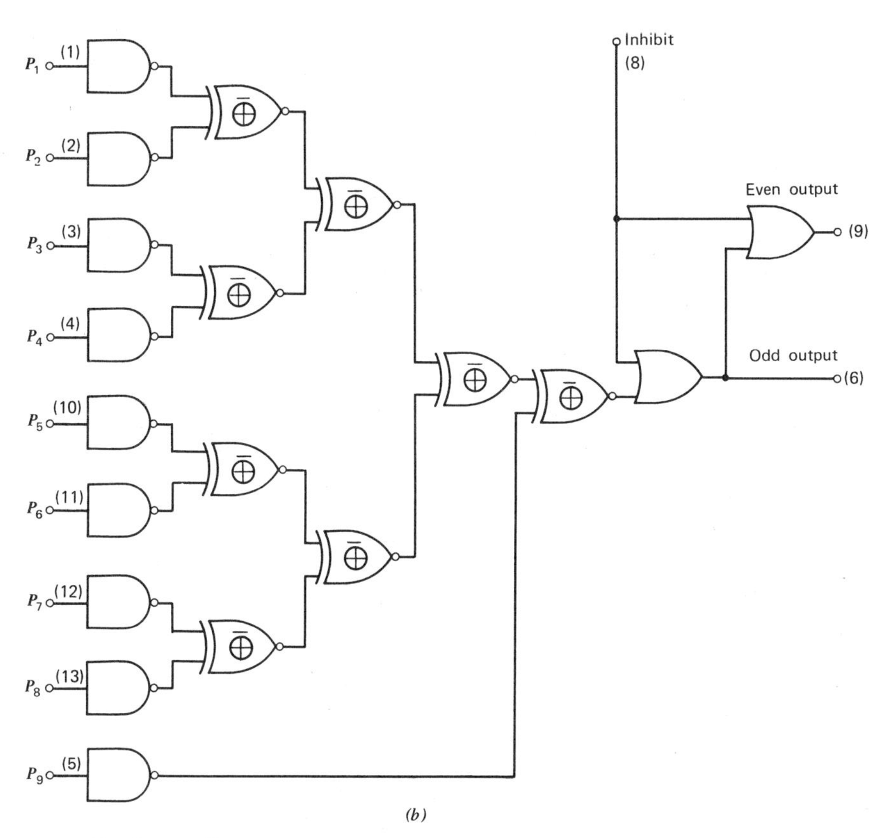

(*b*)

Logic equations:

$$\text{Odd} = P_1 \oplus P_2 \oplus P_3 \oplus P_4 \oplus P_5 \oplus P_6 \oplus P_7 \oplus P_8 \oplus P_9$$

$$\text{Even} = \overline{P_1 \oplus P_2 \oplus P_3 \oplus P_4 \oplus P_5 \oplus P_6 \oplus P_7 \oplus P_8 \oplus P_9}$$

Figure 5.16. Typical MSI parity generators: (*a*) 8 bit parity generator and checker (courtesy of Fairchild Semiconductor); (*b*) 9 bit parity generator and checker (courtesy of Signetics).

In the area of sequence generation, read only memories (ROMs) can be used. In this case, a sequence of binary numbers is permanently entered into a ROM from which it is accessed through direct addressing of the stored information.

Figure 5.17 illustrates the block diagram of a ROM that can be used as a sequence generator. Each of the ROMs memory locations has a unique address where the addresses range from 0...00 to 1...11. Thus, if the address input is changed from 0...00 to 1...11 in an ascending order, the ROM output will provide the content of memory locations 0...00 through

BLOCK DIAGRAM

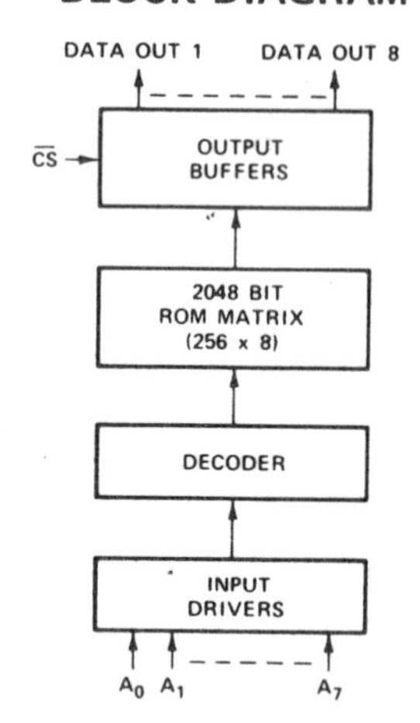

PIN CONFIGURATION

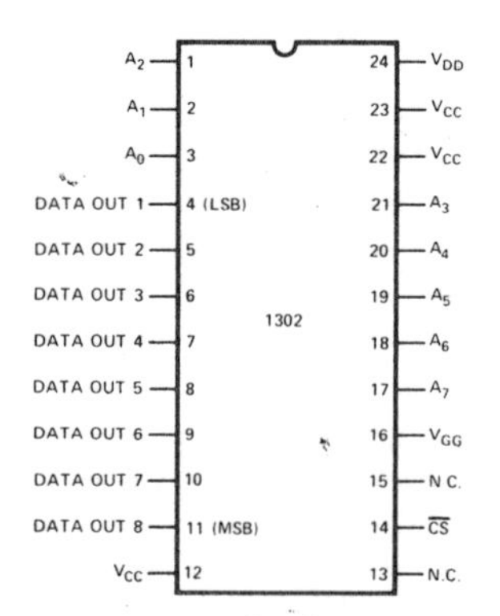

PIN NAMES

A_0–A_7	Address Inputs
$\overline{CS}$	Chip Select Input
D_{OUT1}–D_{OUT8}	Data Outputs

Figure 5.17. A 2048 bit mask programmable ROM (courtesy of Intel).

1...11 in that order. These contents can be a preassigned sequence with the first number of the sequence occupying memory location 0...00 and the last number occupying memory location 1...11.

For example, the arbitrary continuous sequence

$$S: \ldots 1011011101011001 \ldots$$

can be generated by means of a 16 bit ROM of the following content:

Memory location address	Memory location content
0000	1
0001	0
0010	1
0011	1
0100	0
0101	1
0110	1
0111	1
1000	0
1001	1
1010	0
1011	1
1100	1
1101	0
1110	0
1111	1

If the address is made to change linearly, the ROM output will be the sequence that was preset into the ROM.

5.8 APPLICATIONS OF SEQUENTIAL NETWORKS

Sequential networks find wide use in the generation and decoding of bit sequences, and in the generation of pseudorandom digital codes. In this section are three examples of sequential network application.

Example 5.10. Design a pseudorandom generator.

The pseudorandom generator presented here consists of a sequence generator, a binary counter, and antilocking logic, as shown in Figure 5.18. The sequence generator has a six-stage shift register, all outputs of which gated enter a parity network—that is, a chain of exclusive-OR gates that provide a 1 output when the number of 1s in the inputs is odd. The output

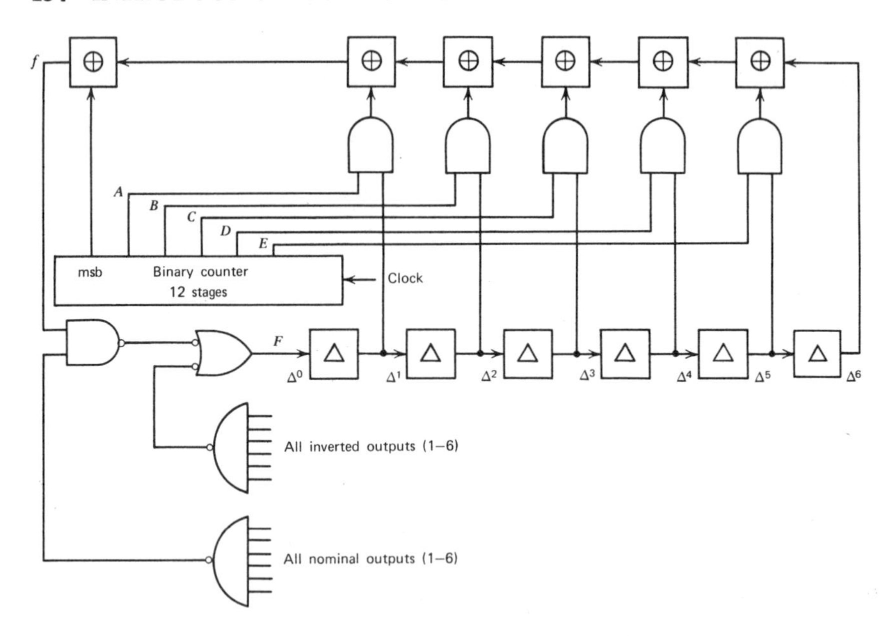

Figure 5.18. General-purpose pseudorandom generator; the network's general expression is $f = S[\Delta^0 \oplus (A \cdot \Delta^1) \oplus (B \cdot \Delta^2) \oplus (C \cdot \Delta^3) \oplus (D \cdot \Delta^4) \oplus (E \cdot \Delta^5) \oplus \Delta^6.]$

of the parity network enters an exclusive-OR gate which is also entered by the MSB of a binary counter.

The binary counter serves as a timing network which every 2^6 clock periods changes the form of the generator's characteristic equation by applying selected shift-register outputs to the parity network. The general form of the generator's characteristic equation is

$$f = S[\Delta_0 \oplus (A \cdot \Delta^1) \oplus (B \cdot \Delta^2) \oplus (C \cdot \Delta^3) \oplus (D \cdot \Delta^4) \oplus (E \cdot \Delta^5) \oplus \Delta^6] \quad (5.74)$$

To prevent the generator from "locking," that is, continuing in an endless sequence of 1s or 0s, protective logic has been included. This logic makes the input to the shift register a 0 when all nominal outputs of the shift register are 1. When all outputs are 0, and consequently the inverted outputs are 1, the logic makes the input to the shift register a 1. With these two provisions included, the characteristic equation of the pseudorandom generator of Figure 5.18 takes the form:

$$F = f(\overline{\Delta^1 \Delta^2 \Delta^3 \Delta^4 \Delta^5 \Delta^6}) + \overline{\Delta^1}\,\overline{\Delta^2}\,\overline{\Delta^3}\,\overline{\Delta^4}\,\overline{\Delta^5}\,\overline{\Delta^6} \quad (5.75)$$

The network of Figure 5.18 provides a pseudorandom sequence that can be accessed from many of the points on the network. More than one such point can be accessed simultaneously in which case a random sequence of parallel binary numbers is obtained.

Example 5.11. Design of an arbitrary sequence generator with a continuous sequence.

$$S: \ldots 7, 6, 4, 5, 2 \ldots$$

In the design of digital systems, it is sometimes required that a certain arbitrary sequence of numbers be generated for the control of a given system, circuit, or component. For the generation of such an arbitrary sequence, a storage register can be used. In order for the stages of the register to change in accordance with the desired sequence, appropriate logic is added that relates the state to come to the presently existing state of the register outputs and generates the appropriate inputs for the register stages. The design process of this sequence generator starts with the listing of the desired sequence both in decimal and in binary numbers.

Desired sequence

Number in counter	Counter outputs
	4 2 1
7	1 1 1
6	1 1 0
4	1 0 0
5	1 0 1
2	0 1 0
7	1 1 1
.	. . .
.	. . .
.	. . .

From this table equations are written that indicate the conditions under which the various synchronous set and synchronous clear inputs of the register must be enabled.

The first stage of the register must be set to 1 after numbers 2 and 4 and should be reset to 0 after numbers 5 and 7. The second stage should be changed to 1 after number 5 and to 0 after number 6. Similarly, the third stage should be changed to 1 after 2 and to 0 after 5.

The conditions necessary for the synchronous set S_s and synchronous reset C_s to be 1 can be expressed by

$$S_{s1} = 4 \text{ and } 2 \tag{5.76}$$

$$C_{s1} = 7 \text{ and } 5 \tag{5.77}$$

$$S_{s2} = 5 \tag{5.78}$$

$$C_{s2} = 6 \tag{5.79}$$

$$S_{s3} = 2 \tag{5.80}$$

$$C_{s3} = 5 \tag{5.81}$$

These expressions show the numbers after which the register stages should be set or cleared to produce the desired sequence. Figure 5.19 illustrates the implementation of the equations. ∎

Example 5.12. Design a sequence generator for the operation of a stepping motor.

Sometimes instead of having a sequence of numbers, as in Example 5.11 a set of waveforms may be given as the basis for the design of a sequence generator. In the case of the stepping motor, the waveforms of Figure 5.20 are required for proper operation. From these waveforms, a sequence of numbers may be written which correspond to the given waveforms. If the HIGH is considered 1 and the LOW considered 0, the sequence then becomes

A	B
0	0
0	1
1	1
1	0
0	0

Thus, the synchronous set and synchronous clear for stages A and B should be enabled as follows:

$$S_{sA} = \bar{A}B \tag{5.82}$$

$$C_{sA} = A\bar{B} \tag{5.83}$$

$$S_{sB} = \bar{A}\bar{B} \tag{5.84}$$

$$C_{sB} = AB \tag{5.85}$$

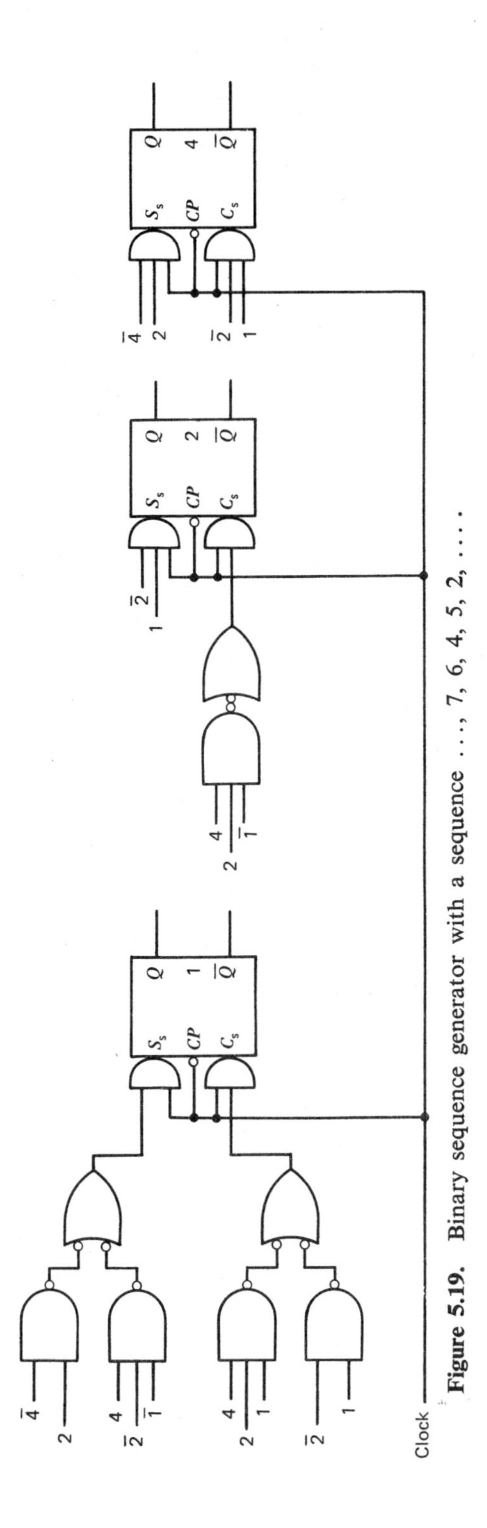

Figure 5.19. Binary sequence generator with a sequence . . ., 7, 6, 4, 5, 2, ...

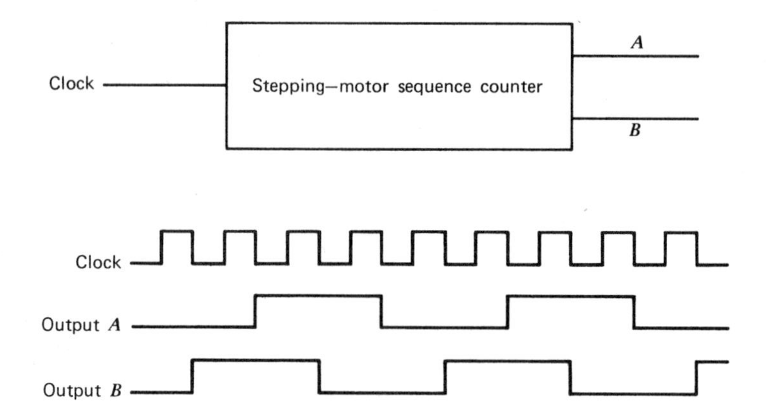

Figure 5.20. Functional diagram and waveforms of the operation of a stepping-motor sequence generator.

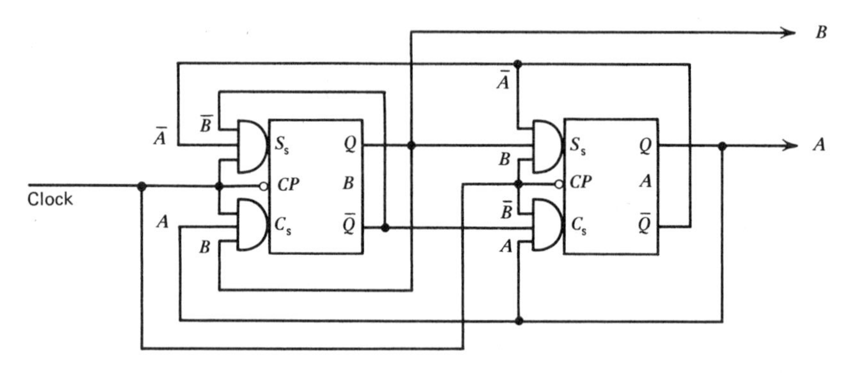

Figure 5.21. Sequence generator for a stepping motor.

Figure 5.21 shows the implementation of this sequence generator. This circuit can be also used as a divide-by-4 counter, since the period of the output waveform is four times that of the input clock frequency. In this case the frequency of both outputs A and B is one-fourth that of the clock, and output B leads output A by one clock period. ∎

PROBLEMS

1. Define a parity-oriented sequence.
2. Define a continuous sequence.

3. Design a sequential network that decodes the sequence S: 1011011.
4. Design a sequential network that decodes the continuous sequence S: . . .001001. . . .
5. Design a sequential network that decodes the continuous sequence S: . . .110011001100. . . .
6. Design a sequence generator producing the sequence S: . . .3, 5, 9, 8, 6, 2, 4, 14, 7. . . .
7. Design a sequence generator implementing the function $F = S(\Delta^0 \oplus \Delta^1 \oplus \Delta^4)$ and determine the resulting continuous sequence.
8. Analyze the sequence generator of Figure 5.9 after replacing the exclusive-OR gates by coincidence gates.
9. Analyze the sequence filter of Figure 5.6 after replacing the exclusive-OR gates by coincidence gates.
10. Design a tunable sequence filter that detects parity-oriented sequences having a 4 bit basic length.

CHAPTER SIX

Counting Networks

6.1 INTRODUCTION

The counter is a device that counts digital pulses. Almost every digital system uses counters, and because of their extensive use counters are available in integrated circuit form in a variety of configurations. Depending on the application, counters may be used to measure a number of functions such as frequency, time, temperature, or they may be used to count the number of operations performed by a given system.

The building block of the counter is the flip-flop. A counter consists of a number of flip-flops connected in cascade where the output of a flip-flop drives the input of the following flip-flop. The specific interconnections that form this cascade determine the properties of the counter.

Counters can be generally classified as either serial or synchronous. An additional cross-classification separates counters into linear and feedback types of counters. In a serial counter each flip-flop stage is triggered by the immediately preceding flip-flop, while in a synchronous counter all stages are triggered by a common clock. Linear counters are those where the state of a flip-flop stage depends only on the state of the stages that precede it, while feedback counters are those where the state of a flip-flop depends on the state of both the preceding as well as the following flip-flops in the cascade.

6.2 SERIAL BINARY COUNTERS

A serial counter, also called a ripple counter, is a chain of trigger flip-flops. A trigger flip-flop changes state each time an input pulse is applied to its trigger input. This change of state may be activated by the positive-going or by the negative-going edge of the input pulse, depending on the type of flip-flop used.

Two input pulses are required for a trigger flip-flop to complete a cycle. If a trigger flip-flop is initially in the 1 state, it will go to the 0 state after the first input pulse and will return to the 1 state after the second input pulse. It thus takes two pulses at the trigger input to generate one pulse at the output and, as a result, the output frequency of a trigger flip-flop is one-half its input frequency. Therefore, if a given frequency f feeds the trigger input of the first flip-flop in a series counter, the output frequency of this flip-flop will be $f/2$.

Figure 6.1 illustrates a four-stage series counter. The output frequency of each stage is expressed as a function of the input frequency to the first stage of the counter. Since the output of the first flip-flop, which is $f/2$, drives the trigger input of the second flip-flop, the output frequency of the second flip-flop will be $f/4$, which is one-half its input frequency, or one-fourth the input frequency to the first flip-flop.

The most commonly used trigger flip-flop is the master-slave type with cross feedback. This type is composed of two flip-flops, the master where the information enters, and the slave to which the information is transferred from the master, and from which that information is available for use.

Normally, information enters the master when the clock input is HIGH and is transferred to the slave when the clock switches to LOW. With cross feedback, the information entering the flip-flop is the opposite of the information the flip-flop already contains. In this configuration, the nominal flip-flop output Q feeds the synchronous clear input C_S, and the inverse flip-flop output $\bar{Q}$ feeds the synchronous set input S_S, as shown in Figure 6.1.

If a LOW initial state for all flip-flop outputs is assumed, the waveforms of the outputs of the serial counter in Figure 6.1(*a*) will be as shown in Figure 6.1(*c*). Here it is assumed, and this is generally the case, that the negative-going edge of the clock pulse activates the flip-flops, and causes their output states to change. From the waveforms in Figure 6.1(*c*), it can be seen that it takes one clock pulse—one negative-going edge of the clock waveform—to change the state of the first flip-flop, two clock pulses to change the state of the second, four to change the state of the third, and eight to change the state of the fourth flip-flop. This property can be used when a number of pulses is to be counted. If the binary weight of the first flip-flop of the counter is 1,

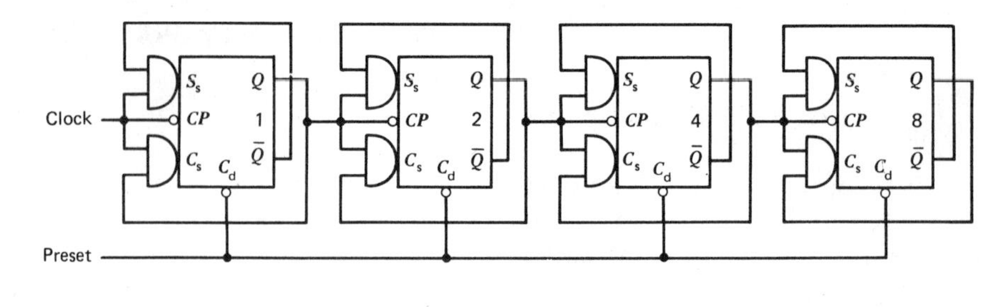

		Counter output
		8 4 2 1
	Preset	0 0 0 0
Number of clock pulses after preset	1	0 0 0 1
	2	0 0 1 0
	3	0 0 1 1
	4	0 1 0 0
	5	0 1 0 1
	6	0 1 1 0
	7	0 1 1 1
	8	1 0 0 0
	9	1 0 0 1
	10	1 0 1 0
	11	1 0 1 1
	12	1 1 0 0
	13	1 1 0 1
	14	1 1 1 0
	15	1 1 1 1
	16	0 0 0 0
	17	0 0 0 1 ← New cycle

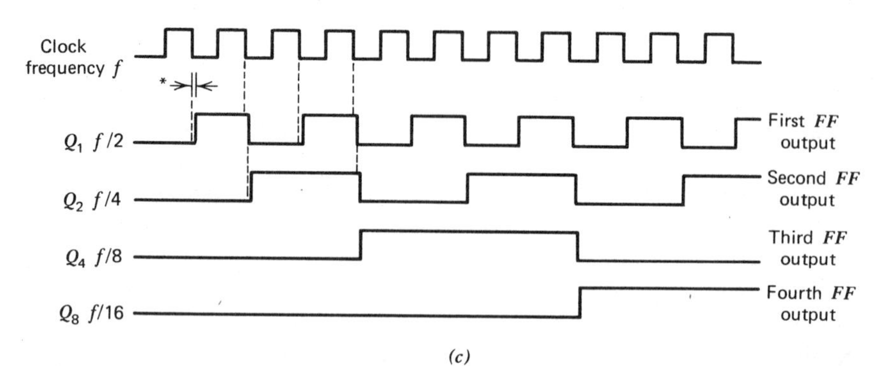

(*c*)

Figure 6.1. Four-stage serial binary counter: (*a*) logic diagram; (*b*) truth table; (*c*) waveforms of operations. (*Note:* *Propagation delay.)

that of the second 2, and that of the third 4, a 3 bit counter can be used to count up to seven pulses.

For a serial binary counter made up of n flip-flops, the output frequency at the nth flip-flop f_n is

$$f_n = 2^{-n} f_{\text{clock}} \tag{6.1}$$

This property can be used to determine the number of flip-flops necessary for specific frequency reductions; for example, given a clock frequency of 16 kHz, and a desired frequency 1 kHz, by substitution to (6.1) we obtain

$$\begin{aligned} 1\text{ kHz} &= 2^{-n}\,(16\text{ kHz}) \\ \tfrac{1}{16} &= 2^{-n} \\ 2^n &= 16 \\ 2^n &= 2^4 \\ n &= 4 \end{aligned}$$

The number of trigger flip-flops required to make this frequency reduction is therefore four.

The relationship between the number of pulses fed into the counter and the counter output is shown on the truth table of Figure 6.1. This table contains the same information as the waveforms of Figure 6.1(c) but it presents it in a different form. The waveforms represent the state of the flip-flops as HIGH or LOW, while the truth table shows the same information as 1 or 0. Truth tables may also be written in terms of HIGHs and LOWs. Unlike waveforms, however, the truth table does not indicate propagation delays.

The choice between truth tables and waveforms, as a means of representing the operation of the flip-flops in a binary series counter, depends on the point of view of the designer. From an operational point of view, truth tables are preferable because they show the flip-flop states directly, while from a design point of view, waveforms are more convenient because they show the time relationships of various outputs as well as propagation delays.

Using waveforms, it is possible to graphically represent the propagation delay in the counting elements thereby clearly showing their effect on the circuit operation. In a series counter, the output of the first flip-flop which is driven by the clock changes state one delay interval after the clock switches from HIGH to LOW. The output of the second flip-flop changes state one delay interval after the output of the first flip-flop switches from HIGH to LOW. Thus the output of the second flip-flop lags behind the activating clock edge by two propagation delay intervals, as shown in Figure 6.1(c). Similarly, the third flip-flop will lag behind the clock by three propagation delay intervals, the fourth by four, and so on. This accumulation of propagation delays can have undesirable results, as we shall see in Example 6.1.

Example 6.1. A 10 bit series counter is driven by a 5 MHz clock. The propagation delay of each of the ten flip-flops used is approximately 25 nsec. The counter starts counting with each of its flip-flops in their 0 state. The desired operation of this circuit is to generate a signal when the counter reaches the binary number 1000000000.

This number will occur when the output of the tenth flip-flop becomes HIGH, and the outputs of the remaining flip-flops become LOW. This number will also indicate that 512 pulses have entered the counter, since the counter was initially reset to 0000000000. The binary number 1000000000 follows 0111111111 which indicates that in the counter the transition from the latter number to the former requires that all flip-flops change state in a chain reaction. Since this is a sequential process, the propagation delays will be cumulative.

Figure 6.2 illustrates a 10 bit counter and its timing waveforms at the pulse which changes the counter output from 0111111111 to 1000000000. Since the propagation delay for each flip-flop is 25 nsec, the transition will require 10×25 nsec, or 250 nsec. But, 200 nsec after the arrival of the activating pulse that will initiate the transition from 0111111111 to 1000000000, the clock pulse corresponding to the number 1000000001 will arrive, and 25 nsec after its arrival, the first flip-flop of the counter will change state. This change will occur before the changes resulting from the clock pulse corresponding to number 1000000000 have been completed.

Thus, the state of the first flip-flop will be correct for number 1000000001 at the same time that the state of the tenth flip-flop will be correct for number 0111111111 and the state of the flip-flops in between will be correct for 1000000000. Since the total propagation delay exceeds the clock frequency period, there is no single instant at which the entire counter output indicates number 1000000000. Therefore, although the flip-flops of the counter change state properly with incoming clock pulses, the counter output cannot always be correctly decoded. ∎

The limits placed on the input frequency of serial counters in cases where the output is to be decoded can be expressed as

$$T_{\text{clock}} - nd > PW_g \tag{6.2}$$

where T_{clock} is the clock frequency period, n is the number of flip-flops which must change state to form the number to be decoded, d is the propagation delay for each flip-flop, and PW_g is the minimum acceptable pulse width expected at the output of the decoding gate.

It is clear that for accurate decoding the total propagation delay nd, the time between the activating edge of the clock pulse and the completion of the last change in the counter initiated by that clock pulse, must be less

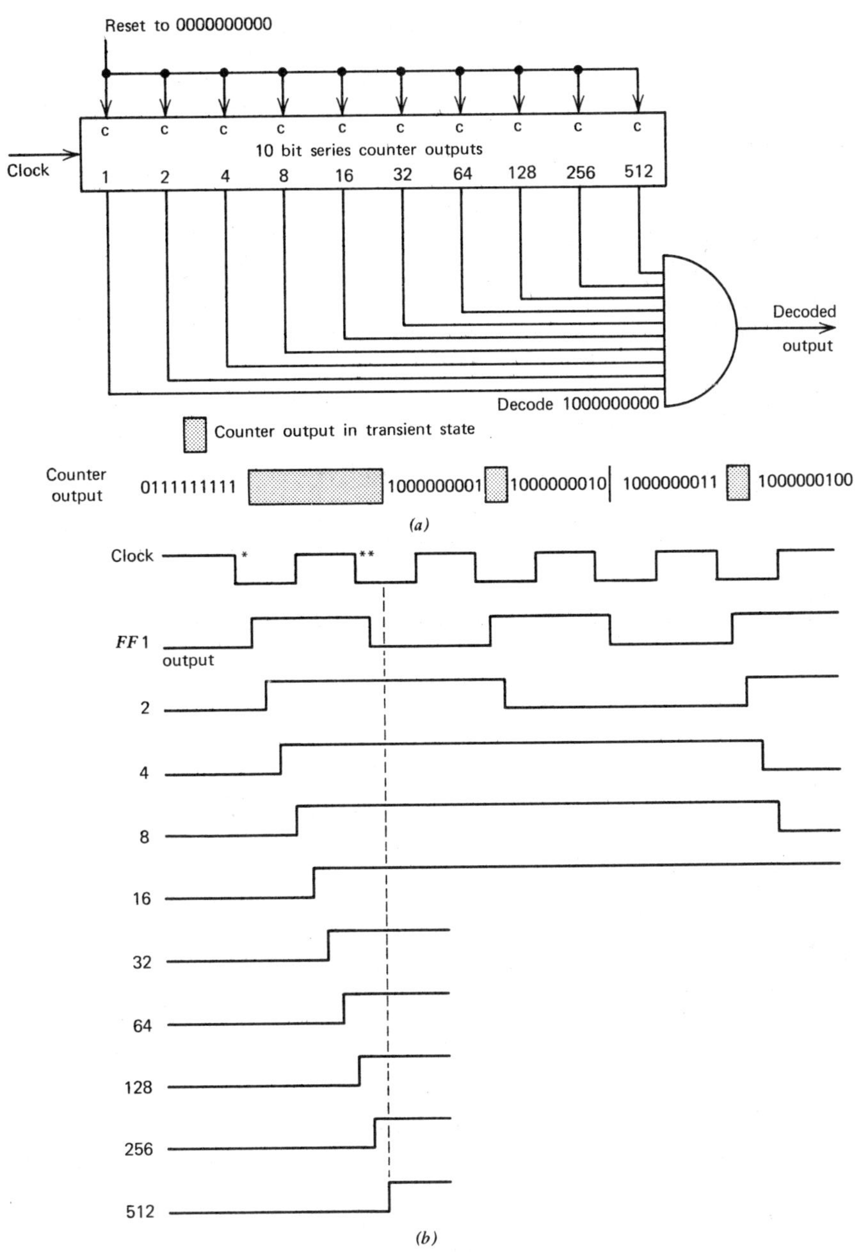

Figure 6.2. (*a*) A 10 bit serial counter having number 1000000000 decoded; (*b*) waveforms indicate timing when the clock changes the counter from 0111111111 to 1000000000. (*Note:* *Activating edge 512 of the clock since the counter was preset to 0000000000. **Activating edge 512 of the clock since the counter was preset to 0000000001.)

than the clock frequency period. It should be also noted that total propagation delay has no effect upon frequency division accomplished by a binary serial counter.

The problem of propagation delay can sometimes be bypassed by decoding an equivalent number, the formation of which requires a change of state in fewer flip-flops than is required by the original number.

In the Example 6.1, decoding the number 1000000000 was an indication that 512 pulses had entered the counter since the counter was preset to 0000000000. If the same counter was initially preset to 0000000001 instead,

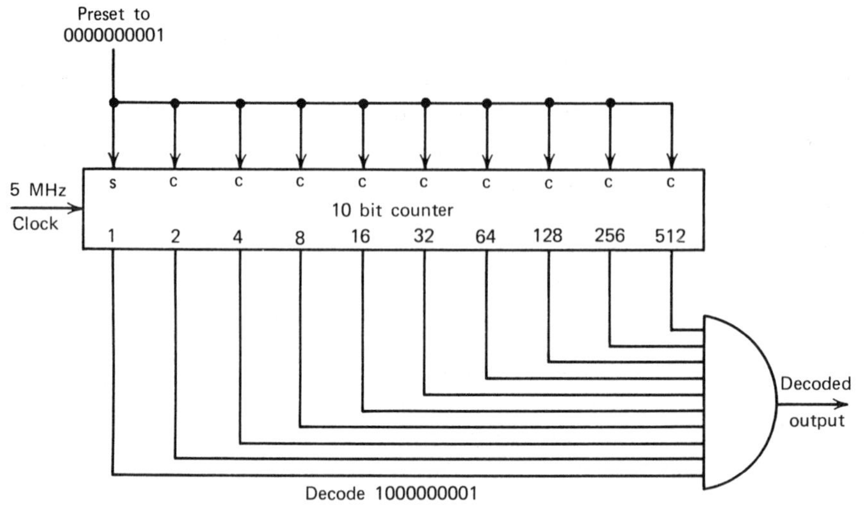

Figure 6.3. A 10 bit serial counter preset to 0000000001.

the result would have been a counter output of 1000000001 at pulse 512 which requires a change of state of only 1 bit. In this case, pulse 512 will change the state of only the first flip-flop of the series counter. This way, an indication of the occurrence of the same number of pulses, 512, can be obtained only two propagation delay intervals after clock pulse 512, while before such indication was impossible to obtain, as shown in Figure 6.2. Figure 6.3 shows a 10 bit serial counter which is preset to 0000000001 and number 1000000001 is decoded from it.

When decoding numbers from serial counters, it should be kept in mind that the formation of odd numbers at the output of a series counter—that is, numbers whose LSB is 1—takes less time than the formation of even numbers.

Serial counters are most effectively used in applications where the counter output does not have to be synchronized with the pulse that originates it, and where time deviation due to propagation delay can be tolerated. When used as a frequency divider and no numbers are decoded from it, the serial counter is limited only by the response of the first flip-flop, which operates at the highest frequency—namely that of the input of the counter. Therefore, a serial counter is as fast as its first flip-flop.

Serial Up-Binary Counter

The serial binary counter is the simplest form of binary counter, and requires less hardware than any other type of counter performing the same function. The serial binary counter is a chain of trigger flip-flops. The first flip-flop is triggered by the clock-input frequency, the second flip-flop is triggered by the nominal output of the first flip-flop, the third flip-flop by the nominal output of the second flip-flop, and so on.

The advantage of this configuration is simplicity from the hardware point of view; while from the performance point of view, the disadvantage is asynchronous operation. By asynchronous operation, we mean that the counter flip-flops are not all directly triggered by a single synchronizing pulse, resulting in a nonsimultaneous change in outputs of the counter flip-flops. Figure 6.1 shows a 4 bit series binary counter that counts upwards—that is, its output, which expresses a binary number, changes in the upward direction by 1 each time the input clock triggers the counter. Also in Figure 6.1 is the counter truth table for 17 clock pulses. The counter has been preset to 0000, and as the input clock "runs," the counter advances by 1 bit per input pulse. The counter output, which is a binary number, represents the number of pulses having triggered the counter since the counter was in the 0000 state.

Serial-Down Binary Counter

When a serial-down binary counter is needed, the circuit of Figure 6.1 can be used with a slight modification—this being that each flip-flop of the counter except the first one will be driven by the inverted output $\bar{Q}$ of the preceding flip-flop and not by the nominal output Q. Such a modified circuit is shown in Figure 6.4; also shown is the truth table of that circuit.

The "weighting" of the flip-flop in the counter is the same as that of the serial-up binary counter. From the truth table it can be seen that the counter output, which represents a binary number, decreases by 1 any time the counter is triggered by a pulse. Serial-down binary counters are often used as subtractors where the minuend is preset into the counter and the subtrahend enters the counter from the clock input as a group of pulses. The weight of each flip-flop in a series counter depends on its location within the

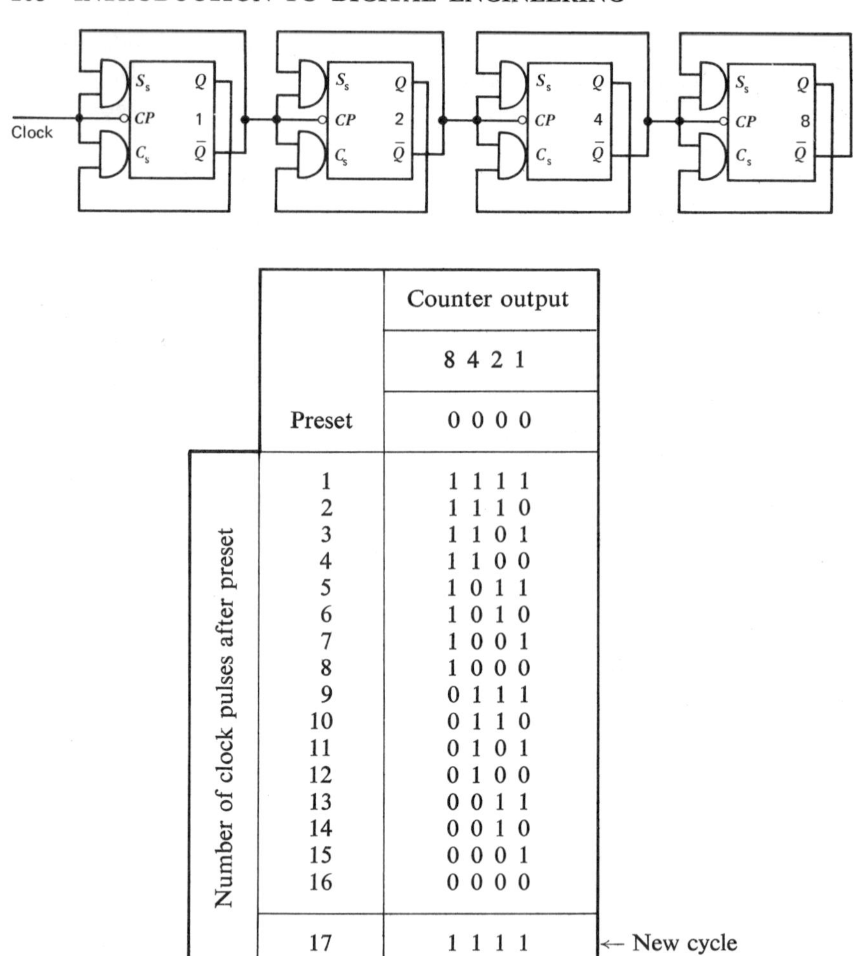

		Counter output
		8 4 2 1
	Preset	0 0 0 0
Number of clock pulses after preset	1	1 1 1 1
	2	1 1 1 0
	3	1 1 0 1
	4	1 1 0 0
	5	1 0 1 1
	6	1 0 1 0
	7	1 0 0 1
	8	1 0 0 0
	9	0 1 1 1
	10	0 1 1 0
	11	0 1 0 1
	12	0 1 0 0
	13	0 0 1 1
	14	0 0 1 0
	15	0 0 0 1
	16	0 0 0 0
	17	1 1 1 1 ← New cycle

Figure 6.4. Series-down binary counter: (*a*) logic diagram; (*b*) truth table.

counter. The first flip-flop—that is, the flip-flop which is driven directly by the input clock—represents the LSB of the number expressed by the output of the counter and has a weight of 1. The second flip-flop, which is driven by the output of the first flip-flop represents the next higher bit and has a weight of 2. The third flip-flop represents the next bit, and has a weight of 4; the fourth flip-flop has a weight of 8; and so on. In the counter shown on Figure 6.4, the flip-flop weighting is, from left to right, 1, 2, 4, 8.

The weight of a flip-flop in a binary counter may be expressed as 2^{n-1}, where n is the number of that flip-flop, counting the flip-flops of that counter from left to right. For example, the weight of the fifth flip-flop in a binary counter is 2^{5-1}, which is 16, and the weight of the eleventh flip-flop is 2^{11-1} which is 1024.

6.3 SYNCHRONOUS BINARY COUNTERS

The cumulative propagation delay, which constitutes one of the major limitations of the serial counter, can be avoided by using a synchronous counter. In a synchronous counter all stages are triggered simultaneously by a single clock. After each trigger pulse at least one flip-flop of the counter changes state. Change depends on the states of the flip-flops that precede it. Figures 6.5 and 6.6 show the general form of two synchronous counters which differ only in the way their carry signal is gated.

The carry signal of a flip-flop in a synchronous counter indicates that the conditions, under which that flip-flop should change state, have been met. That is, the carry signal dictates whether the flip-flop will change state upon arrival of the clock pulse. The carry signal enables the trigger input of a flip-flop only when all preceding flip-flops are in the 1 state.

In a synchronous counter, the first flip-flop changes state with each clock pulse. The second flip-flop changes state when two conditions are met—namely, when it is triggered by the clock and when the output of the first flip-flop is 1. The third flip-flop changes state when it is triggered by the clock and the output of the first flip-flop as well as that of the second are both 1. In other words, in a synchronous binary counter, a given flip-flop changes state when it is triggered by the clock while all flip-flops that precede it are in the 1 state.

Following are the equations of the carry, or enable signals for the first 4 bits of the n bit synchronous counters shown in Figures 6.5 and 6.6:

$$C_{FF1} = 1 \qquad \text{(all the time)} \qquad (6.3)$$

$$C_{FF2} = Q_1$$

$$C_{FF3} = Q_1 \cdot Q_2 = C_{FF2} \cdot Q_2$$

$$C_{FF4} = Q_1 \cdot Q_2 \cdot Q_3 = C_{FF3} \cdot Q_3$$

where FF stands for flip-flop. From these specific equations, two equivalent general equations can be written for the carry input of the nth flip-flop:

$$C_{FFn} = Q_1 \cdot Q_2 \cdot Q_4 \cdots Q_{(n-1)} \qquad (6.4)$$

and

$$C_{FFn} = C_{FF(n-1)} \cdot Q_{(n-1)} \qquad (6.5)$$

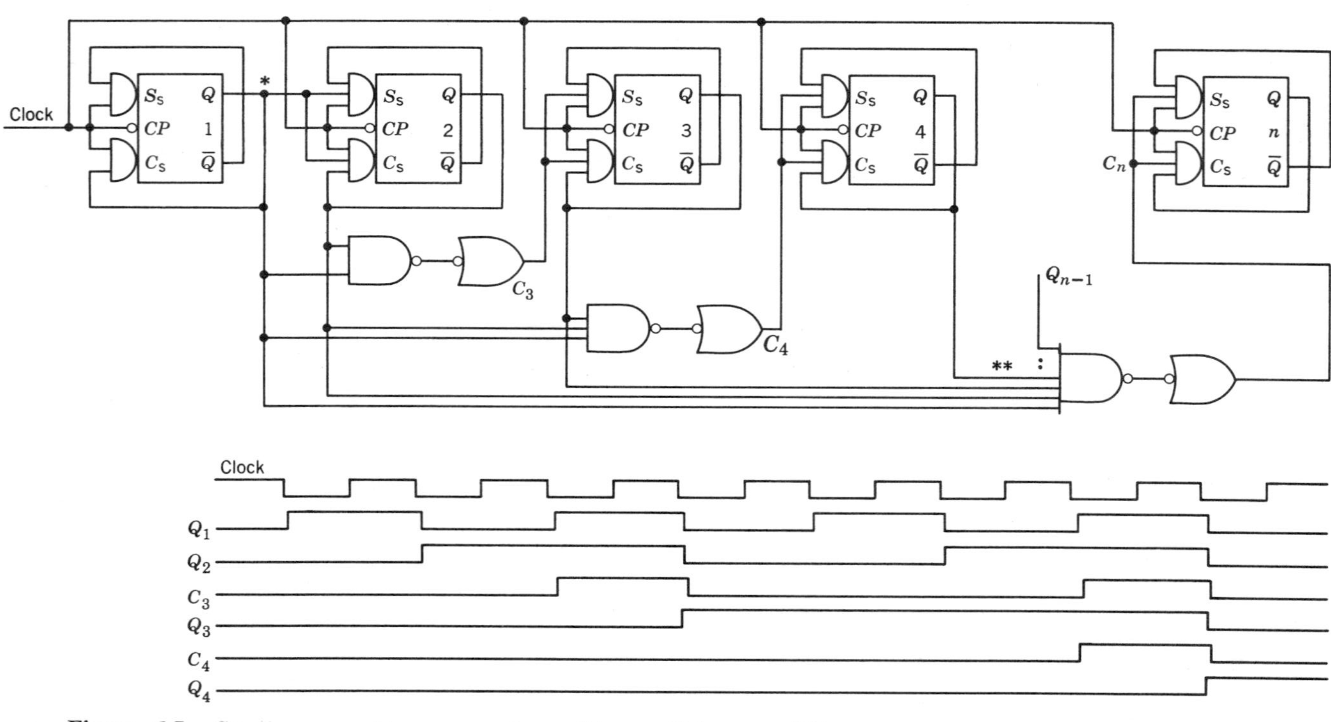

Figure 6.5. Synchronous binary counter with parallel-carry. (*Note:* * High fan-out requirement. ** High fan-in requirement.)

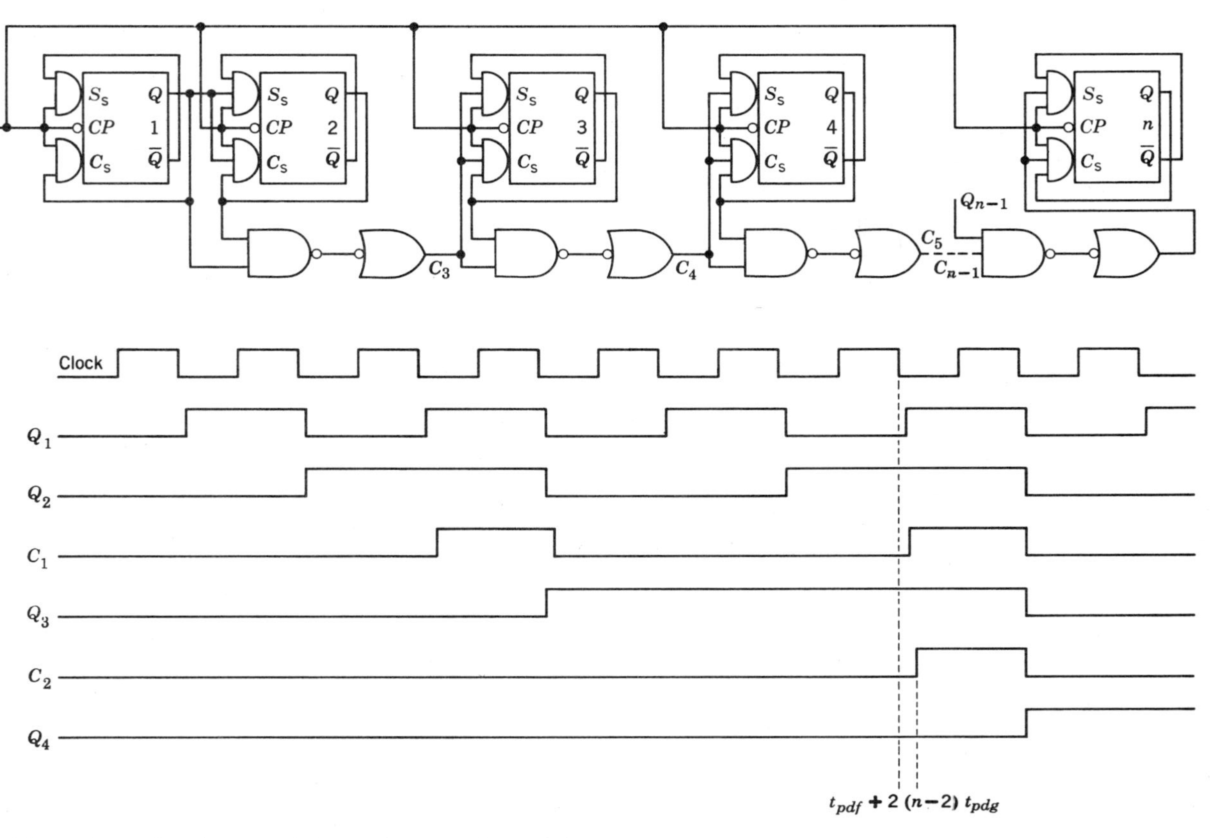

Figure 6.6. Synchronous binary counter with serial-carry.

There are two methods of implementing the above equations. One is to employ a separate gate for each carry signal (Fig. 6.5). This type of counter is called a parallel-carry synchronous counter. The advantage of this implementation is speed. Here, the output of each flip-flop is generated only one propagation delay period after the activating edge of the clock pulse. Thus, the speed of a parallel-carry synchronous counter is limited only by the propagation delay of its flip-flops and not by the number of flip-flop stages as is the case with serial counters.

This limitation can be expressed by

$$T_{\text{clock}} > 2t_{\text{pdg}} + t_{\text{pdf}} + t_{\text{in}} \tag{6.6}$$

where T_{clock} is the period of the clock frequency, t_{pdg} is the average propagation delay of the type of gate used, t_{pdf} is the average propagation delay of one flip-flop, and t_{in} is the minimum time required by the flip-flop to read synchronous information. In this counter configuration the output pulse width of any decoded number always equals the clock period T_{clock}.

The disadvantages of the parallel-carry synchronous counter are its high fan-out requirements for the first flip-flops, and high fan-in requirements of the carry gates of the last flip-flops. Figure 6.5 is the direct implementation of equation 6.4. The implementation of equation 6.5 is shown in Figure 6.6. This figure illustrates a serial-carry synchronous counter. Its advantage is that it requires minimal flip-flop fan-out and gate fan-in. Its disadvantage is the cumulative delay created by the carry signal gates, which limits the speed of the counter.

This limitation can be expressed by the following inequality:

$$T_{\text{clock}} > 2(n - 2)(t_{\text{pdg}}) + (t_{\text{pdf}}) + (t_{\text{in}}) \tag{6.7}$$

where n is the number of flip-flop stages in the counter. The coefficient in the inequality is 2 because two gates are used for each stage (Fig. 6.6). Between clock pulses the output of the counter is stable, and when numbers are decoded from the counter, the pulse width of the decoded information is

$$PW = T_{\text{clock}} - 2(m - 2)t_{\text{pdg}}$$

where m is the number of counter stages that changed state due to the most recent clock pulse. An approach to counter selection is illustrated in Example 6.2.

Example 6.2. An 8 bit binary counter is to be designed using master-slave flip-flops which have a propagation delay of 50 nsec and a read-in time of 60 nsec and the clock frequency of the system is 4 MHz. The counter output is to be decoded and the decoding pulse must have a minimum pulse width of 200 nsec. Gates having an average propagation delay of 25 nsec are available.

It should be kept in mind that in any design one should start with the simplest and least expensive configuration and proceed to more complex approaches only when simpler methods have proven to be inadequate. The simplest and least expensive binary counter is the serial counter but its disadvantage is that it cannot operate effectively at high speeds. An equation can be derived from the inequality (6.2) to express the minimum pulse width of a decoded number in a series counter in terms of the clock period, the number of stages in the counter, and the propagation delay of the flip-flops used.

$$PW = T_{\text{clock}} - nd \tag{6.8}$$

where n is the number of flip-flops in the counter, and d is the propagation delay of the flip-flops used. In this example, the minimum acceptable pulse width PW of the decoded output must be greater than 200 nsec. Replacing T_{clock} by 250 nsec n by 8, and d by 50 nsec, we obtain

$$\begin{aligned} PW_{\min} &= 250 \text{ nsec} - 8(50 \text{ nsec}) \\ &= 250 \text{ nsec} - 400 \text{ nsec} \\ &= -150 \text{ nsec} \end{aligned}$$

The negative pulse width means that the counter output cannot be decoded. Consequently, the serial counter cannot be used and the designer should now proceed to consider the counter which is next in speed and cost. This is the synchronous counter with serial-carry (Fig. 6.6). This counter has the advantage of speed over the serial counter, and the disadvantage of an additional two gates per stage. For reliable operation in a synchronous counter with serial carry the following expression should be satisfied:

$$T_{\text{clock}} > 2(n - 2)(t_{\text{pdg}}) + (t_{\text{pdf}}) + (t_{\text{in}}) \tag{6.9}$$

Substituting in the given numerical values we obtain

$$\begin{aligned} &250 \text{ nsec} > 2(8 - 2)(25 \text{ nsec}) + (50 \text{ nsec}) + (60 \text{ nsec}) \\ &250 \text{ nsec} > (300 + 50 + 60)\text{nsec} \\ &250 \text{ nsec} > 410 \text{ nsec} \end{aligned}$$

Since the inequality is not satisfied, a serial-carry synchronous counter cannot be used. The fact that the carry propagation delay, which is the right-hand expression of the inequality, is greater than the clock period indicates that the clock pulse will arrive at the last flip-flops of the counter before the carry signal has reached the trigger enable input of those flip-flops.

The counter next in order of speed and cost, is the parallel-carry synchronous counter. The following inequality expresses the relationship that must be satisfied to ensure proper operation of this counter:

$$T_{\text{clock}} > 2(t_{\text{pdg}}) + (t_{\text{pdf}}) + (t_{\text{in}}) \tag{6.10}$$

Substituting in the given numerical values we obtain

$$250 \text{ nsec} > 2(25 \text{ nsec}) + 50 \text{ nsec} + 60 \text{ nsec}$$

$$250 \text{ nsec} > 160 \text{ nsec}$$

Satisfaction of this inequality indicates that a parallel-carry synchronous counter meets the requirements of the example. The 200 nsec minimum pulse width requirement is automatically satisfied since the outputs of a synchronous counter remain in each state for a full clock period, which in this case is 250 nsec. ▮

It is worth noting that in almost all cases of digital design, an increase in speed is accompanied by an increase in complexity and hardware.

The synchronous counter configurations illustrated in Figures 6.5 and 6.6 depict up-counting devices. Down-counting can be accomplished by connecting to the input of the gates the inverted outputs of the flip-flops rather than the nominal. Figure 6.7 shows two 6 bit synchronous binary counters with serial-carry in the up-counting and down-counting configurations.

Reversible Synchronous Binary Counter

In the two preceding sections, the synchronous binary counter was considered as either a count-up or a count-down device. Some applications, however, require that the counter perform both functions, count up as well as count down, and have the ability to switch from one mode to the other without altering the number contained in the counter.

Figure 6.8(a) illustrates an n bit reversible synchronous binary counter. In this circuit, when the "direction-control" input is HIGH, the U gates are enabled and the D gates are disabled. With the U gates enabled, the counter counts up. When the direction-control input is LOW, the U gates are disabled and the D gates are enabled allowing the counter to count down.

The need often arises for a reversible counter that accepts two trains of noncoincident pulses, where the pulses of one train should be added to the content of the counter while those of the other pulse train should be subtracted. In this configuration the counter has two clock inputs. One clock is gated to the nominal outputs of the counter and increments the counter, and the other is gated to the inverted outputs and decrements the counter.

Figure 6.8(b) illustrates a synchronous reversible binary counter that uses

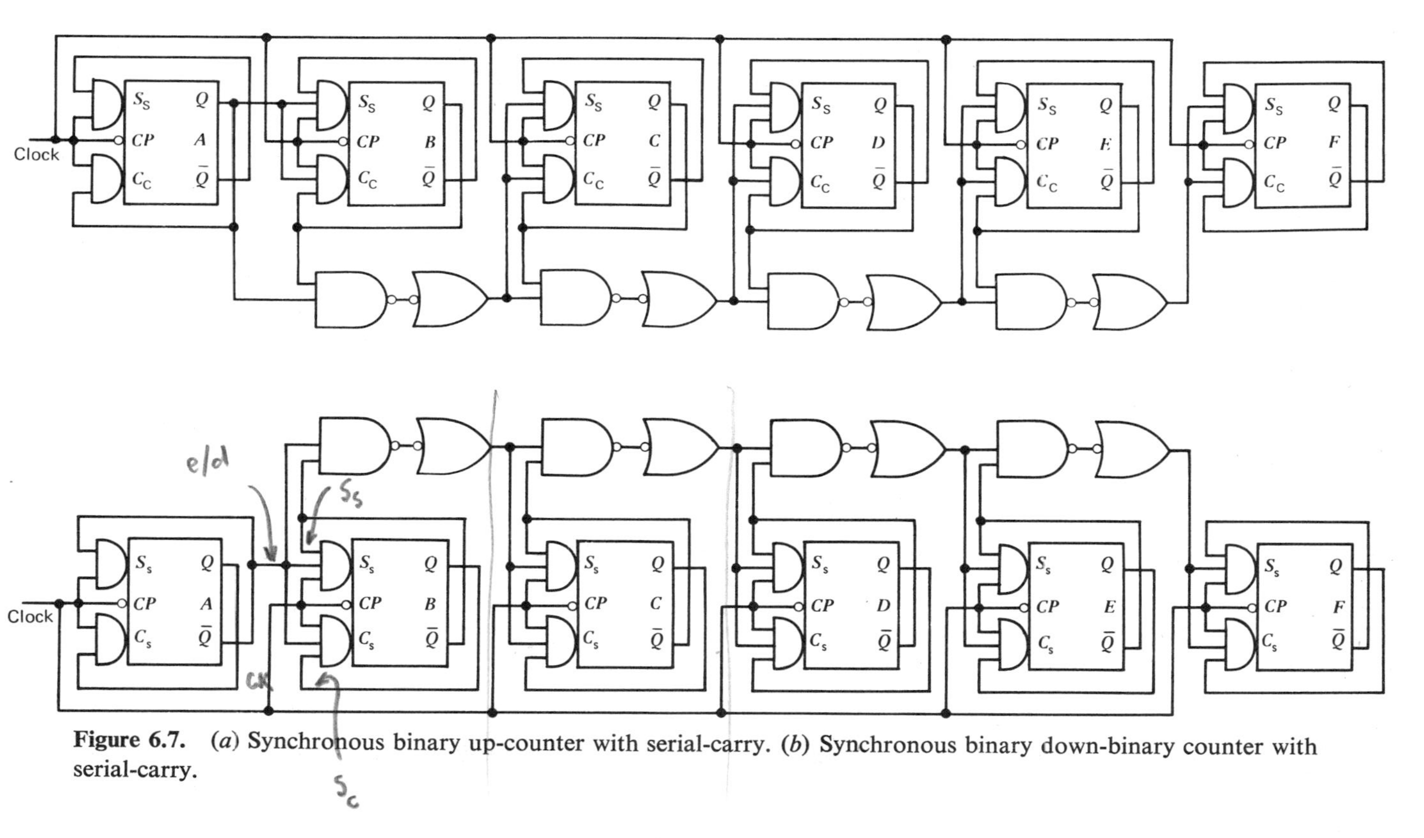

Figure 6.7. (*a*) Synchronous binary up-counter with serial-carry. (*b*) Synchronous binary down-binary counter with serial-carry.

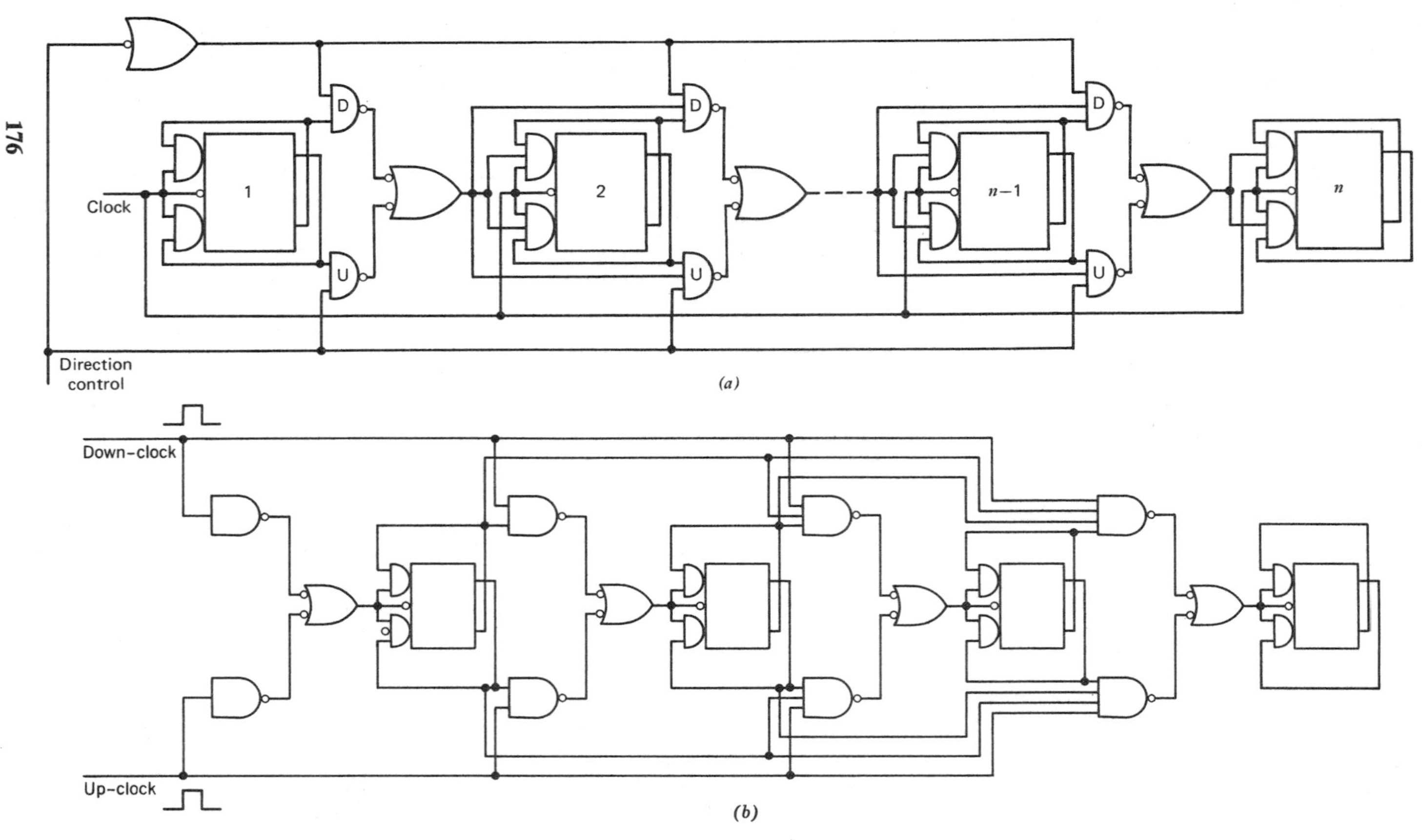

Figure 6.8. Reversible synchronous binary counters: (*a*) using serial-carry with single clock and direction-control input; (*b*) using parallel-carry and dual clock.

parallel-carry and has separate clock inputs. The up-clock activates the clock input of a stage when the nominal output of all preceding stages is 1, while the down-clock activates it when the inverted output of all preceding stages is 1.

6.4 FEEDBACK COUNTERS

The configurations examined in the preceding sections represent counters that count in the pure binary system—that is, each of the counter flip-flops changes stage only when all preceding flip-flops are in the 1 state. In the binary counters, the first flip-flop has the weight of 1, the second flip-flop the weight of 2, the third flip-flop the weight of 4, and the *n*th flip-flop the weight of 2^{n-1}. A counter of this type counts from 0 to $2^n - 1$ and then goes back to zero. Its counts can be from 0 to 3, 7, 15, 31, 63, and so on. If a count other than $2^n - 1$ is needed, provision should be made for the counter to return to the 0 state after it has reached the desired number.

Figure 6.9 shows an *n* bit binary counter that counts from 0 to 2^{n-1} and then returns to 0. A truth table and critical waveforms illustrate the counter operation. This circuit consists of a binary counter, one flip-flop that operates as a shift register, one control gate, and one inverter. The binary counter can be either a series of a synchronous counter. The counter is free-running, with the exception of its first flip-flop, the synchronous set of which is inhibited when the shift register $\bar{Q}$ output is LOW as shown in Figure 6.9. If we consider this circuit to be an *n* bit counter, its last flip-flop, which is a shift register, will then hold the 2^{nth} bit of the binary number contained in the counting network. The NAND gate allows the clock to trigger the shift register only when $\bar{Q}_{(n-1)}$ is HIGH; that is, only when $Q_{(n-1)}$ is LOW; $Q_{(n-1)}$ is the output of the flip-flop located immediately before the shift register. The inverter provides the clock to the NAND gate with the proper phase.

The feedback counter—that is, the entire circuit of Figure 6.9—starts with all its flip-flops at 0 and operates as a typical binary counter until it reaches the binary number that corresponds to 2^{n-1}. It then returns to 0 and starts counting all over again. At 0 all counter outputs are in the 0 state; while at 2^{n-1}, output Q_n is 1 and outputs $Q_1 \ldots Q_{(n-1)}$ are 0; Q_n is the output of the last flip-flop which is used as a shift register.

After counting to 2^{n-2}, $Q_{(n-1)}$ becomes 1, and its output enters the slave portion of flip-flop *n*. As long as $Q_{(n-1)}$ is 1, the clock is inhibited from triggering the shift register flip-flop, because of the NAND gate, and the Q_n output remains in the 0 state. When the counter reaches $2^{n-1} - 1$, all outputs are 1 except that of the last flip-flop *n*. Now, the arrival

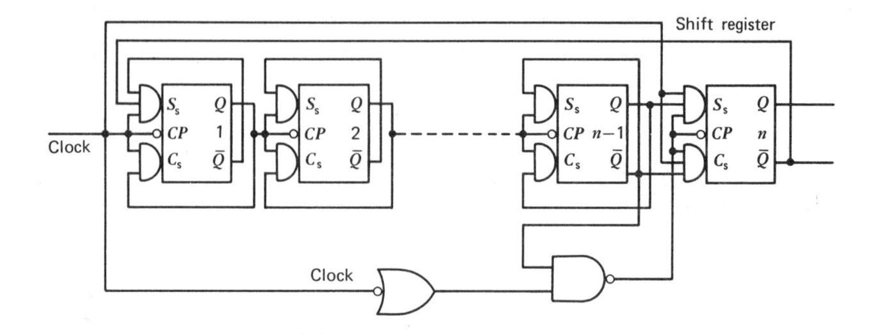

Output in Decimal	Counter output					
	n	$n-1$	1. . .	3	2	1
0	0	0	0. . .0	0	0	0
1	0	0	0. . .0	0	0	1
2	0	0	0. . .0	0	1	0
3	0	0	0. . .0	0	1	1
4	0	0	0. . .0	1	0	0
5	0	0	0. . .0	1	0	1
⋮	⋮	⋮	⋮ ⋮	⋮	⋮	⋮
$2^{n-1}-2$	0	1	1. . .1	1	1	0
$2^{n-1}-1$	0	1	1. . .1	1	1	1
2^{n-1}	1	0	0. . .0	0	0	0
New cycle	0 ⋮	0 ⋮	0. . .0 ⋮ ⋮	0 ⋮	0 ⋮	0 ⋮

(*b*)

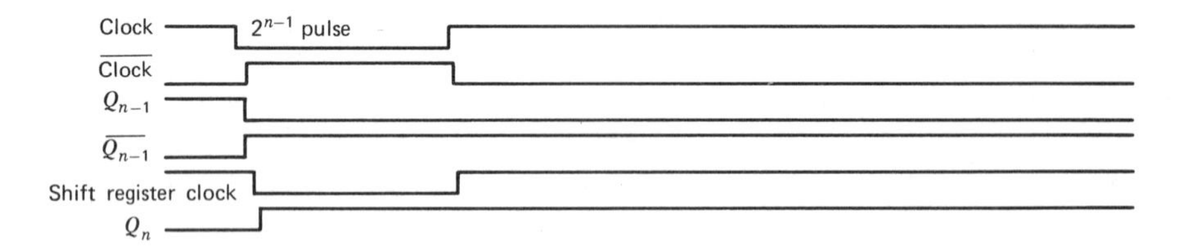

Figure 6.9. Feedback counter: (*a*) logic diagram; (*b*) truth table; (*c*) timing waveforms. The waveforms show operation at the 2^{n-1} pulse.

of a clock pulse will change all the outputs of the binary counter from 1 to 0, including that of $Q_{(n-1)}$. When $Q_{(n-1)}$ goes to 0, the NAND gate which is controlled by $Q_{(n-1)}$ will be activated and the clock will trigger the shift register flip-flop, switching it from 0 to 1 (see waveforms, Fig. 6.9). The counter output is now 2^{n-1}. In this state all counter outputs, excluding that of the shift register which represents 2^{n-1}, are 0. During the 2^{n-1} period, $\bar{Q}_n$ output which now becomes LOW inhibits the first flip-flop of the counter from becoming 1. And the output of $Q_{(n-1)}$ enters a 0 into the shift register flip-flop.

At the arrival of the next clock pulse, the first flip-flop remains in the 0 state, and the 0 that was entered into the shift register flip-flop is now transferred to the output of the shift register. Thus, the counter output has returned to 0 and is ready to start counting again—that is, all flip-flops of the counter are in the 0 state.

The type of feedback counter discussed above and shown in Figure 5.6 may count from 0 to 2^{n-1}, where n is the number of stages in the entire counter. The counter configuration of Figure 6.9 may be used to count from 0 to 2, 4, 8, 16, 32, and so on. For example, if a 17 state binary counter is needed, a 5 bit feedback counter of the type discussed above may be used. This counter will count from 0 to 16, which corresponds to 17 states. Figure 6.10 shows the counter diagram. The number of stages was determined as follows:

$$\text{Counter number of states } 2^{n-1} + 1 = 17$$

Solving for n, we have

$$2^{n-1} = 16$$

$$2^{n-1} = 2^4$$

$$n - 1 = 4$$

$$n = 5$$

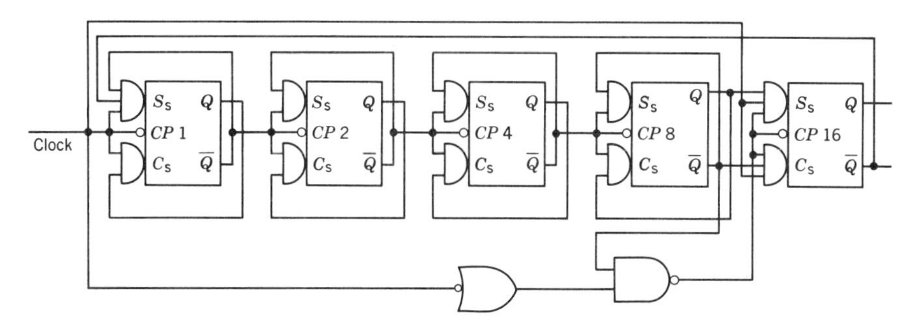

Figure 6.10. A 5 bit feedback counter counting from 0 to 16.

It should be noted that at the beginning of their operation the feedback counters must be cleared.

6.5 COMBINATION COUNTERS

The feedback counter in Figure 6.9 counts from 0 to any 2^{n-1}. If a free-running counter is connected in series with it, the new counter will count from 0 to $2^{n-1} \cdot 2^m$, or from 0 to 2^{n+m-1}, where m is the number of stages in the free-running counter. To obtain a binary count from 0 to 2^{n+m-1}, the free-running counter of the combination counter must precede the feedback counter in the series as shown in Figure 6.11.

The counts obtained using a combination counter of Figure 6.11 are binary multiples of the counts, which can be obtained with the feedback counter when it is used alone. For example, if a feedback counter has four stages and a binary counter has three, the combination counter made by connecting the two counters in series will count from 0 to 2^{4+3-1}, or from 0 to 64.

The maximum count of a combination counter is thus a function of the number of stages of the binary and feedback counters. This is illustrated in Figure 6.12. If the counter starts from 0, its maximum count can be obtained by subtracting 1 from the total number shown. To obtain counts that are not included in the figure and still maintain the binary code, a general-purpose counter, discussed in the following section, should be used.

Example 6.3. A binary counter is to be designed that counts in pure binary and has a total of 72 counts.

The counter needed must count from 0 to 71 and then return to 0. Since 72 is not a power of 2, a free-running binary counter alone does not meet the requirements. If a combination counter is to be used, the number of stages, for each of its two parts, should be determined. This is done by successively dividing the number of counts by 2 until the division gives an odd number—that is,

$$\frac{72}{2} = 36 \qquad \frac{36}{2} = 18 \qquad \frac{18}{2} = 9$$

Since three divisions were needed to reach an odd number, the number of stages in the free-running counter is 3, and the number of counts in the feedback counter should be 9. In the expression of the combination counter $2^{n-1} \cdot 2^m$, m is 3. To find n, which is the number of stages in the feedback counter, the general expression for the number of counts in the feedback

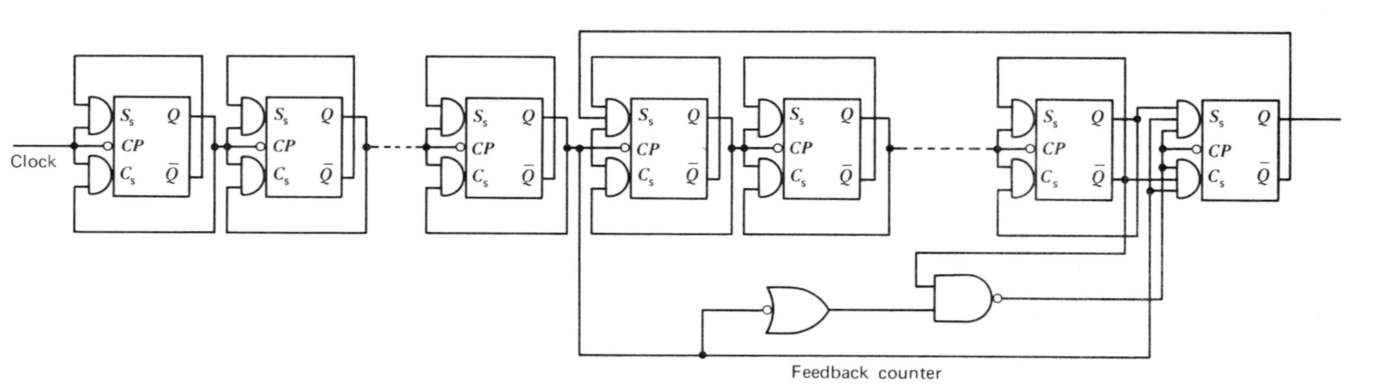

Figure 6.11. Combination counter.

Number of stages in the binary counter	Number of stages in the feedback counter							
	0	1	2	3	4	5	...	n
0			3	5	9	17	...	$2^{n-1}+1$
1	2		6	10	18	34	...	$2(2^{n-1}+1)$
2	4		12	20	36	68	...	$4(2^{n-1}+1)$
3	8		24	40	72	136	...	$8(2^{n-1}+1)$
4	16		48	80	144	272	...	$16(2^{n-1}+1)$
5	32		96	160	288	544	...	$32(2^{n-1}+1)$
⋮	⋮		⋮	⋮	⋮	⋮	...	⋮
m	2^m		.	.	.	.	...	$2^m(2^{n-1}+1)$

Figure 6.12. Number of counts in a binary combination counter as a function of the number of stages in the binary and feedback counters of which it is composed.

counter $(2^{n-1}+1)$ should be used. Placing 9 for the number of states, we obtain

$$9 = 2^{n-1} + 1$$

$$8 = 2^{n-1}$$

$$8 = \frac{2^n}{2}$$

$$16 = 2^n$$

$$n = 4$$

The feedback component of the combination counter should thus have four stages. The binary combination counter for this example is shown in Figure 6.13. The case where n is not an integer is discussed later in the section on multiple-feedback counters. ∎

6.6 GENERAL-PURPOSE COUNTERS

When a count N cannot be obtained using the binary counters covered in the preceding sections, a general-purpose counter should be used. In this configuration the desired count N may be obtained by decoding $N - 1$ from a typical binary counter, delaying it for one-half clock period, and then using it to reset the counter. This is a straightforward approach that can make a binary counter count to any number. Its disadvantage is that, in addition to the counter, it requires an additional flip-flop and a multi-input gate. Figure 6.14 shows this type of counter in its general form.

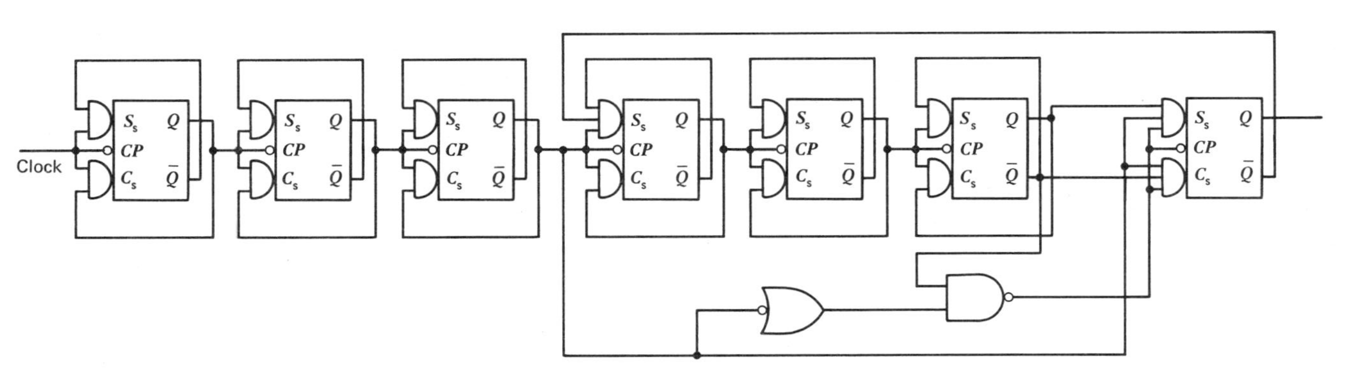

Figure 6.13. Binary combination counter counting from 0 to 71.

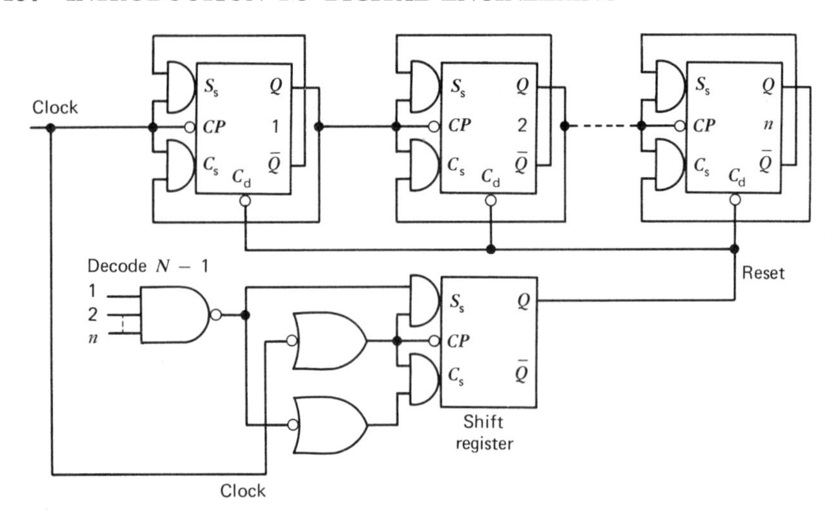

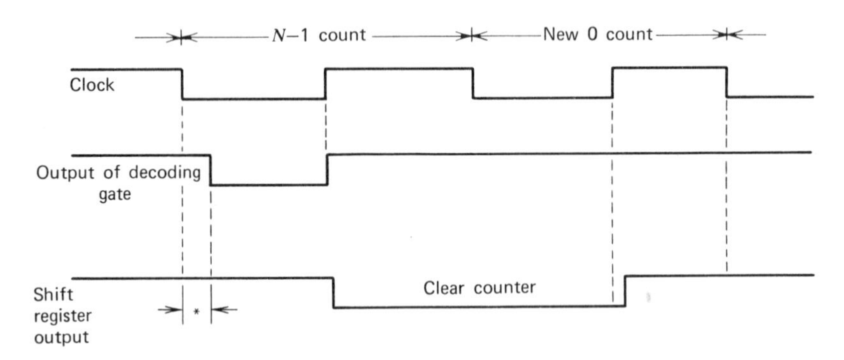

Figure 6.14. General-purpose binary counter. (*Note:* N = desired number of states in the counter. *Counter and decode gate delay.)

The binary counter starts from 0 and counts to $N - 1$. When the counter reaches $N - 1$, all inputs to the multi-input gate are HIGH and the gate output is LOW. This LOW state is entered into the master portion of the shift register flip-flop. When the clock switches to HIGH, after the Nth pulse, the LOW signal is transferred from the master into the slave portion of the shift register making the shift register output a LOW which clears the counter.

Example 6.4. If a binary counter that has 23 counts and starts from 0 is required, the type described above can be used. If $N = 23$, $N - 1$ is 22. Since 22 in binary is the 5 bit number 10110, a five-stage counter will be

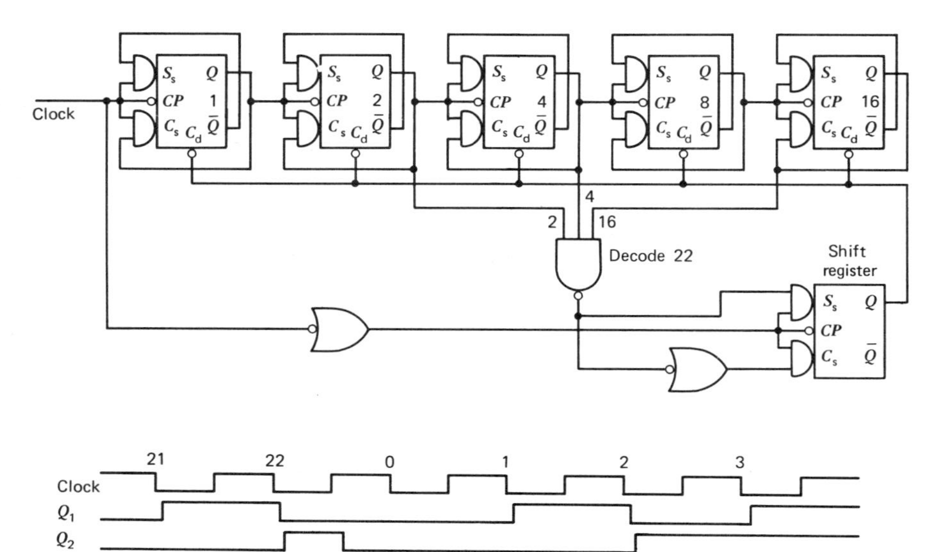

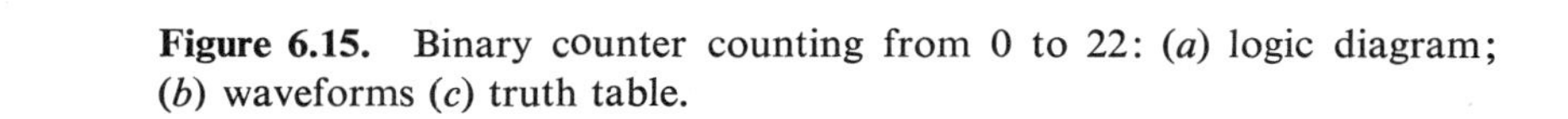

Figure 6.15. Binary counter counting from 0 to 22: (*a*) logic diagram; (*b*) waveforms (*c*) truth table.

Decimal	Binary
0	00000
1	00001
2	00010
3	00011
4	00100
5	00101
6	00110
7	00111
8	01000
9	01001
10	01010
11	01011
12	01100
13	01101
14	01110
15	01111
16	10000
17	10001
18	10010
19	10011
20	10100
21	10101
22	10110
0	00000
1	00001
⋮	⋮

(*c*)

required. In binary, number 22 consists of 16 + 4 + 2; consequently to decode this number the output of flip-flops 16, 4, and 2 must be ANDed. The output of the AND gate will be delayed through a shift register with the output of which the binary counter will then be cleared. This counter with its truth table and timing diagram is shown in Figure 6.15. ▮

In this section, the design of counters that count from 0 to a number N was examined. Some applications, however, may require counters that count

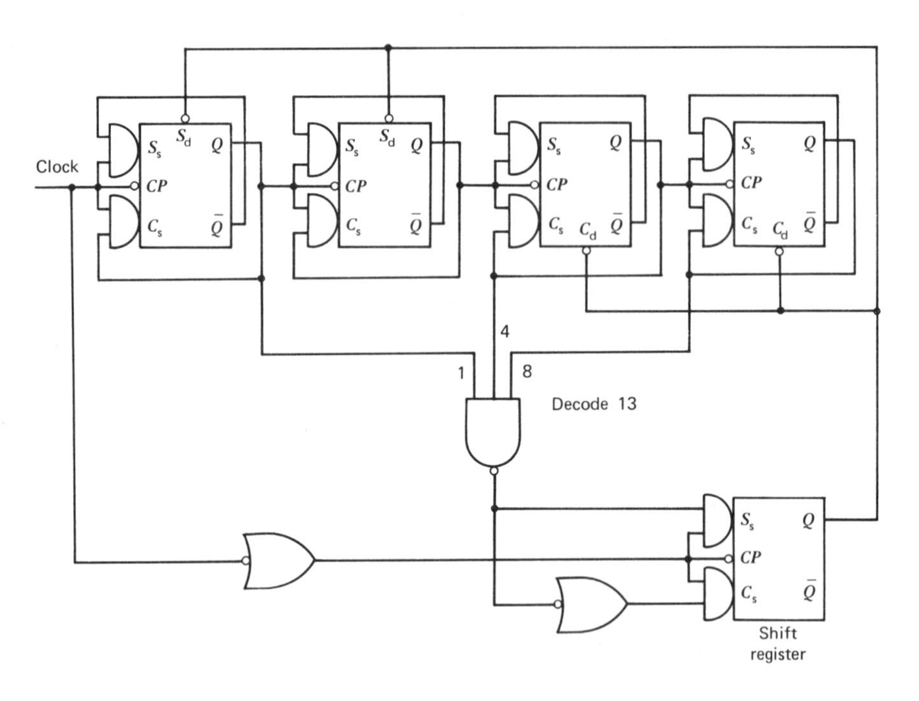

Figure 6.16. Binary counter producing the sequence 3, 4, 5, 6, 7, 8, 9, 10, 11, 12, 13, 3, ...

from a number M to another number N, go back to M, and start counting again. In this case, the decoded number is N. The preset line, however, instead of clearing the counter, will set it to M. If, for example, a 4 bit counter is to count from 3 to 13, a general-purpose binary counter may be used where the decoding gate decodes number 13 and uses the decoded signal to preset the counter to 3 where it will start counting again. This counter is illustrated in Figure 6.16.

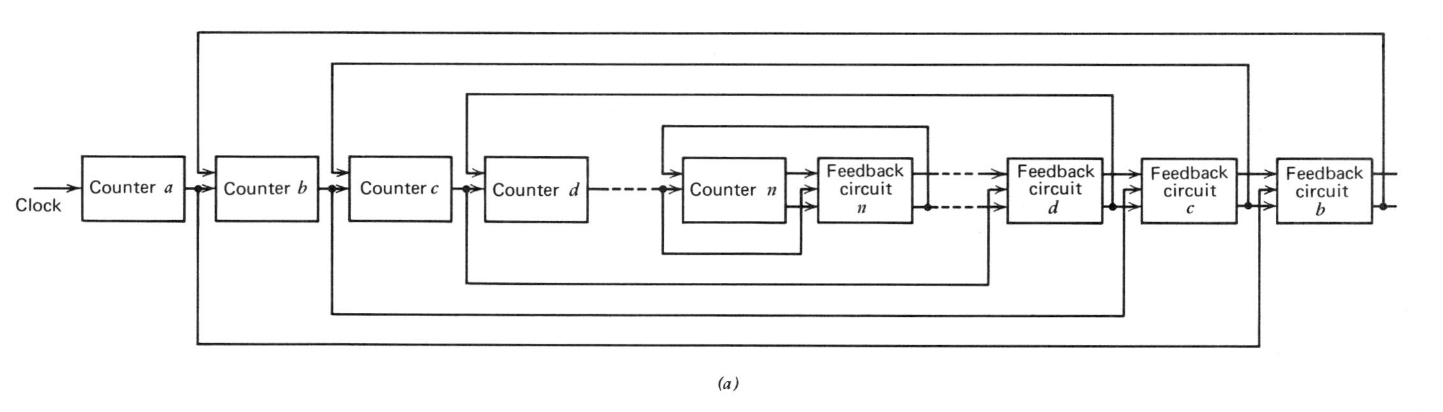

Figure 6.17. Nonbinary multiple-feedback counter: (*a*) block diagram; (*b*) schematic.

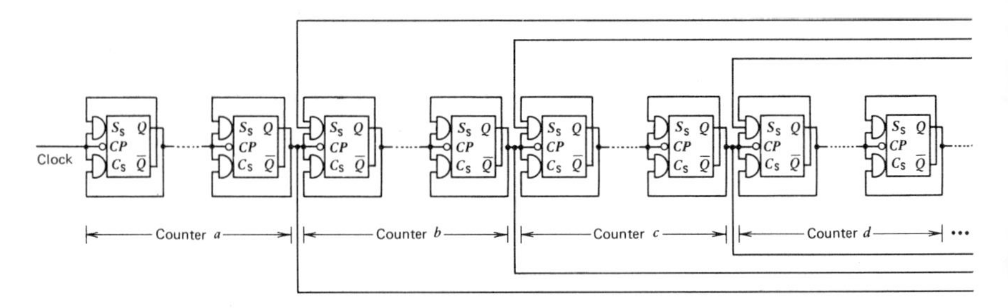

6.7 MULTIPLE-FEEDBACK COUNTERS

Another way of generating counts that are not powers of 2 is by means of Multiple-feedback counters. These counters consist of feedback counters that contain other typical feedback counters. Figure 6.17 shows a counter of this type in general block form and in schematic form. The number of counts or number of different states a multiple-feedback counter of this type has is determined by

$$N = [[[[\ldots[(2^n + 1)2^{n-1} + 1]\ldots]2^d + 1]2^c + 1]2^b + 1]2^a \qquad (6.11)$$

where N is the number of counts in the counter, a is the number of stages in binary counter a, b is the number of stages in binary counter b, c is the number of stages in binary counter c, d is the number of stages in binary counter d, and n is the number of stages in the nth or last binary counter. In the design of multiple-feedback counters, the values of $a, b, c, d, \ldots, n$ are computed from the following equations:

$$a = \log_2 \frac{N}{K_1} \qquad (6.12)$$

$$b = \log_2 \frac{K_1 - 1}{K_2} \qquad (6.13)$$

$$c = \log_2 \frac{K_2 - 1}{K_3} \qquad (6.14)$$

$$d = \log_2 \frac{K_3 - 1}{K_4} \qquad (6.15)$$

$$\vdots$$

where K is the odd number that makes the logarithm an integer. This process continues until the denominator becomes 1. It should be stated that multiple-feedback counters count in arbitrary code. The following examples illustrate the use of the above equations.

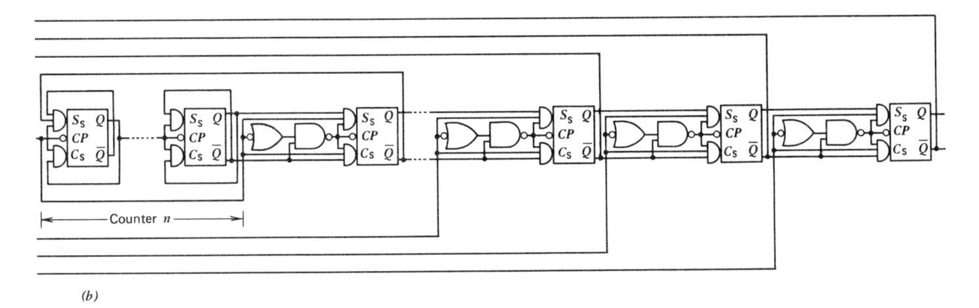

Example 6.5. Design a counter that generates a sequence of 32 different numbers.

$$a = \log_2 \frac{N}{K_1} \qquad (N = 32)$$

$$= \log_2 \frac{32}{K_1} \qquad (K_1 = 1)$$

$$= \log_2 \frac{32}{1}$$

$$= 5$$

The number 1 was chosen for K_1, because 1 is the only odd integer that makes the fraction $32/K_1$ a power of 2. Counter a should therefore have five stages. The five-stage free-running binary counter shown in Figure 6.18 will satisfy the requirements of this example. ∎

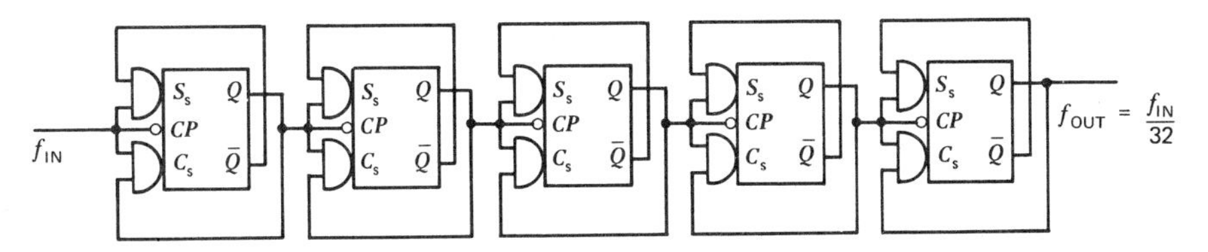

Figure 6.18. Five-stage free-running counter of Example 6.5.

Example 6.6. Design a counter that has nine states. Here, N is 9 and the number of stages of the a counter can be determined using the same equation used in Example 6.5; K_1 is the odd number that makes the fraction N/K_1 a power of 2. Since the numerator is odd, the only power of 2 that can be generated is 2^0 when the fraction has the value of 1. This will be the case when K_1 is the same as the numerator which is 9; therefore,

$$a = \log_2 \frac{9}{K_1} \qquad (K_1 = 9)$$

$$= \log_2 \frac{9}{9}$$

$$= \log_2 1$$
$$= 0$$

Thus, the counter required by this example has no a component. To determine the b component of the counter,

$$b = \log_2 \frac{K_1 - 1}{K_2} \qquad (6.16)$$

where K_2 is the odd number that makes the fraction $(K_1 - 1)/K_2$ a power of 2. Substituting its numerical value for K_1 in (6.16), we obtain

$$b = \log_2 \frac{9 - 1}{K_2}$$

$$b = \log_2 \frac{8}{K_2}$$

The odd number denominator that will make the fraction $8/K_2$ a power of 2 is 1—that is, K_2 should be 1. Solving now for b, we have

$$b = \log_2 \frac{8}{1}$$

$$= \log_2 8$$
$$= 3$$

Therefore, the b component of the counter should have three stages. When K becomes 1, it is an indication that the computations are completed and that the counter has no components other than those computed thus far. In this case, the desired counter has only a b component which consists of three stages. This circuit is shown in Figure 6.19. ▮

Example 6.7. Design a counter with an output signal period of 22 clock pulses. Here N is 22. The number of stages in the a counter can be determined by

$$a = \log_2 \frac{N}{K_1} \qquad (6.17)$$

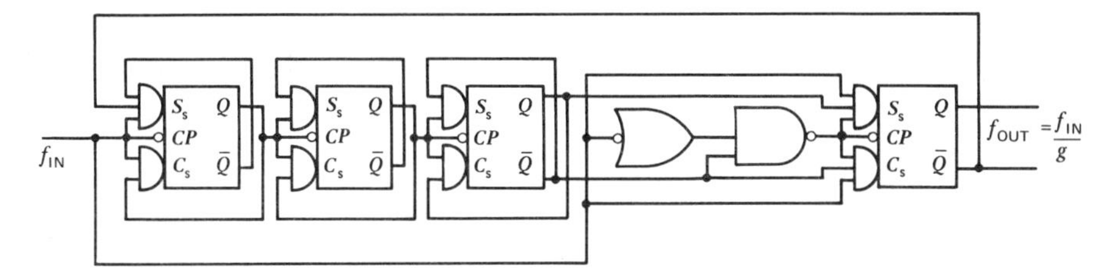

Figure 6.19. Nine-state counter of Example 6.6.

where K_1 is the odd number that makes the fraction N/K_1 a power of 2. Substituting its numerical value for N in (6.17) we obtain

$$a = \log_2 \frac{22}{K_1}$$

The odd number denominator K_1 that will make the fraction $22/K_1$ a power of 2 is 11. Solving for a, we have

$$a = \log_2 \frac{22}{11} \qquad (K_1 = 11)$$

$$= \log_2 2$$

$$= 1$$

Therefore, the a component of the counter will have one stage. To determine the b component,

$$b = \log_2 \frac{K_1 - 1}{K_2} \qquad (6.18)$$

Substituting its numerical value for K_1, we obtain

$$b = \log_2 \frac{11 - 1}{K_2}$$

$$= \log_2 \frac{10}{K_2}$$

The odd number denominator K_2 that will make the fraction $10/K_2$ a power of 2 is 5. Solving for b, we have

$$b = \log_2 \frac{10}{5} \qquad (K_2 = 5)$$

$$= \log_2 2$$

$$= 1$$

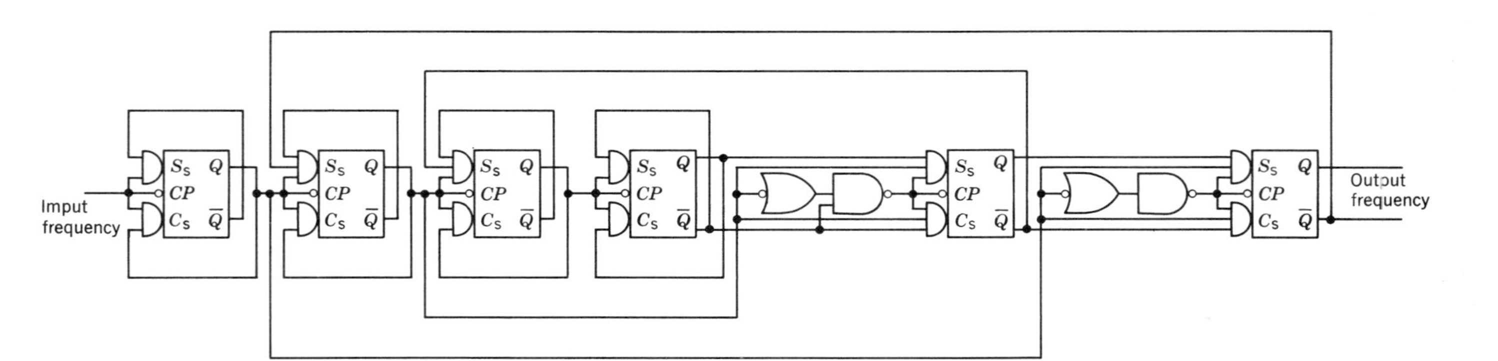

Figure 6.20. Circuit of Example 6.7. The counter has an output frequency period equal to 22 times that of the input frequency.

Therefore, the b component of the divide-by-22 counter has one stage.

Since we have not reached a K of unity yet, the computations continue. Next we must determine component c of the counter.

$$c = \log_2 \frac{K_2 - 1}{K_3} \tag{6.19}$$

Substituting the numerical value of K_2 into the above equation, we obtain

$$c = \log_2 \frac{5 - 1}{K_3}$$

$$= \log_2 \frac{4}{K_3}$$

The odd number K_3 that will make the fraction $4/K_3$ a power of 2 is 1. Solving for c, we have

$$c = \log_2 \frac{4}{1} \qquad (K_3 = 1)$$

$$= \log_2 4$$

$$= 2$$

Therefore, the c component of the divide-by-22 counter has two stages.

Since we have reached a K equal to 1, we may conclude that the desired counter has an a component of 1, a b component of 1, and a c component of 2. The schematic of this counter is shown in Figure 6.20. ▮

6.8 RING COUNTERS

The ring counter is a shift register, the output of which feeds its input directly—that is, the Q output of the last stage of the shift register feeds the synchronous set input and the inverted $\bar{Q}$ output feeds the synchronous reset input of the first flip-flop stage of the shift register. This way, a closed-loop shift register is formed, as shown in Figure 6.21(a). In this circuit the clock triggers all stages simultaneously thus shifting the information contained in each stage to the stage that follows. The ring counter does not generate new information, as do the circuits discussed in the previous sections of this chapter, but it "circulates" the information that was preset into it.

The advantages of the ring counter configuration are: First, speed because there is no cumulative propagation delay; second, simplicity because no decoding circuits are required; and third, fewer components. This last advantage occurs when the ring counter is replacing a regular counter where more than 50% of the counter states are decoded.

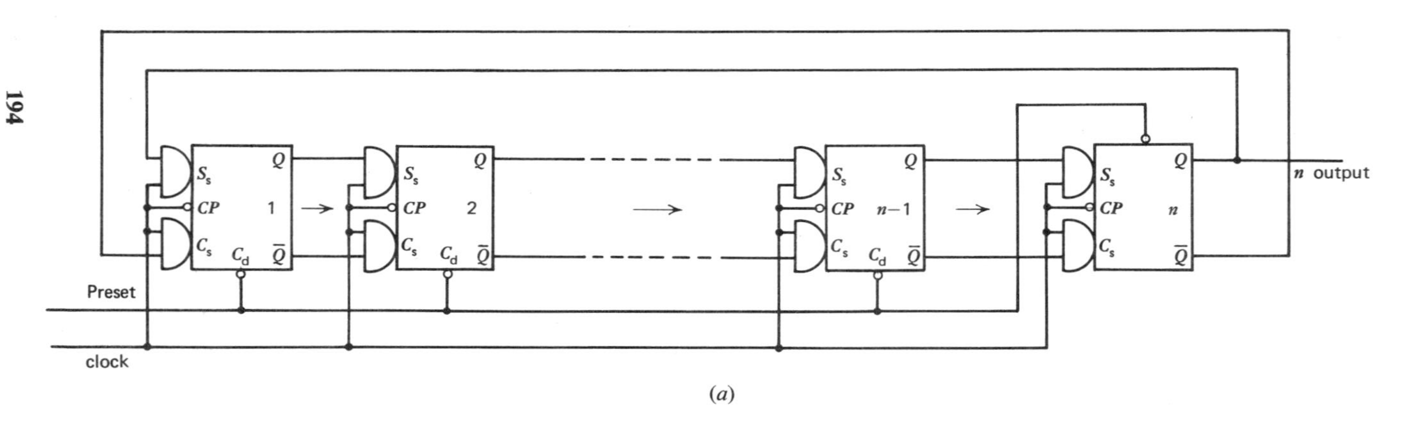
n output
Q
S_s
CP
C_s
C_d
Q̄
1
2
n−1
n
Preset
clock
(a)

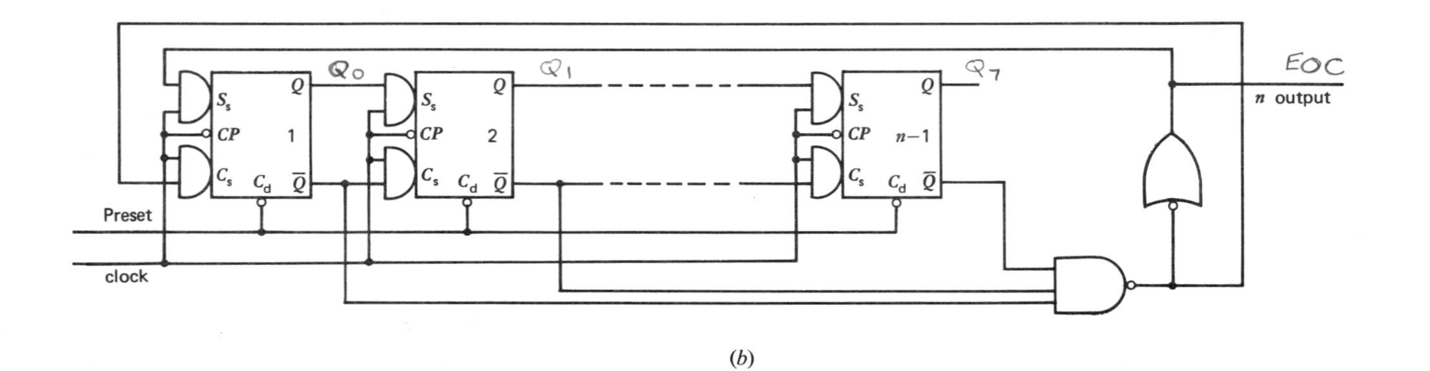
EOC
n output
Q0
Q1
Q7
Q
S_s
CP
C_s
C_d
Q̄
1
2
n−1
Preset
clock
(b)

Outputs after clock pulse	Ring counter outputs 1 2 3 4...$n-1$ n
	0 0 0 0... 0 1
1	1 0 0 0... 0 0
2	0 1 0 0... 0 0
3	0 0 1 0... 0 0
⋮	⋮ ⋮ ⋮ ⋮ ⋮ ⋮
$n-1$	0 0 0 0... 0 0
n	0 0 0 0... 1 0
$n+1$	0 0 0 0... 0 1

(*c*)

Figure 6.21. Ring counters: (*a*) conventional (arrows show the direction of signal flow) and (*b*) modified; and (*c*) truth table.

The disadvantages of the ring counter configuration are: First, there is no binary readout; and second, it needs to be initially preset—that is, a ring counter cannot properly work without being preset. The number of unique states a ring counter may have equals the number of its stages. Ring counters are mainly used when both counting and decoding is required.

Ring counters should be always preset because without presetting their operation is undetermined, and the manner in which the counter flip-flops will change is not predictable. If not preset, the initial output of a four-stage ring counter can be any of the 16 numbers that can be formed with 4 bits (see Fig. 6.22, first column). The generated sequence, after the clock starts "running," is given in the row corresponding to the initial state. If the counter is preset to one of the 16 states, its operation will be predictable.

The typical use of the ring counter is to sequentially enable a group of circuits where each circuit is enabled by one of the outputs of the shift register. In this case, one of the stages of the ring counter is set to 1, and the remaining are set to 0. As the clock triggers the ring counter, the 1 will be transferred to its adjacent stage, thus sequentially enabling the corresponding circuits.

Circuit (*a*) of Figure 6.21 is a conventional ring counter. In addition to the conventional ring counter, there is a modified version. Use of the modified version in many cases results in hardware gains. Circuit (*b*) in Figure 6.21 is the general form of the modified version of a ring counter. In this circuit, the last flip-flop is replaced by a multi-input NAND gate and an inverter, thus forming a noninverting AND gate. This gate produces an output of 1 whenever all inverted outputs of the flip-flops are 1. This output is the same as that of the nth flip-flop of the conventional ring counter.

Initial state	Ring Counter Output				
	After clock pulse 1	After clock pulse 2	After clock pulse 3	After clock pulse 4	After clock pulse 5
0000	0000	0000	0000	0000	0000
0001	1000	0100	0010	0001	1000
0010	0001	1000	0100	0010	0001
0011	1001	1100	0110	0011	1001
0100	0010	0001	1000	0100	0010
0101	1010	0101	1010	0101	1010
0110	0011	1001	1100	0110	0011
0111	1011	1101	1110	0111	1011
1000	0100	0010	0001	1000	0100
1001	1100	0110	0011	1001	1100
1010	0101	1010	0101	1010	0101
1011	1101	1110	0111	1011	1101
1100	0110	0011	1001	1100	0110
1101	1110	0111	1011	1101	1101
1110	0111	1011	1101	1110	0111
1111	1111	1111	1111	1111	1111

Figure 6.22. Output sequences generated by a four-stage ring counter.

The information generated by this noninverting AND gate feeds the first flip-flop in the same way that the output of the nth flip-flop feeds the first flip-flop in the conventional ring counter. The advantage of the modified ring counter over the conventional ring counter is that it is self-starting—that is, even if it is not preset, it will go into its normal sequence after a maximum period of time KT, where K is the number of clock pulses necessary to clear the counter, and T is the clock period. The maximum value for K is the number of flip-flops in the counter.

6.9 SHIFT COUNTERS

The shift counter, like the ring counter, is a shift register with feedback. In the case of the shift counter, the feedback is crossed. Here the Q output of the last flip-flop feeds the synchronous clear of the first flip-flop, and the $\bar{Q}$ output of that flip-flop feeds the synchronous set of the first flip-flop. An n bit shift counter and its truth table are shown in Figure 6.23.

The shift counter is similar to the ring counter in that it requires presetting. When all flip-flops are initially preset to the same state, either all 1 or all 0, the number of different states of the shift counter is twice the number of stages in the counter. Therefore, a shift counter of the configuration of Figure 6.23 always has an even number of states, and the relationship between the number of states and the number of stages is

$$N = 2n$$

where N is the number of states and n the number of stages in the counter.

An eight-state shift counter is illustrated in Figure 6.24. To decode a specific state of a shift counter the uniqueness of that state should be recognized and detected. The third column of the truth table in Figure 6.24 shows the uniqueness of each of the eight states of the four-stage shift counter. If number four, for example, is to be decoded, the nominal outputs of the first and last flip-flops should be ANDed. These two outputs are both 1 only in the fourth state, counting from the 0000 presetting. Similarly, the seventh state can be decoded by detecting the coincidence of the 1 in $\bar{Q}_3$ and Q_4, where $\bar{Q}_3$ is the inverted output of the third flip-flop, and Q_4 is the nominal output of the fourth flip-flop.

The arrangement shown in Figure 6.23 can be used only in the design of a counter having an even number of states. When an odd number of states is necessary, the circuit in Figure 6.25 should be used. This figure illustrates a $(2n - 1)$-state shift counter and its truth table. The equation for the relationship between the number of states and the number of stages for the counter is

$$N = 2n - 1$$

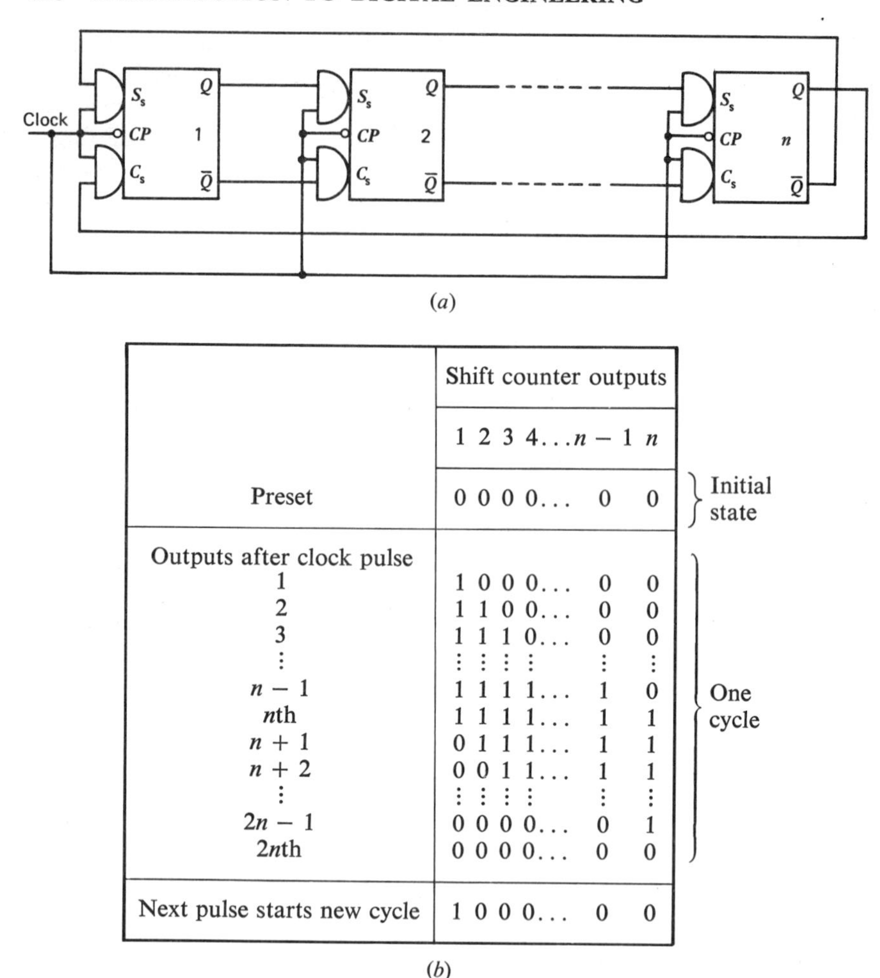

	Shift counter outputs	
	1 2 3 4. . . $n-1$ n	
Preset	0 0 0 0. . . 0 0	Initial state
Outputs after clock pulse		One cycle
1	1 0 0 0. . . 0 0	
2	1 1 0 0. . . 0 0	
3	1 1 1 0. . . 0 0	
⋮	⋮ ⋮ ⋮ ⋮ ⋮ ⋮	
$n-1$	1 1 1 1. . . 1 0	
nth	1 1 1 1. . . 1 1	
$n+1$	0 1 1 1. . . 1 1	
$n+2$	0 0 1 1. . . 1 1	
⋮	⋮ ⋮ ⋮ ⋮ ⋮ ⋮	
$2n-1$	0 0 0 0. . . 0 1	
$2n$th	0 0 0 0. . . 0 0	
Next pulse starts new cycle	1 0 0 0. . . 0 0	

(b)

Figure 6.23. Shift counter with $2n$ states: (*a*) logic diagram; (*b*) truth table.

where N is the number of states and n is the number of stages in the counter.

In this circuit, the first $(n-1)$ stages work as a typical cross-feedback shift counter until the arrival of the $2n-2$ clock pulse. During the $2n-2$ period, immediately after the $2n-2$ clock pulse, the $(n-1)$ stage is 0 and feeds a 1 to the first stage. The first stage, however, does not enter it because its synchronous set input is inhibited by the $\bar{Q}$ output of the nth stage. During that period the only stage in 1 is the nth and at the arrival of the $2n-1$ pulse the nth stage will change to 0 making all outputs 0, which is the initial state.

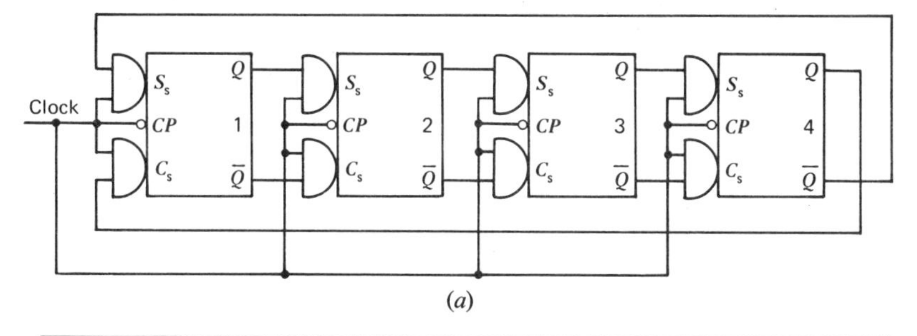

(a)

	Shift counter output	Output for decoding
	1 2 3 4	
Preset	0 0 0 0	$\bar{1}$ $\bar{4}$
After clock pulse		
1	1 0 0 0	1 $\bar{2}$
2	1 1 0 0	2 $\bar{3}$
3	1 1 1 0	3 $\bar{4}$
4	1 1 1 1	1 4
5	0 1 1 1	$\bar{1}$ 2
6	0 0 1 1	$\bar{2}$ 3
7	0 0 0 1	$\bar{3}$ 4
8	0 0 0 0	$\bar{1}$ $\bar{4}$
Start of new cycle	1 0 0 0	1 $\bar{2}$

(b)

Figure 6.24. Eight-state shift counter: (a) logic diagram; (b) truth table.

In the $2n - 1$ period the first stage is not inhibited by stage n anymore and it enters the inverse of $n - 1$ stage which is 1. The arrival of the next pulse will make the output of the first stage 1, thus starting a new cycle.

Example 6.8. If a seven-state shift counter is required, equation $N = 2n - 1$, can be used to determine the number of stages in that counter. Solving for n, we obtain

$$n = \frac{N+1}{2}$$

$$= \frac{7+1}{1} \tag{6.20}$$

$$= 4$$

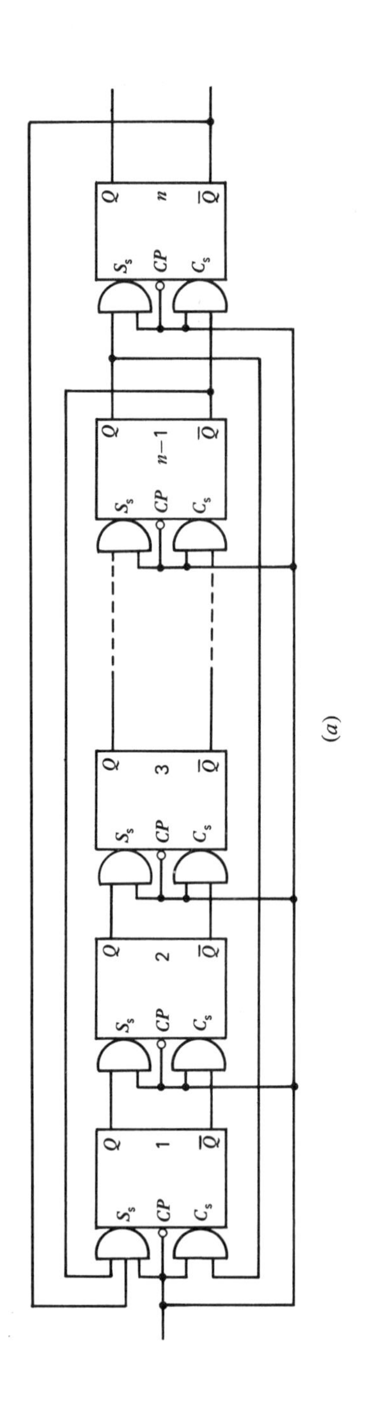

(a)

	Shift counter outputs	
	1 2 3 4. . . $n - 1$ n	
Preset	0 0 0 0. . . 0 0	Initial state
Outputs after clock pulse		One cycle
1	1 0 0 0. . . 0 0	
2	1 1 0 0. . . 0 0	
3	1 1 1 0. . . 0 0	
⋮	⋮ ⋮ ⋮ ⋮ ⋮ ⋮	
$n - 1$	1 1 1 1. . . 1 0	
$n + 1^n$	0 1 1 1. . . 1 1	
$n + 1$	0 0 1 1. . . 1 1	
$n + 2$	0 0 0 1. . . 1 1	
⋮	⋮ ⋮ ⋮ ⋮ ⋮ ⋮	
$2n - 3$	0 0 0 0. . . 1 1	
$2n - 2$	0 0 0 0. . . 0 1	
$2n - 1$	0 0 0 0. . . 0 0	
Next pulse starts new cycle	1 0 0 0. . . 0 0	

(*b*)

Figure 6.25. A $(2n - 1)$-state shift counter: (*a*) logic diagram; (*b*) truth table.

The shift counter should thus have four stages. Figure 6.25 illustrates the general form of a shift counter having an odd number of states, and Figure 6.26 illustrates a three-state, a five-state, and a seven-state shift counter. ▮

The shift counter has the advantage of speed over the serial counter, and the advantage of simplicity over a synchronous counter. Its disadvantage however, in both cases is that the output is not available in binary form and there is an increase in hardware of an additional flip-flop when compared to the series counter.

When shift counters are used in place of serial or synchronous counters in applications where all counter states are decoded, there is a possible hardware gain. The following example illustrates this point.

Example 6.9. Design a 12-state counter with maximum circuit speed where all states of the counter will be individually decoded.

Figures 6.27 and 6.28 show two configurations, the overall function of which is the same. The circuit of Figure 6.27 is a binary counter that counts

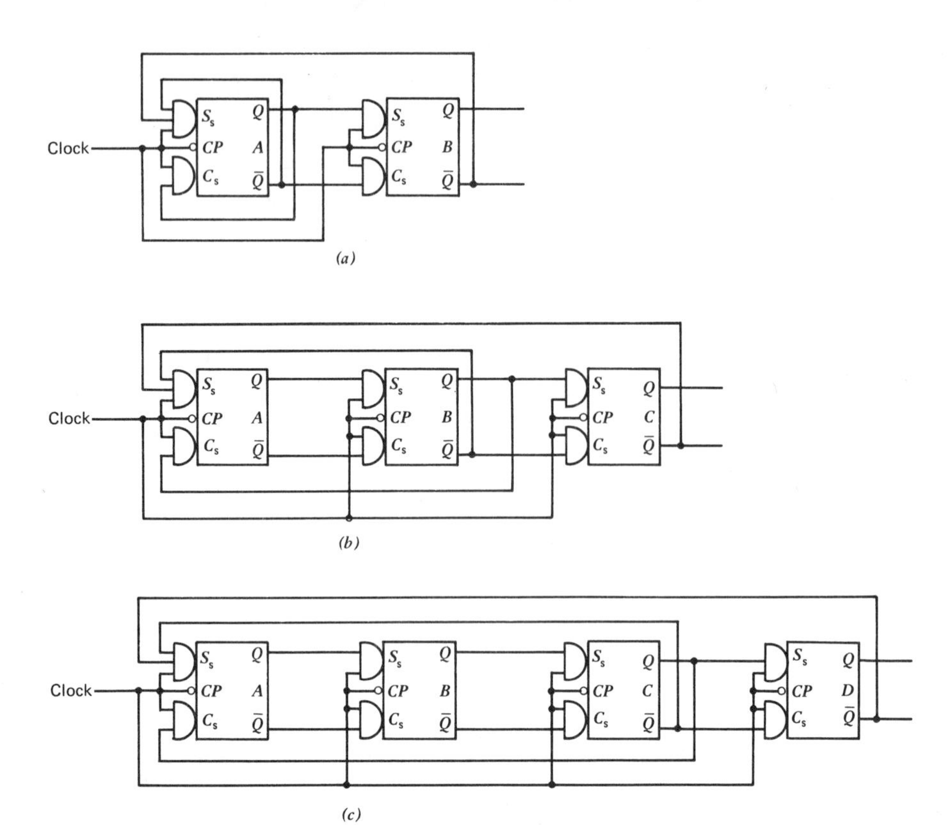

Figure 6.26. (*a*) Three-state shift counter. (*b*) Five-state shift counter. (*c*) Seven-state shift counter.

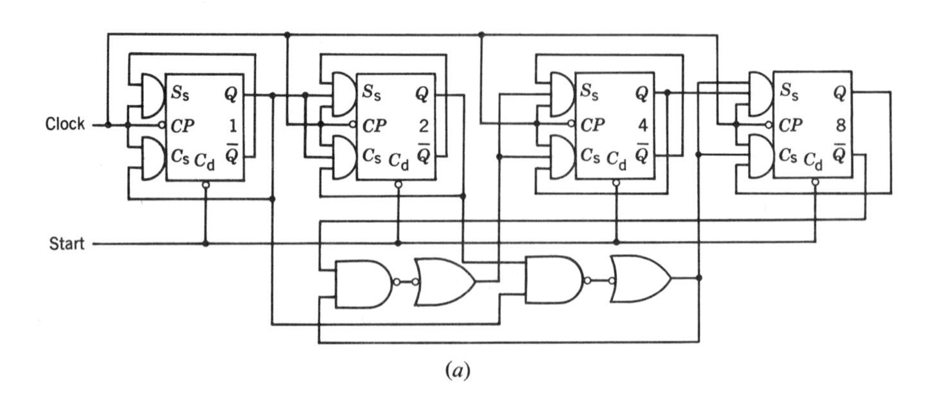

(*a*)

Decimal	Shift counter output 1 2 4 8	Outputs required for decoding
0	0 0 0 0	$\bar{8}\ \bar{4}\ \bar{2}\ \bar{1}$
1	0 0 0 1	$\bar{8}\ \bar{4}\ \bar{2}\ 1$
2	0 0 1 0	$\bar{4}\ 2\ \bar{1}$
3	0 0 1 1	$\bar{4}\ 2\ 1$
4	0 1 0 0	$4\ \bar{2}\ \bar{1}$
5	0 1 0 1	$4\ \bar{2}\ 1$
6	0 1 1 0	$4\ 2\ \bar{1}$
7	0 1 1 1	$4\ 2\ 1$
8	1 0 0 0	$8\ \bar{2}\ \bar{1}$
9	1 0 0 1	$8\ \bar{2}\ 1$
10	1 0 1 0	$8\ 2\ \bar{1}$
11	1 0 1 1	$8\ 2\ 1$
New cycle	0 0 0 0	

(*b*)

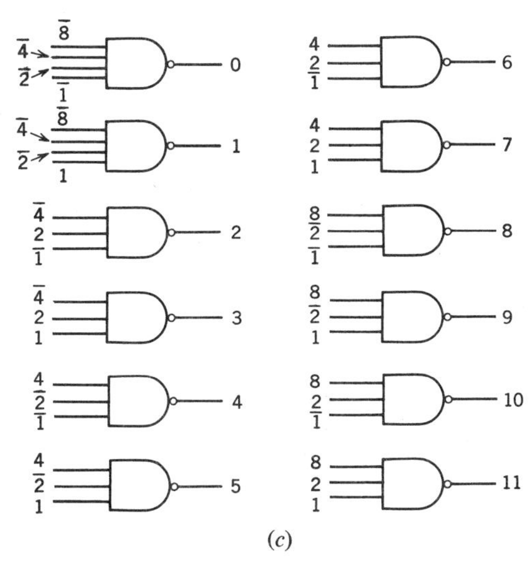

(*c*)

Figure 6.27. Synchronous 12-state counter: (*a*) logic diagram; (*b*) decoding chart; (*c*) decoding gates.

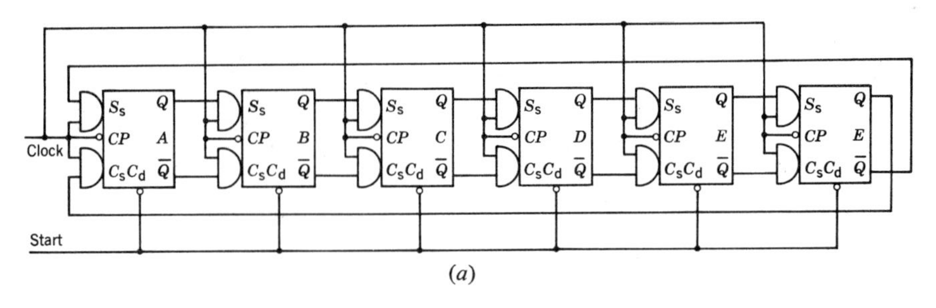

(*a*)

Decimal	Shift counter outputs $A\ B\ C\ D\ E\ F$	Outputs required for decoding
0	0 0 0 0 0 0	$\bar{A}\ \bar{F}$
1	1 0 0 0 0 0	$A\ \bar{B}$
2	1 1 0 0 0 0	$B\ \bar{C}$
3	1 1 1 0 0 0	$C\ \bar{D}$
4	1 1 1 1 0 0	$D\ \bar{E}$
5	1 1 1 1 1 0	$E\ \bar{F}$
6	1 1 1 1 1 1	$A\ F$
7	0 1 1 1 1 1	$\bar{A}\ B$
8	0 0 1 1 1 1	$\bar{B}\ C$
9	0 0 0 1 1 1	$\bar{C}\ D$
10	0 0 0 0 1 1	$\bar{D}\ E$
11	0 0 0 0 0 1	$\bar{E}\ F$
New cycle	0 0 0 0 0 0	

(*b*)

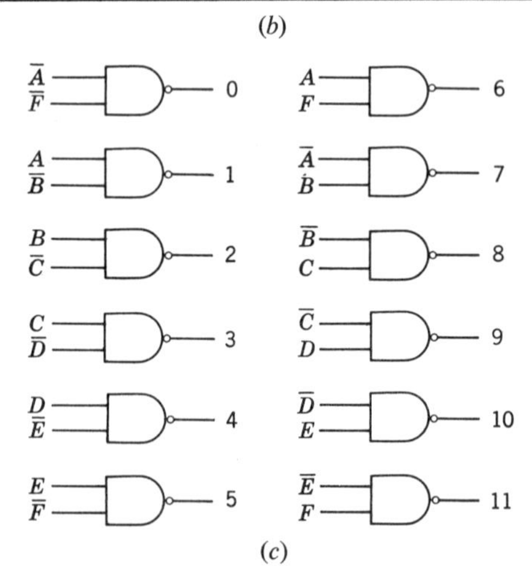

(*c*)

Figure 6.28. A 12-state shift counter: (*a*) logic diagram; (*b*) decoding chart; (*c*) decoding gates.

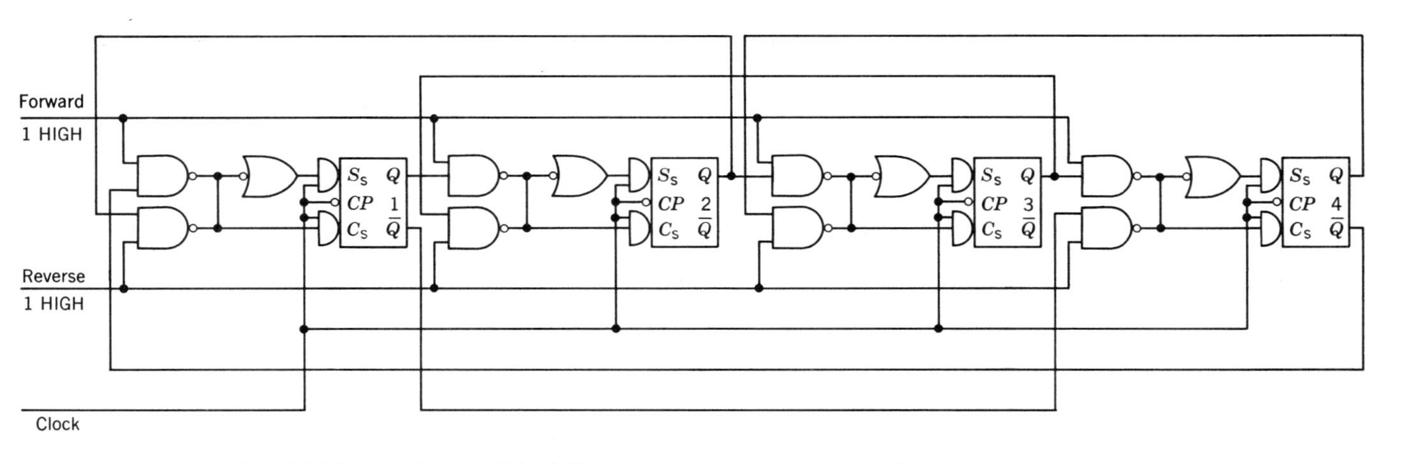

Figure 6.29. Octal (eight-state) reversible shift counter.

from 0 to 11. The output of this counter is a binary number. Its minimum hardware requirements are four master-slave flip-flops, one dual four-input gate, four triple three-input gates, and one quad two-input gate. Four different types of microcircuit packages are needed. The circuit of Figure 6.28 is a 12-state shift counter. This circuit has no carry-propagation delay, and can operate at frequencies higher than those of the circuit of Figure 6.27. Its minimum hardware requirements are six master-slave flip-flops and three quad two-input gates. Therefore only two different types are needed. Although the difference in hardware of one microcircuit package may seem insignificant, it can become a major cost and space consideration if a system uses large numbers of such circuits.

In addition to the increase in hardware, the digital designer should be concerned with the increase in packaging time and volume, and the decrease in system reliability and maintainability that accompany any increase in hardware. In choosing between these two counter circuits, the designer must decide between less hardware in the case of the shift counter, and binary code output in the case of the binary counter. This choice will be determined by whether the output of that counter is needed in binary code elsewhere in the system.

The shift counters are normally unidirectional devices—that is, in these devices information flows only in one direction. In the case of the 12-state shift counter (Fig. 6.28) information flows from flip-flop *A* to flip-flop *B*, from *B* to *C*, from *C* to *D*, from *D* to *E*, from *E* to *F*, and from flip-flop *F* to flip-flop *A*, thus closing the loop. In this configuration, the shift counter counts up. ∎

With some additional circuitry, a shift counter can become reversible. In a reversible shift counter, information may travel from each stage to either of its adjacent stages. An octal eight-stage reversible shift counter is shown in Figure 6.29. This circuit counts up when the "forward" input is HIGH, and down when the "reverse" input is HIGH. It should be noted that for proper operation the forward and reverse direction-control inputs must always be complementary—that is, when one input is HIGH, the other should be LOW.

6.10 GRAY CODE COUNTERS

Gray code is a digital code, similar to the binary code, in which two successive numbers differ only by 1 bit. In binary code, two successive numbers can differ by 1 or more bits.

In Gray code, the bits of the numbers, or states of the stages of the counter, carry no numerical weight as they do in binary code where the first bit has the weight of 1, the second of 2, the third of 4, the fourth of 8, and so on. Gray code is derived from binary code and the two codes have the following relationship:

$$G_n = B_n \oplus B_{n+1} \tag{6.21}$$

$$= B_n\bar{B}_{n+1} + \bar{B}_n B_{n+1} \tag{6.22}$$

where, G_n is the nth-order bit of the Gray number, B_n is the nth-order bit of the binary number, and B_{n+1} is the $(n + 1)$-order bit of the binary number.

Figure 6.30 illustrates the decimal, binary, and Gray codes. It can be seen that when counting from 15 to 16 in binary code, all 5 bits of the number change, while in Gray code, only 1 bit changes. For example, binary number 01111 (15 in decimal) is 01000 in Gray code. This is obtained in the following manner:

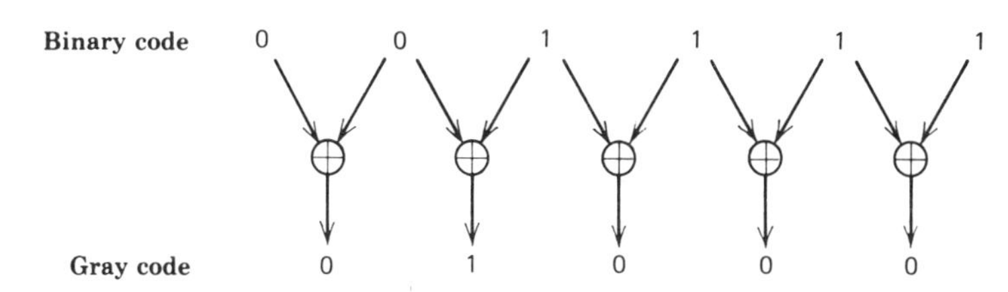

In Figure 6.30 we observe that the extreme right column of the Gray code is similar to that of the weight-2 column of the binary code. With the addition of an auxiliary column of alternating 1s and 0s that starts with 1, the right column of the Gray code can be considered as being the weight-2 bit of a 2 bit binary counter starting at 01. From this observation we may conclude that the bit of the right column of the Gray code must change state each time the auxiliary bit of the preceding row is 1; thus,

$$E_1 = A \tag{6.23}$$

where E_1 is the signal that enables the right column bit of the Gray code to change and A is the nominal output of the auxiliary signal.

We also observe that in the Gray code the bit of the second column from the right changes state whenever the bit of the right Gray column in the preceding row is 1 and the auxiliary bit is 0. The expression for the enable signal of this column then becomes

$$E_2 = G_1\bar{A} \tag{6.24}$$

where G_1 is the right column bit in the Gray code.

From a similar observation we may conclude that in Gray code the bit of

Decimal code	Binary code	Gray code	Auxiliary bit
0	00000	00000	1
1	00001	00001	0
2	00010	00011	1
3	00011	00010	0
4	00100	00110	1
5	00101	00111	0
6	00110	00101	1
7	00111	00100	0
8	01000	01100	1
9	01001	01101	0
10	01010	01111	1
11	01011	01110	0
12	01100	01010	1
13	01101	01011	0
14	01110	01001	1
15	01111	01000	0
16	10000	11000	1
17	10001	11001	0
18	10010	11011	1
19	10011	11010	0
20	10100	11110	1
21	10101	11111	0
22	10110	11101	1
23	10111	11100	0
24	11000	10100	1
25	11001	10101	0
26	11010	10111	1
27	11011	10110	0
28	11100	10010	1
29	11101	10011	0
30	11110	10001	1
31	11111	10000	0

Figure 6.30. Decimal, binary, and Gray codes.

the third column from the right changes state whenever G_2 is 1, G_1 is 0, and A is 0 in the preceding row, resulting in the following expression:

$$E_3 = G_2\overline{G}_1\overline{A} \tag{6.25}$$

The above observations may be expressed by

$$E_n = G_{n-1}\overline{G}_{n-2}\overline{G}_{n-3}\cdots\overline{A} \tag{6.26}$$

where E_n is the signal that enables the Gray code bit G_n to change state. In

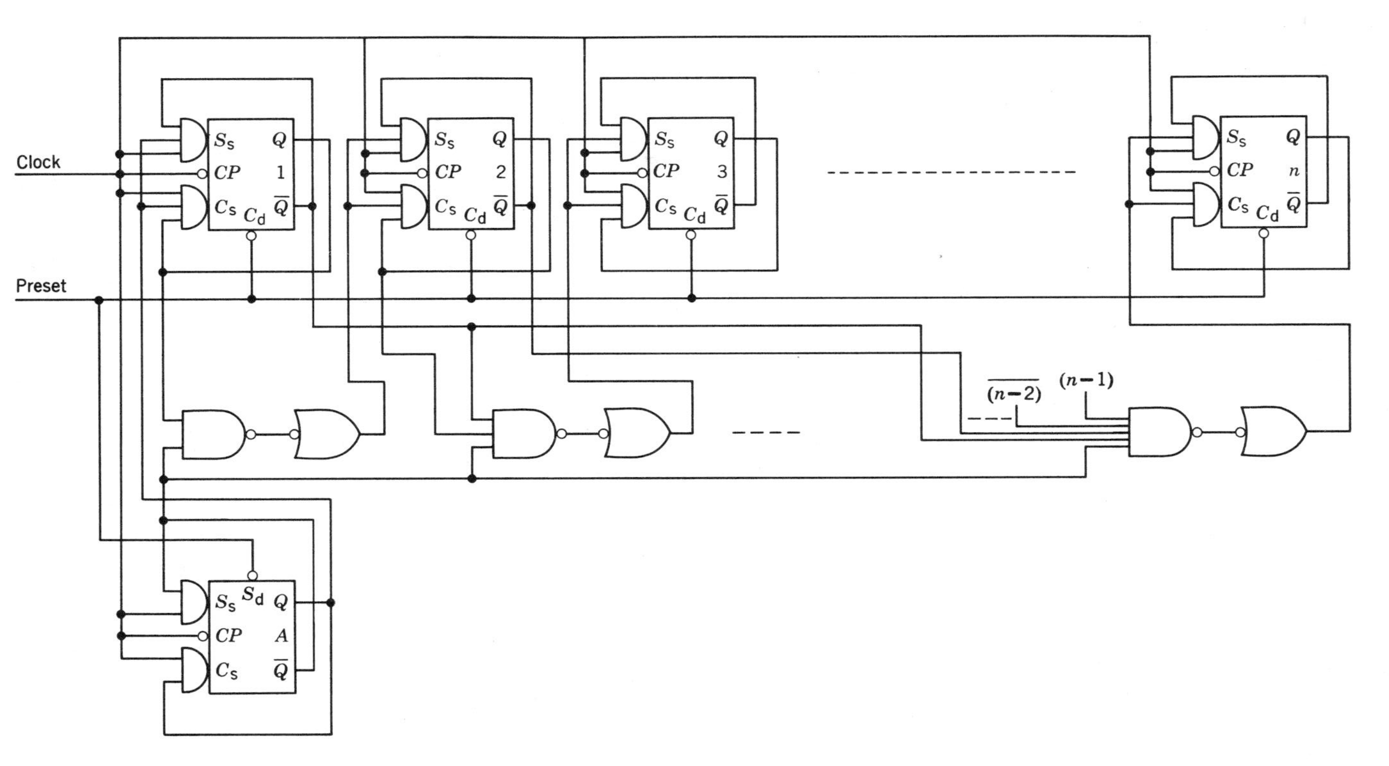

Figure 6.31. General form of a Gray code counter.

general, the nth bit of the Gray code changes state when the $n - 1$ bit is 1 and all preceding bits, including the auxiliary one, are 0.

Based on the above derived information, a digital counter can be designed that counts in Gray code. Figure 6.31 illustrates the general form of such a counter.

6.11 COMPLEMENTS OF A COUNTER OUTPUT

In most binary counters each stage provides a nominal as well as an inverted output. Thus, the 1's complement of the counter output is readily available.

The 2's complement of a counter output OUT_n is the same as the 1's complement of the counter output of one clock period before. Thus, if

$$\text{1's complement of } \mathrm{OUT}_n = \overline{\mathrm{OUT}}_n$$

then,

$$\text{2's complement of } \mathrm{OUT}_n = \overline{\mathrm{OUT}}_n + 1$$

As the counter output increases, its 1's complement decreases. Therefore,

$$\mathrm{OUT}_n = \mathrm{OUT}_{n-1} + 1$$

and,

$$\overline{\mathrm{OUT}}_n = \overline{\mathrm{OUT}}_{n-1} - 1$$

Substituting in the above equation for the 2's complement we have

$$\begin{aligned} \text{2's complement of } \overline{\mathrm{OUT}}_n &= \mathrm{OUT}_n + 1 \\ &= \overline{\mathrm{OUT}}_{n-1} - 1 + 1 \\ &= \overline{\mathrm{OUT}}_{n-1} \end{aligned}$$

Therefore, the 2's complement of a counter output can be obtained from the output of a storage register triggered by the counter clock and fed by the 1's complement of the counter output. Such a configuration appears in Figure 6.32.

6.12 MEDIUM-SCALE INTEGRATION COUNTERS

Digital counting networks are available in MSI mainly in 4-bit modules. Some are pure binary counters, either serial or parallel, while others consist of two parts. In the latter case one stage is used to divide by 2 and the other

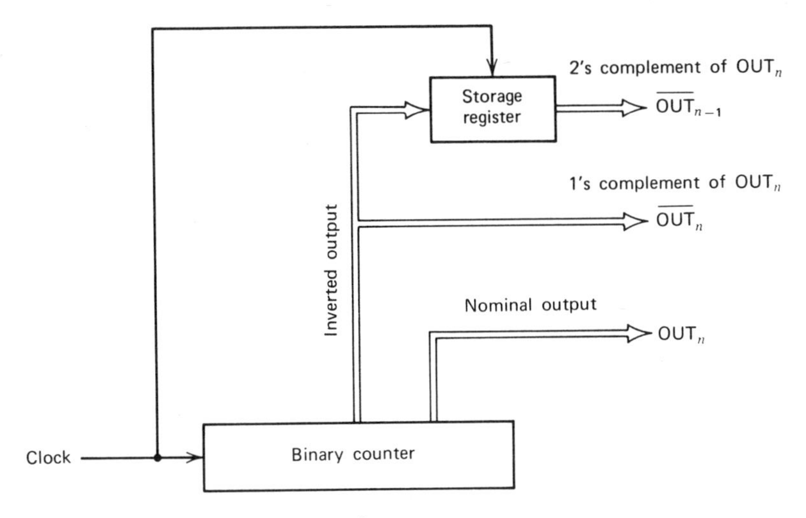

Figure 6.32. Complements of a counter output.

three are connected so that they divide either by 5 or 6. Division by 2 along with division by 5 makes a division by 10, while division by 2 followed by division by 6 makes a divide-by-12 counter. Figure 6.33 illustrates three typical configurations of binary MSI counters.

6.13 APPLICATION OF COUNTERS

Counters are widely used in a variety of applications, but one application that stands out is frequency-scaling. Most digital systems require more than one frequency for their operations. Having a separate generator for each frequency is, however, impractical in most cases.

Normally, only one frequency is generated by an oscillator and the remaining frequencies are provided by properly dividing that frequency. It is, therefore, required that the initial frequency be equal to or greater than the maximum desired frequency of the system.

By dividing a frequency by a number N a new frequency is generated, the period of which is N times longer than the period of the original frequency—that is, for every N pulses or cycles of the original frequency, one pulse or cycle of the new frequency occurs. Such frequency-scaling is widely implemented by means of counters. For example, if a given frequency is to be divided by 16, the binary counter of Figure 6.1 can be used; if it is to be

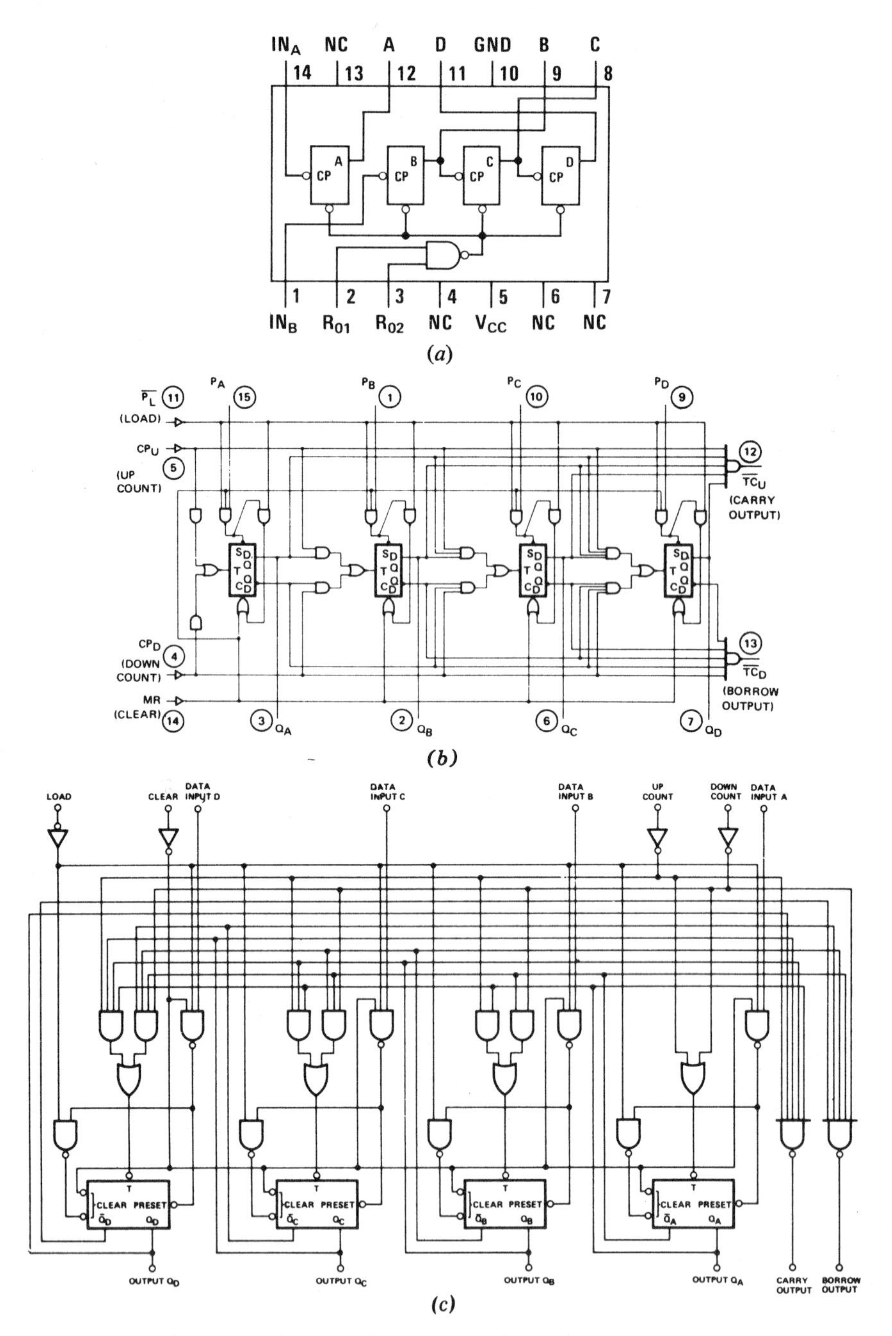

Figure 6.33. Typical configurations of binary MSI counters: (*a*) 4 bit serial counter (courtesy of National Semiconductor); (*b*) 4 bit synchronous counter (courtesy of Signetics); (*c*) 4 bit up-down counter (courtesy of Fairchild Semiconductor).

divided by 17, the feedback counter of Figure 6.10 can be used; or, if a frequency is to be divided by 22, the multiple-feedback counter of Figure 6.20 can be used.

The counters discussed in the preceding sections of this chapter can be used in their present form only to divide a given frequency by an integer. If N is not an integer, other methods must be used to generate the required frequency.

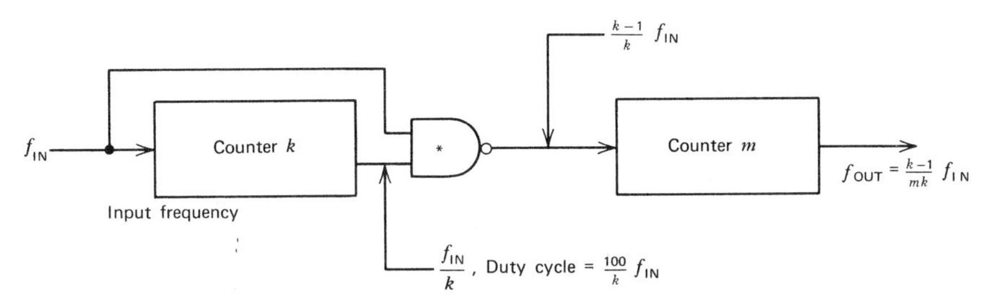

Figure 6.34. General form of a frequency-division counter where the frequency divider is a noninteger number. (*Note:* *In this gate one pulse out of every k pulses is inhibited.)

Figure 6.34 shows a divide-by-N network where N is not an integer. The relationship of output frequency to input frequency in this network is

$$f_{OUT} = \left(\frac{1}{N}\right) f_{IN} \tag{6.27}$$

$$= \left(\frac{k-1}{k}\right)\left(\frac{1}{m}\right) f_{IN} \tag{6.28}$$

where N is the desired noninteger frequency divider, k is the number of states of the first counter, and m is the number of states of the second counter. Terms k and m are, consequently, integers. The fraction $(k - 1)/k$ represents the operation performed by the counter k and the AND gate (Fig. 6.34) is where one pulse out of every k pulses is inhibited. Thus, the frequency at the output of the AND gate is $[(k - 1)/k] f_{IN}$. The fraction $1/m$ indicates the integral division performed by counter m. Division by m is the major division of the frequency-scaling while the fraction $(k - 1)/k$ acts as a correction factor.

To design such a network we proceed as follows: First, we determine N from the equation

$$N = \frac{f_{IN}}{f_{OUT}} \tag{6.29}$$

Having determined N, m which is the closest integer below N can be identified as

$$m < N < m + 1 \tag{6.30}$$

Expressions (6.27) and (6.28) indicate that

$$\frac{1}{N} = \left(\frac{k-1}{k}\right)\left(\frac{1}{m}\right) \tag{6.31}$$

Solving for k, we have

$$k = \frac{N}{N-m} \tag{6.32}$$

With m and k known the desired frequency-scaling network can be designed in more detail. It should be noted that the output of the k counter equals the period of the input frequency, so that it inhibits only one pulse at the AND gate. Thus, for every k pulses that are applied to the AND gate, $k - 1$ go through. This method is illustrated by means of Example 6.10.

Example 6.10. Assume that the frequency of 175 kHz $\pm$ 0.5 kHz is to be generated from a 2 MHz clock. First, we compute divisor N

$$N = \frac{f_{\text{IN}}}{f_{\text{OUT}}} \tag{6.33}$$

$$= \frac{2\text{ MHz}}{175\text{ kHz}}$$

$$= 11.42$$

Next, we determine whether the required frequency can be generated within the allowed tolerance by dividing the input frequency by either of the integers above and below N. These two integers are 12 and 11, respectively. Division by 12 gives 166.66 kHz, and division by 11 gives 181.81 kHz. It is obvious that neither of these frequencies satisfies the requirement of 175 kHz $\pm$ 0.5 kHz. Since division of the input frequency by an integer does not provide the required accuracy, the counters discussed in the preceding chapter cannot be used in their present form.

The approach of Figure 6.34 should be now employed to design a circuit generating 175 kHz $\pm$ 0.5 kHz from 2 MHz. First, the values of m and k are determined. Since the input to output frequency ratio is 11.42, m should be 11 to satisfy the relationship

$$m < N < m + 1 \tag{6.34}$$

The term k can be now computed from equation $k = N/(N - m)$. Substituting numerical values for m and N, and solving for k we obtain

$$k = \frac{11.42}{11.42 - 11}$$

$$= 27.027$$

When k is not an integer, it is rounded off to the nearest integer, which in this case is 27. The k counter should therefore have 27 states. The output of this counter should inhibit the original clock frequency f_{IN} for one cycle out of every 27 cycles.

To determine the accuracy of the designed frequency divider, the produced output frequency is computed using the equation on which the design was based

$$f_{OUT} = \left(\frac{k - 1}{mk}\right) \tag{6.35}$$

Substituting the values of m and k in the above equation, we obtain

$$f_{OUT} = \frac{27 - 1}{(11)(27)} 2 \text{ MHz}$$

$$= \frac{26}{297} 2 \text{ MHz}$$

$$= 174.42 \text{ kHz}$$

If the accuracy obtained is not sufficient, the input frequency must be increased, so that m or k will increase and k will more closely approach an integral value. In this example, however, the output frequency is within the tolerance specified by the requirements and therefore the approach of Figure 6.34 can be used. Figure 6.35 shows the implementation of the above example using general-purpose counters. ▮

The disadvantage of this type of frequency division is that the obtained frequency is not a pure frequency but a simulated one. That is, the pulses of the new frequency are of the correct number, but they are not always spaced the same. Although the overall frequency is correct, close examination of the waveform will show that there is no uniform pulse density. To minimize the nonuniform regions the original clock frequency should be as high as possible. Whether the nonuniform pulse density is immaterial or harmful, of course, depends on the specific application. More counter applications are found throughout the following chapters.

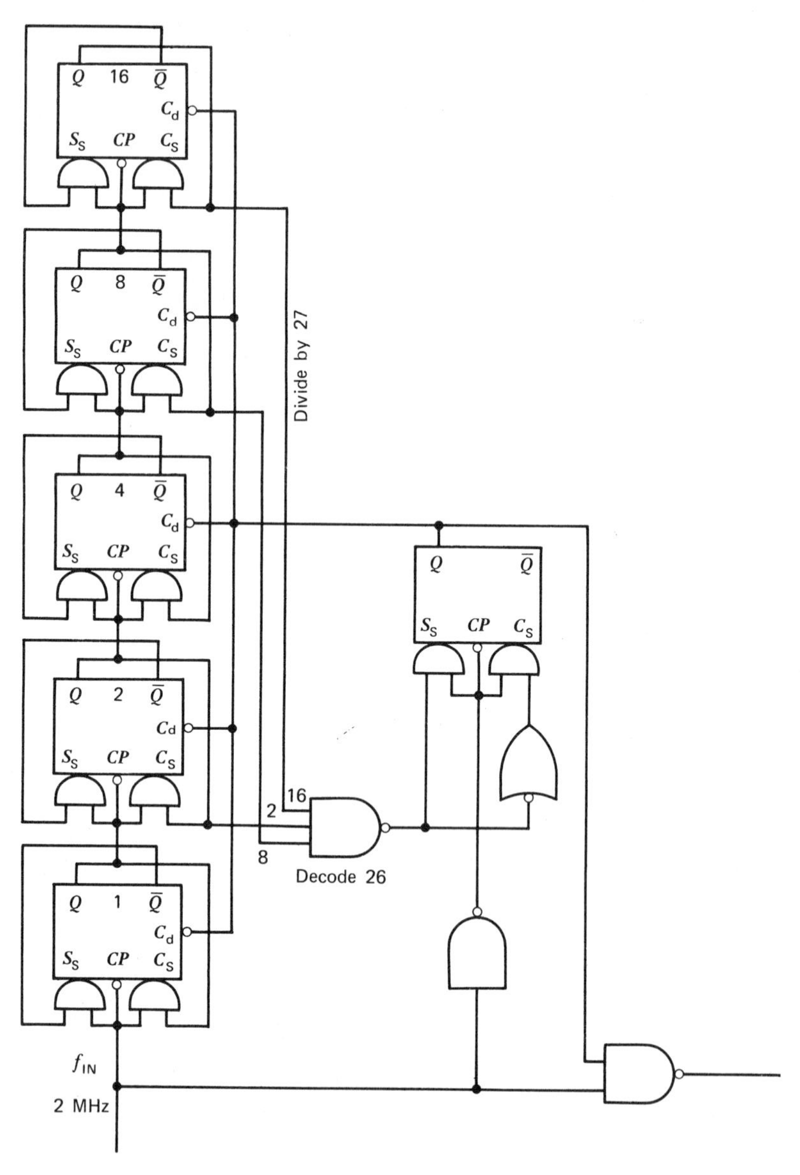

Figure 6.35. Frequency divider generating 175 ± 0.5 kHz from 2 MHz.

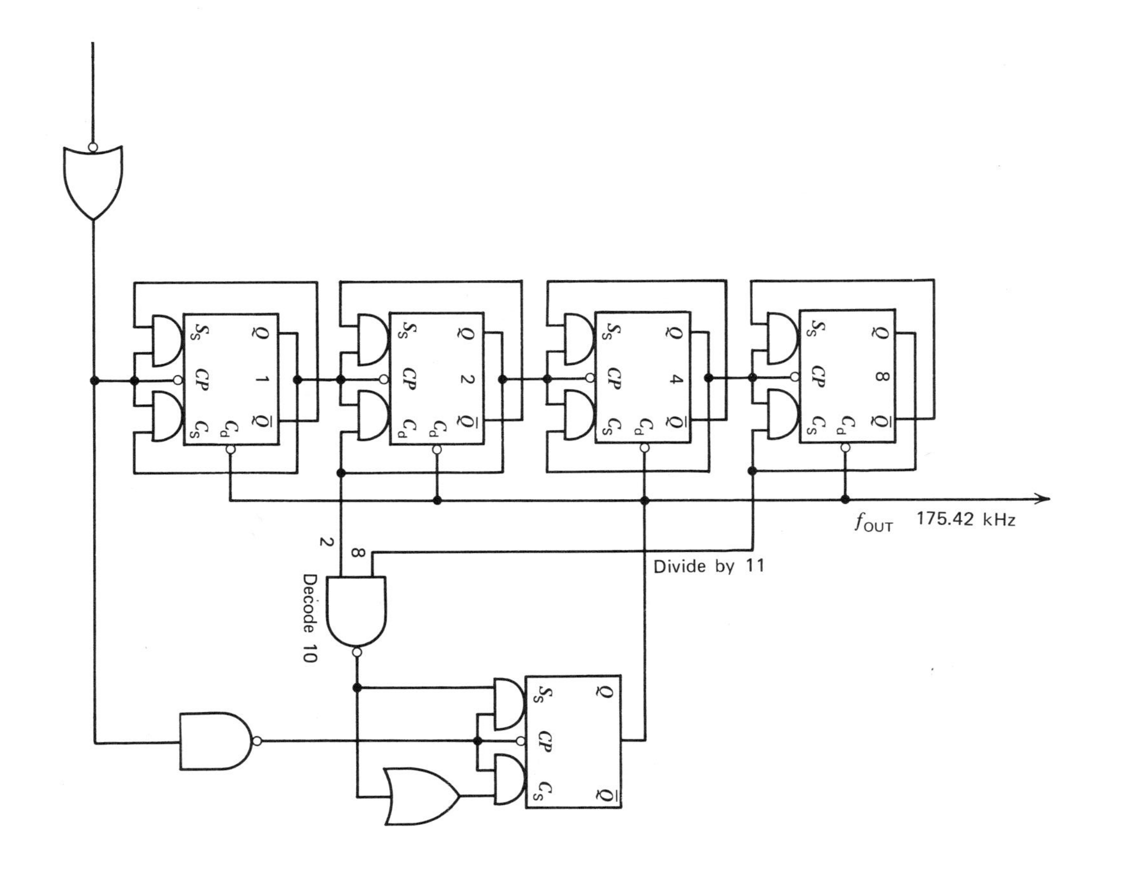

fOUT 175.42 kHz
Divide by 11
Decode 10
Q 8 Q̄
Q 4 Q̄
Q 2 Q̄
Q 1 Q̄
Cd
SS CP CS
8
2

PROBLEMS

1. State the advantages and disadvantages of:
   (*a*) Serial and synchronous counters. (*b*) Serial- and parallel-carry.
   (*c*) Multiple-feedback counters. (*d*) Ring and shift counters.
   (*e*) Gray counters.
2. Using master-slave flip-flops, design a 5 bit series counter, and determine the time it takes the counter to change from number 15 to number 16. Also, compute the time interval for which the counter output is 16 and 31. Assume a flip-flop propagation delay for 30 nsec and a clock frequency of 1 MHz.
3. In a 6 bit series counter, number 32 is to be decoded and provide a minimum pulse width of 600 nsec. If the propagation delay per counter stage is 20 nsec, determine the maximum input frequency at which the counter can be used.
4. Design a 5 bit synchronous counter for a 10 MHz operation using master-slave flip-flops having a propagation delay t_{pdf} of 30 nsec and a read-in time t_{in} of 5 nsec. The gates used have a propagation delay t_{pdg} of 10 nsec. Determine whether a serial-carry or a parallel-carry counter should be used.
5. Determine the maximum frequency of operation of a 6 bit serial-carry synchronous counter employing components with the following characteristics:

$$\text{Flip-flops: } t_{\text{pdf}} = 30 \text{ nsec} \qquad t_{\text{in}} = 15 \text{ nsec}$$
$$\text{Gates: } t_{\text{pdg}} = 20 \text{ nsec}$$

6. Design a counter which counts from number 7 to 23.
7. Design an 81-state multiple-feedback counter.
8. Determine the number of states of a multiple-feedback counter with the following characteristics: $a = 3$, $b = 2$, $c = 1$, $d = 2$. Draw the network.
9. Design a ring counter and a shift counter each having six states, all of which are decoded. Use currently available hardware and compare requirements.
10. Design a frequency dividing network that generates 345 kHz $\pm$ 2 kHz from a 3 MHz clock frequency.

PART TWO

DIGITAL IMPLEMENTATION OF BINARY MATHEMATICS

CHAPTER SEVEN

Addition of Binary Numbers

7.1 INTRODUCTION

The most commonly performed binary operation is addition. Addition of binary numbers is performed on a bit-by-bit basis, and it is governed by the two boolean expressions

$$\begin{aligned} \text{Sum} &= A\bar{B} + \bar{A}B \\ &= A \oplus B \end{aligned} \tag{7.1}$$

$$\text{Carry} = AB \tag{7.2}$$

that is, the sum output is 1 when either A or B is 1 and not both. The carry output, on the other hand, is 1 when both A and B are 1. In the addition of 3 bits, the sum and carry expressions take the form

$$\begin{aligned} \text{Sum} &= A\bar{B}\bar{C} + \bar{A}B\bar{C} + \bar{A}\bar{B}C + ABC \\ &= A \oplus B \oplus C \end{aligned} \tag{7.3}$$

$$\text{Carry} = AB\bar{C} + A\bar{B}C + \bar{A}BC + ABC \tag{7.4}$$

The sum output is 1 when the number of bits in the 1 state is an odd number. In this case, we obtain a sum of 1 when either bit or all 3 bits are 1. The carry output is obtained when 2 or more bits are 1.

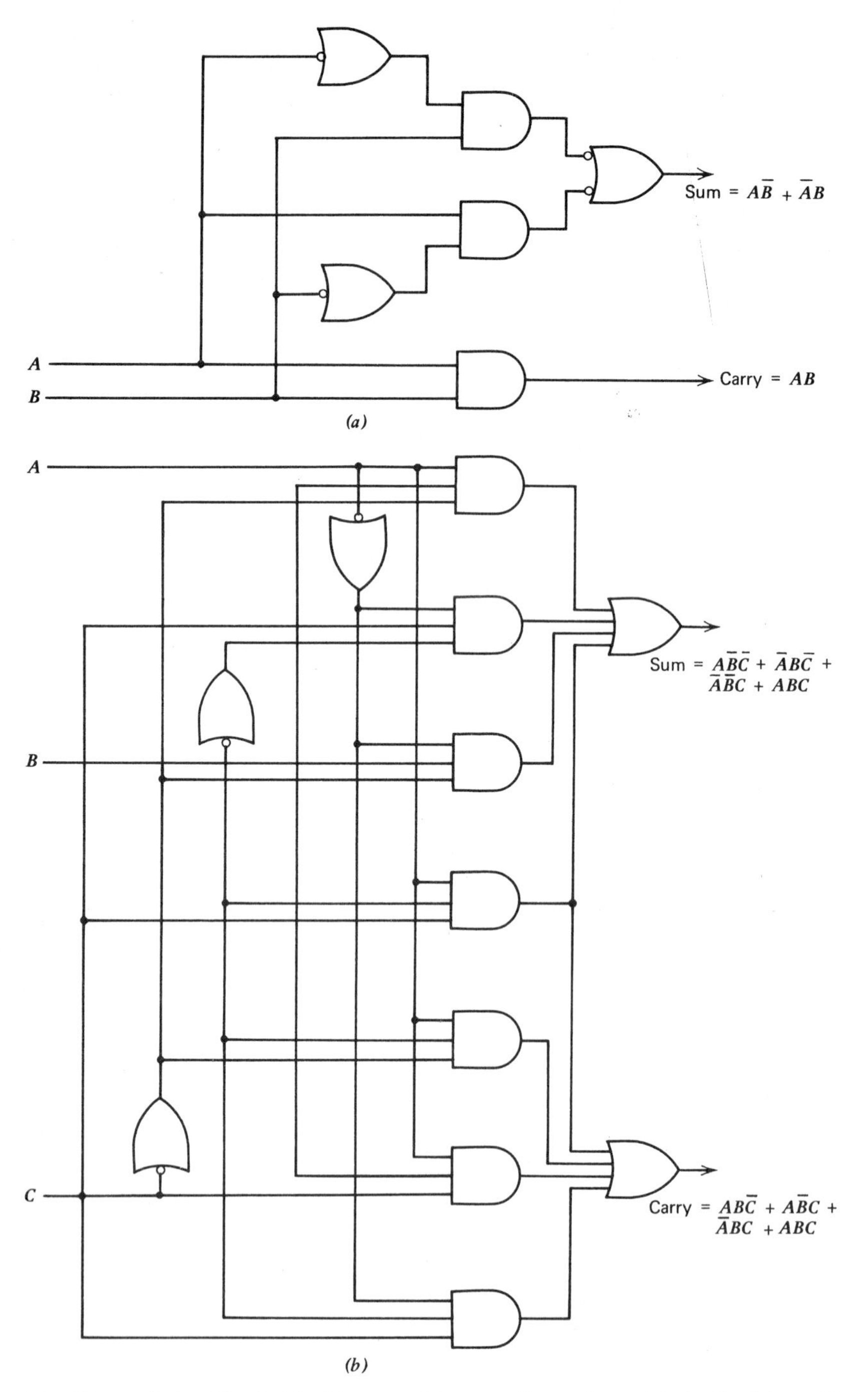

Figure 7.1. Single-bit binary adder configuration: (*a*) half-adder; (*b*) full-adder.

The device that implements (7.1) and (7.2) is called "half-adder" and its logical diagram is illustrated in Figure 7.1(*a*). It is called half-adder because it has only two inputs and does not provide for a carry input.

In the addition of two multibit numbers, the carry of each single bit addition is added to the next-in-significance bit. This process requires that the adder used in that addition handle three inputs. Such a circuit is called a full-adder and satisfies (7.3) and (7.4). The logical diagram of the full-adder is illustrated in Figure 7.1(*b*).

7.2 PARALLEL-PARALLEL ADDITION

Addition of parallel multibit numbers if performed by means of full-adders connected in a chain fashion where the carry produced by each 2 bit addition is added with the next-in-significance bits. Figure 7.2 shows an n bit adder

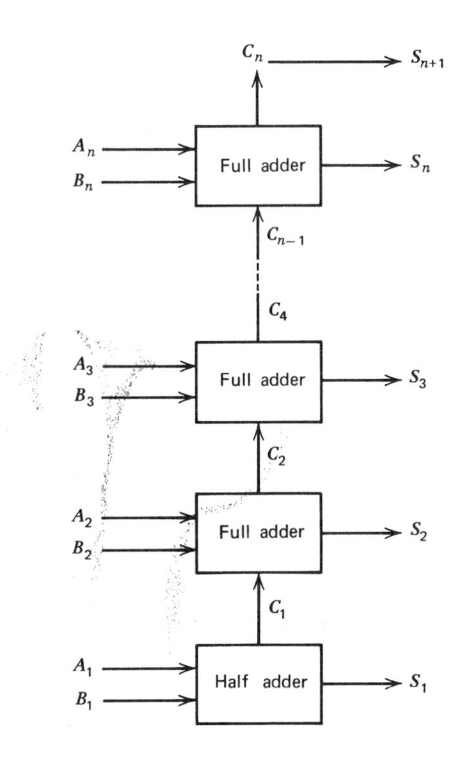

Figure 7.2. General configuration of a parallel n bit adder.

where both input numbers are available in parallel form. The operation performed by the n bit adder is

$$\begin{array}{r} A_n \ldots A_3 A_2 A_1 \\ B_n \ldots B_3 B_2 B_1 \\ \hline S_{n+1}\, S_n \ldots S_3 S_2 S_1 \end{array}$$

The addition of all bits, except that of the LSB, is performed by a full-adder. The addition of the LSB can be performed with a half-adder since there is no carry input from any lower order addition. The carry output resulting from the addition that generates sum S_n can be considered as the S_{n+1} sum output since A_{n+1} and B_{n+1} are 0.

Full-adder circuits are available in integrated form in single-adder, dual-adder, and quad-adder packages. Using these circuits, one does not have to design an adder. In systems where there are not enough adders used, however, it is preferred to build the adders out of gates if this keeps the number of different integrated circuits used to a minimum.

In the adder configuration of Figure 7.2 the output is available after the individual carry signals have propagated through all higher order adders. Because of this ripple type of carry propagation, the adder configuration of Figure 7.2 is called the ripple-carry adder.

The delay due to the carry propagation through the adder stages can be minimized when a carry-look-ahead scheme is used. This technique computes the carry of an adder group independently of the normal binary addition performed in that group. The general expression for the carry can be derived as follows:

First Stage

$$C_1 = A_1B_1 + A_1C_{IN} + B_1C_{IN}$$
$$= A_1B_1 + (A_1 + B_1)C_{IN} \tag{7.5}$$

Second Stage

$$C_2 = A_2B_2 + (A_2 + B_2)C_1 \tag{7.6}$$

Third Stage

$$C_3 = A_3B_3 + (A_3 + B_3)C_2 \tag{7.7}$$

Fourth Stage

$$C_4 = A_4B_4 + (A_4 + B_4)C_3 \tag{7.8}$$

Similarly, the expression for the carry of the addition of the nth stage will be

***n*th stage**

$$C_n = A_nB_n + (A_n + B_n)C_{n-1} \tag{7.9}$$

In (7.9), the term A_nB_n is called carry-generate since it independently generates a carry; and $A_n + B_n$ is called carry-propagate, since it merely propagates carry C_{n-1}. These two terms will be denoted by G_n and P_n, respectively. Employing this notation, (7.5) through (7.9) can be written as

$$C_1 = G_1 + P_1C_{IN} \tag{7.10}$$

$$C_2 = G_2 + P_2C_1$$
$$= G_2 + P_2(G_1 + P_1C_{IN}) \tag{7.11a}$$
$$= G_2 + P_2G_1 + P_2P_1C_{IN} \tag{7.11b}$$

$$C_3 = G_3 + P_3C_2$$
$$= G_3 + P_3[G_2 + P_2(G_1 + P_1C_{IN})] \tag{7.12a}$$
$$= G_3 + P_3G_2 + P_3P_2G_1 + P_3P_2P_1C_{IN} \tag{7.12b}$$

$$C_4 = G_4 + P_4C_3$$
$$= G_4 + P_4(G_3 + P_3[G_2 + P_2(G_1 + P_1C_{IN})]) \tag{7.13a}$$
$$= G_4 + P_4G_3 + P_4P_3G_2 + P_4P_3P_2G_1 + P_4P_3P_2P_1C_{IN} \tag{7.13b}$$

$$C_n = G_n + P_nC_{n-1}$$
$$= G_n + P_n(G_{n-1} + P_{n-1}[G_{n-2} + \cdots + P_2(G_1 + P_1C_{IN})\cdots]) \tag{7.14a}$$
$$= G_n + P_nG_{n-1} + P_nP_{n-1}G_{n-2} + \cdots + P_nP_{n-1}\cdots P_2G_1 + P_nP_{n-1}\cdots P_2P_1C_{IN} \tag{7.14b}$$

In the above equations the carry is expressed in two forms. The first form (equations a) indicates the way the carry is generated inside a ripple-carry adder, and its implementation for 4 bits appears in Figure 7.3(*a*). From this figure it can be seen that in the carry generation there is a cumulative delay which is proportional to the number of bits in the adder.

The second form (equations b) relates the carry directly to the carry-generate and carry-propagate, rather than to the immediately preceding

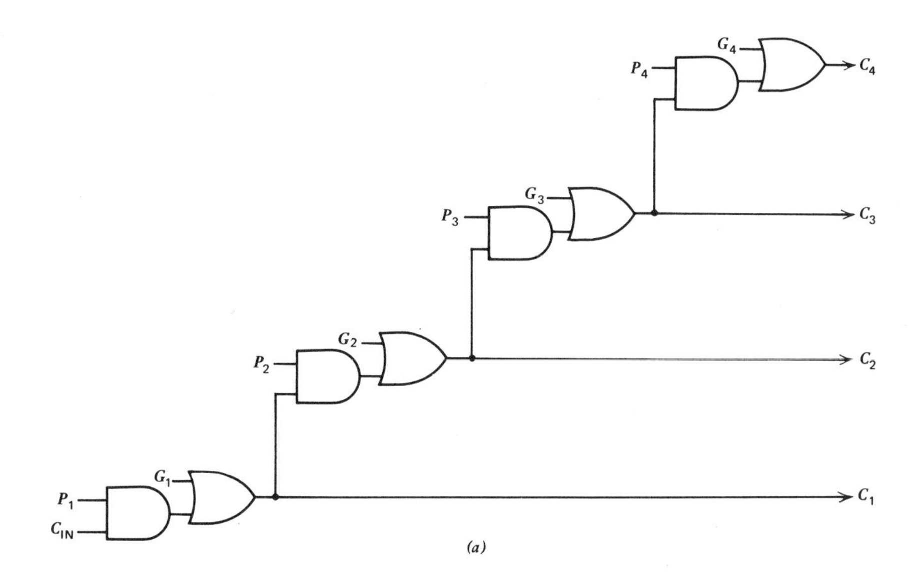

(*a*)

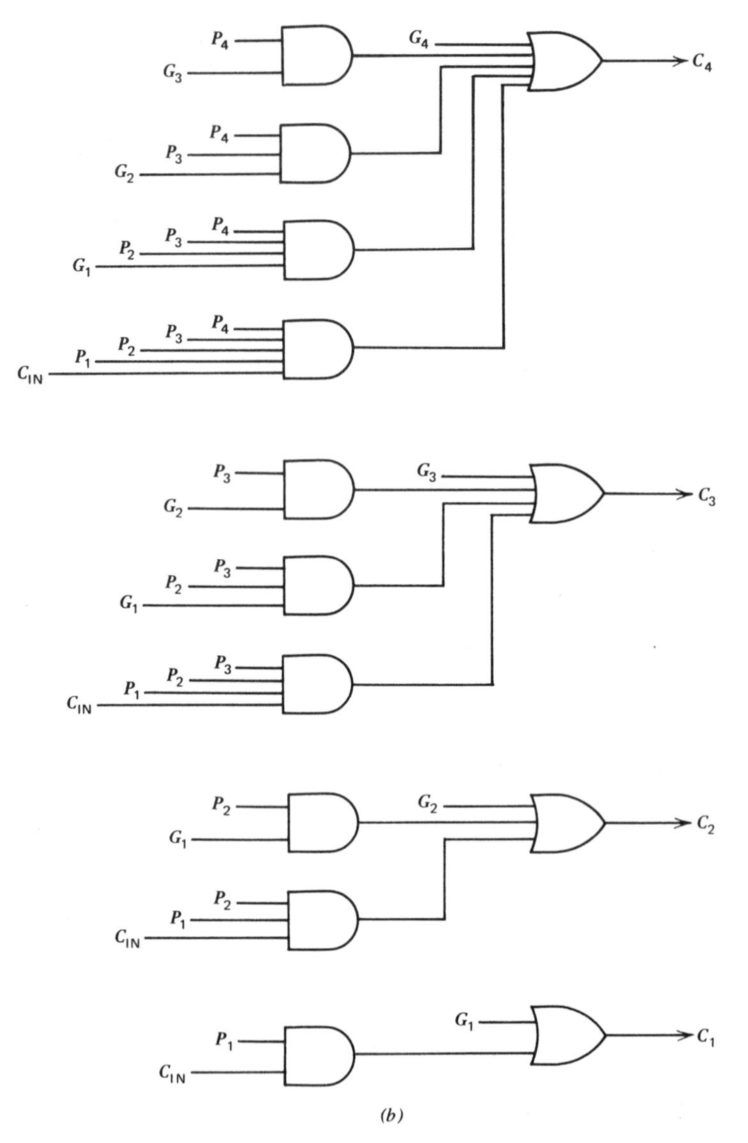

(b)

Figure 7.3. Carry-generation in a 4 bit adder: (*a*) in a ripple-carry configuration; (*b*) in a carry-look-ahead configuration.

carry. The implementation of this approach for a 4 bit addition is illustrated in Figure 7.3(*b*). The generation of any of the four carries takes only two gate delays, while in the carry-ripple scheme the delay varies from two to eight gate delays. A comparison of the two approaches clearly indicates that additional hardware has been the price for speed.

The connection of the carry-look-ahead circuit to the adders is shown in Figure 7.4. Figure 7.5 shows the logical configuration of a full-adder that provides sum, carry-generate, and carry-propagate. The carry-look-ahead circuit of Figure 7.4 is the direct implementation of (7.10)–(7.13).

Equation (7.14b), which is the general expression for the carry, indicates that as the size of the binary adder increases the fan-in requirements of the

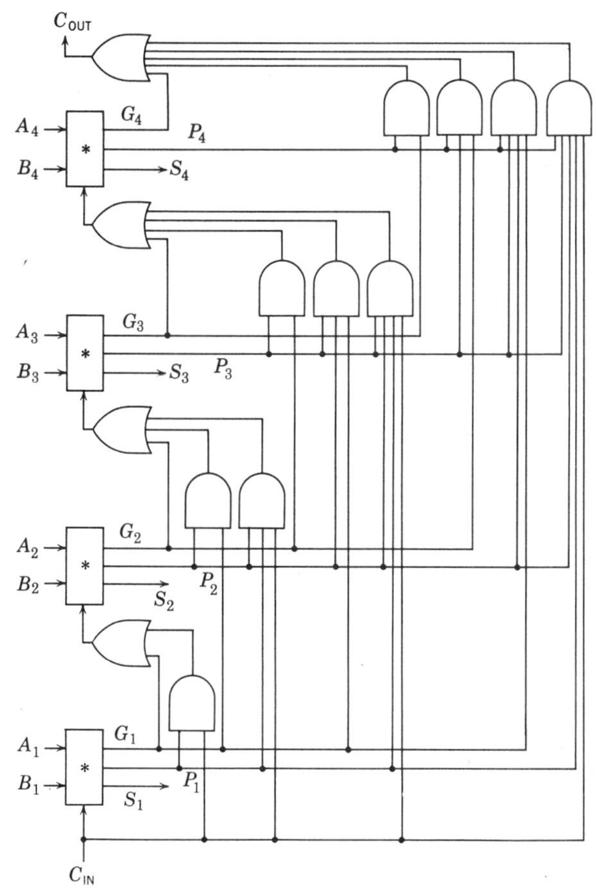

Figure 7.4. A 4 bit adder with carry-look-ahead circuit. (*Note:* * Full-adder providing G and P outputs.)

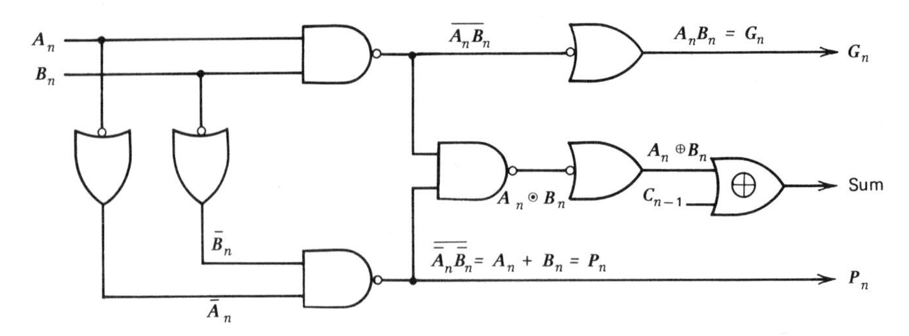

Figure 7.5. Full-adder providing sum, carry-generate, and carry-propagate. It is obvious that the generation of G_n and P_n is independent of C_{n-1}.

AND gates and the fan-out requirements of the G and P inputs considerably increase. To minimize that effect, the binary adder is partitioned into m-adder modules, where each module has its own carry-look-ahead circuit and carries ripple through from module to module, as shown in Figure 7.6. The expressions of the intermodule carries are

$$C_{m1} = (G_4 + P_4G_3 + P_4P_3G_2 + P_4P_3P_2G_1) + (P_4P_3P_2P_1C_{IN}) \tag{7.15}$$

$$C_{m2} = (G_8 + P_8G_7 + P_8P_7G_6 + P_8P_7P_6G_5) + (P_8P_7P_6P_5C_{m1}) \tag{7.16}$$

$$C_{m3} = (G_{12} + P_{12}G_{11} + P_{12}P_{11}G_{10} + P_{12}P_{11}P_{10}G_9) + (P_{12}P_{11}P_{10}P_9C_{m2}) \tag{7.17}$$

$$C_{m4} = (G_{16} + P_{16}G_{15} + P_{16}P_{15}G_{14} + P_{16}P_{15}P_{14}G_{13}) + (P_{16}P_{15}P_{14}P_{13}C_{m3}) \tag{7.18}$$

In each of the above expressions a carry-generate and a carry-propagate part can be identified. By labeling these parts G_m and P_m, respectively, the intermodule carry expressions can be written as

$$C_{m1} = G_{m1} + P_{m1}C_{IN} \tag{7.19}$$

$$C_{m2} = G_{m2} + P_{m2}C_{m1} \tag{7.20}$$

$$C_{m3} = G_{m3} + P_{m3}C_{m2} \tag{7.21}$$

$$C_{m4} = G_{m4} + P_{m4}C_{m3} \tag{7.22}$$

Expressing the intermodule carries in terms of G_m, P_m, and C_{IN}, we have

$$C_{m1} = G_{m1} + P_{m1}C_{IN} \tag{7.23}$$

$$C_{m2} = G_{m2} + P_{m2}(G_{m1} + P_{m1}C_{IN}) \tag{7.24}$$

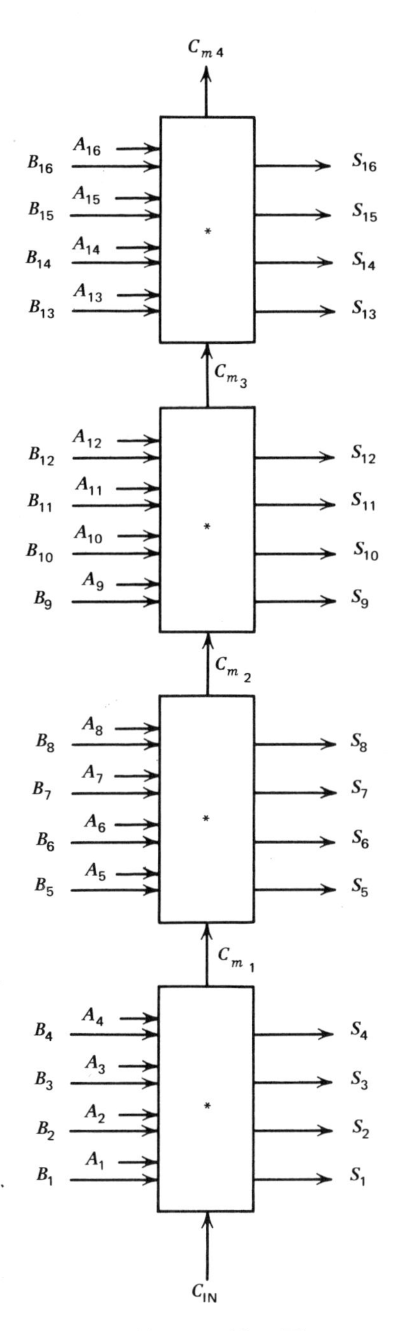

Figure 7.6. Four-module binary adder. The carry ripples from module to module. (*Note:* *4 bit adder with carry-look-ahead.)

$$C_{m3} = G_{m3} + P_{m3}[G_{m2} + P_{m2}(G_{m1} + P_{m1}C_{IN})] \quad (7.25)$$

$$C_{m4} = G_{m4} + P_{m4}(G_{m3} + P_{m3}[G_{m2} + P_{m2}(G_{m1} + P_{m1}C_{IN})]) \quad (7.26)$$

It is the direct implementation of these expressions that takes place when the carry ripples through modules as in Figure 7.6. In this case, the delay in the intermodule carry-generation varies from two gate delays in C_{m1}, to eight gate delays in C_{m4}.

The carry-look-ahead concept used to propagate the carry between full-adders can be equally used to propagate carry between adder modules. Expanding (7.23)–(7.26) we have

$$C_{m1} = G_{m1} + P_{m1}C_{IN} \quad (7.27)$$

$$C_{m2} = G_{m2} + P_{m2}G_{m1} + P_{m2}P_{m1}C_{IN} \quad (7.28)$$

$$C_{m3} = G_{m3} + P_{m3}G_{m2} + P_{m3}P_{m2}G_{m1} + P_{m3}P_{m2}P_{m1}C_{IN} \quad (7.29)$$

$$C_{m4} = G_{m4} + P_{m4}G_{m3} + P_{m4}P_{m3}G_{m2} + P_{m4}P_{m3}P_{m2}G_{m1} + P_{m4}P_{m3}P_{m2}P_{m1}C_{IN} \quad (7.30)$$

These equations are identical to (7.10)–(7.13b) with the only difference being the subscript m. Thus, the carry-look-ahead circuit of Figure 7.3(*b*) developed for (7.10)–(7.13b) can be equally applied for the implementation of (7.27)–(7.30).

Figure 7.7 illustrates a four-module binary adder employing carry-look-ahead at the bit adder as well as at the module level. In this configuration the propagation time of the intermodule carries is four gate delays, two for the generation of the respective G_m and P_m and two for the generation of C_m.

In the 16 bit adder of Figure 7.7 incorporation of carry-look-ahead reduces the addition time from 33 gate delays to only 6. Justifiably, the concept of carry-look-ahead is widely used in applications where speed considerations are of prime importance.

Parallel addition of more than two numbers is accomplished in stages. The first stage consists of partial addition, while the stages that follow perform complete additions.

Partial addition performs bit per bit addition of three numbers without rippling through the produced carry. This operation is performed with a multibit adder that does not have interadder carry connections and it is called carry-save addition.

The output of this configuration is two binary numbers, one consisting of the carry outputs and the other of the preliminary sum outputs. These numbers are then added in a fully interconnected multibit binary adder to produce the sum of the three initial numbers. Figure 7.8(*a*) illustrates this technique where the addition of three 8 bit numbers is implemented. For

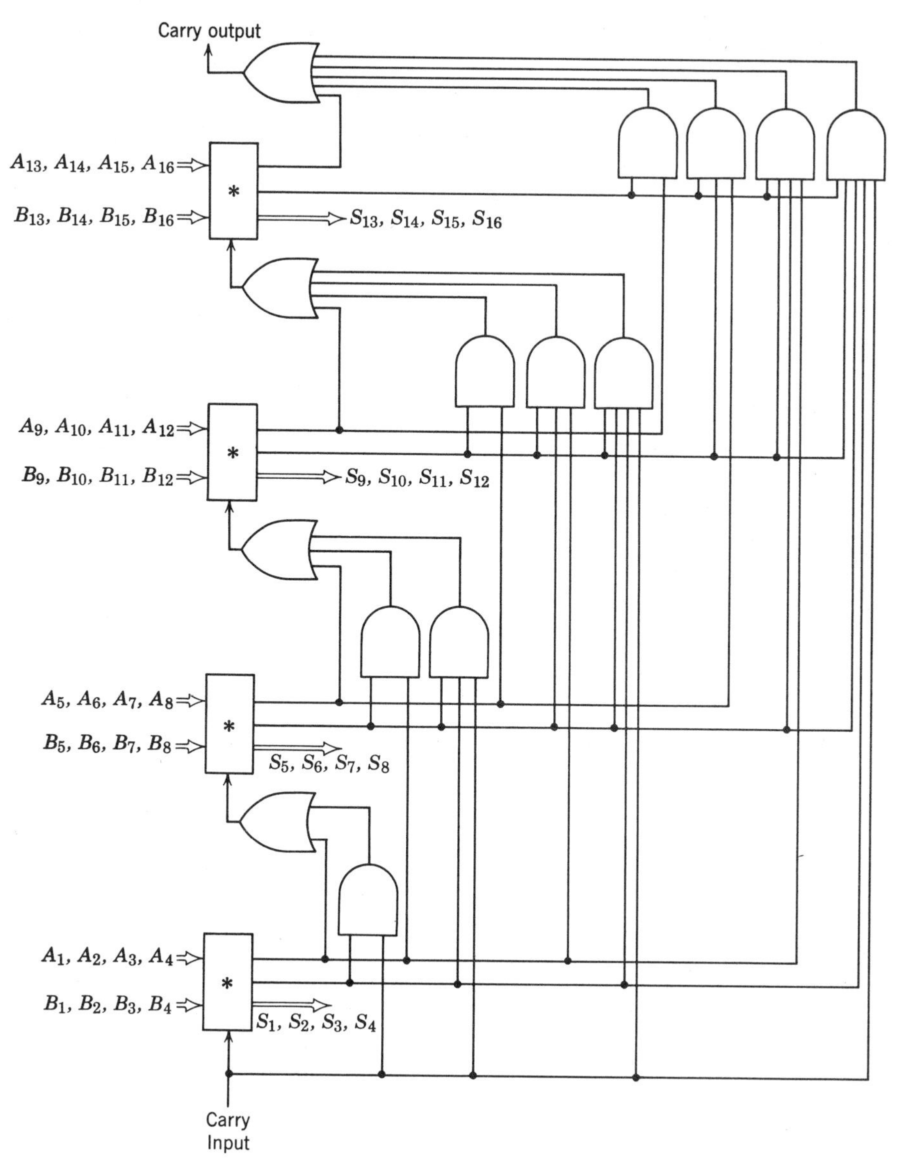

Figure 7.7. A 16 bit adder employing two levels of carry-look-ahead. (*Note:* *4 bit adder with carry-look-ahead that provides G_m and P_m outputs.)

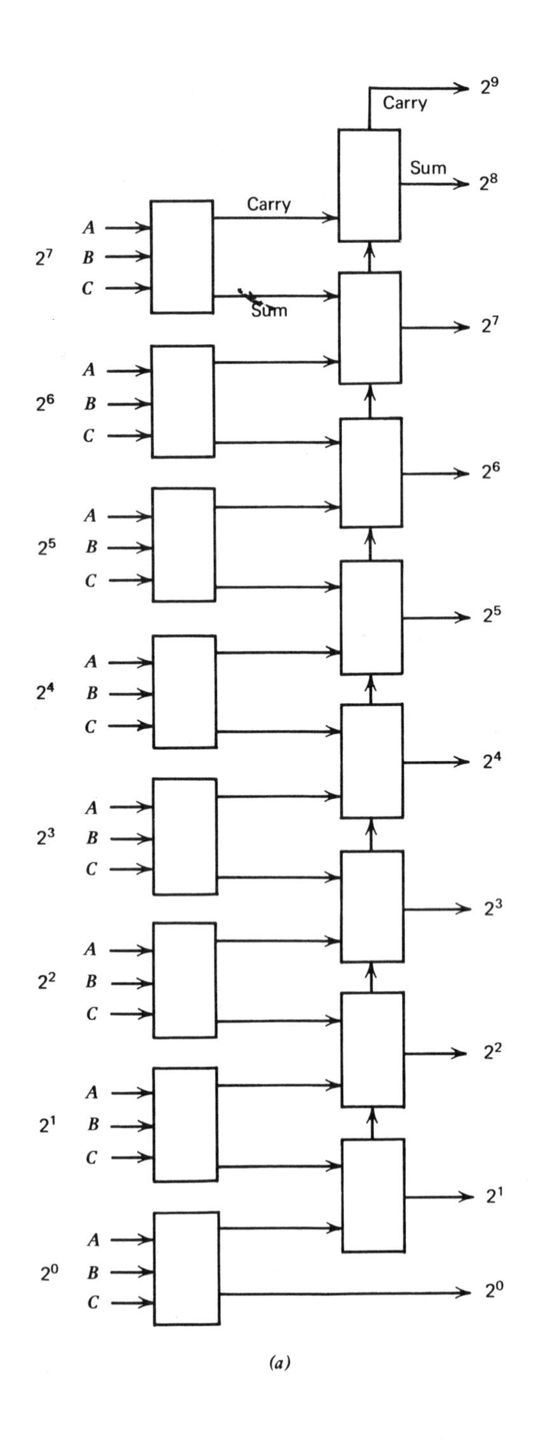

Carry
2^9
Sum
2^8
Carry
A
2^7
B
C
Sum
2^7
A
2^6
B
C
2^6
A
2^5
B
C
2^5
A
2^4
B
C
2^4
A
2^3
B
C
2^3
A
2^2
B
C
2^2
A
2^1
B
C
2^1
A
2^0
B
C
2^0

(a)

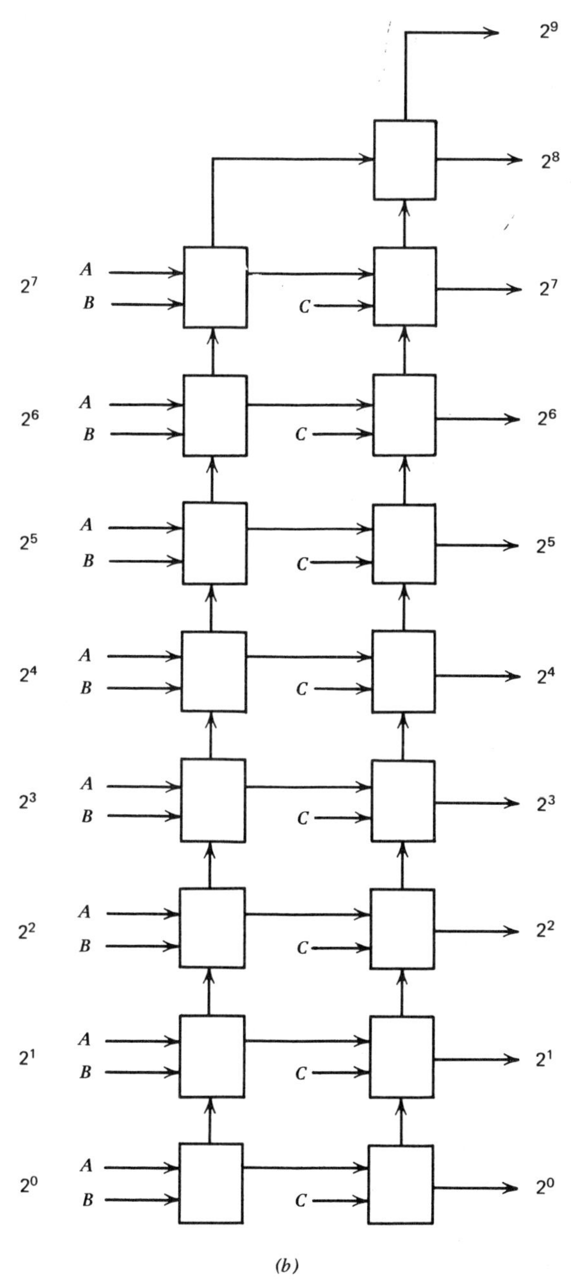

(b)

Figure 7.8. Methods of three number addition: (*a*) three number addition employing the carry-save technique; (*b*) conventional three number addition. (*Note:* Each block represents a binary adder.)

comparison, Figure 7.8(*b*) shows the conventional three-number addition. The advantage of this approach is simplicity and slightly less hardware, with the disadvantage being that no intermediate sum is available. The need for such a sum is usually a rare case. Considerable savings in hardware are realized when adders with carry-look-ahead are used. In this case, the first stage of the adder does not need carry-look-ahead circuits because no carry-propagation takes place.

7.3 SERIAL-SERIAL ADDITION

Addition of multibit serial numbers is performed using a full-adder and a flip-flop. The concept used is that the bits of the two numbers to be added are made available to the adder synchronously and in ascending order of significance—that is, the serial numbers must be available with their LSB first, their MSB last, and with those in between ordered so that the binary weight of the available bits increases with time.

After the addition of the two LSBs a sum and a carry are produced. The sum output is the LSB of the sum of the two input numbers and its binary weight is 2^0. The carry output produced from the addition of the two LSBs has a binary weight of 2^1, and it is stored in the flip-flop to be added in the next clock period to the next bits. The flip-flop must be cleared before each serial addition.

The produced sum is available in serial form with the LSB first and, if it is not used as soon as it becomes available, it must be stored in a shift register. From the shift register, the sum may be fed back into the adder to be added with a third serial number. Care must be exercised so that the size of the selected shift register is enough to hold the generated serial sum or sums.

The number of stages of the shift registers should be one more than the number of bits of the largest of the two input numbers. The additional stage is for the carry that results from the addition of the MSBs of the two numbers. A binary adder for the addition of two serial numbers is illustrated in Figure 7.9.

7.4 PARALLEL-SERIAL ADDITION

The case often arises in the design of digital systems where two numbers must be added where one is in parallel while the other is in serial. This requirement can be accommodated by means of a serial adder and a shift register. Figure 7.10 illustrates this configuration.

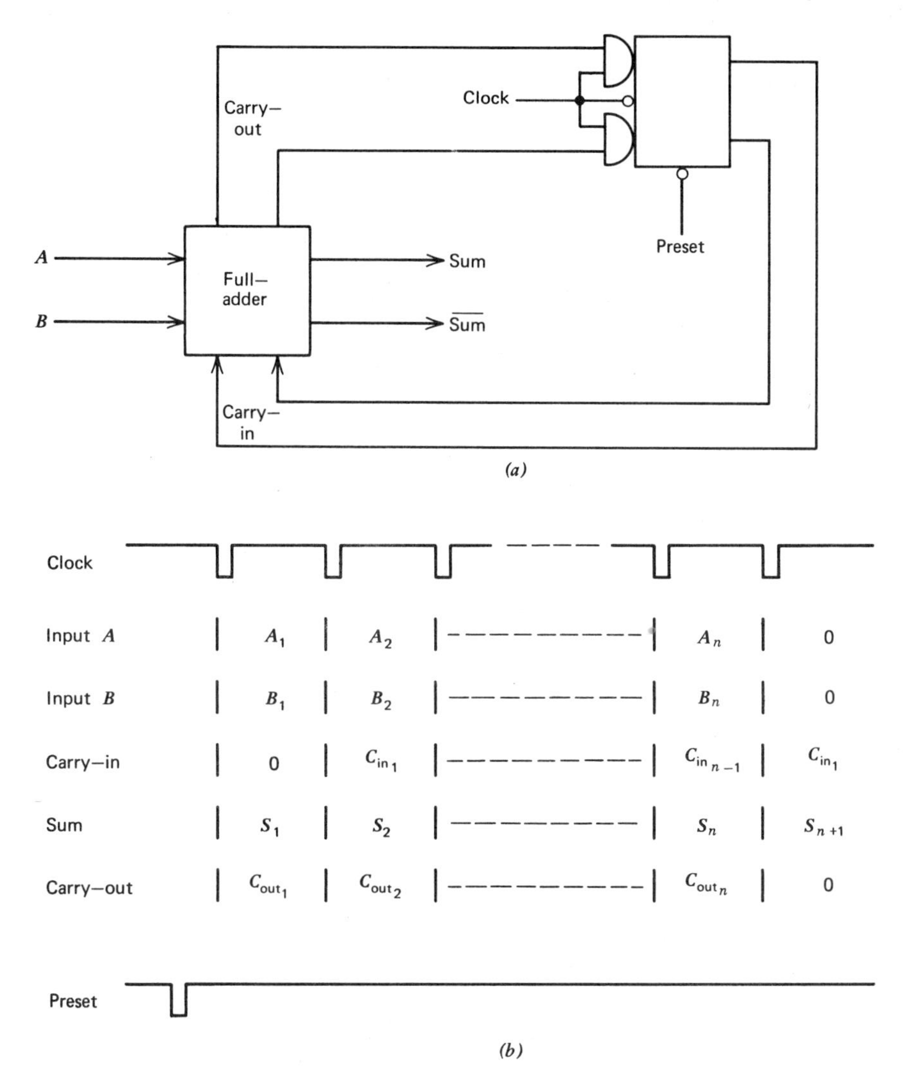

Figure 7.9. Serial adder: (*a*) logic diagram; (*b*) timing diagram.

The parallel number is applied to the shift register and it becomes available to the adder 1 bit at a time in synchronization with the serially-inputted number. After the addition, the produced sum is in the shift register. The sum will also be available in serial form at the output of the adder immediately after it is produced.

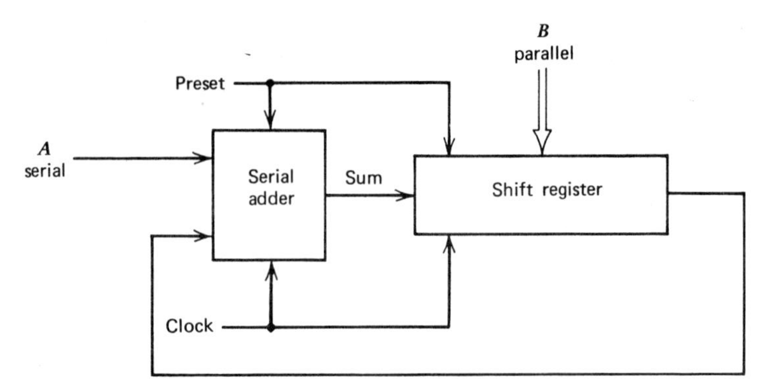

Figure 7.10. Binary adder for the addition of a parallel number to a serial number.

7.5 PARALLEL PULSE-TRAIN ADDITION

Binary addition can be also performed by means of binary counters. In this case, the numbers to be added enter the counter one after the other at the clock input. For proper addition, these numbers must be in pulse-train form, where the number of pulses represents the magnitude of the number.

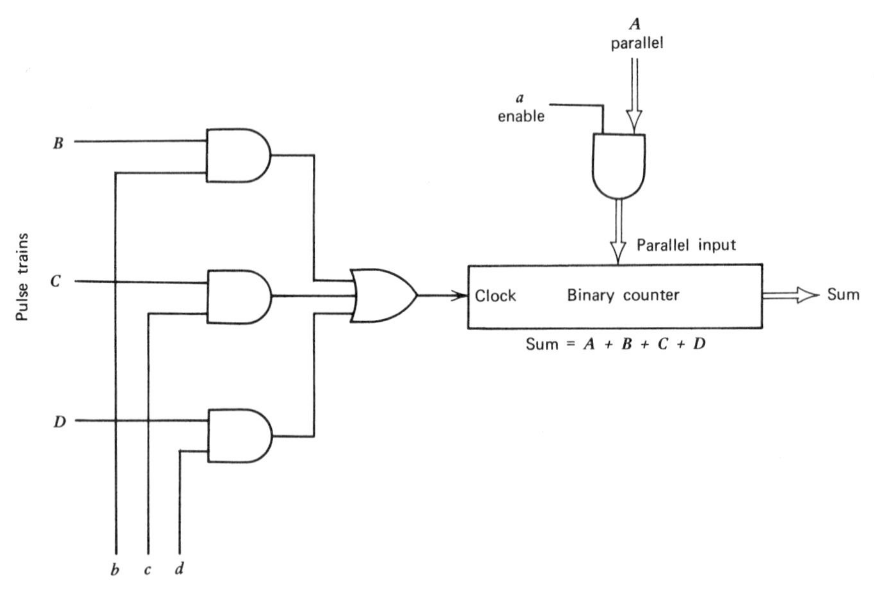

Figure 7.11. Binary counter performing the addition $A + B + C + D$ where A is in parallel form and B, C, and D, are in pulse-train form.

One number may be available in parallel, in which case it is preset into the counter by means of the counter's parallel input. Figure 7.11 illustrates a binary counter that performs addition $A + B + C + D$, where A is in parallel form and B, C, and D are in pulse-train form. Parallel input A must be preset into the counter first and inputs B, C, and D should follow. It should be noted that the enable-A signal must be released before the pulse-train inputs are enabled; otherwise, the counter will be fixed at A and will

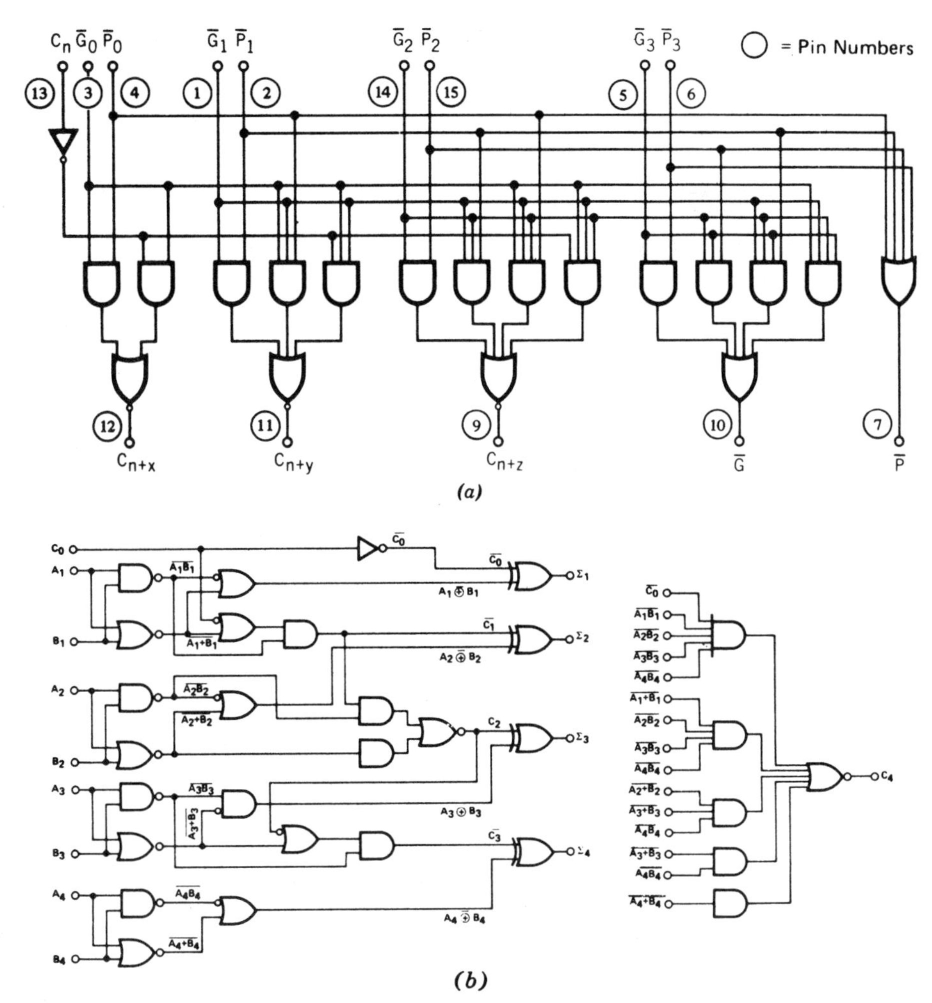

Figure 7.12. MSI circuits for binary addition: (*a*) carry-look-ahead generator (courtesy of Fairchild Semiconductor); (*b*) 4 bit binary full-adder with carry-look-ahead (courtesy of Signetics).

not be affected by the input pulse trains. The sum that results from the addition is available in parallel form at the output of the counter.

7.6 MEDIUM-SCALE INTEGRATION ADDERS

The convenience of MSI is also extended to adders as well as to carry-look-ahead circuits. Adders can be found in 1, 2, or 4 bit packages with and without the carry-look-ahead feature. Figure 7.12 illustrates typical binary adders and carry-look-ahead circuits available in MSI form.

PROBLEMS

1. Design a full-adder using standard TTL hardware.
2. State the advantages and disadvantages of carry-look-ahead adders.
3. Using the carry-look-ahead concept, design a logical network that determines whether the sum of two 6 bit numbers is less than 64.
4. Design a carry-look-ahead circuit that services a 6 bit adder.
5. Design a 9 bit carry-look-ahead adder that consists of three 3 bit adder modules. Employ two levels of carry-look-ahead.
6. Design a logical network that will accumulate (compute the summation of) ten 6 bit binary numbers. The numbers are available at different time intervals, all in parallel form.
7. State the advantages and disadvantages of the adder illustrated in Figure 7.8(*a*).
8. Design a logical network that computes the summation of a wattage reading once every hour. The reading is a 7 bit serial number where the LSB represents 0.5 W. The measured energy is not expected to exceed 2 kWh for the monitored interval.
9. Design a logical network that simultaneously adds three serial numbers.
10. Design a logical network that adds numbers A, B, and C, where A and B are 7 bit serial numbers and C is a 6 bit parallel number.

CHAPTER EIGHT

Subtraction of Binary Numbers

8.1 INTRODUCTION

Binary subtraction is a very commonly performed operation in digital systems. Subtraction $A - B$, where A and B are single-bit numbers, is governed by the following two expressions:

$$\textbf{Difference} = A\bar{B} + A\bar{B} \tag{8.1}$$
$$= A \oplus B$$
$$\textbf{Borrow} = \bar{A}B \tag{8.2}$$

that is, the difference is 1 when 1 of the 2 bits A and B is 1 while the other is 0, which is the same as the computation of the sum of two single-bit numbers. The borrow is 1 only when the minuend A is 0 and the subtrahend B is 1. This operation is called half-subtraction, and it does not consider borrow inputs that result from the subtraction of lower order bits. Its implementation is shown in Figure 8.1.

In the subtraction of multibit binary numbers, however, a borrow input which is generated in the subtraction of the bits of immediately lower significance must be considered along with the minuend and subtrahend bits. That is, in the subtraction of multibit numbers (full-subtraction) 3 bits are

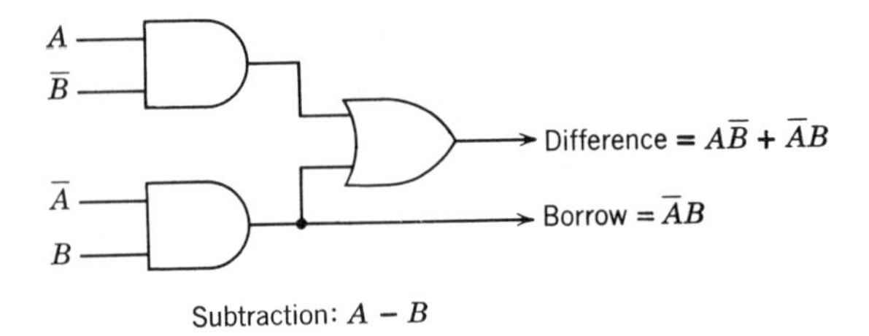

Figure 8.1. Logical diagram of a half-subtracter.

involved—namely, the minuend A, the subtrahend B, and the borrow input which is added to the subtrahend. Full-subtraction results in a difference and a borrow output, and it is governed by the following two expressions:

$$\textbf{Difference} = \bar{A}\bar{B}B_{\text{IN}} + A\bar{B}\bar{B}_{\text{IN}} + \bar{A}B\bar{B}_{\text{IN}} \times ABB_{\text{IN}} \tag{8.3}$$

$$= A \oplus B \oplus B_{\text{IN}}$$

$$\textbf{Borrow} = \bar{A}\bar{B}B_{\text{IN}} + \bar{A}B\bar{B}_{\text{IN}} + \bar{A}BB_{\text{IN}} + ABB_{\text{IN}} \tag{8.4}$$

where B_{IN} stands for borrow input. Here, we see that the difference is 1 any time the number of inputs in the 1 state is odd, which was also the case with the sum bit of the full-addition. The borrow is 1 any time binary sum $B + B_{\text{IN}}$ is greater than A. Figure 8.2 illustrates the logic diagram of a full-subtracter.

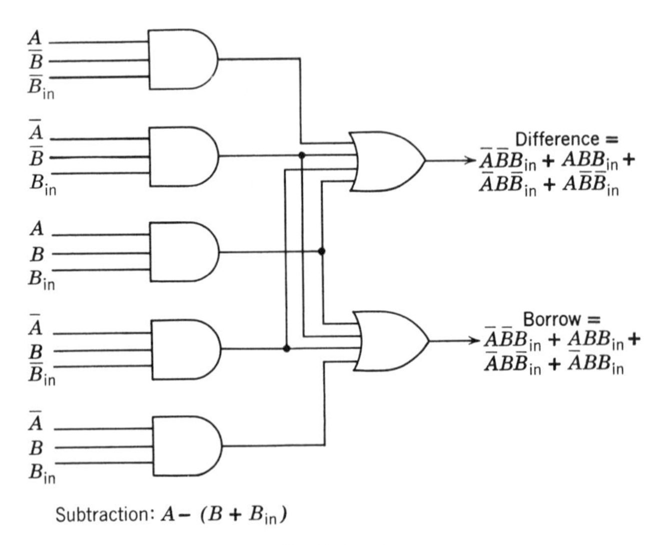

Figure 8.2. Logic diagram of a full-subtracter.

8.2 PARALLEL-PARALLEL SUBTRACTION

A multibit subtracter can be formed by using a full-subtracter for each single-bit subtraction. Figure 8.3 shows an n bit subtracter where all inputs are available in parallel form. The operation performed is

$$\begin{array}{rl} & A_n \ldots A_3 A_2 A_1 \\ & B_n \ldots B_3 B_2 B_1 \\ \hline (\text{Borrow})_n & D_n \ldots D_4 D_2 D_1 \end{array}$$

The borrow output that results from the subtraction that generates the difference D_n indicates the polarity of the obtained difference. When that borrow output is 0, the difference is positive and when it is 1, the difference is negative.

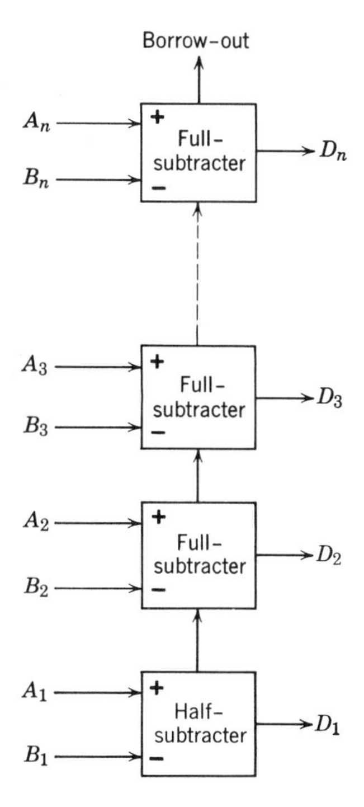

Figure 8.3. An n bit subtracter. (*Note:* When the borrow-out is 1, the difference is negative and in 2's complement form; when it is 0, the difference is positive.)

Subtraction of binary numbers can be also achieved indirectly by addition. This can be done by considering the following relationship:

$$-(B_n \ldots B_3 B_2 B_1) = (\bar{B}_n \ldots \bar{B}_3 \bar{B}_2 \bar{B}_1) + (0 \ldots 001) \tag{8.5}$$

This relationship means that a negative number can be expressed as its l's complement plus one LSB. The validity of (8.5) can be proven as follows: Bringing all terms on one side, we obtain

$$(0_n \ldots 0_3 0_2 0_1) = (\bar{B}_n \ldots \bar{B}_3 \bar{B}_2 \bar{B}_1) + (B_n \ldots B_3 B_2 B_1) + (0 \ldots 001) \tag{8.6}$$

Performing the addition of the first two terms of the right side we have

$$\begin{array}{r} B_n \ldots B_3 B_2 B_1 \\ \underline{\bar{B}_n \ldots \bar{B}_3 \bar{B}_2 \bar{B}_1} \\ 1 \;\ldots 1\;1\;1 \end{array}$$

Substituting sum into (8.6), we obtain

$$(0_n \ldots 0_3 0_2 0_1) = (1 \ldots 1\;1\;1) + (0 \ldots 001) \tag{8.7}$$

Performing the addition indicated on the right side of (8.7), we have

$$\begin{array}{rr} & 0_n \ldots 1_3 1_2 1_1 \\ & \underline{0_n \ldots 0_3 0_2 1_1} \\ (1_{n+1}) & 0_n \ldots 0_3 0_2 0_1 \end{array}$$

Substituting this into (8.6), we obtain

$$0_n \ldots 0_3 0_2 0_1 = 1_{n+1}\, 0_n \ldots 0_3 0_2 0_1 \tag{8.8}$$

The binary system considered here has been assumed to have n bits and the generated $n + 1$ bit carries no significance. Therefore, the original assumption of (8.5) is valid. The implementation of this approach does not require subtracters, but only adders. Figure 8.4 shows an n bit subtracter that uses full-adders and performs the following addition:

$$\begin{array}{rr} & A_n \ldots A_3 A_2 A_1 \\ & \bar{B}_n \ldots \bar{B}_3 \bar{B}_2 \bar{B}_1 \\ & \underline{0 \;\ldots 0\;0\;1} \\ (C_n) & S_n \ldots S_3 S_2 S_1 \end{array}$$

which is equivalent to the subtraction

$$\begin{array}{rr} & A_n \ldots A_3 A_2 A_1 \\ & \underline{B_n \ldots B_3 B_2 B_1} \\ (\text{Borrow}) & D_n \ldots D_3 D_2 D_1 \end{array}$$

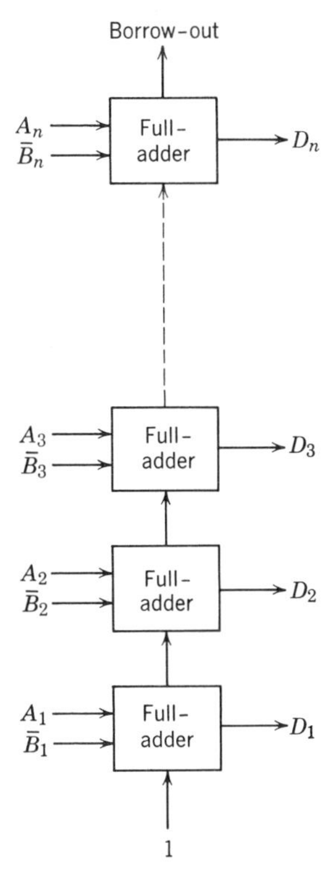

Figure 8.4. An n bit subtracter using full-adders where $D = A - B$. (*Note:* When the borrow-out is 0, the difference is negative and in 2's complement form; when it is 1, the difference is positive.)

In the addition the n-carry (C_n) which is the final borrow indicates the polarity of the difference. When C_n is 1, the difference is positive and when it is 0, the difference is negative, which is the opposite of what is normally obtained in a pure subtraction. This is, of course, immaterial considering the convenience offered by the use of adders. Adders are readily available, while subtracters are not available at all. Subtracters are not manufactured because digital systems designers prefer to use adders in order to minimize the number of different logic circuits used.

8.3 SERIAL-SERIAL SUBTRACTION

Subtraction of multibit serial binary numbers can be performed using a full-subtracter and a storage element. Figure 8.5 illustrates a binary subtracter

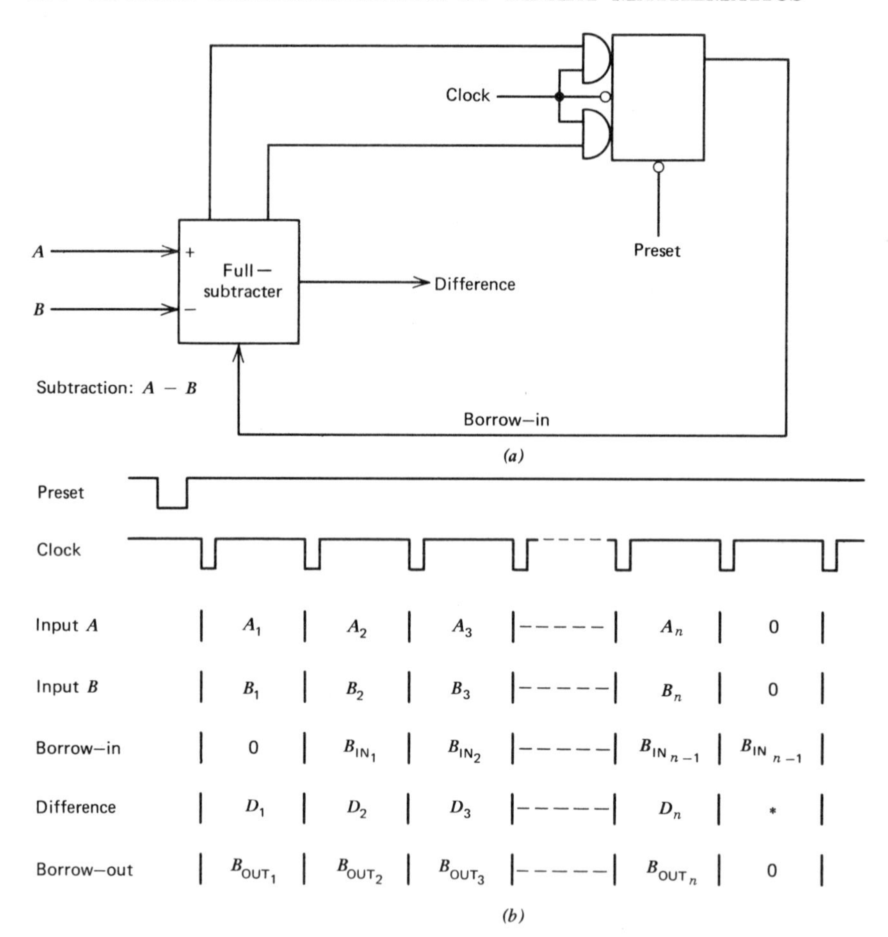

Figure 8.5. Serial subtracter: (*a*) logic diagram; (*b*) timing diagram. (*Note:* *Indicates polarity of difference where 0 = positive and 1 = negative.)

for serial-serial subtraction where both the minuend and the subtrahend are in serial form. Also shown in this figure is the circuit's timing diagram.

A serial subtracter, like a serial adder, requires that the minuend and the subtrahend enter the circuit synchronously and with their LSB first. The difference produced by a serial subtracter is in serial form; it is available with its LSB first; and its last bit indicates the polarity of that difference. The difference is positive when that bit is 0 and negative when it is a 1, in which case it is a 2's complement.

If the obtained difference is not immediately used in its initial serial form, it must be stored in an $(n + 1)$ bit shift register from where it will be later used in either serial or parallel form. It should be noted that before each serial subtraction the storage flip-flop must be cleared.

Sequential subtraction can be also accomplished by means of a serial adder. When a serial adder is used for serial-serial subtraction, the subtrahend must enter the circuit as a 1's complement and the storage flip-flop must be preset to 1 before each subtraction. It is preset to 1 in order to provide an LSB of 1 for the addition.

$$\begin{array}{l} A_n \ldots A_3 A_2 A_1 \\ \bar{B}_n \ldots \bar{B}_3 \bar{B}_2 \bar{B}_1 \\ 0 \ \ldots 0 \ 0 \ 1 \qquad \text{(1 preset into storage f-f)} \\ \hline D_n \ldots D_3 D_2 D_1 \end{array}$$

The presetting of the storage flip-flop must take place once before each subtraction. The output produced is an $(n + 1)$ bit serial word, where the first n bits express the magnitude of the obtained difference and the $(n + 1)$ bit is the polarity of the difference. When the $(n + 1)$ bit is 1, the difference is positive and when it is 0, the difference is negative and is expressed as a 2's complement. A serial adder used for serial-serial subtraction is shown in Figure 8.6.

8.4 PARALLEL-SERIAL SUBTRACTION

In digital systems design, the case often arises where binary subtraction needs to be performed where one of the operands is in serial form while the other is in parallel form. Such operation can be performed by means of a serial subtracter and a shift register, as shown in Figure 8.7.

In this configuration the parallel input is preset into the shift register with its LSB as the last flip-flop of the register. When the serial input becomes available, the clock starts triggering the shift register and the storage flip-flop of the binary subtracter, thus converting the parallel input to serial and updating the storage flip-flop of the serial subtracter.

The computed difference of the two input operands is provided in serial form at the output of the serial subtracter where it is available for use. If the difference output of the serial subtracter is connected to the serial input of the shift register, it will enter the register and at the completion of the subtraction it will be available in parallel form.

Figure 8.7 also illustrates the input logic that enables selection of minuend and subtrahend, so that there are no restrictions as to which of the two operands will be in parallel or serial form.

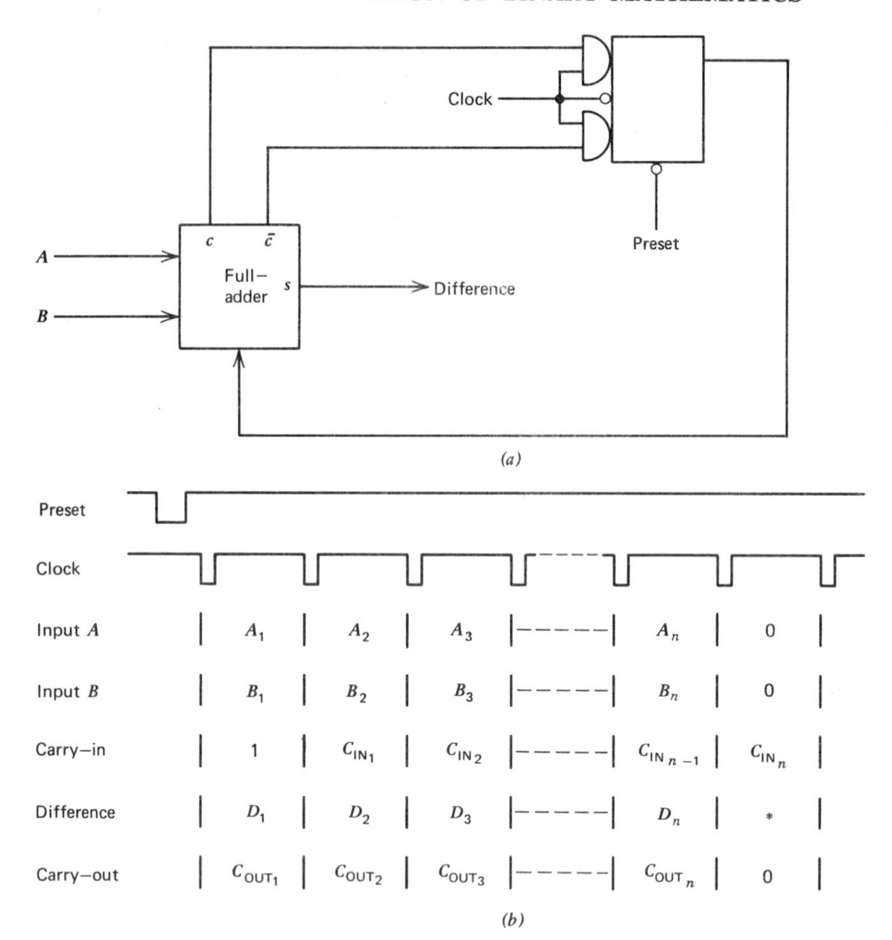

Figure 8.6. Serial adder used as a serial subtracter: (*a*) logic diagram; (*b*) timing diagram. (*Note:* *Indicates polarity of difference where 1 = positive and 0 = negative.)

8.5 PARALLEL PULSE-TRAIN SUBTRACTION

When one of the operands is in parallel form and the other is in pulse-train form, a binary counter can be used to perform the subtraction. The operand available in parallel form is preset into the counter, while the other enters at the clock input of the counter.

When the minuend A is in parallel form and the subtrahend B is available as a pulse train, a down-counter is used, where the minuend is preset into the

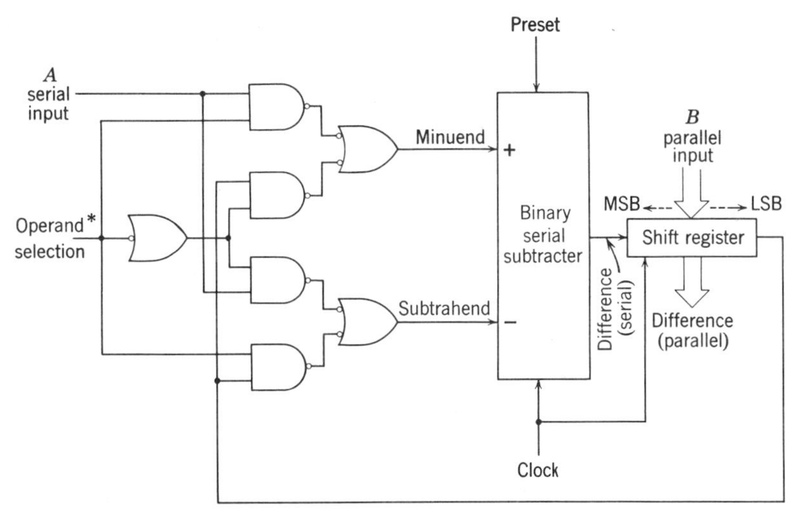

Figure 8.7. Logic network performing parallel-serial subtraction. (*Note:* *HIGH: $A - B$; LOW: $B - A$.)

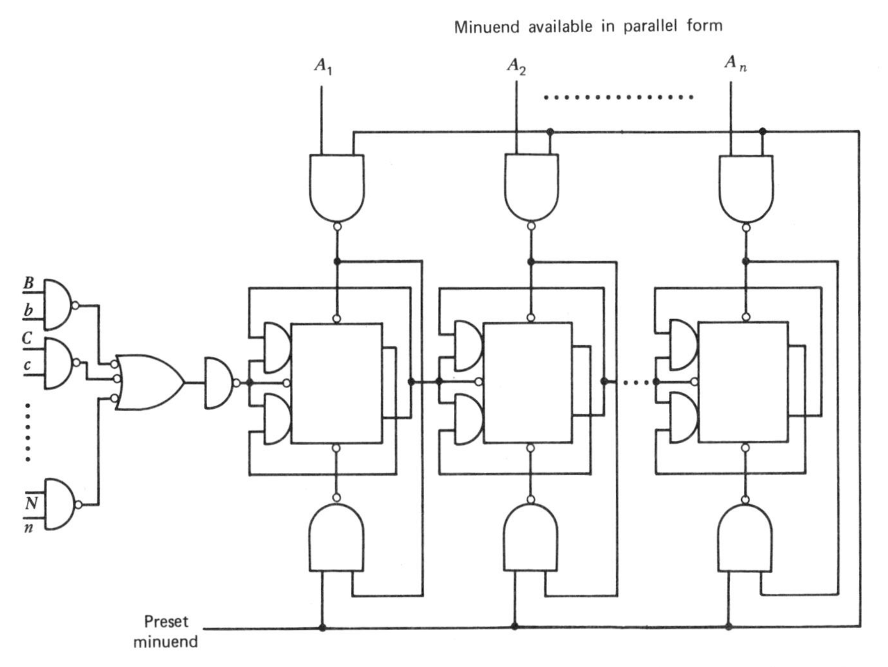

Figure 8.8. An n bit multisubtrahend counter-subtracter where the counter output $= A - B - C - \cdots - N$.

counter and the subtrahend enters at the counter's clock input. The performed operation is $A - B$.

When the subtrahend B is in parallel form and the minuend A is available as a pulse train, an up-counter is used, where the 2's complement of the subtrahend is preset into the counter and the minuend enters the counter at the clock input. The performed operation is $-B + A$, which is the same as $A - B$.

When the minuend is in parallel form, the subtraction may have one or more subtrahends all in pulse-train form. Figure 8.8 illustrates a counter-subtracter configuration where the minuend is available in parallel form and the subtrahend input is accessed by more than one operand—all available in pulse-train form of nonoverlapping pulses.

8.6 ABSOLUTE VALUE OF A BINARY DIFFERENCE

In a binary subtraction the obtained difference can be positive or negative; where the polarity is indicated by the value of the most significant borrow resulting from the subtraction. In the subtraction $A - B$, if the most significant borrow is 0, it means that A is greater than B, but if it is 1, it means that A is less than B, as shown

$$\begin{array}{rl} & A_n \ldots A_4 A_2 A_1 \\ & B_n \ldots B_4 B_2 B_1 \\ \hline D_{n-1} & D_n \ldots D_4 D_2 D_1 \end{array}$$

↑ **Most significant borrow**

When D_{n+1} is 0, it means that A is greater than B and that the obtained difference is the true difference $A - B$. But, when D_{n+1} is 1, it means that A is less than B and the obtained difference is the 2's complement of the true difference $A - B$. To determine the true difference between A and B, we compute the 2's complement of the obtained difference.

The implementation of the computation of the true difference, which is the absolute value of the difference, is shown in Figure 8.9, and it consists of a multibit subtracter, a complementing circuit, and gating that selects either the subtracter output (when $A > B$) or the output of the complementing circuit (when $A < B$).

The complementing circuit produces the 2's complement of the subtracter output D by adding 0...01 to the 1's complement of D. The gating circuit selects the appropriate output and it is controlled by the most significant borrow. When the borrow is 0, the circuit output is the output of the multibit

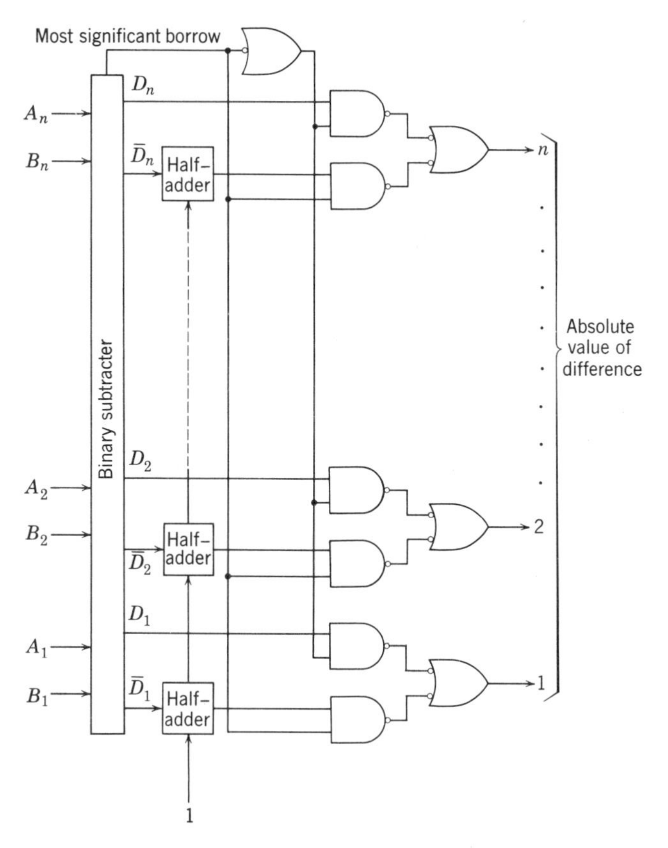

Figure 8.9. Computation of the absolute value of binary difference.

subtracter and when it is 1, the gating circuit output is the output of the complementing circuit.

PROBLEMS

1. State the difference between a half-subtracter and a full-subtracter Also, explain how a parallel multibit adder can be used for binary subtraction.
2. Design a logical network that implements the expression $(A + B) - (C + D)$, where all numbers are in parallel form and have 8 bits.
3. Design a network that computes the 2's complement of binary serial numbers.

4. State the reason for which the storage element of a serial adder should be preset to 1 when the network is used for serial subtraction.
5. Design a network that implements the expression $A + B - C$, where A and B are in serial form and C is parallel, having 8 bits each.
6. Design a network that implements the expression $A + B - C - D$, where all operands are in serial form.
7. Design a digital unit to be used in a parking lot to light the "sorry—full" sign. The parking lot has a capacity of 400 cars, and there are three entrances and one exit to it. At each of the lot's ports there is a sensor that provides one 0.5 μsec pulse for each car that crosses that port. The possibility of overlapping pulses may be neglected.
8. Design a logical network that implements expression $A = B - C$, where B is fixed and C is variable. An "update A" signal is available. The operands B and C are in parallel form; B has 16 bits and C has 4.
9. Design a logical network that performs binary subtraction where the minuend is in pulse-train form and the subtrahend is in parallel form. The minuend's magnitude is always less than 2^9, and the subtrahend is never greater than the minuend.
10. Design a logical network that implements the expression

$$\left|\frac{A - B}{32}\right|$$

where A and B are 10 bit parallel binary numbers.

CHAPTER NINE

Multiplication of Binary Numbers

9.1 INTRODUCTION

The purpose of this chapter is to provide the reader with a repertoire of digital implementation techniques of binary multiplication. There are several methods of binary multiplication with each one having its own merits. Selection of the most suited method, however, depends on the form in which the multiplicand and multiplier are available, and on the time allowed for the performance of the operation. The multiplication algorithms presented here, treat multiplication of binary numbers that are available in parallel, serial, or pulse-train form.

9.2 PARALLEL-PARALLEL MULTIPLICATION

In this section three methods of parallel-parallel multiplication are presented.

Method One is a completely parallel approach and is the direct implementation of the binary multiplication algorithm described in Chapter One. The implementation offers very fast operation, the speed of which is limited only by the propagation delay of the logic circuits used. The design requires no clock, and it should be used in applications where speed is the prime factor.

Its drawback is that it requires more hardware per bit of multiplication than the slower methods do.

Method Two uses binary counters for the conversion of the multiplicand from parallel to pulse-train form and for the accumulation of the product. It is a relatively slow approach, but it compensates in terms of less hardware requirements.

Method Three also uses counters. It requires less hardware than Method Two but it is slower.

Method One

The method presented here is the most straightforward approach to binary multiplication. The multiplication to be performed is accomplished by a number of additions, the final sum of which is the desired product. The number of terms of the sum equals the number of bits of the multiplier. The terms to be added are the multiplicand multiplied by each of the bits of the multiplier—that is, in the multiplication of $A \times B$, where $A = A_8A_4A_2A_1$ and $B = B_8B_4B_2B_1$, the following operation takes place:

Multiplicand				A_8	A_4	A_2	A_1	
Multiplier				B_8	B_4	B_2	B_1	
				A_8	A_4	A_2	A_1	multiplied by B_1
			A_8	A_4	A_2	A_1		multiplied by B_2
		A_8	A_4	A_2	A_1			multiplied by B_4
	A_8	A_4	A_2	A_1				multiplied by B_8
P_{128}	P_{64}	P_{32}	P_{16}	P_8	P_4	P_2	P_1	**Product**

For the bits of the multiplier that are 1 the corresponding term in the addition is the multiplicand, while for the bits that are 0 the term is zero. For example if the multiplicand is 9 (binary 1001) and the multiplier is 12 (binary 1100) the multiplication will be

9		**Multiplicand**		1001
12		**Multiplier**		1100
	1001		0	0000
Multiplicand	1001	Multiplied by	0	0000
	1001	multiplier bits	1	1001
	1001		1	1001
108		**Product** =		1101100

The multiplication of the multiplicand by each bit of the multiplier can be easily achieved with the use of AND gates, as shown in Figure 9.1, where

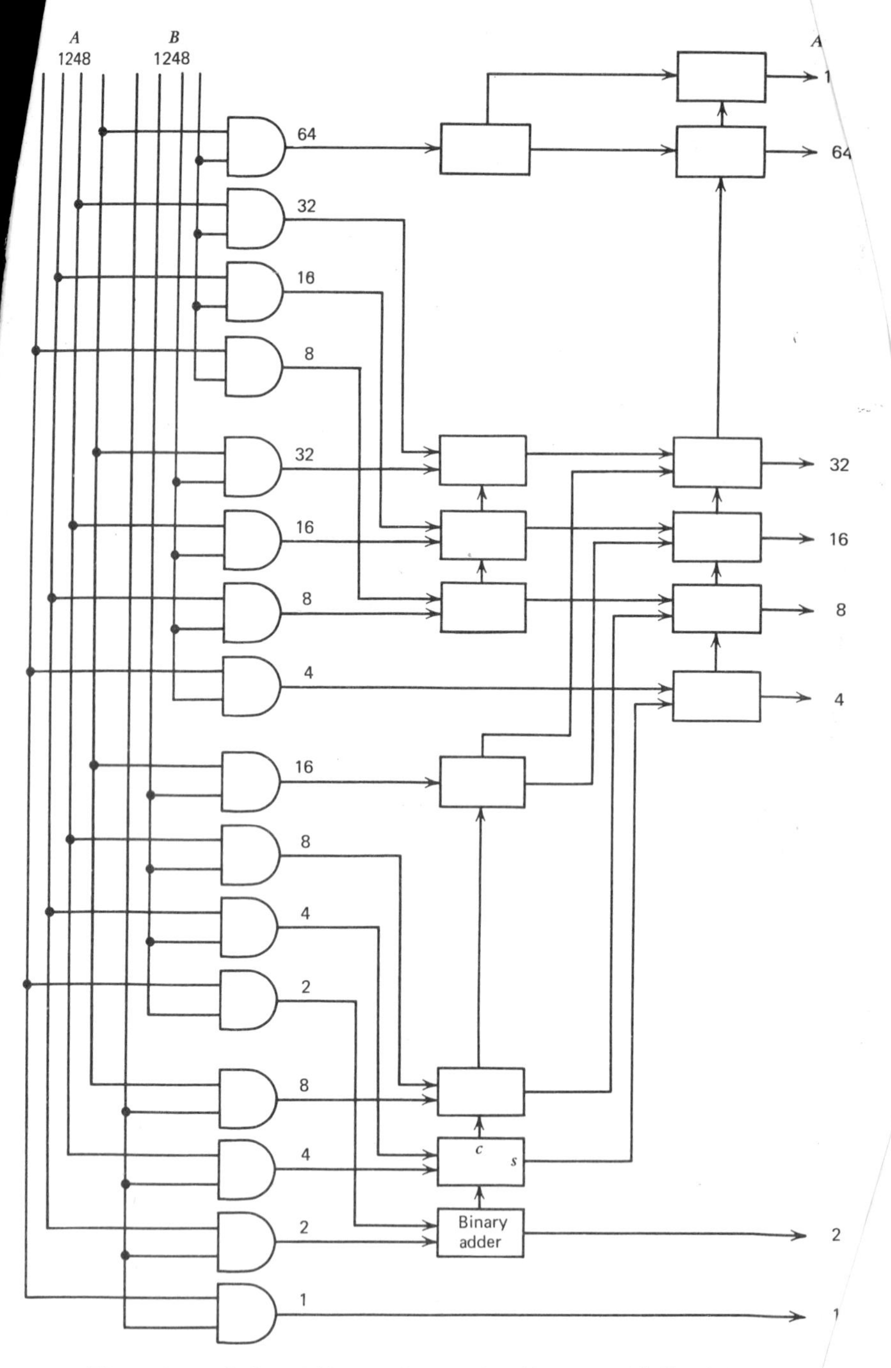

Figure 9.1. A 4 × 4 bit parallel-parallel binary multiplier (Method O

a 4 × 4 bit parallel-parallel binary mutliplier is shown. The output o AND gates provides the multiplicand when the multiplier bit is 1, ar ZERO when the multiplier bit is 0. The partial products provided by gates are fed into adders, the final sum of which is the product of the multiplication. The maximum number of bits a binary product may have equals the sum of the bits of the multiplicand and the multiplier.

Method Two

The amount of hardware needed for the implementation of Method One increases exponentially with an increase in the number of bits of the operands. The hardware implementation of a 10 bit multiplier, for example, requires 100 AND gates, another 100 adders and close to 1000 element interconnections. This requirement is impractical for many applications, expecially those where hardware instead of speed is the prime factor.

For such applications binary counters can be used with the binary operands treated as pulse-train groups rather than bits of binary weight. In this case, the multiplicand is converted into a pulse train once for each bit of the multiplier. For the multiplier bits that are 1, the multiplicand pulse train feeds a product counter, triggering it at the stage, the binary weight of which equals that of the referenced multiplier bit.

Figure 9.2 illustrates the general design of a parallel-parallel multiplier where the multiplicand is converted to pulse trains, and it is multiplied by the multiplier which has remained in parallel binary form. The pulse trains are ANDed with the multiplier bits and the enable signals and serve as clock inputs to the product counter.

The operation of the multiplier of Figure 9.2 is as follows: The START pulse initializes the timing unit, enters the multiplicand into the down-counter and clears the product counter. Instead of being cleared the product counter may be preset to any binary number which will be added to the product.

After the removal of the START pulse the timing unit enables the down-counter and gate 2^0. The down-counter is now counting until its output becomes 0. The number of pulses counted until then equals the magnitude of the multiplicand.

If the 2^0 bit of the multiplier is 1, the pulses will clock the product counter at the input of stage 2^0. If the 2^0 bit of the multiplier is 0, the pulses will be blocked, and the content of the product counter will remain unchanged.

When the down-counter reaches 0, the zero decode circuit will generate a pulse which will preset the down counter to the value of the multiplicand and will trigger the timing unit. As a result, the timing unit will disable gate 2^0 and enable gate 2^1.

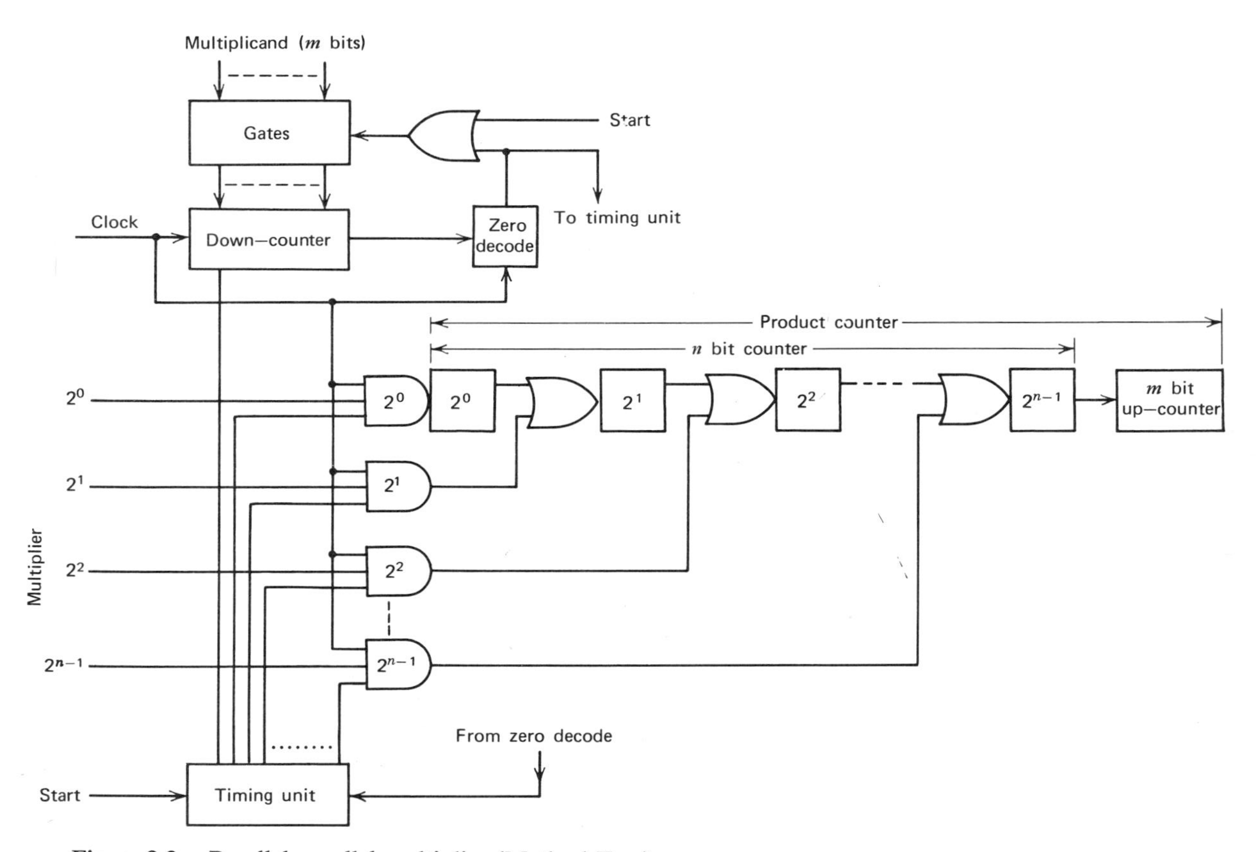

Figure 9.2. Parallel-parallel multiplier (Method Two).

The down-counter will start counting again. This time, if the 2^1 bit of the multiplier is 1, the clock through gate 2^1 will trigger the product counter at stage 2^1. If the 2^1 bit of the multiplier is 0, the clock will be disabled at gate 2^1 and it will not alter the content of the product counter.

When the down-counter reaches 0, the zero decode circuit will generate a pulse which will initiate the pulse-train generation for the third multiplier bit. The process will repeat until all multiplier bits have been enabled to allow or block a multiplicand pulse train to enter the appropriate stage of the product counter.

For the same size of operands, the hardware implementation of this method requires considerably less parts and interconnections than the implementation of Method One. The price paid is time. The maximum time requirements of the multiplier of Figure 9.2 are $n(2^m - 1)$ clock pulses, where n is the number of bits of the multiplier and m is the number of bits of the multiplicand.

Method Three

Further hardware simplification can be obtained using counters, by converting both operands of the multiplication to pulse trains.

Figure 9.3 illustrates this approach the operation of which is as follows: The START pulse presets the multiplicand and the multiplier in their respective down-counters and clears the product counter. Should a binary number

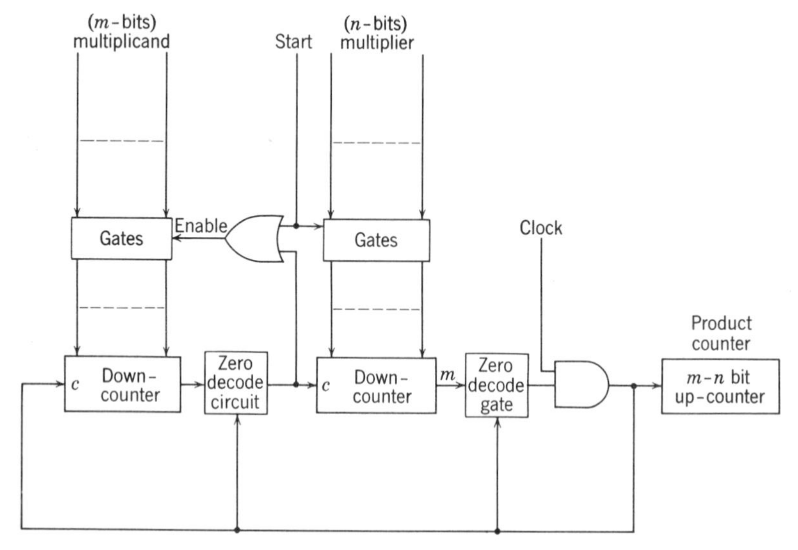

Figure 9.3. Parallel-parallel multiplier (Method Three).

be preset in the product counter, that number will be added to the multiplicand-multiplier product.

After the removal of the START pulse, the multiplicand down-counter starts counting down. The clock pulses that trigger that counter also trigger the product counter which is counting up. When the multiplicand down-counter reaches 0 the zero decode circuit generates a pulse which reenters the multiplicand into its down-counter, and it triggers the clock input of the multiplier down-counter, decrementing it by one.

The number of clock pulses counted from the removal of START until the generation of the decode zero signal equals the binary number of the multiplicand. At this time the content of the product counter is its initial value plus value of the multiplicand.

After the activation of the zero decode signal, the multiplicand down-counter will start counting down again, while the product counter will be counting up. After a number of pulses equal to the multiplicand, the multiplicand down-counter will reach 0 and its zero decode circuit will again generate a signal that will clock the multiplier down-counter, and preset the multiplicand into its down counter. The content of the product counter is now its initial value plus twice that of the multiplicand.

This process continues until the multiplier down-counter reaches 0. This will take place immediately after the multiplicand down-counter has converted the multiplicand to pulse train a number of times equal to the value of the multiplier. Then the zero decode gate disables the clock and the multiplier network stops counting. The final content of the product counter is its initial value plus the multiplicand-multiplier product.

Considering the same size of operands, the hardware implementation of this method requires less parts and interconnections than the preceding two methods. The maximum time requirements of the multiplication of Figure 9.3 are $(2^n - 1)\cdot(2^m - 1)$ clock pulses, where n is the number of bits of the multiplier and m is the number of bits of the multiplicand.

9.3 SERIAL-SERIAL MULTIPLICATION*

To arrive at an algorithm for the multiplication of serial binary numbers, the binary multiplication process is carefully analyzed. In this analysis the formation of the product is related to the availability of the bits of the serial operands.

Figure 9.4 illustrates the multiplication of binary numbers A and B where each number has 4 bits. Also shown is the timing relationship between

* The material in Sections 9.3 and 9.4 originally appeared by the author in *Digital Design*, April 1973, pp. 34–39.

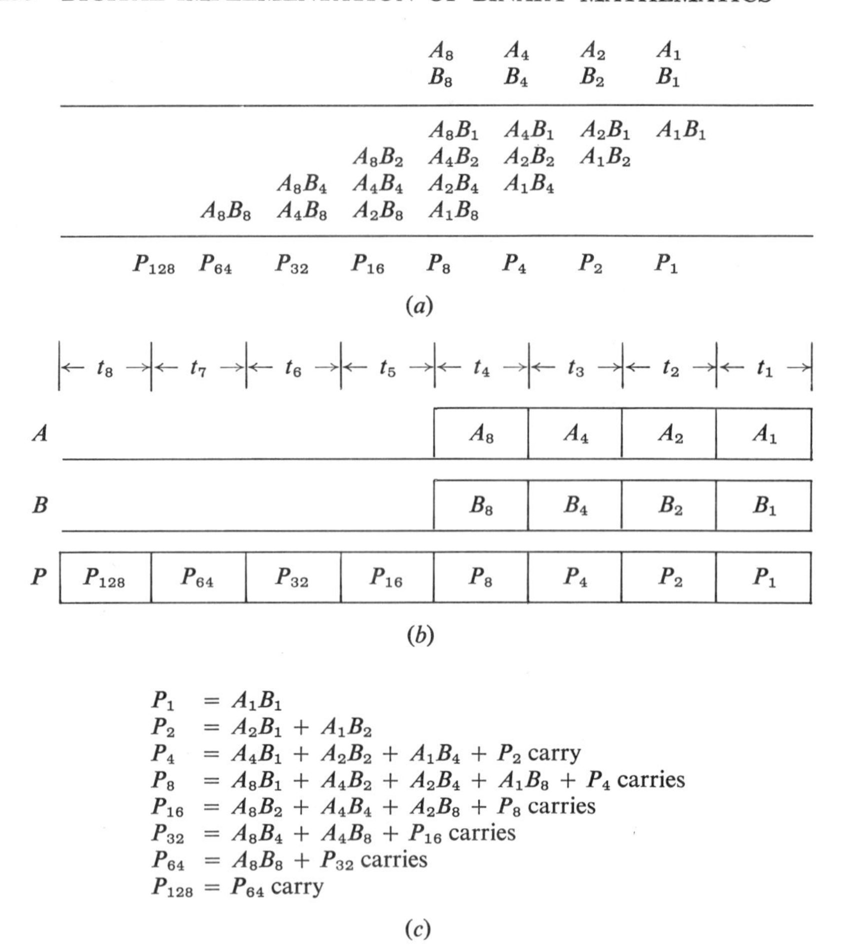

Figure 9.4. (*a*) Multiplication of serial binary numbers A and B; (*b*) timing; (*c*) expression of product bits.

serial numbers A and B. An examination of both presentations indicates that multiplication may start immediately with the arrival of the two serial numbers and for every clock period thereafter 1 product bit will be produced. In time interval t_1, multiplication of A_1 and B_1 gives P_1, the LSB of the product. In time interval t_2, newly arrived A_2 and B_2, together with bits A_1 and B_1 which are already available, may provide P_2 since $P_2 = A_1B_2 + A_2B_1$. That is, 1 product bit can be computed during each clock period, starting with the first period that provides LSBs A_1 and B_1. The expressions for each product bit are listed in Figure 9.4(*c*).

The next step is to develop from the above algorithm a hardware implementation for serial-serial binary multiplication. A second examination of the 4 × 4 bit multiplication of Figure 9.4(*a*), may give one the impression that numbers *A* and *B* slide vertically, one next to the other. The bits of coincidence are then ANDed, and the AND results are added together. Figure 9.5

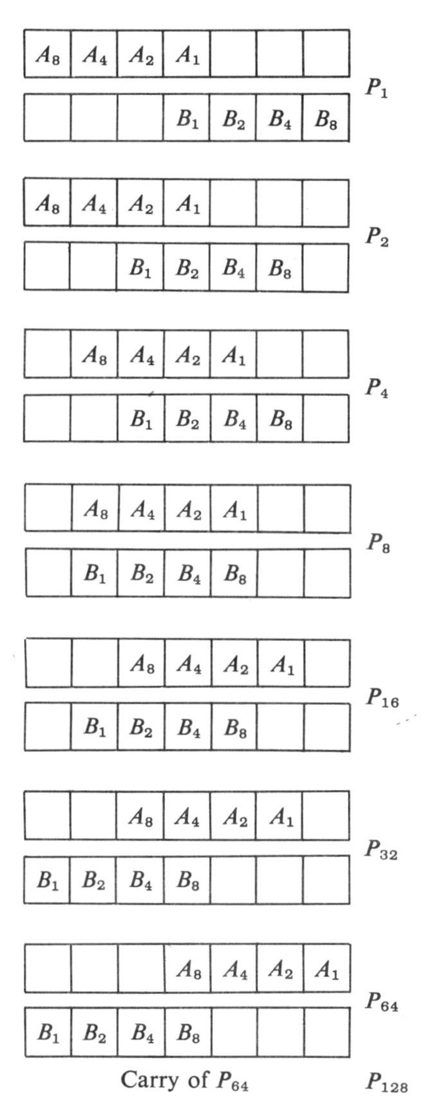

Figure 9.5. Pictorial presentation of a 4 bit binary multiplication.

illustrates this observation. Intuitively, one would think that two shift registers, one for number A and one for number B, would simulate this dynamic bit coincidence. This, however, is not exactly so because in each step in Figure 9.5 only one number shifts.

If each bit of numbers A and B is followed by a space, then both numbers may slide synchronously in opposite directions and form the bit coincidences necessary for the computation of the product bits. Figure 9.6 is a pictorial

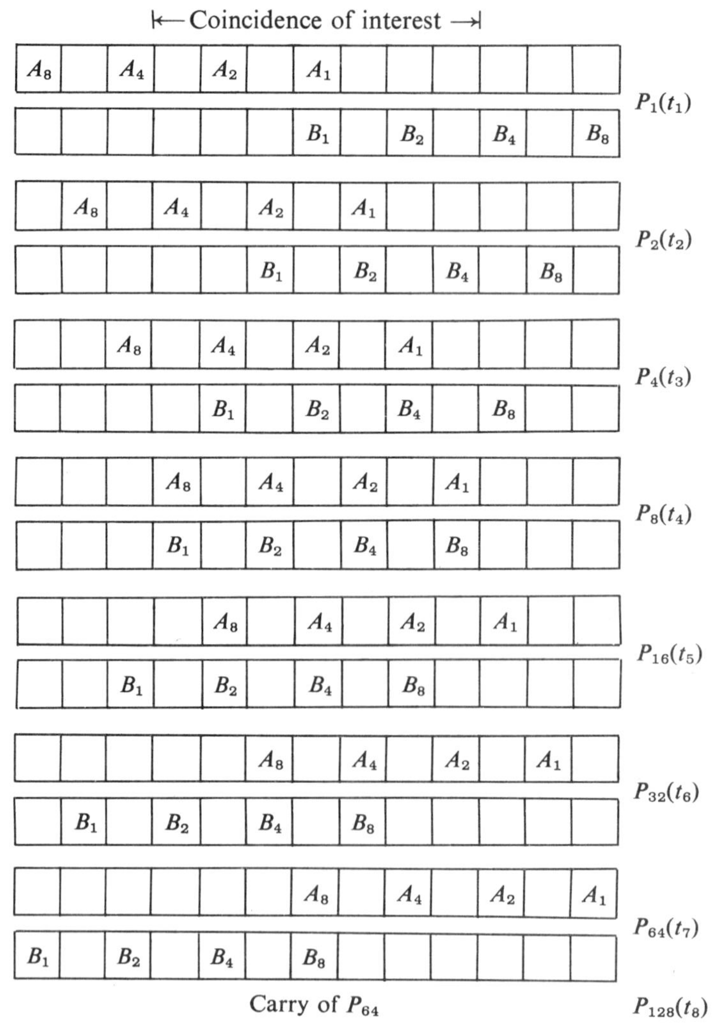

Figure 9.6. Pictorial presentation of a 4 bit binary multiplication where each bit is followed by a space.

representation of a 4 bit binary serial-serial multiplication, where each bit is followed by a space.

If serial binary numbers A and B enter individual shift registers, and after each bit-entry a 0 enters the registers, the bit coincidence illustrated in Figure 9.4 can be achieved. The parallel outputs of the two registers can be ANDed bit by bit, and the AND products then added to form the appropriate bit of product $A \times B$.

Figure 9.7 shows the logic diagram of a serial-serial multiplier employing the above-described method. Operands A and B enter serially, and their bits are placed into selected shift register stages so that the necessary spacing between bits is obtained. The selection of those stages was derived from Figure 9.6, which indicates the clock period during which operand bits are coincident for the first time.

9.4 SERIAL-PARALLEL MULTIPLICATION*

In the design of special-purpose digital equipment, the need often arises for binary multiplication where one operand is available in serial form while the other is in parallel form. Described below are three methods of serial-parallel multiplication.

Method One

The serial-serial multiplication algorithm can be modified to accommodate multiplications where one of the operands is available in parallel form. Figure 9.8 shows the logic diagram of a serial-parallel binary multiplier, where the parallel operand has 4 bits and the size of the serial operand is unrestricted.

Serial operand A enters the multiplier shift register from where it is available for ANDing with the parallel operand B. The operation of the serial-parallel multiplier of Figure 9.8 appears in Figure 9.9(*b*), while Figure 9.9(*a*) illustrates a $4 \times n$ bit multiplication. Serial number A moves 1 bit per clock period in front of the stationary parallel operand B. The overlapping bits are ANDed and the results summed along with carries generated in the preceding periods.

The product is available in serial form with the LSB appearing first. This bit becomes available when the LSB of the serial operand, A, is applied to the multiplier network. The remainder of the product follows serially 1 bit per clock period in ascending binary order.

The network of Figure 9.8 provides a simple way of multiplying serial and parallel binary numbers. A more elaborate method follows that has provisions

* See footnote to Section 9.3.

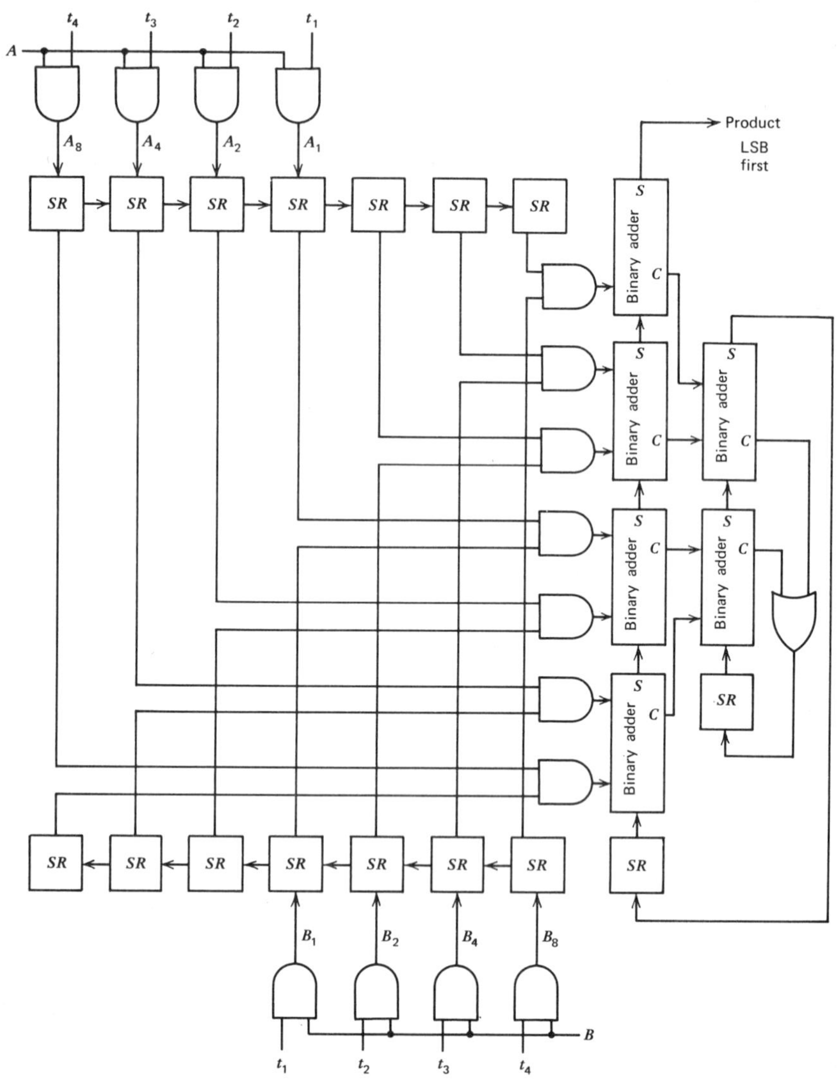

Figure 9.7. Logic diagram of a serial-serial binary multiplier. (*Note:* All shift registers (*SR*) are initially cleared and are driven by a common clock line synchronous with the serial operands.) (Courtesy of *Digital Design.*)

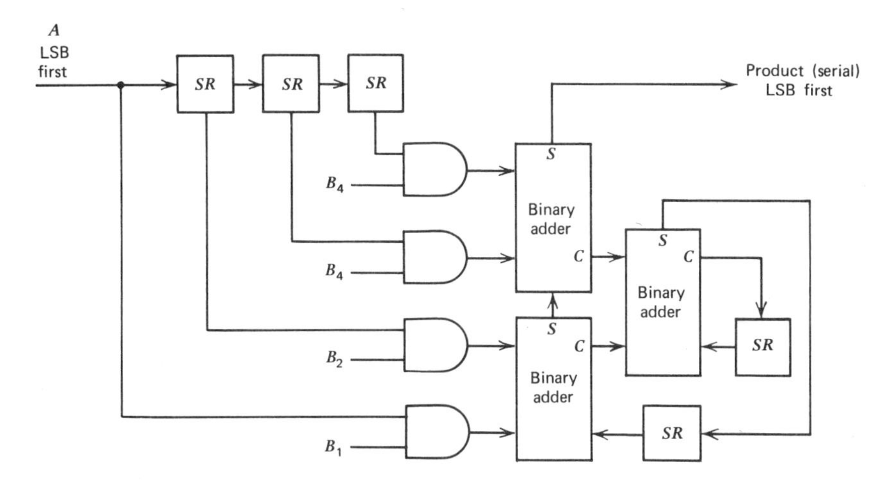

Figure 9.8. Logic diagram of a 4 bit serial-parallel binary multiplier (Method One). (*Note:* All shift registers (SR) are initially cleared and are driven by a common clock line synchronous with the serial operand A.) (Courtesy of *Digital Design.*)

for two parallel constants to be simultaneously added to the serial product of the multiplication. This property is naturally accompanied by an increase in hardware requirements.

Method Two

Binary multiplication can be viewed as a series of additions, rather than one summation of terms having the multiplicand's value properly shifted. An algorithm is presented here where the multiplication terms are added one at a time. This algorithm can be best illustrated by means of the following general, but simple, example. Let the multiplicand be $A_8A_4A_2A_1$ and the multiplier be $B_n \ldots B_8B_4B_2B_1$. The conventional binary multiplication of operands A and B appears in Figure 9.10, where the product of this multiplication is the following summation:

$$P(A \times B) = (A_8A_4A_2A_1)B_1 + (A_8A_4A_2A_1)B_2 + (A_8A_4A_2A_1)B_4 + (A_8A_4A_2A_1)B_8 + \cdots + (A_8A_4A_2A_1)B_n$$

This summation can be rewritten in the following manner:

$$P(A \times B) = [[\cdots[[[[0 + (A_8A_4A_2A_1)B_1] + (A_8A_4A_2A_1)B_2] + (A_8A_4A_2A_1)B_4] + (A_8A_4A_2A_1)B_8] + \cdots] + (A_8A_4A_2A_1)B_n]$$

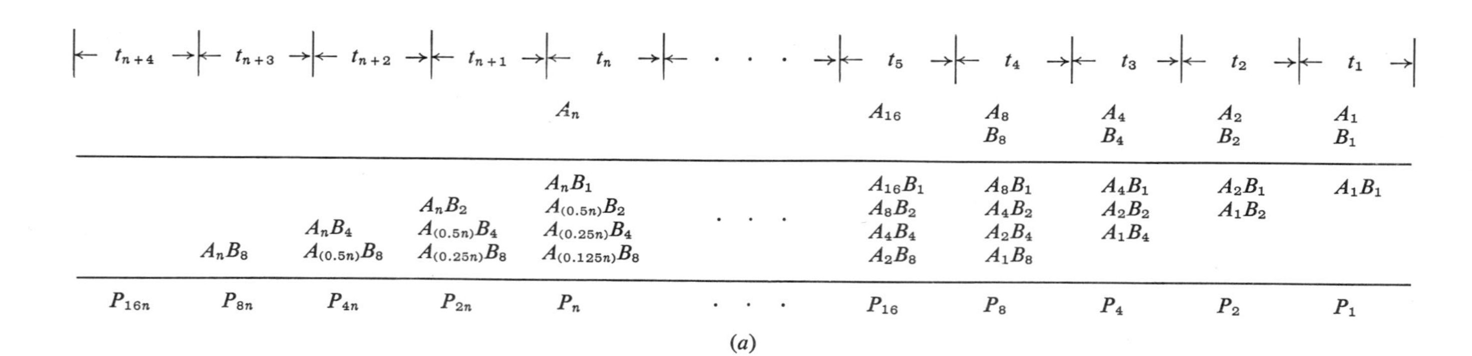

Figure 9.9. *(a)* A 4 × *n* bit multiplication. *(b)* Operation of the serial-parallel multiplier of Figure 9.8. (Courtesy of *Digital Design.*)

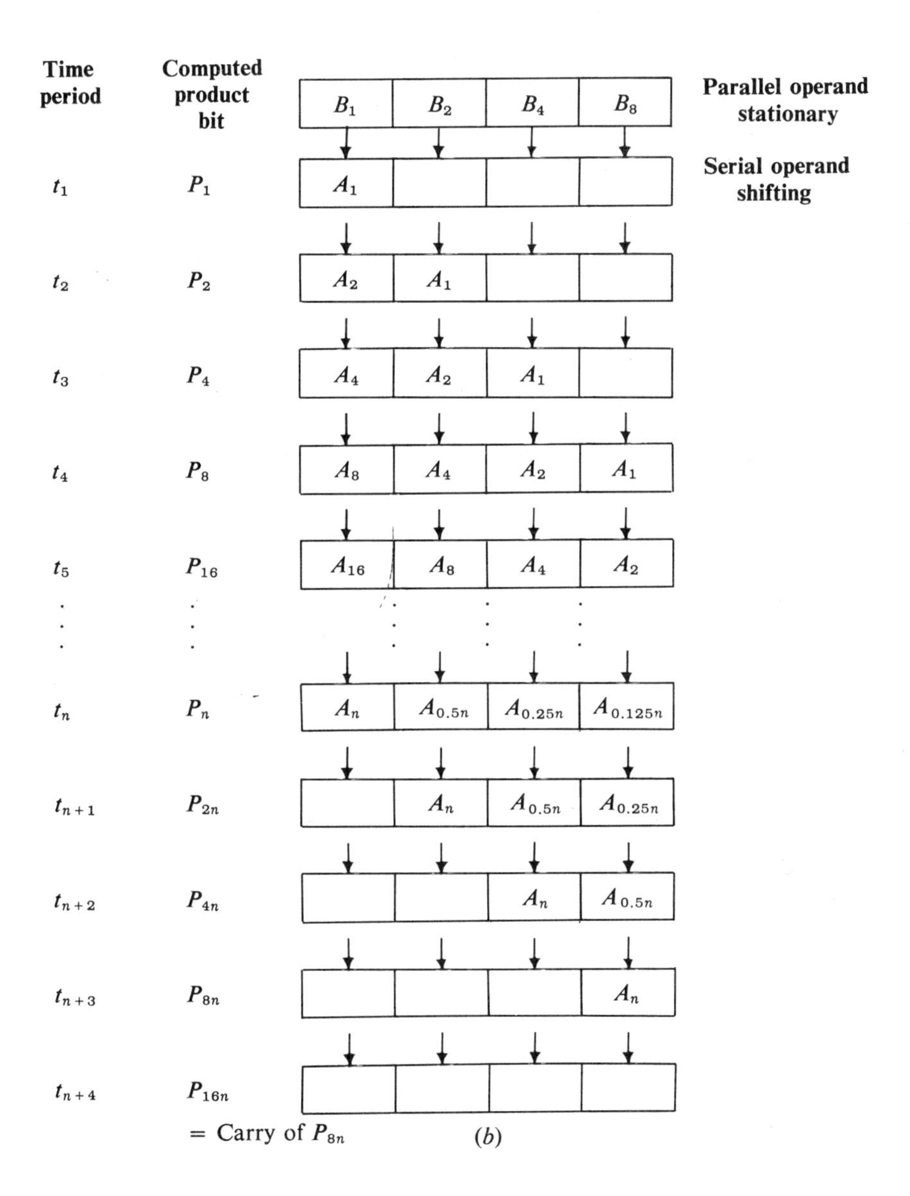
Time period
Computed product bit
B_1
B_2
B_4
B_8
Parallel operand stationary
Serial operand shifting
t_1
P_1
A_1
t_2
P_2
A_2
A_1
t_3
P_4
A_4
A_2
A_1
t_4
P_8
A_8
A_4
A_2
A_1
t_5
P_{16}
A_{16}
A_8
A_4
A_2
t_n
P_n
A_n
$A_{0.5n}$
$A_{0.25n}$
$A_{0.125n}$
t_{n+1}
P_{2n}
A_n
$A_{0.5n}$
$A_{0.25n}$
t_{n+2}
P_{4n}
A_n
$A_{0.5n}$
t_{n+3}
P_{8n}
A_n
t_{n+4}
P_{16n}
= Carry of P_{8n}

(*b*)

									A_8	A_4	A_2	A_1
				B_n	⋯				B_8	B_4	B_2	B_1
									A_8B_1	A_4B_1	A_2B_1	A_1B_1
								A_8B_2	A_4B_2	A_2B_2	A_1B_2	
							A_8B_4	A_4B_4	A_2B_4	A_1B_4		
						A_8B_8	A_4B_8	A_2B_8	A_1B_8			
					⋮							
	A_8B_n	A_4B_n	A_2B_n	A_1B_n								
P_{16n}	P_{8n}	P_{4n}	P_{2n}	P_n		P_{64}	P_{32}	P_{16}	P_8	P_4	P_2	P_1

Figure 9.10. A 4 × n binary multiplication. (Courtesy of *Digital Design*.)

The above expression is a repetitive operation of the form

$$K + L \cdot M$$

where K is the result of the preceding operations, L is always the parallel operand A, and M is the appropriate bit of serial operand B.

During the first time period, when B_1 is available, the operations indicated in the innermost brackets are performed. During the second time period, when B_2 is available, the result obtained from the operations of the first time period is the K term of the $K + L \cdot M$ expression to be computed in the second time period; M, for that computation, is B_2, and L remains the parallel multiplier A. During the n time period, when B_n is available, the result obtained in time period $n - 1$ is added to product $(A_8A_4A_2A_1)B_n$, forming the final product of the multiplication.

Figure 9.11 illustrates the digital implementation of the above serial-parallel multiplication algorithm. The design consists of a set of m AND gates that perform the 4×1 bit multiplication, of an m bit binary adder, which performs the $K + (L \cdot M)$ addition, and of an $m + 1$ bit shift register where m is the number of bits of multiplicand A.

The hardware requirements of the parallel-serial multiplier of Figure 9.11 are restricted only by the bit size of the parallel operand.

The advantage of this algorithm, over that of Method One, is that a parallel m bit constant C may be preset into the shift register making the final result of the operation $A \times B + C$. Also, a serial constant D of bit size B or less may be added to the product; D will enter the network at the unused input of the LSB of the binary adder. The final result will then be $A \times B + C + D$, where A and C are parallel and B and D are serial binary numbers.

The disadvantages of this method over the preceding one are a small increase in hardware requirements and a speed limitation due to the carry-propagation of the m bit adder. Carry-propagation can be minimized if carry-look-ahead circuits are used.

Method Three

In the method presented here the binary multiplication is treated as being an equation of nested operations as in Method Two, where the product is represented by the following expression:

$$P(A \times B) = [[\cdots[[[[0 + (A_8A_4A_2A_1)B_1] + (A_8A_4A_2A_1)B_2] + (A_8A_4A_2A_1)B_4] \\ + (A_8A_4A_2A_1)B_8] + \ldots] + (A_8A_4A_2A_1)B_n]$$

Here the speed limitation, mentioned in the description of Method Two, is overcome by performing the successive additions on a bit-by-bit basis, storing away the sum and carry outputs. This way, no carry-propagation

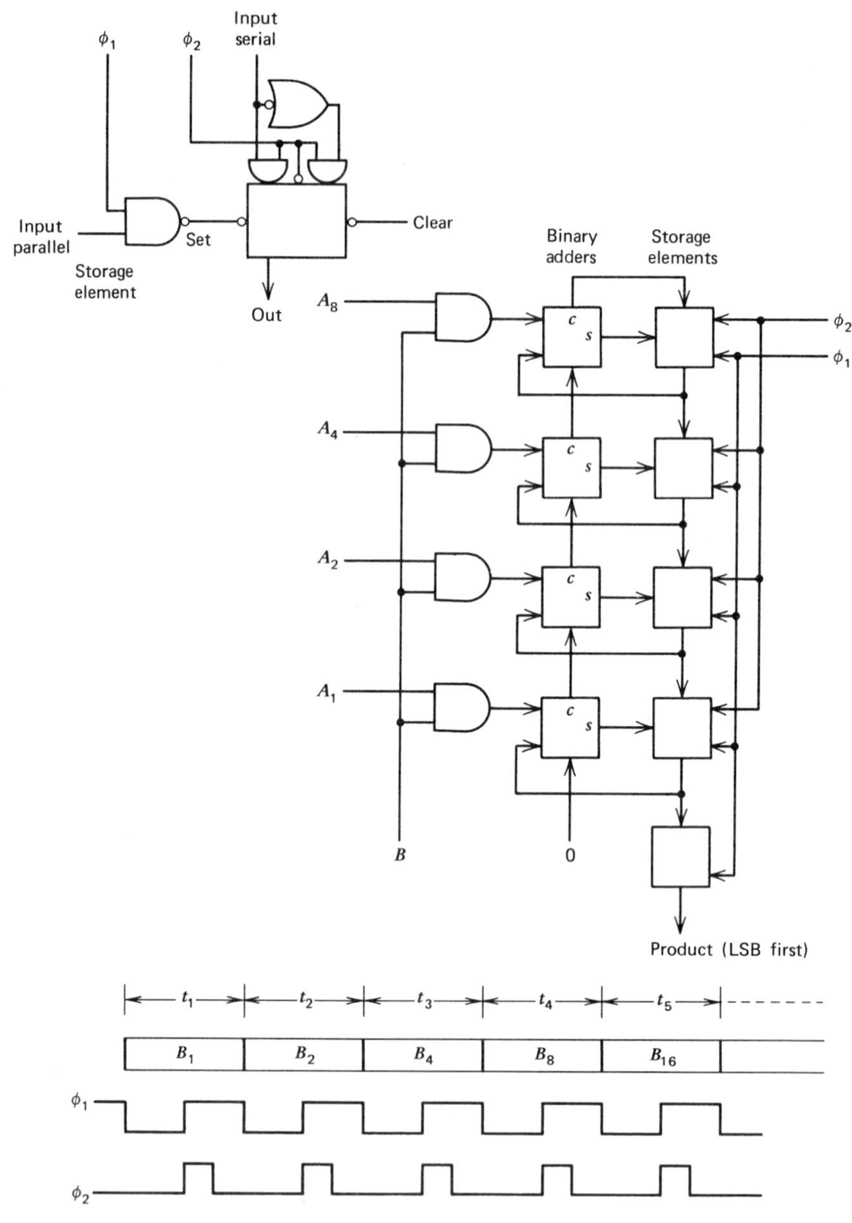

Figure 9.11. Parallel-serial multiplier (Method Two). (Courtesy of *Digital Design*.)

takes place and the network may operate at speeds higher than those of the preceding methods. In addition, this approach has provision for two parallel constants to be added to the product during the multiplication process.

Figure 9.12 illustrates the general form of a serial-parallel multiplier employing this technique. As it was the case with the implementation of the preceding two methods of serial-parallel multiplication, the hardware requirements of the design of Figure 9.12 are directly proportional to the

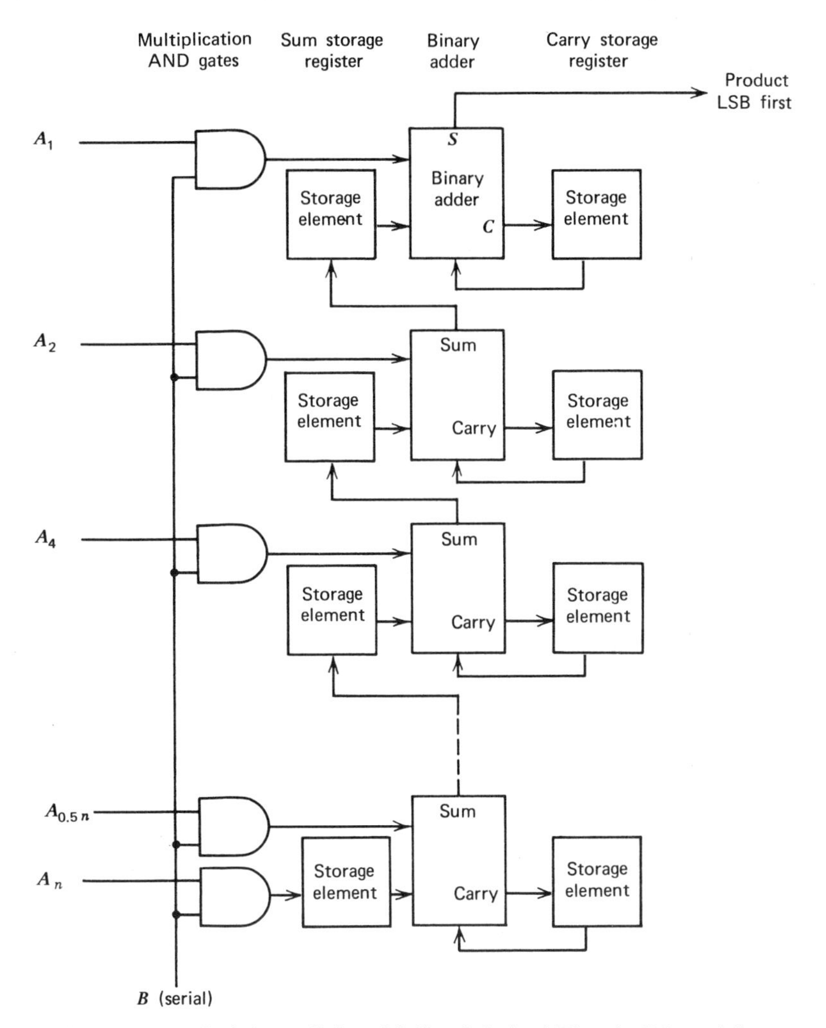

Figure 9.12. Serial-parallel multiplier (Method Three). (*Note:* All storage elements are initially cleared and are driven by a common clock line synchronous with the serial operand.) (Courtesy of *Digital Design.*)

bit size of the parallel operand while the serial operand is of unlimited size.

The algorithm and its implementation can be best explained by means of the following example, where both operands are assumed to have 4 bits each. The necessary network for the serial-parallel 4 × 4 multiplication employing the bit-by-bit addition technique appears on Figure 9.13.

In the multiplier of Figure 9.13 the parallel operand A is ANDed with each bit of the serial operand B, and the generated product enters a special

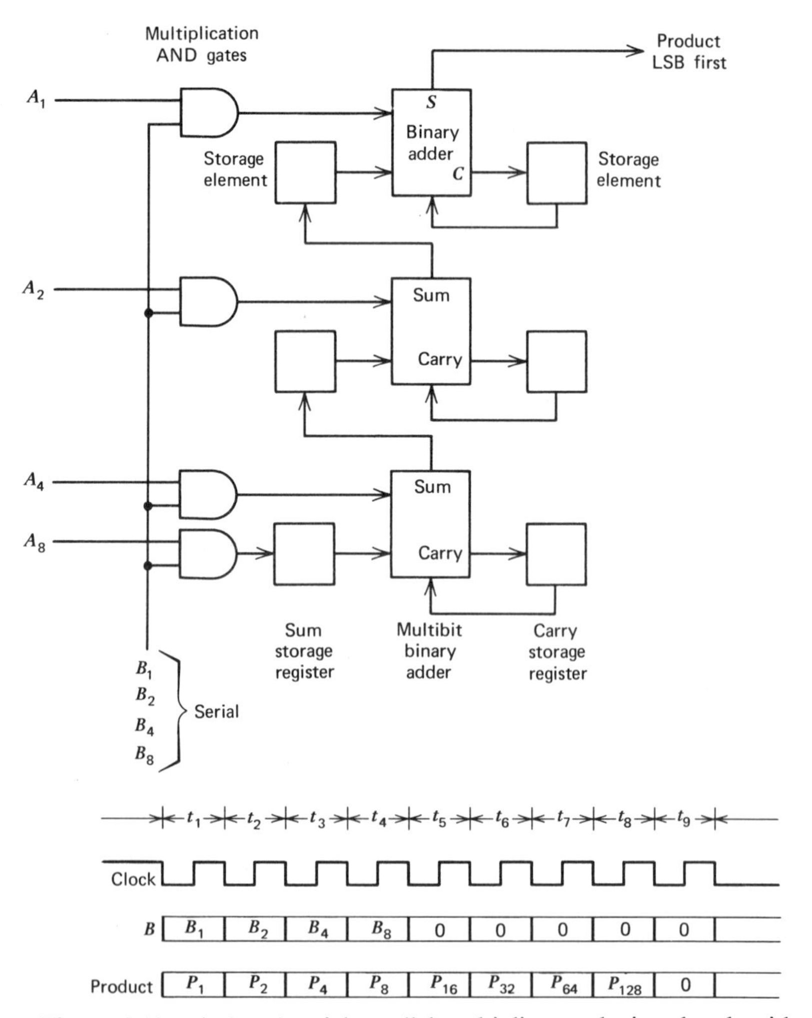

Figure. 9.13. A 4 × 4 serial-parallel multiplier employing the algorithm of Method Three. (Courtesy of *Digital Design.*)

type of multibit adder. Serial operand B enters the multiplier with its LSB first. When the B bit is 1, parallel operand A is applied to the multibit adder; while a 0 B bit applies zero to the adder.

The multibit adder consists of a chain of full-adders, the sum and carry outputs of which are delayed for one clock period before being used. The delayed sum output feeds the adder which is next in least significance, while the delayed carry output feeds back the adder that has produced it. Before the multiplication starts, the two storage registers are cleared, or preset by one or two binary numbers that will be added to the product during the multiplication process.

For synchronization to exist in the serial-parallel multiplication process the updating clock of the two storage registers is the same as that of the serial operand. Figure 9.14 illustrates the detailed timing of the multiplier of Figure 9.13 when a 4×4 multiplication is performed.

During the first clock period parallel number A is multiplied by the LSB of serial input B. This bit is available at the serial input of the multiplier for the first clock period only, and serves as an enable signal to the AND gates.

The product of this multiplication enters the multibit adder as $A_8A_4A_2A_1$. or as 0000, depending on the value of B_1, which is the LSB of serial operand B. The output of the gates feeds the multibit adder where it is added to the outputs of the sum and carry storage registers.

The storage registers are initially cleared, and since this is the first clock period, the output of the AND gates will appear at the sum output of the multibit adder. Consequently, the multiplier output is the same as the output of the uppermost gate, which is A_1B_1. Indeed, the serial product output of this clock period is P_1 and should be the product of A_1 and B_1. At the end of the first clock period the sum and carry outputs of the multibit adder are stored in their respective storage registers from where they are available in the next clock period.

During the second clock period parallel number A is multiplied by B_2, which is the next bit in least significance of serial number B. The $(A_8A_4A_2A_1) \times B_2$ multiplication is accomplished by the AND gates where the parallel number A is enabled by means of input B_2.

At the multibit adder the output of the gates is added to the outputs of the sum and carry storage registers that now hold the results of the addition that took place in the preceding clock period. The output of the multiplier during the second clock period is P_2 of the product and it is the sum output of the uppermost adder stage. The inputs to that stage are A_1B_2 from the gates, A_2B_1 from the sum storage register, and 0 from the carry storage register; P_2 is indeed $A_1B_2 + A_2B_1$, as illustrated in Figure 9.10.

During the third clock period parallel number A is multiplied by B_4 and

Serial input	Clock period	Bit-by-bit additions				
		Adder inputs	A_4B_1	A_2B_1	A_1B_1	Output of AND gates
			0	0	0	Output of sum register
B_1	1		0	0	0	Output of carry register
		Adder outputs	A_4B_1	A_2B_1	P_1	$= A_1B_1$
			0	0	0	
			A_4B_2	A_2B_2	A_1B_2	
			A_8B_1	A_4B_1	A_2B_1	
B_2	2		0	0	0	
			$S_{2,1}$	$S_{2,2}$	P_2	$= A_1B_2 + A_2B_1$
			$C_{2,1}$	$C_{2,2}$	$C_{2,3}$	
			A_4B_4	A_2B_4	A_1B_4	
			A_8B_2	$S_{2,1}$	$S_{2,2}$	
B_4	3		$C_{2,1}$	$C_{2,2}$	$C_{2,3}$	
			$S_{3,1}$	$S_{3,2}$	P_4	$= A_1B_4 + A_2B_2 + A_4B_1$
			$C_{3,1}$	$C_{3,2}$	$C_{3,3}$	+ carry
			A_4B_8	A_2B_8	A_1B_8	
			A_8B_4	$S_{3,1}$	$S_{3,2}$	
B_8	4		$C_{3,1}$	$C_{3,2}$	$C_{3,3}$	
			$S_{4,1}$	$S_{4,2}$	P_8	$= A_1B_8 + A_2B_4 + A_4B_2$
			$C_{4,1}$	$C_{4,2}$	$C_{4,3}$	$+ A_8B_1 +$ carries
			A_40	A_20	A_10	
			A_8B_8	$S_{4,1}$	$S_{4,2}$	
0	5		$C_{4,1}$	$C_{4,2}$	$C_{4,3}$	
			$S_{5,1}$	$S_{5,2}$	P_{16}	$= A_2B_8 + A_8B_2 + A_4B_4$
			$C_{5,1}$	$C_{5,2}$	$C_{5,3}$	+ carries
			A_40	A_20	A_10	
			A_80	$S_{5,1}$	$S_{5,2}$	
0	6		$C_{5,1}$	$C_{5,2}$	$C_{5,3}$	
			$S_{6,1}$	$S_{6,2}$	P_{32}	$= A_4B_8 + A_8B_4 +$ carries
			0	$C_{6,2}$	$C_{6,3}$	
			A_40	A_20	A_10	
			A_80	$S_{6,1}$	$S_{6,2}$	
0	7		0	$C_{6,2}$	$C_{6,3}$	
			0	$S_{7,2}$	P_{64}	$= A_8B_8 +$ carries
			0	0	$C_{7,3}$	
			A_40	A_20	A_10	
			A_80	0	$S_{7,2}$	
0	8		0	0	$C_{7,3}$	
			0	0	P_{128}	= Final carry
			0	0	0	

Figure 9.14. Timing of a 4 × 4 serial-parallel multiplier (Method Three).

the product is applied to the multibit adder, where it is added to the sums and carries formed in the preceding (second) clock period. The output of the multiplier during the third clock period represents bit P_4 of the product.

The operation continues until all bits of serial operand B have entered the multiplier and the four additional clock periods have passed which are necessary for the information left in the storage registers after the end of B to be processed. The total number of clock periods required for the completion of the multiplication equals the sum of bits of the two operands while the number of stages in the multiplier equals the number of bits of the parallel operand less one.

9.5 PARALLEL PULSE-TRAIN MULTIPLICATION

In specialized designs of digital equipment the need often arises for multiplication hardware that multiplies two numbers where one is available in parallel form and the other is represented by the number of pulses of a pulse train. This type of multiplication can be accomplished by means of an accumulator network which consists of a multibit adder and a storage register.

Figure 9.15 illustrates a multiplier network where the multiplicand is a parallel binary number and the multiplier is a train of pulses. Here the output of the adder enters the storage register and the output of the storage register feeds back into the adder. The parallel operand enters the multiplier as an input to the multibit adder while the pulse-train operand serves as the updating clock of the storage register.

Initially, the storage register is cleared, or it is preset to a binary number that will be added to the computed product. Before the arrival of the first pulse, the output of the adder is parallel operand A, assuming that the storage register was initially cleared. Otherwise, it is A plus the number that was preset in the register.

After the arrival of the first pulse, the output of the adder, which is also the input to the storage register, enters that register and becomes its output. The adder inputs are now the parallel operand A and the output of the storage register which is also A. The adder output is consequently $2A$, $A + A$.

After the arrival of the second pulse, the $2A$ output of the adder enters the storage register making the output of that register $2A$. This output feeds back into the binary adder, the sum of which now becomes $3A$, which is the parallel operand A, plus the content of the storage register which is $2A$.

After the arrival of the third pulse, the register output becomes $3A$, after the fourth pulse $4A$, and so on, assuming that the storage register was initially

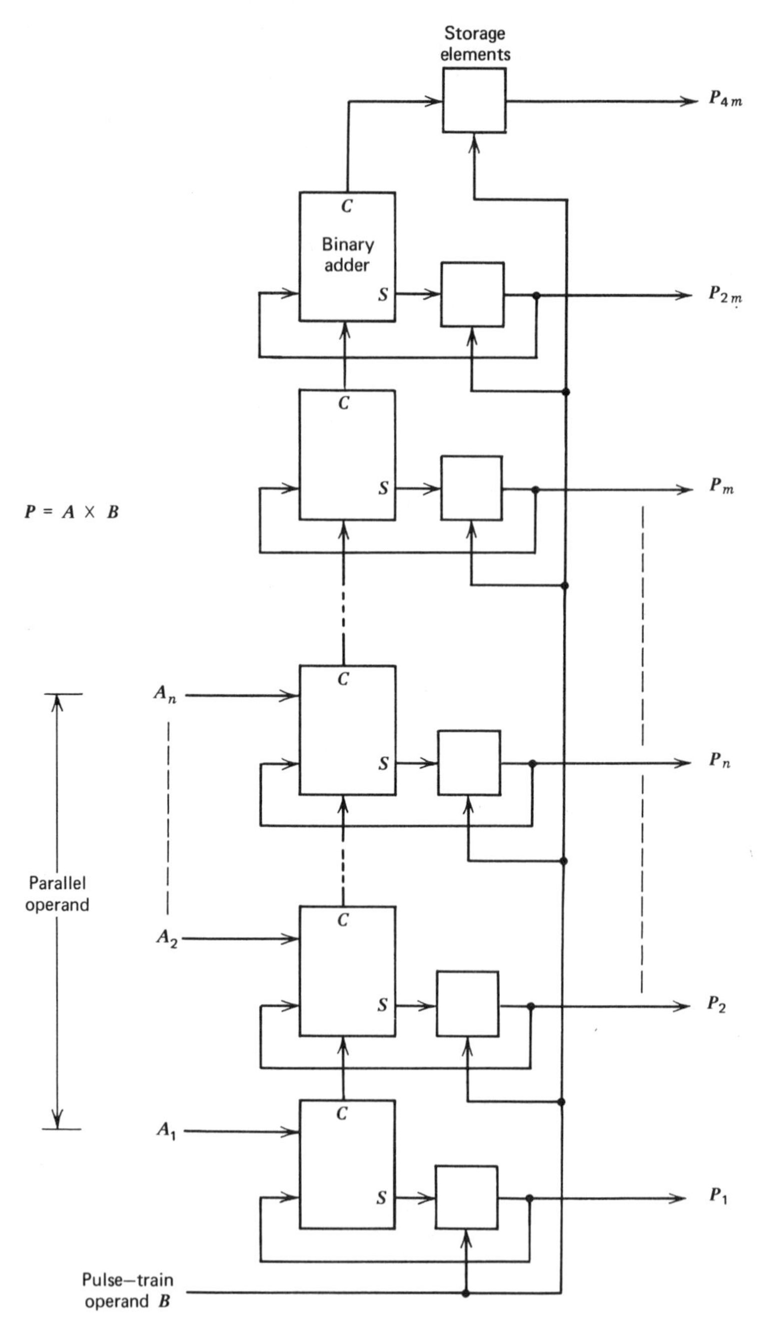

Figure 9.15. Parallel pulse-train multiplier.

Pulse-train operand after pulse	Adder inputs		Adder output	Product Register output
	Parallel operand	Register output		
	A	0	A	0
1	A	A	$2A$	A
2	A	$2A$	$3A$	$2A$
3	A	$3A$	$4A$	$3A$
4	A	$4A$	$5A$	$4A$
⋮	⋮	⋮	⋮	⋮
n	A	nA	$(n + 1)A$	nA

Figure 9.16. Timing of a parallel pulse-train multiplier.

cleared and it can handle the size of product nA where n is the number of pulses of the pulse-train operand and A is the binary value of the parallel operand. Figure 9.16 illustrates the timing of the multiplier of Figure 9.15.

This method of multiplication finds wide application in instrumentation, where the pulse train represents elapsed time and the parallel operand is a variable parameter that should be integrated such as temperature, light, flow, etc.

9.6 MEDIUM-SCALE INTEGRATION MULTIPLICATION

Binary multipliers find wide use in general-purpose as well as special-purpose computers, where speed is of greatest importance in most cases.

Multibit multipliers are formed by the appropriate connection of multiplier modules that perform 4×4 or 4×2 bit multiplications. To facilitate modular construction of multibit multipliers, multiplier modules provide for the addition of a constant, thus performing function

$$P = X \cdot Y + K$$

Use of this expression is made after the performed multiplication has been expanded and then expressed in a form that uses iterative operations consisting of one multiplication and one addition

$$\begin{aligned}(X_{0,1,2,3}) \cdot (Y_{0,1,2,3,4,5,6,7}) + K &= (X_{2,3}) \cdot (Y_{4,5,6,7}) + (X_{0,1}) \cdot (Y_{4,5,6,7}) \\ &\quad + (X_{2,3}) \cdot (Y_{0,1,2,3}) + (X_{0,1}) \cdot (Y_{0,1,2,3}) + K \\ &= [(X_{2,3}) \cdot (Y_{4,5,6,7}) + [(X_{0,1}) \cdot (Y_{4,5,6,7}) + [(X_{2,3}) \cdot (Y_{0,1,2,3}) \\ &\quad + [(X_{0,1}) \cdot (Y_{0,1,2,3}) + K]]]]\end{aligned}$$

Thus, for the implementation of the above 4 × 8 bit multiplication, four 2 × 4 bit multiplier modules are needed. Figure 9.17 illustrates a 4 × 8 bit multiplier consisting of 2 × 4 bit multiplier modules.

Several types of binary multipliers are available in MSI form, two of which appear in Figures 9.18 and 9.19. Each of the figures illustrates the building block as well as use of that block for the construction of multibit multipliers.

PROBLEMS

1. Design a parallel-parallel multiplier employing Method One (Fig. 9.1), where A has 3 bits and B has 7 bits.
2. Determine the minimum clock frequency of a parallel-parallel multiplier, employing the technique of Method Two (Fig. 9.2), where the multiplicand has 6 bits, the multiplier has 7 bits, and the available time is 10 msec.
3. Repeat Problem 2 using Method Three (Fig. 9.3). Compare the result with that obtained in Problem 2.
4. Modify the serial-serial multiplier of Figure 9.7, so that two 4 bit serial numbers, C and D, are added to the product $A \times$ B while the multiplication is performed.
5. Design a serial-parallel multiplier employing Method One (Fig. 9.8), where A has 6 bits and B has 10 bits. Compute the maximum number of bits the product may have, and the required number of clock pulses.
6. Draw the detailed timing diagram of the serial-parallel multiplier of Figure 9.11 for a 5 × 5 bit multiplication.
7. Using the information in Figure 9.14, perform the binary multiplication where $A = 1010$ and $B = 1011$. Verify the result by computing the product using the conventional multiplication algorithm.
8. Design an instrument of digital readout that will compute the daily total volume of flow through a pipe. The rate of flow is available in binary parallel form (6 bits). The volume reading should be incremented every 20 sec.
9. Design an instrument of digital readout that will compute the instantaneous pressure-temperature product of a boiler, where both parameters are 6 bit numbers. The pressure is available in parallel form while the temperature is in serial form.
10. Using currently available hardware, design a high-speed 8 × 12 bit multiplier.

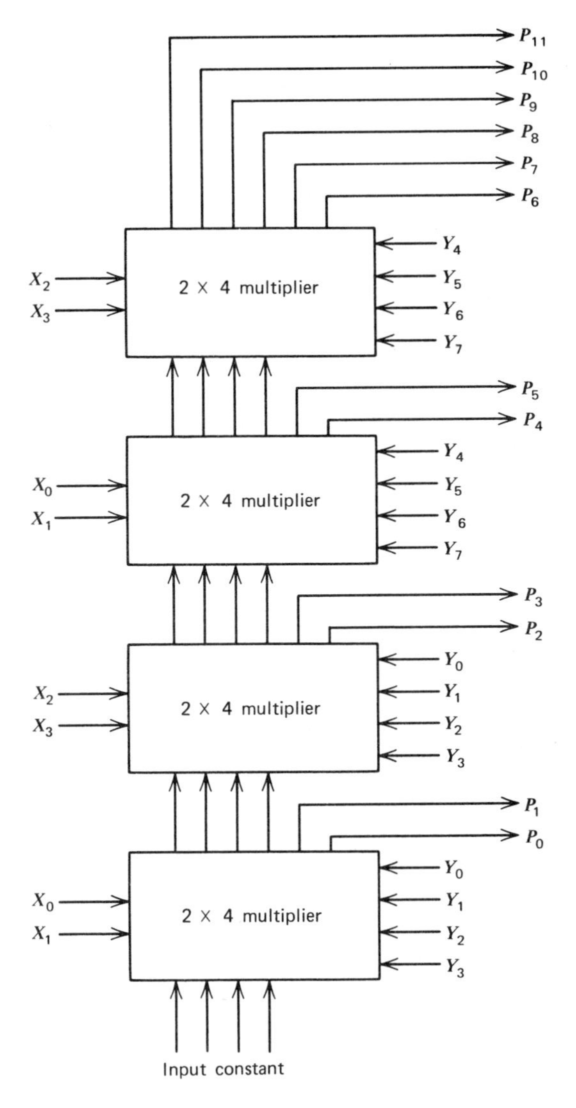

Figure 9.17. (*a*) A 4 × 8 bit multiplier consisting of four 2 × 4 bit multiplier modules (module characteristics: $P = X \cdot Y + K$).

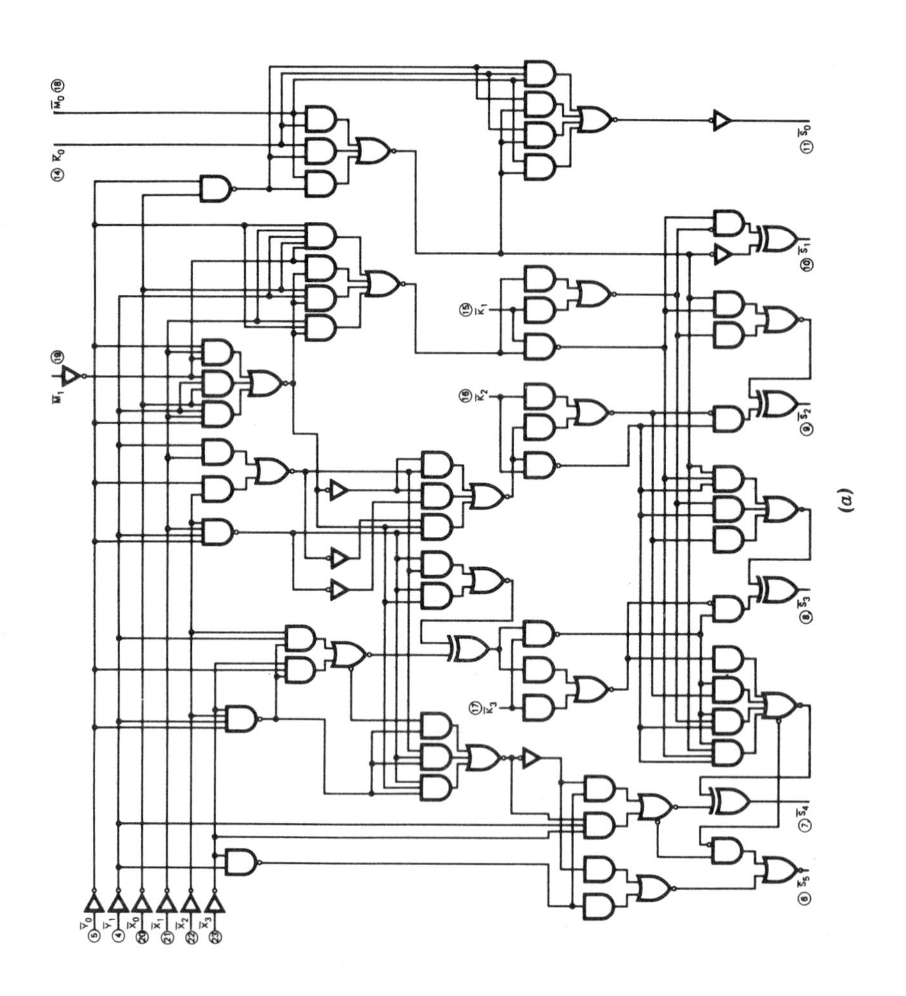
14 $\bar{K}_0$
$\bar{M}_0$ 18
$\bar{M}_1$ 19
15 $\bar{K}_1$
16 $\bar{K}_2$
17 $\bar{K}_3$
11 $\bar{S}_0$
10 $\bar{S}_1$
9 $\bar{S}_2$
8 $\bar{S}_3$
7 $\bar{S}_4$
6 $\bar{S}_5$
5 $\bar{Y}_0$
4 $\bar{Y}_1$
20 $\bar{X}_0$
21 $\bar{X}_1$
22 $\bar{X}_2$
23 $\bar{X}_3$

(*a*)

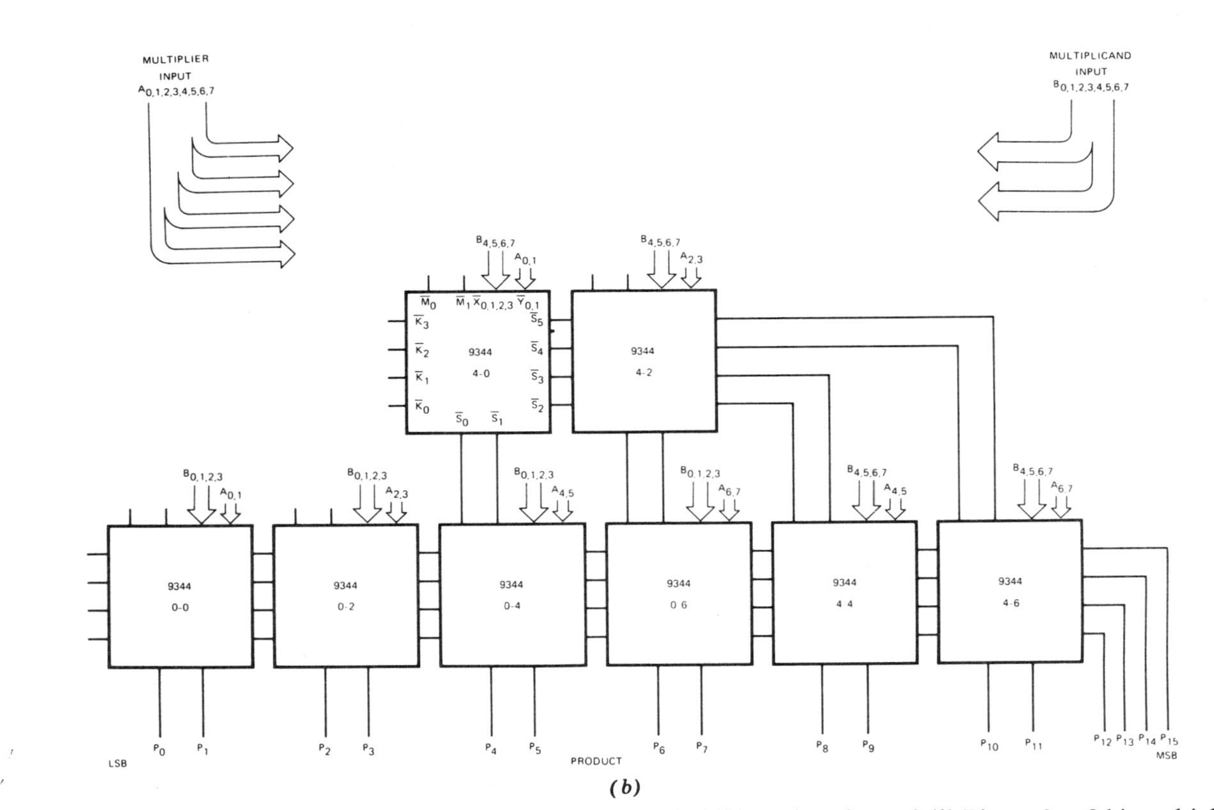

(*b*)

Figure 9.18. (*a*) Binary 4 × 2 bit multiplier. (Courtesy of Fairchild Semiconductor.) (*b*) Binary 8 × 8 bit multiplication. (Courtesy of Fairchild Semiconductor.)

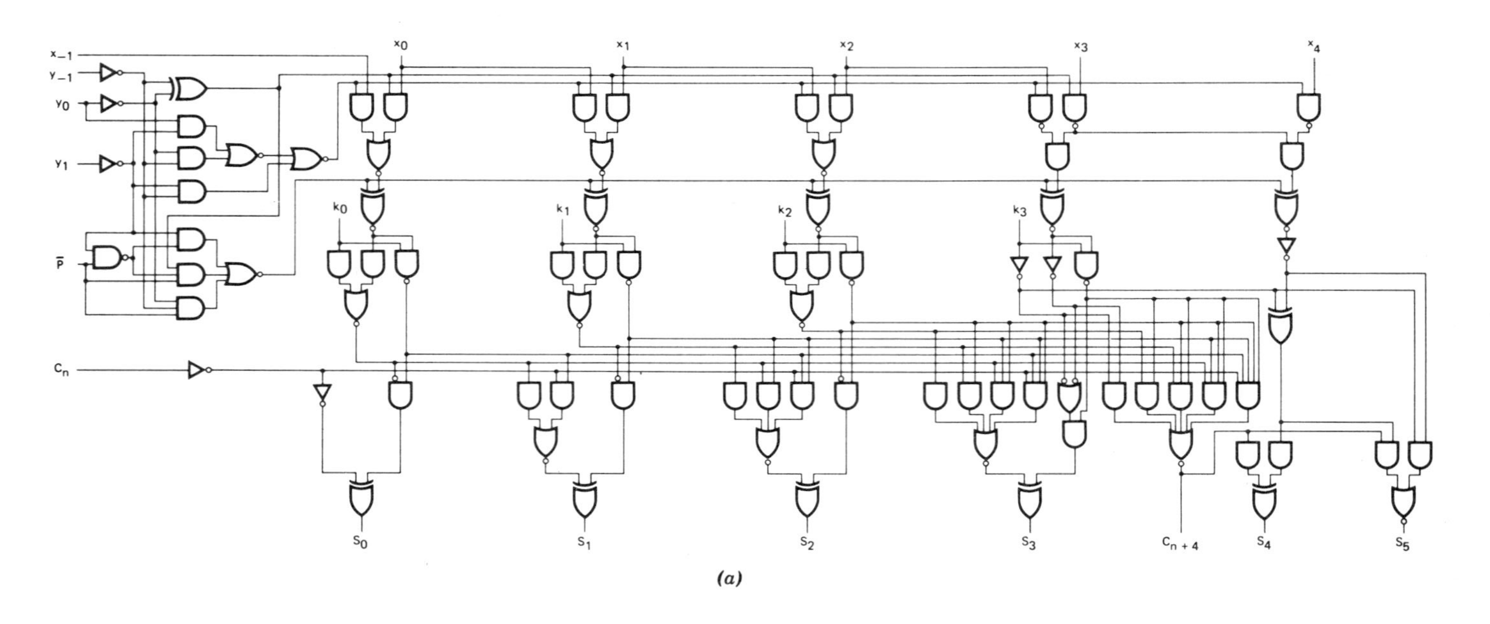

(a)

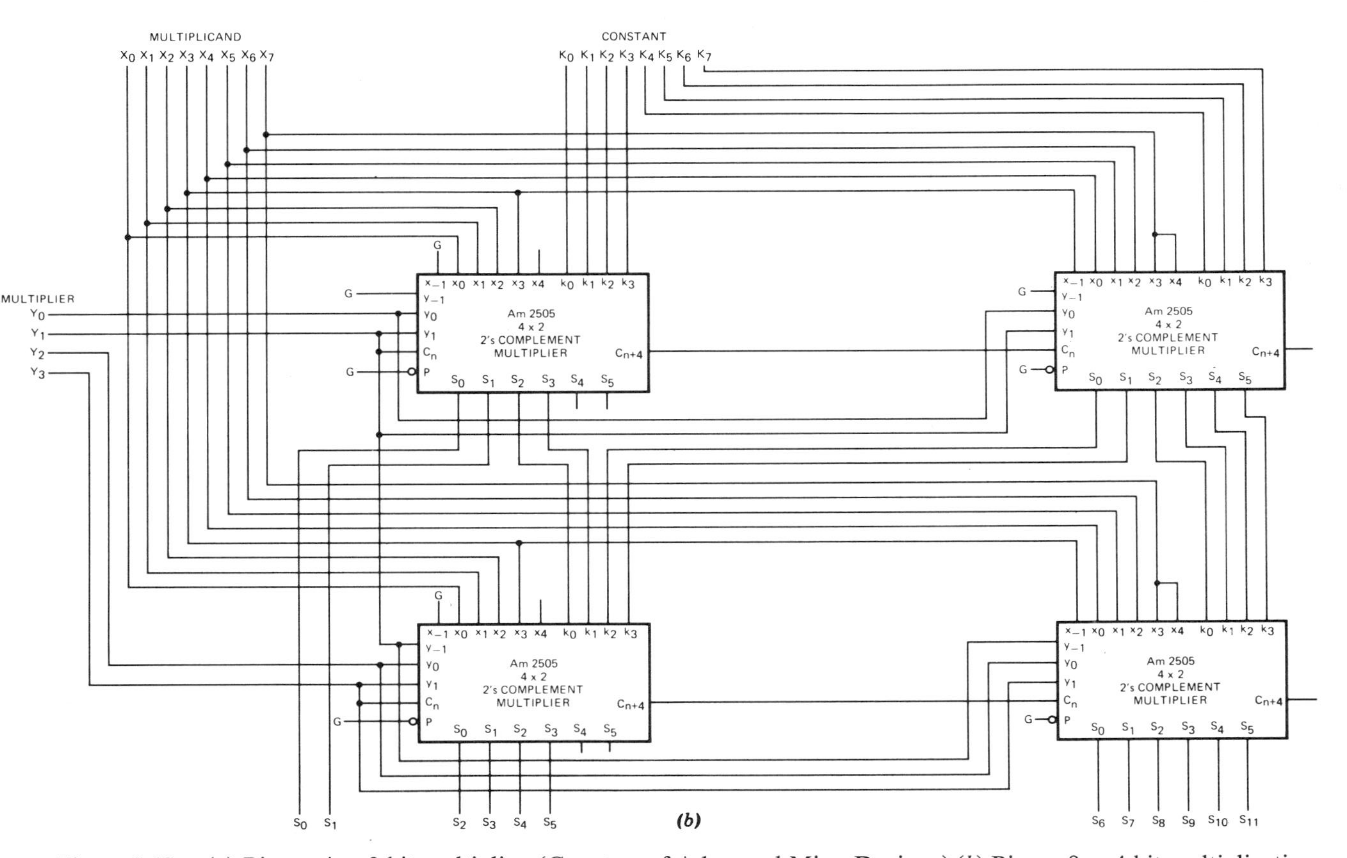

Figure 9.19. (*a*) Binary 4 × 2 bit multiplier. (Courtesy of Advanced MicroDevices.) (*b*) Binary 8 × 4 bit multiplication. (Courtesy of Advanced MicroDevices.)

CHAPTER TEN

Division of Binary Numbers

10.1 INTRODUCTION

In the digital implementation of equations, digital division is very often required. The best approach to dividing binary numbers is by successive subtractions. In this method the divisor is compared to a certain portion of the dividend and if found to be greater, then the quotient for that comparison is 0, while if it is found to be less or equal, the divisor is subtracted from the dividend and the quotient for that comparison is 1. This comparison is repeated for as many times as the number of bits needed in the quotient.

10.2 PARALLEL-PARALLEL DIVISION

Method One

The parallel-parallel division is a direct implementation of the principle of binary division that was described in Chapter One. Figure 10.1 illustrates the block diagram, functional diagram, and logic diagram of a parallel-parallel divider.

The divider consists of a number of subtract-and-enable circuits, the number of which determines the number of bits of the quotient. Each subtract-and-enable circuit consists of a multibit parallel subtracter and of an output-select circuit. A chain of such comparison circuits comprising a 7 bit divisor is shown in Figure 10.1(c).

The first subtract-and-enable circuit is fed by the dividend as well as by the divisor, where the divisor is subtracted from the dividend. The most significant borrow of the subtraction determines if the divisor is greater than the dividend. If the most significant borrow output is 0, then the divisor is equal to or smaller than the dividend, and if it is 1, the divisor is greater than the dividend. The inverse of this borrow bit represents the MSB of the division's quotient.

Example 10.1. If the dividend is $D = 0111011$ and the divisor is $d = 1011011$, the most significant borrow can be obtained by subtracting d from D as follows:

D	0111011
d	1011011
Difference	1100000
Borrow	1000000

The borrow of 1 indicates that the divisor is greater than the dividend, which is actually the case, because $D = 59$ and $d = 91$, and the quotient bit resulting from this comparison is consequently 0. If $D = 1011011$ and $d = 0010110$, then the most significant borrow would be 0:

D	1011010
d	0010110
Difference	1000100
Borrow	0000100

The borrow of 0 indicates that the divisor is less than the dividend, which is actually the case, because $D = 90$ and $d = 22$, and the quotient bit resulting from this comparison is 1. ∎

When the borrow is 0, indicating that the divisor is less than or equal to the dividend, the obtained difference is transferred to the next subtract-and-enable circuit, where it is again compared to the divisor. The transfer is achieved by the gating, and the difference is routed through the AND gates which are enabled by the inverse of the borrow output which is 1.

When the borrow is 1, indicating that the divisor is greater than the dividend, the difference obtained is not transferred but the dividend itself is transferred to the next subtract-and-enable circuit. The transfer is achieved by the gating where the dividend is routed through the gates which are enabled by the nominal borrow output which is now 1.

From each subtraction two outputs result. One is the inverse of the most significant borrow, which is the quotient bit for this comparison, and the other is a multibit parallel number, which is either the minuend input of the subtraction or the difference of the subtraction, depending on the value of the obtained borrow. The remainder of the division appears at the multibit output of the last subtract-and-enable circuit of the network.

The parallel-parallel divider of Figure 10.1(c) handles dividend operands having up to 10 bits, and provides a 7 bit quotient. The size of the operands can be increased by increasing the number of circuits within each subtract-and-enable stage, and the number of bits of the quotient can be increased by increasing the number of the subtract-and-enable stages.

In the parallel-parallel divider of Figure 10.1(c) where the divisor is a 4 bit number, the subtraction performed at the first subtract-and-enable stage is

$$\begin{array}{lllllllllll} & A_{512} & A_{256} & A_{128} & A_{64} & A_{32} & A_{16} & A_8 & A_4 & A_2 & A_1 \\ & B_8 & B_4 & B_2 & B_1 & & & & & & \\ \hline b & D_{512} & D_{256} & D_{128} & D_{64} & & & & & & \end{array}$$

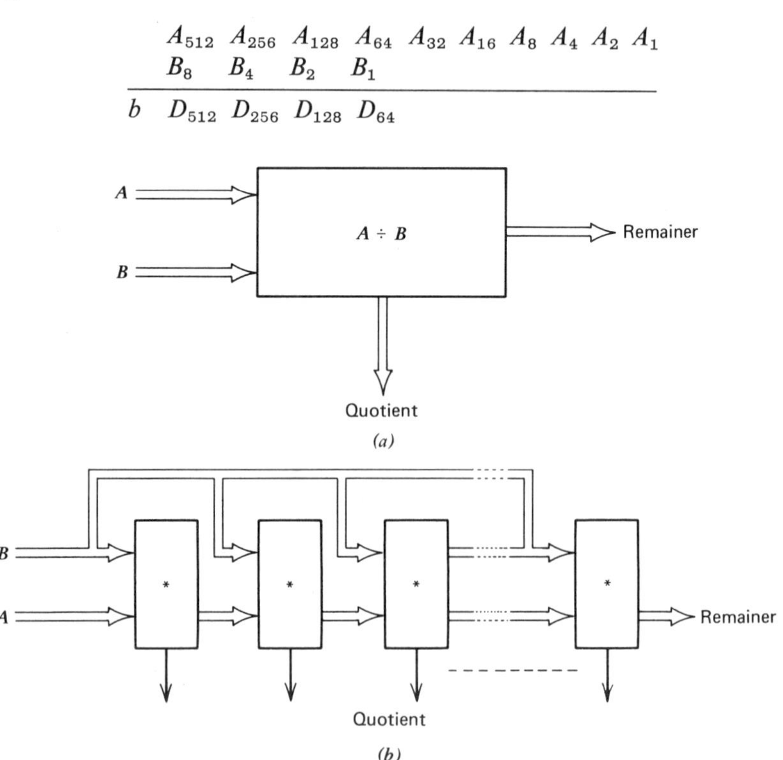

Figure 10.1. Parallel-parallel binary divider: (a) block diagram; (b) functional diagram. (*Note:* *multibit subtract-and-enable network); (c) logic diagram; (d) basic element of logic diagram. (*Note:* **binary subtracter).

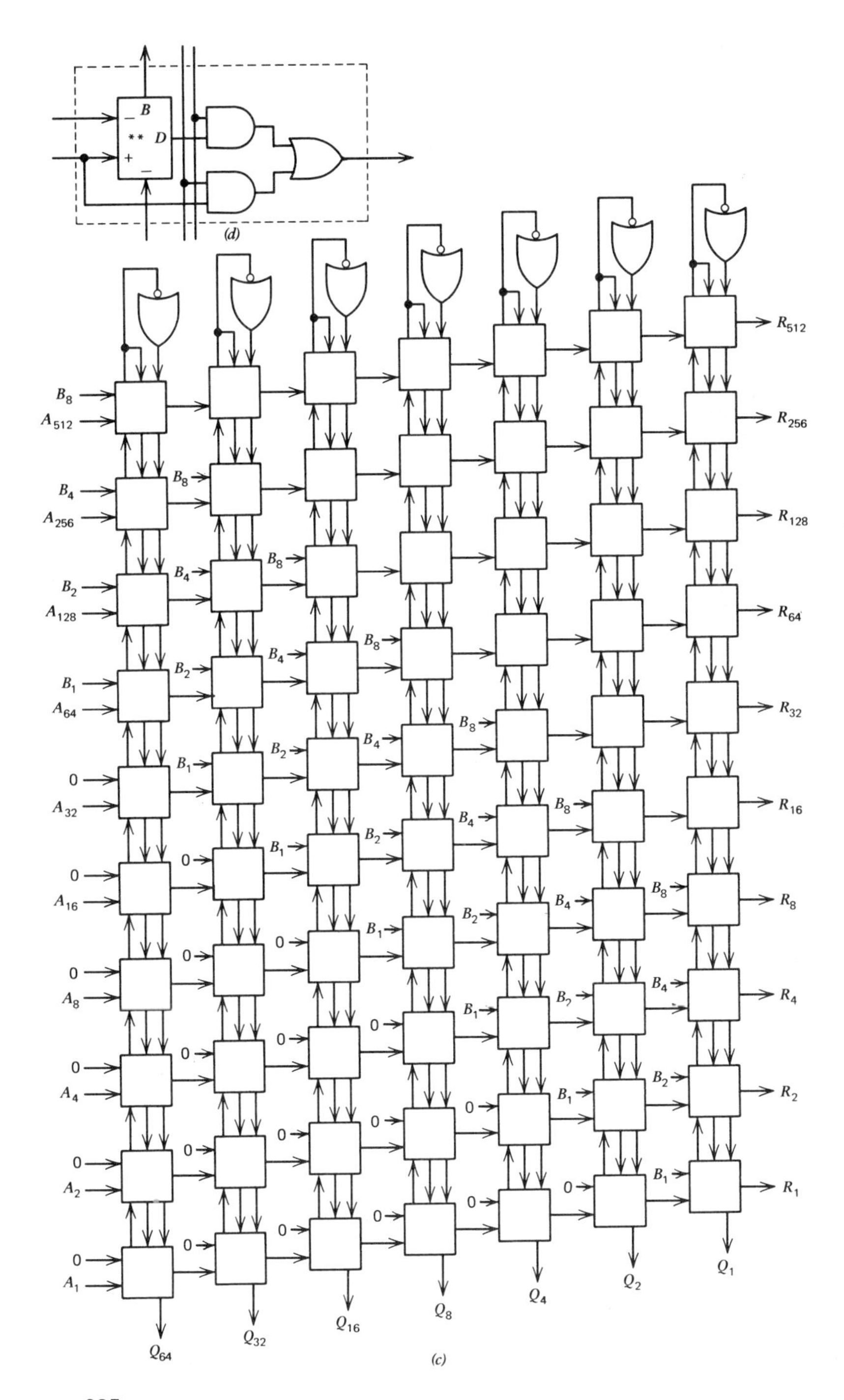
B
**
D
+
(d)
B_8
A_{512}
B_4
A_{256}
B_2
A_{128}
B_1
A_{64}
0
A_{32}
0
A_{16}
0
A_8
0
A_4
0
A_2
0
A_1
R_{512}
R_{256}
R_{128}
R_{64}
R_{32}
R_{16}
R_8
R_4
R_2
R_1
Q_{64}
Q_{32}
Q_{16}
Q_8
Q_4
Q_2
Q_1
(c)

where b is the borrow bit resulting from the first subtraction. The inverse of this bit is a quotient bit, the weight of which is the weight of the A bit right above B_1, which is 64—that is, the most significant borrow of this subtraction is $\overline{Q_{64}}$, which is the inverse of the most significant bit of the quotient of the division.

The second stage of the divider performs subtraction

$$\begin{array}{cccccccccc} D_{512} & D_{256} & D_{128} & D_{64} & A_{32} & A_{16} & A_8 & A_4 & A_2 & A_1 \\ & B_8 & B_4 & B_2 & B_1 & & & & & \\ \hline \end{array} \quad \textbf{Output of first stage}$$

the result of which provides Q_{32} of the quotient and the subtrahend input for the next stage. Similarly, the third, fourth, fifth, sixth, and seventh stages provide Q_{16}, Q_8, Q_4, Q_2, and Q_1, respectively. The multibit output of the last stage is always the remainder of the division. ∎

For a better understanding of the operation of the parallel-parallel divider, the following example is provided:

Example 10.2. Using the network of Figure 10.1(*c*) divide 001110101 by 1011.

First, the divisor is subtracted from the dividend in the first subtract-and-enable circuit, where the minuend is 0001. The subtraction performed is

$$\begin{array}{cccccl} A_{512} & A_{256} & A_{128} & A_{64} & 0001 & \textbf{Partial dividend} \\ B_8 & B_4 & B_2 & B_1 & 1011 & \textbf{Divisor} \\ \hline & & & & 10110 & \end{array}$$

Final borrow ↑ ↑ **Difference**

giving a final borrow of 1. The final borrow of 1 indicates that the subtrahend is greater than the minuend and enables the gates fed by the dividend (see Fig. 10.1(*d*)). The inverse of the final borrow at the same time disables the output of the subtractor; thus the output of the OR gates is 0001. The inverse of the final borrow also indicates the value of the quotient bit for this comparison, which is 0. Therefore, $Q_{64} = 0$. It is Q_{64} because B_1 is under A_{64}.

Next, the divisor is subtracted from the new minuend, which is 0011. The subtraction performed is

$$\begin{array}{rl} 0011 & \textbf{Partial dividend} \\ 1011 & \textbf{Divisor} \\ \hline 11000 & \end{array}$$

Final borrow ↑ ↑ **Difference**

giving a final borrow of 1 which indicates that the divisor is greater than the portion of the dividend to which it is compared. The borrow of 1 enables the gates fed by the minuend of this subtraction, so that this minuend will feed the third subtract-and-enable circuit for another comparison. The quotient bit resulting from the second comparison is the inverse of the final borrow of the second subtraction, which is 1. Therefore, $Q_{32} = 0$.

In the third stage the minuend is 0111 and the subtraction performed is

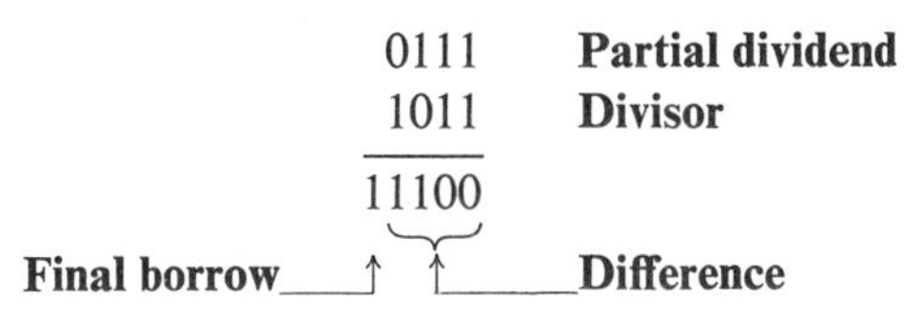

giving a final borrow of 1 which indicates that the divisor is again greater than the portion of the dividend to which it is compared. The borrow of 1 enables the gates fed by the minuend of this subtraction, so that this minuend will feed the fourth subtract-and-enable circuit for another comparison. The quotient bit Q_{16} resulting from the third comparison is the inverse of the final borrow of the third subtraction, which is 1. Therefore, $Q_{16} = 0$.

In the fourth stage the minuend will be 1110 and the subtraction performed is

1110 **Partial dividend**
1011 **Divisor**
00011
Final borrow ↑ ↑ **Difference**

giving a final borrow of 0 which indicates that the divisor is smaller than the portion of the dividend to which it is compared. The borrow of 0 disables the gates fed by the minuend of this subtraction and enables the difference 0011 to feed the fifth subtract-and-enable stage. The quotient bit Q_8 resulting from the fourth comparison is the inverse of the final borrow of the fourth subtraction which is 0, making $Q_8 = 1$.

In the fifth stage the minuend is 00111, which is the output of the fourth stage with A_4 as the LSB, and the subtrahend is the divisor 1011. The subtraction performed is

00111 **Output of preceding (fourth) stage and A_4**
1011 **Divisor**
11100
Final borrow ↑ ↑ **Difference of fifth stage**

giving a final borrow of 1, indicating that the difference obtained in the preceding stage, which is the minuend of this stage, is smaller than the divisor which is the subtrahend. The borrow of 1 enables the gates fed by the minuend of this subtraction so that this minuend will feed the sixth stage.

The output of this stage is 0111. The quotient bit derived from the operation of this stage is Q_4 and its value is 0, which is the inverse of the final borrow.

The minuend of the sixth stage is the output of the fifth stage with A_2 as the LSB. The divisor is subtracted from this number, yielding

```
                    01110    Output of preceding (fifth) stage with A2
                     1011    Divisor
                    -----
                    00011
                    ↑ ↑
Final borrow________↑ ↑______Difference of sixth stage
```

This subtraction gives a final borrow of 0 which indicates that $Q_2 = 1$ and that the obtained difference should be transferred to the next stage for further processing. Thus, the output of this stage is 0011.

In the seventh stage the minuend is the output of the sixth stage with A_1 as the LSB. The subtraction performed here is

```
                    00111    Output of preceding (sixth) stage with A1
                     1011    Divisor
                    -----
                    11100
                    ↑ ↑
Final borrow________↑ ↑______Difference of seventh stage
```

This subtraction yields a final borrow of 1 which indicates that $Q_1 = 0$. The output of stage seven is its own minuend, because the divisor 1011 was found to be greater than the minuend of that stage. That is, the output of stage seven is 0111 and is also the remainder of the division, since the seventh stage is the last stage of divider used.

To summarize, the division performed was

$$0001110101 \div 1011$$

which yielded a quotient of

$$Q = 0001010$$
$$= \text{decimal } 10$$

and a remainder of

$$R = 111$$
$$= \text{decimal } 7$$

In decimal this division is

```
      10
11 ⎡ 117
       7
```

which verified the obtained results. ∎

A parallel-parallel divider of Figure 10.1 is a high-speed device limited only by the total propagation delay of the components used.

Method Two

The preceding method of binary division offers high speed at the expense of considerable hardware. For applications where some reduction in speed can be accommodated, the divider of Figure 10.1(*c*) can be reduced to that of Figure 10.2.

The new divider consists of a binary subtracter, a selection circuit, a storage register, and two shift registers. In this configuration the two parallel operands are initially preset into their respective registers from which they are accessed for the performance of the division. The quotient and remainder produced are available in parallel form with the quotient also available in serial form.

The division is performed in steps where each step produces 1 bit of the quotient. At start of operation, the dividend is preset into a storage register, the output of which serves as the minuend of a binary subtracter. The subtrahend input is the output of a shift register where the divisor has been initially preset.

After the presetting of the dividend and divisor, the binary subtracter computes the difference between dividend and divisor with their MSBs

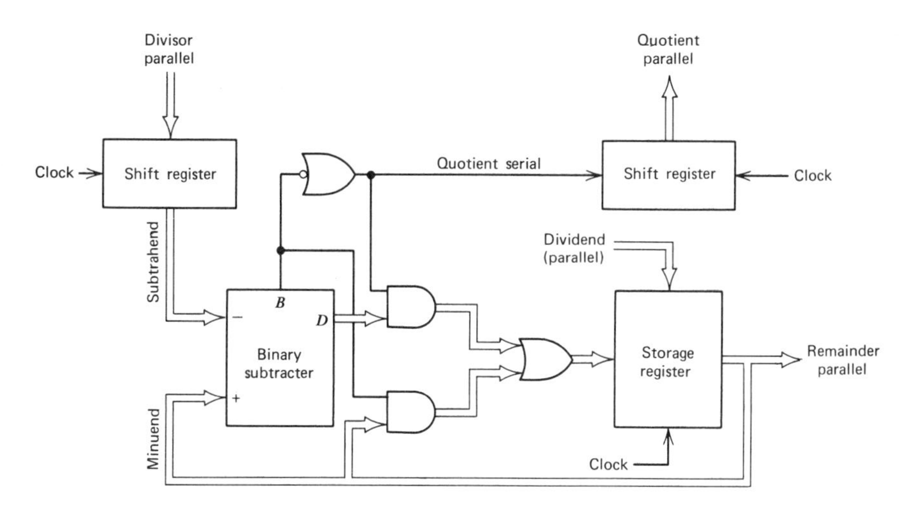

Figure 10.2. Binary division of parallel numbers producing serial quotient.

aligned—that is, if the dividend is 10111011 and the divisor is 1101, the performed subtraction will be

```
                 10111011
               − 1101
               ──────────
                111101011
Final borrow ______↑
```

The inverse of the final borrow of the performed subtraction is the quotient's MSB.

When the final borrow is 1, the minuend is reentered into the dividend storage register to be used again in the computation of the next quotient bit. When the final borrow is 0, the produced difference is entered into the dividend storage register to serve as the minuend in the computation of the next quotient bit.

The first clock pulse enters the generated MSB of the quotient into the quotient shift register and shifts the divisor one binary order down so that it will be ready for the computation of the next quotient bit. During the first clock period the divisor is subtracted from the number stored in the dividend storage register and from this subtraction the second MSB of the quotient is determined. This bit is again the inverse of the final borrow, which at the second clock pulse enters into the quotient shift register.

Similarly, the remaining bits of the quotient are determined and stored in the quotient shift register from where they are available in parallel form.

The number of bits of the quotient equals the number of clock pulses that will update the divider's storage elements.

10.3 PARALLEL-SERIAL DIVISION

Sometimes, in the implementation of binary division, the divisor is in parallel form and the dividend in serial form. The performance of this division cannot be achieved with the circuit discussed in Section 10.2 because that circuit requires that both operands be in parallel form.

Figure 10.3 illustrates a parallel-serial divider employing a single subtract-and-enable stage and four storage elements. This network handles divisors having up to 4 bits, and dividends with an unlimited number of bits. The divisor must be in parallel form and the dividend is serial with its MSB entering the network first. The produced quotient is available in serial form and the remainder is in parallel as a 4 bit number. The number of bits of the quotient equals the number of bits of the dividend.

It is important to note that the amount of hardware required for a parallel-serial divider is a function only of the divisor's number of bits. The operation of the parallel-serial divider of Figure 10.3 is as follows:

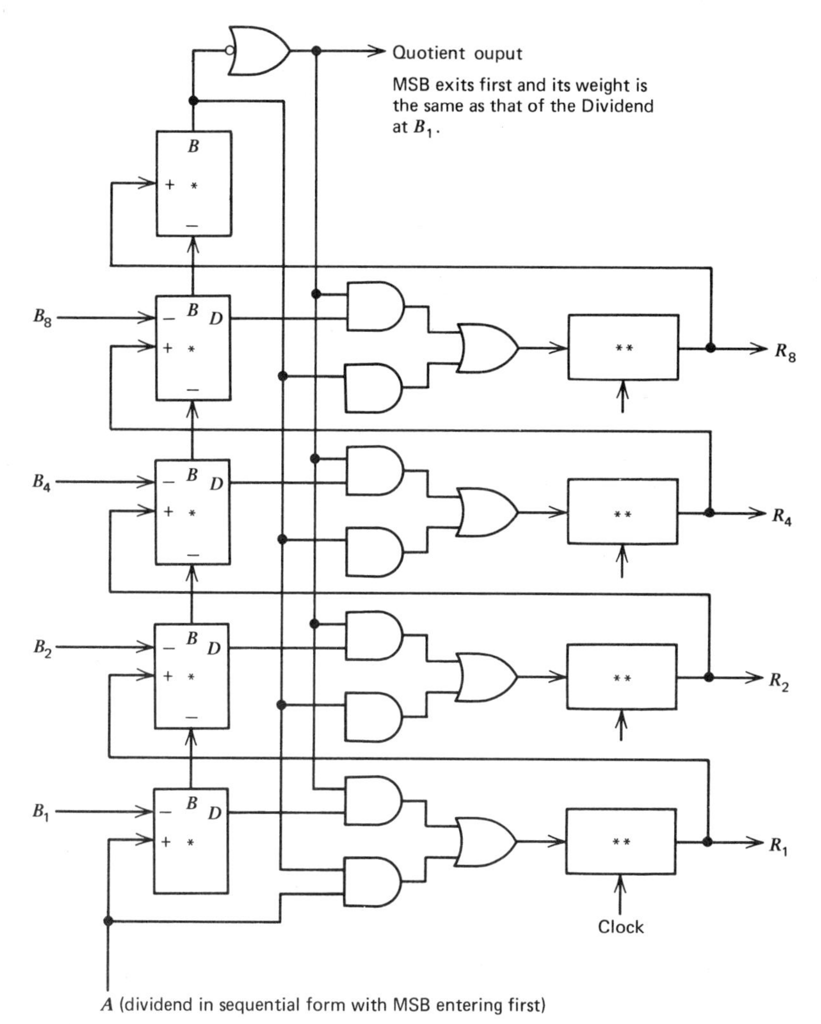

Figure 10.3. Parallel-serial divider. (*Note:* * Binary subtracter. ** Storage register flip-flop initially cleared. At the quotient output the MSB exits first and its weight is the same as that of the dividend at B_1.)

Initially, all storage elements are cleared by means of the "start operation" pulse. During the first clock period, the MSB of the dividend appears at the A input. In this clock period, the minuend of the multibit subtracter is $0000A_{\text{MSB}}$ and the subtrahend is the divisor. The obtained final borrow from the performed subtraction determines the minuend of the subtraction of the next clock period. The inverse of the final borrow is the quotient bit and has

Step	Clock period						
	1	2	3	4	5	6	7
Numbers in Multibit Subtracter	0 0 0 A_{64} $\underline{B_8B_4B_2B_1}$ Difference 1	ST_{OUT2} A_{32} $\underline{B_8B_4B_2B_1}$ Difference 2	ST_{OUT3} A_{16} $\underline{B_8B_4B_2B_1}$ Difference 3	ST_{OUT4} A_8 $\underline{B_8B_4B_2B_1}$ Difference 4	ST_{OUT5} A_4 $\underline{B_8B_4B_2B_1}$ Difference 5	ST_{OUT6} A_2 $\underline{B_8B_4B_2B_1}$ Difference 6	ST_{OUT7} A_1 $\underline{B_8B_4B_2B_1}$ Difference 7
Storage input	(F.B.) (Minuend) or $\overline{\text{(F.B.)}}$ (Difference 1)	(F.B.) (Minuend) or $\overline{\text{(F.B.)}}$ (Difference 2)	(F.B.) (Minuend) or $\overline{\text{(F.B.)}}$ (Difference 3)	(F.B.) (Minuend) or $\overline{\text{(F.B.)}}$ (Difference 4)	(F.B.) (Minuend) or $\overline{\text{(F.B.)}}$ (Difference 5)	(F.B.) (Minuend) or $\overline{\text{(F.B.)}}$ (Difference 6)	(F.B.) (Minuend) or $\overline{\text{(F.B.)}}$ (Difference 7)
Storage output	0000	ST_{IN1}	ST_{IN2}	ST_{IN3}	ST_{IN4}	ST_{IN5}	ST_{IN6}
Quotient	Q_{64}	Q_{32}	Q_{16}	Q_8	Q_4	Q_2	Q_1
Dividend input	A_{64}	A_{32}	A_{16}	A_8	A_4	A_2	A_1

Figure 10.4. Timing diagram of the parallel-serial divider of Figure 10.3 for a 7 bit dividend. (*Note:* ST_{OUT2} = storage output during clock period 2; F.B. = final borrow; *remainder; ST_{IN1} = storage input of clock period 1.)

Step	Clock period							Decimal number
	1	2	3	4	5	6	7	
Number in multibit subtracter	0 0 0 1 1 0 1 1 1 0 1 1 0	0 0 0 1 1 1 0 1 1 1 1 0 0 0	0 0 1 1 1 1 0 1 1 1 1 1 0 0	0 1 1 1 0 1 0 1 1 0 0 0 1 1	0 0 1 1 1 1 0 1 1 1 1 1 0 0	0 1 1 1 0 1 0 1 1 0 0 0 1 1	0 0 1 1 1 1 0 1 1 1 1 1 0 0	divider 11
Storage input	$(1)\cdot(0001)$	$(1)\cdot(0011)$	$(1)\cdot(0111)$	$(\bar{0})\cdot(0011)$	$(1)\cdot(0111)$	$(\bar{0})\cdot(0011)$	$(1)\cdot(0111)$	Remainder 7
Storage output	0 0 0 0	0 0 0 1	0 0 1 1	0 1 1 1	0 0 1 1	0 1 1 1	0 0 1 1	
Quotient	0	0	0	1	0	1	0	Quotient 10
Dividend input	1	1	1	0	1	0	1	Dividend 117
Binary weight	64	32	16	8	4	2	1	

Figure 10.5. Timing diagram of synchronous binary divider where $A = 1110101$ and $B = 1011$.

the same weight as the sequential input A for that time period. During the first period the quotient bit has the weight of A_{MSB}.

Figure 10.4 illustrates the general form of the timing diagram of the parallel-serial divider of Figure 10.3 for a 7 bit dividend. An example illustrating the operation of the divider of Figure 10.3 appears in Figure 10.5 where all the steps of division 1110101 ÷ 1011 are shown.

PROBLEMS

1. State the advantages and disadvatages of the parallel-parallel binary dividers of Figures 10.1(*c*) and 10.2.
2. Design a parallel-parallel binary divider similar to that of Figure 10.2 that handles 6 bit operands.
3. Design a parallel-serial binary divider similar to that of Figure 10.3, capable of handling divisors having up to 8 bits. Use currently available hardware.
4. Draw the timing diagram for a parallel-serial binary divider for the division 1011011 ÷ 1011.
5. Design a parallel-serial binary divider similar to that of Figure 10.3, where the dividend has 10 bits and the divisor has 5.
6. Modify the parallel-parallel divider of Figure 10.1(*c*), so that it computes the reciprocal of a given number inputed at the divisor inputs.

CHAPTER ELEVEN

Powers and Roots of Binary Numbers

11.1 INTRODUCTION

Digital systems design often involves the implementation of mathematical expressions which may contain terms that should be raised to a certain exponent. This exponent may be an integer, the reciprocal of an integer, or a noninteger number.

There are various ways of obtaining the powers and roots of binary numbers. The selection of the most suitable approach, however, depends on the following considerations:

1. Form of binary number (parallel, serial, train of pulses).
2. Maximum time allowed for the execution of the desired mathematical operation.
3. Available digital hardware.

Some of the presented methods can be partially implemented by means of MSI circuits and, hopefully, in the future these and many more functions will be available in MSI or LSI form. Regardless of this availability, for the optimum selection of methods and hardware the digital designer needs to be quite familiar with the implemented algorithms he is considering. This way, the advantages and disadvantages offered by the various algorithms will be more appreciated.

In this chapter many methods of exponentiation are presented. It is possible that some of these methods coincide with a designer's needs. It is also possible that none of the presented methods exactly matches one's requirements. A study of these methods, however, will provide the engineer with sufficient background to enable him to appropriately modify the presented methods, or to invent new ones that meet the constraints of his problem.

11.2 SQUARE OF BINARY NUMBERS

Method One

The simplest way to obtain the square of a binary number is by means of a binary multiplier where both inputs to the multiplier are driven by the binary number. Figure 11.1 shows the basic configuration of this method, where the block may represent any of the parallel-parallel multipliers or the serial-serial multiplier of Chapter Nine. The method is simple in principle, but it does not take advantage of the identity of the two multiplication operands.

Method Two

An analysis of the multiplication process of two identical numbers leads to some interesting observations that considerably simplify the digital design of a squaring network for parallel binary numbers. Shown below is the squaring of binary number $A_8A_4A_2A_1$ where the bits of the square are designated by letter S and the plus sign indicates addition.

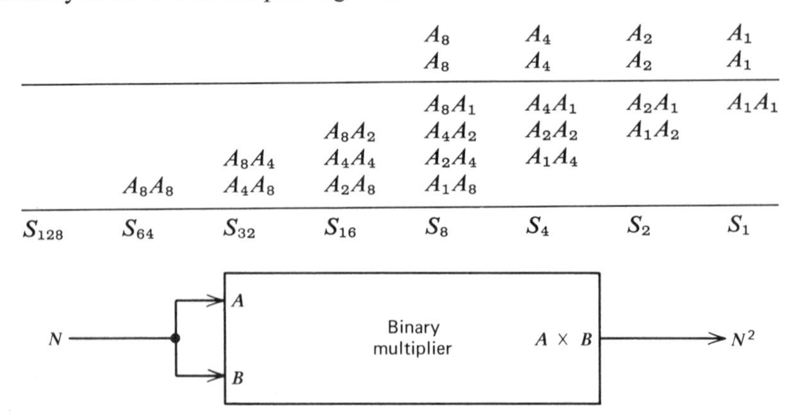

				A_8	A_4	A_2	A_1
				A_8	A_4	A_2	A_1
				A_8A_1	A_4A_1	A_2A_1	A_1A_1
			A_8A_2	A_4A_2	A_2A_2	A_1A_2	
		A_8A_4	A_4A_4	A_2A_4	A_1A_4		
	A_8A_8	A_4A_8	A_2A_8	A_1A_8			
S_{128}	S_{64}	S_{32}	S_{16}	S_8	S_4	S_2	S_1

Figure 11.1. Squaring by means of a binary multiplier.

Now, let us examine how each of the bits of the square is formed. Since S_1 is the AND product of the bit A_1 with itself,

$$S_1 = A_1 \cdot A_1$$
$$= A_1$$

and S_2 is the sum of the terms A_2A_1 and A_1A_2. The two terms are identical and their sum is always 0. Therefore, at all times,

$$S_2 = A_2A_1 + A_1A_2$$
$$= 0$$

The carry of this addition, however, is the product of these two terms, which is $A_2A_1 \cdot A_1A_2$, or A_1A_2. This term should be added in the formation of the next order bit.

The sum of $[(A_4A_1 + A_1A_4) + A_2A_2] + A_1A_2$ is S_4. The last term is the carry from the previous bit. The pair in the parentheses is always 0, producing a carry of A_1A_4 for the next bit, and the second term in the brackets reduces to A_2. Therefore,

$$S_4 = A_2 + A_1A_2$$

Any produced carry is added to the next bit.

The sum of $[(A_8A_1 + A_1A_8) + (A_4A_2 + A_2A_4) + A_1A_4]$ + carry of S_4 is S_8. The pairs in parentheses are always 0 producing two carries A_1A_8 and A_2A_4 for the next bit. Therefore,

$$S_8 = A_1A_4 + \text{carry of } S_4$$

The sum of $[(A_2A_8 + A_8A_2) + A_4A_4] + A_1A_8 + A_2A_4$ + carry of S_8 is S_{16}. The pair in parentheses is always 0, producing a carry of A_2A_8 for the bit next in significance, and the second term in the brackets reduces to A_4. Therefore,

$$S_{16} = A_4 + A_1A_8 + A_2A_4 + \text{carry of } S_8$$

From this addition two carries are produced. Their weights are 32 and 64.

The sum of $(A_4A_8 + A_8A_4)$ + carry of S_{16} (weight 32) is S_{32}. The pair in the parentheses is always 0, producing a carry of A_4A_8 for the bit next in significance. Therefore,

$$S_{32} = \text{carry of } S_{16} \text{ (weight 32)}$$

The sum of A_8A_8 + carry of S_{16} (weight 64) is S_{64}. The first term reduces to A_8. Therefore,

$$S_{64} = A_8 + \text{carry of } S_{16} \text{ (weight 64)}$$

The carry produced from the S_{64} addition is the MSB of the multiplication.

Therefore,

$$S_{128} = \text{carry of } S_{64}$$

The maximum value of $A_8A_4A_2A_1$ however is 1111 (decimal 15) and its square is 125. Therefore,

$$S_{128} = 0$$

and S_{64} does not produce a carry. Consequently, the expression for S_{64} reduces to

$$S_{64} = A_8 \quad \text{or} \quad \text{carry of } S_{16} \text{ (weight 64)}$$

Starting with the LSB, the algorithm can be implemented with binary arithmetic units. Figure 11.2 illustrates the block diagram of a 4 bit squaring network, which is greatly simpler than the 4 bit parallel multiplier of Figure 9.1.

Method Three

Numbers are sometimes expressed as pulse trains where the number of pulses is the actual number. To use any of the preceding squaring methods, the number must be first identified in binary form. This can be achieved by entering the pulses into a binary counter. At the end of the entry, the output of the binary counter expresses the number of pulses in binary parallel form. After this conversion has been accomplished, Method One or Method Two can be used.

The network of Figure 11.3 computes the square of numbers available in pulse-train form and expresses the square in parallel binary form. This network is based on the following equality:

$$X^2 = (X - 1 + 1)^2 \tag{11.1}$$

In an upward counting binary counter, if X_n is the output at time n, $X_n - 1$ will be the output at time $n - 1$. Thus,

$$X_{n-1} = X_n - 1$$

Applying this property to (11.1), we obtain

$$X_n^2 = (X_{n-1} + 1)^2 \tag{11.2}$$

Expanding the right side of (11.2) we obtain

$$X_n^2 = X_{n-1}^2 + 2X_{n-1} + 1 \tag{11.3}$$

Equation (11.3) can be easily implemented as shown by the block diagram of Figure 11.3. A detailed schematic appears in Figure 11.4, where the exact

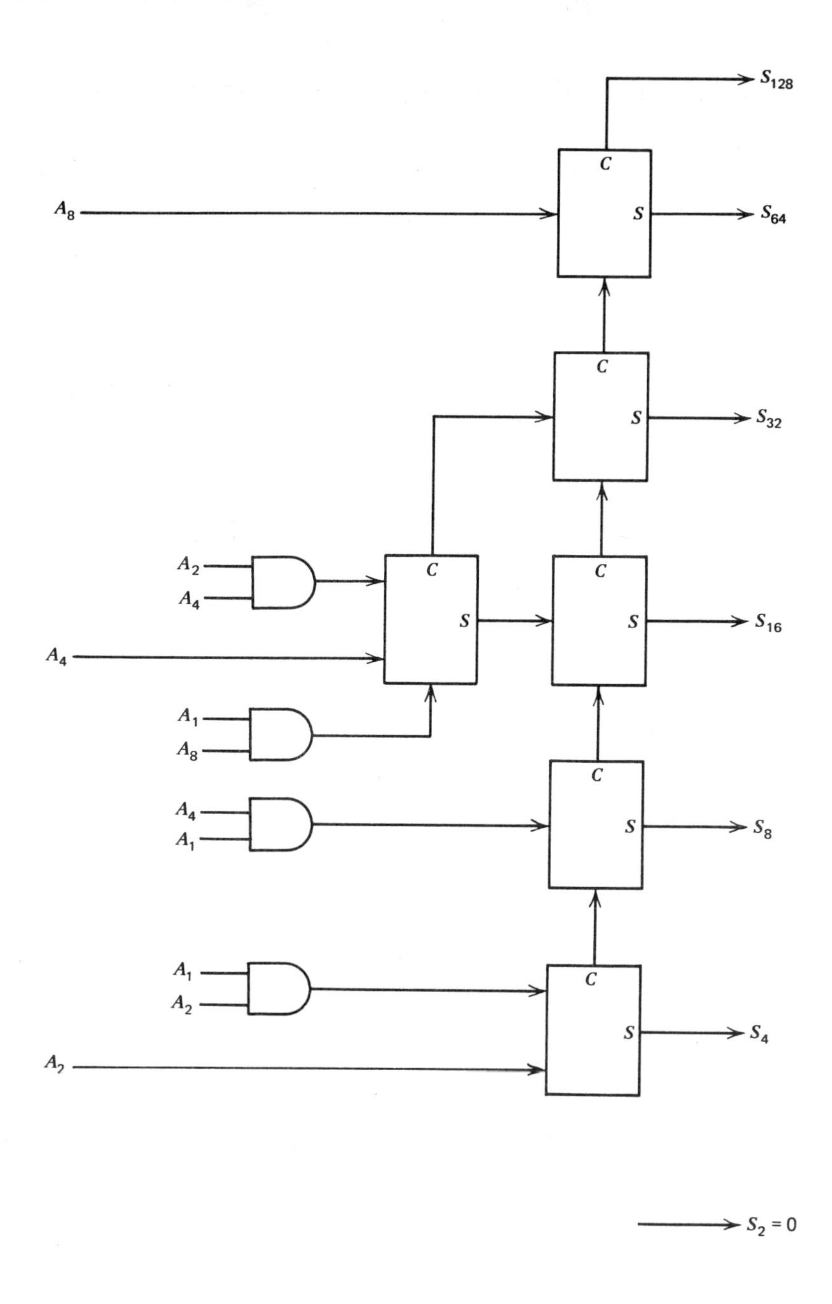

Figure 11.2. A 4 bit parallel squaring network. The blocks represent binary adders.

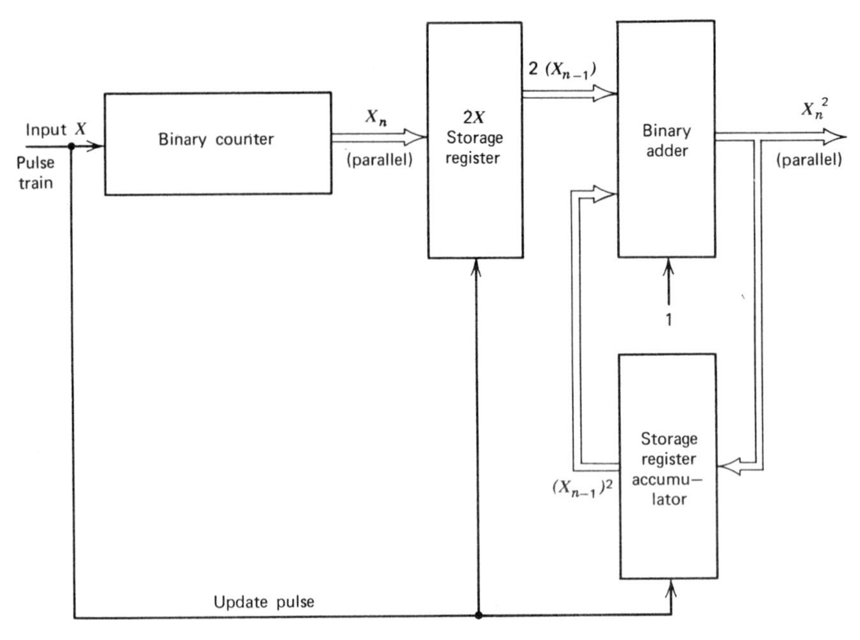

Figure 11.3. Block diagram of digital network used for the computation of the square of numbers available in pulse-train form.

content of the blocks as well as the connections between them are illustrated. The operation of this network is as follows: The number for which the square is needed, enters the counter in pulse-train form. For example, if the number is 15, 15 pulses will enter into the counter. These pulses constitute the clock of the squaring network. In addition to triggering the counter, function of this clock is to update the storage registers. There are three terms that contribute to the output of the squaring network. These are

1. Output of the accumulator which contains X_{n-1}^2.
2. Output of the storage register which represents $2X_{n-1}$; this information enters the register directly from the counter output.
3. Output of a single storage element which provides a 1 of least significance.

When the counter output is initially 0...00, which is the first number in the operation, the output of the squaring network is also 0...00. This is because all inputs to the multibit adder are 0.

At the arrival of the first pulse the counter output changes from 0...00 to 0...01. The first clock pulse also sets the single storage element to 1.

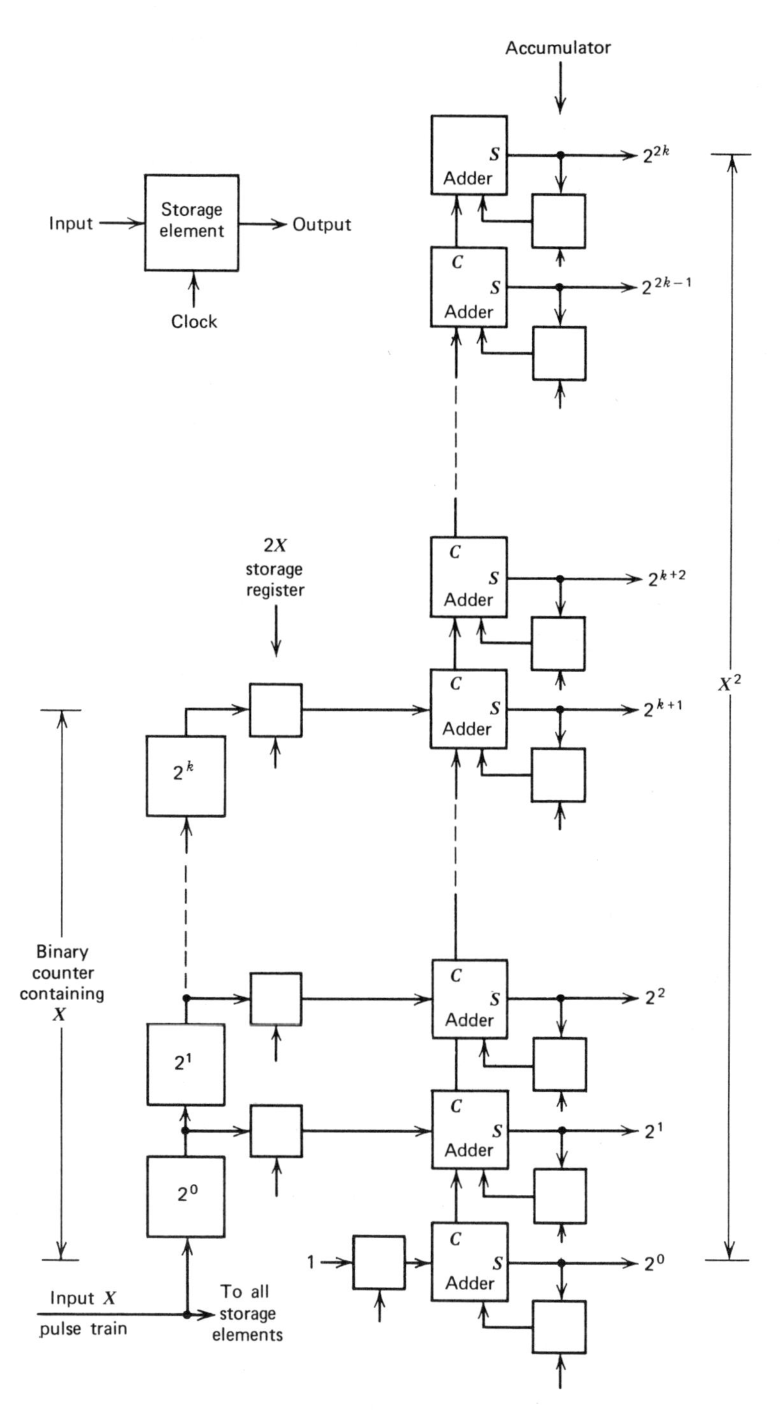

Figure 11.4. Logic diagram of Figure 11.3. (*Note:* All storage elements are initially cleared.)

This storage element will remain in the 1 state until the squaring network is cleared to start a new computation.

After the first clock pulse, the $2X_{n-1}$ storage register is still cleared, holding 2(0. . .00). The accumulator is also cleared, holding $(0\ldots00)^2$, and the single storage element is 1. These three numbers enter the multibit adder and provide the square of the number presently in the counter—that is,

$$X_n^2 = X_{n-1}^2 + 2X_{n-1} + 1$$

is directly implemented. The addition performed in the multibit adder is

Accumulator output	0. . .00	X_{n-1}^2	$X_{n-1} = 0\ldots00$
Storage register output	0. . .0	$2X_{n-1}$	$X_n = 0\ldots01$
Single element output	1	1	X_n = counter output
Multibit adder output	0. . .01	X_n^2	

At the arrival of the second pulse, the counter output changes from 0. . .001 to 0. . .010. This pulse also updates the storage registers. Now, the output of the accumulator is 0. . .01 which is X_{n-1}^2; the output of the $2X_{n-1}$ register is 0. . .01 where the least significant bit has a weight of 2; and the output of the single storage element remains unchanged in the state of 1. These three numbers constitute the input to the multibit adder where the following addition is performed:

0. . .0001	X_{n-1}^2	X_n	(counter output) = 0. . .010
0. . .001	$2X_{n-1}$		
1	1		
0. . .0100	X_n^2		**Adder output**

The sum is 4. The counter output during this period is 2, and the adder output is its square.

At the arrival of the third pulse the counter output changes from 0. . .010 to 0. . .011. This pulse also updates the storage registers. Now, the output of the accumulator is 0. . .0100, the output of the $2X_{n-1}$ storage register is 0. . .010, and the output of the single flip-flop remains 1. These three numbers are added in the multibit adder in the following manner:

0. . .00100	X_{n-1}^2	X_n	(counter output) = 0. . .011
0. . .0010	$2X_{n-1}$		
1	1		
0. . .01001	X_n^2		**Adder output**

The sum is 9, and it is the square of the counter output during that clock period which is 3. Figure 11.5 shows the output at various points of the

					X_n^2								=	X_{n-1}^2							+	$2X_{n-1}$			+ 1	
Number of pulses entering the counter	Counter output				Adder output								Accumulator output								2*X* storage register output				Single element output	Sum output in decimal
	2^3	2^2	2^1	2^0	2^7	2^6	2^5	2^4	2^3	2^2	2^1	2^0	2^7	2^6	2^5	2^4	2^3	2^2	2^1	2^0	2^4	2^3	2^2	2^1	2^0	
0	0	0	0	0	0	0	0	0	0	0	0	0	0	0	0	0	0	0	0	0	0	0	0	0	0	0
1	0	0	0	1	0	0	0	0	0	0	0	1	0	0	0	0	0	0	0	0	0	0	0	0	1	1
2	0	0	1	0	0	0	0	0	0	1	0	0	0	0	0	0	0	0	0	1	0	0	0	1	1	4
3	0	0	1	1	0	0	0	0	1	0	0	1	0	0	0	0	0	1	0	0	0	0	1	0	1	9
4	0	1	0	0	0	0	0	1	0	0	0	0	0	0	0	0	1	0	0	1	0	0	1	1	1	16
5	0	1	0	1	0	0	0	1	1	0	0	1	0	0	0	1	0	0	0	0	0	1	0	0	1	25
6	0	1	1	0	0	0	1	0	0	1	0	0	0	0	0	1	1	0	0	1	0	1	0	1	1	36
7	0	1	1	1	0	0	1	1	0	0	0	1	0	0	1	0	0	1	0	0	0	1	1	0	1	49
8	1	0	0	0	0	1	0	0	0	0	0	0	0	0	1	1	0	0	0	1	0	1	1	1	1	64
9	1	0	0	1	0	1	0	1	0	0	0	1	0	1	0	0	0	0	0	0	1	0	0	0	1	81
10	1	0	1	0	0	1	1	0	0	1	0	0	0	1	0	1	0	0	0	1	1	0	0	1	1	100
11	1	0	1	1	0	1	1	1	1	0	0	1	0	1	1	0	0	1	0	0	1	0	1	0	1	121
12	1	1	0	0	1	0	0	1	0	0	0	0	0	1	1	1	1	0	0	1	1	0	1	1	1	144
13	1	1	0	1	1	0	1	0	1	0	0	1	1	0	0	1	0	0	0	0	1	1	0	0	1	169
14	1	1	1	0	1	1	0	0	0	1	0	0	1	0	1	0	1	0	0	1	1	1	0	1	1	196
15	1	1	1	1	1	1	1	0	0	0	0	1	1	1	0	0	0	1	0	0	1	1	1	0	1	225

Figure 11.5. Operation of squaring circuit of Figure 11.3.

squaring network from the time the counter is cleared to start the computation until it reaches 15.

The approach of Figure 11.3 can be used to provide the square of any continuous linear function. If the function does not start from 0, the starting number should be preset into the counter and proper numbers simulating previous operation should be preset into the storage registers.

If the starting number is K, K should be preset into the counter, $2(K - 1)$ into the $2X_{n-1}$ register, $(K - 1)^2$ into the accumulator, and 1 into the single storage element. For example, if the starting number is 21, the counter should be preset to 21, the $2X_{n-1}$ register to 40, the accumulator to 400, and the single storage element to 1. The output of the multibit adder will then be $400 + 40 + 1$, which is 441 or $(21)^2$. After the arrival of the first pulse, the summation takes the form $441 + 42 + 1$, which is 484 or $(22)^2$.

11.3 *n*th INTEGRAL POWER OF BINARY NUMBERS

Method One

The most straightforward approach of raising a binary number to an integer exponent is by continuous multiplication of the given binary number using separate multipliers. The block diagram of this method is shown on Figure 11.6, where the multiplier used is a parallel-parallel binary multiplier similar to that of Figure 9.1. The advantage of the network of Figure 11.6 is speed. Its operation is asynchronous and as a result it needs no clock. The time delay between the application of input and the availability of output depends only on the total propagation delay of the circuits used. The disadvantage of this circuit is the excessive amount of required hardware.

Method Two

A binary number K can be raised to a given integer exponent n by means of a single multiplier, a storage register, and the necessary control network.

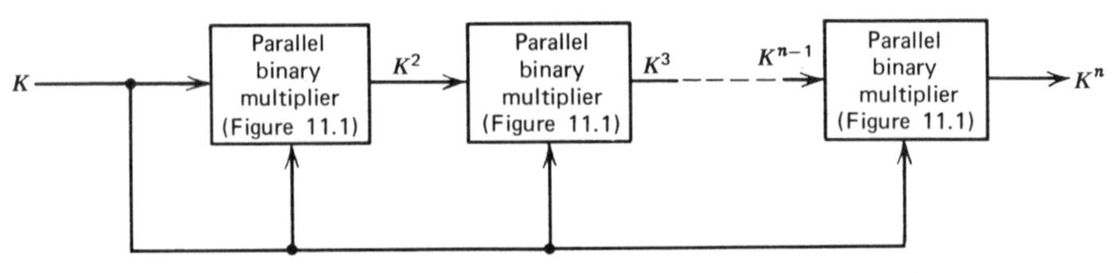

Figure 11.6. Cascaded multipliers used for the generation of K^n.

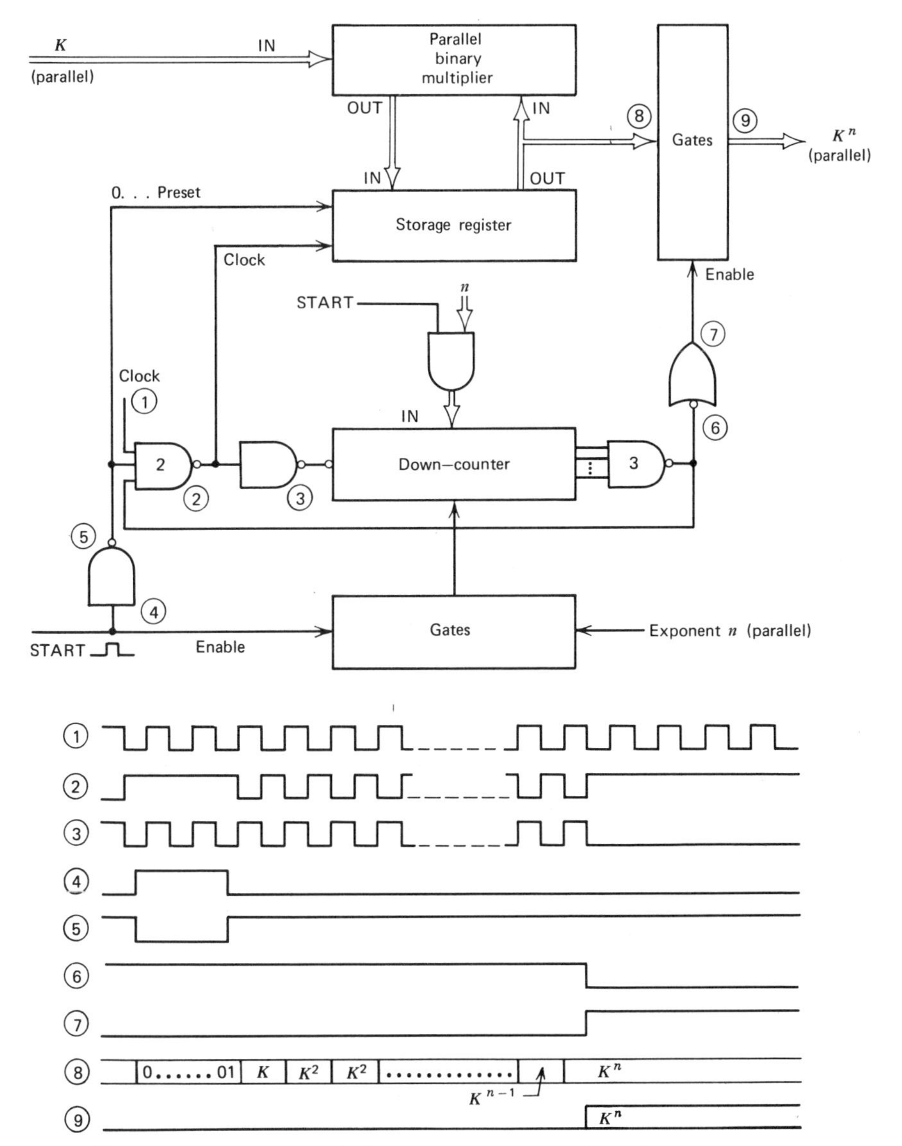

Figure 11.7. Exponentiating network (Method Two).

This approach is illustrated in Figure 11.7. The operation of this network proceeds as follows:

The START pulse presets number 0...01 into the storage register and enters the exponent n into the down-counter. The exponent n must be in parallel form and available during the START pulse. The output of the down-counter determines whether the clock should be used and, if it should, for how long.

When $n = 0$, all the inverted outputs of the down-counter are 1, which is HIGH, and the output of gate 3 is LOW. This output inhibits the clock at gate 2, so that the storage register is not updated. Inverted by gate 4 this output also enables the output gates of the exponentiating network. Consequently, the storage register does not enter the output of the multiplier, which is K, but remains with its preset value of 0...01. Thus, when the exponent n is 0, the output of the exponentiating network is 0...01.

When $n = 1$, after the START pulse the output of gate 3 is HIGH. This is because the inverse output of the LSB of the down-counter is LOW. With $\overline{\text{START}}$ and the output of gate 3 both HIGH, gate 2 passes the clock and updates the storage register. Before the first clock pulse, the output of the multiplier is K. This is the result of the multiplication $(K)\cdot(0...01)$. The first term is the number that should be exponentiated and the second term is the initial output of the storage register.

At the arrival of the first clock pulse the storage register is updated, becoming K or K^1. At the same time the output of the down-counter changes from 0...01 to 0...00. The new output makes the output of gate 3 LOW which in turn disables gate 2. With gate 2 disabled, the clock stops from further updating the storage register. Thus, the computation ends with the output of the exponentiating network being K which is the last number in the storage register.

When $n = 2$, the storage register is updated twice, and its final output is K^2. Similarly, when the exponent is n, the storage register is updated n times and its final output is K^n.

Since the number of bits of the output increases with each multiplication, the multiplier as well as the storage register used must have sufficient bit capability to handle the requirements. For example, if the 4 bit number 1011 is raised to the fourth power, it becomes 11100100110001, which is a 14 bit number. To perform this operation, a 14 bit multiplier and storage unit should be used.

The advantage of Method Two over Method One is less hardware while the disadvantage is lower speed. The time requirements of Method Two are n clock pulses, and the bit size requirement of the multiplier and storage unit is nm where n is the exponent to which the given number K should be raised, and m is the bit size of K.

Method Three

Another way to raise a binary number to a given integer exponent is by means of a shift register and a parallel-serial multiplier, similar to those of Figures 9.12 and 9.13.

The shift register serves two purposes. One is to convert the given binary number from parallel to serial form—this is necessary for the first multiplication, which will provide the square of the input binary number. The other purpose is to provide storage for the output of the parallel-serial multiplier. An exponentiating network employing this approach is illustrated in Figure 11.8. The operation of this network proceeds as follows:

A START pulse clears the storage elements of the parallel-serial multiplier and presets the input binary number 00...01 into the shift register. The LSB 1 is preset into the uppermost storage element of the shift register. Upon removal of the START pulse, the clock starts triggering the shift register and the storage elements of the parallel-serial multiplier.

After s pulses, where s is the number of stages of the shift register, the input binary number K will be in the shift register and be available in parallel form with its LSB occupying the uppermost storage element of the shift register. If the clock continues counting for another s pulse, K, which is in the shift register, will "pass through" the parallel-serial multiplier, be multiplied by K and form K^2; K^2 will end up in the shift register with its LSB occupying the uppermost storage element of the shift register.

If the clock continues counting for another s pulses, the third power of the input number will be generated. This process must continue for ns pulses, where n is the exponent to which input binary number K should be raised and s is the number of stages in the shift register.

The size of the shift register must be such that it can accommodate the products of the performed multiplications. If K has b bits, for K^2 the shift register should have a $2b$ bit length. For the computation of K^3 the shift register should have a $3b$ bit-length, and for the computation of K^n the shift register should have an nb bit-length.

The advantages of this method over the preceding ones are: First, its implementation requires less hardware because the parallel-serial multiplier consists of less elements than a parallel-parallel multiplier; and second, this approach produces the answer K^n in parallel as well as in sequential form.

The disadvantage of this method is time. This network requires ns clock pulses for the exponentiation of K to the n power, while Method One is limited only by the propagation delay of the circuits used, and Method Two requires only n clock pulses for the same computation.

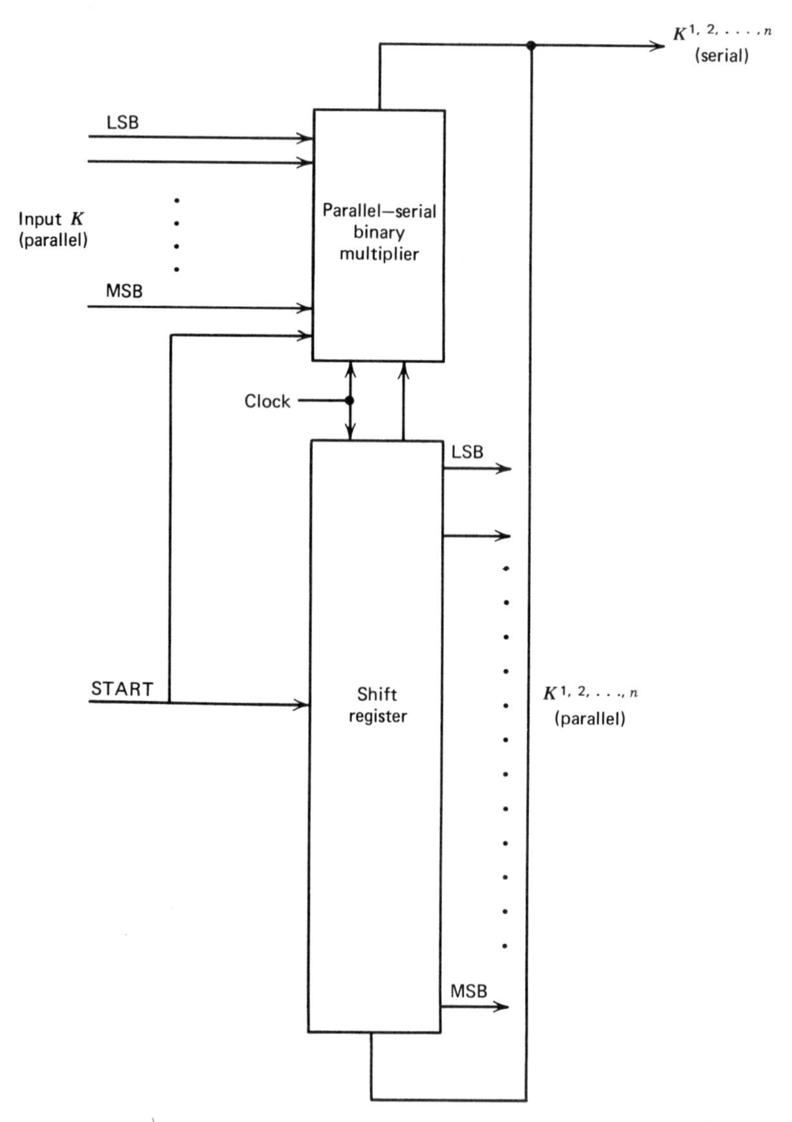

Figure 11.8. Parallel-serial multiplier used for the generation of K^n.

11.4 SQUARE ROOT OF BINARY NUMBERS*

The technique for the computation of the square root of binary numbers presented here has been derived from the general algorithm for the computation of the square root and offers a novel approach to the computation of the square root of binary numbers. The derivation of this technique is illustrated in Appendix A. It is a simple and straightforward approach applicable to long-hand operations as well, and it can be software-implemented even in computers where the arithmetic repertoire includes only subtraction.

In the computation, no correction factors are required, and the binary numbers may have a fractional part, as well as an integral, while the accuracy may be extended to any number of bits. The algorithm for the computation of the square root of binary numbers is

Step 1. Starting from binary point, partition the number into 2 bit groups.

Step 2. Let the first bit of the square root be 1.

Step 3. Subtract 01 from the first 2 bit group of the number.

Step 4. Write the next 2 bit group of the number to the right of the computed difference; call the grouping M.

Step 5. To the right of the square root R obtained thus far, write 01 forming $R01$.

Step 6. If $R01$ is greater than M, the root bit resulting from this iteration is 0. Return to Step 4 and compute the next bit of the square root.

Step 7. If $R01$ is equal to or less than M, the next root bit resulting from this iteration is 1. Compute difference $M - R01$ and return to Step 4 to compute the next bit of the square root.

Figure 11.9 illustrates the above algorithm in flow diagram form. The simplicity of the algorithm can be best illustrated by the following two examples:

Example 11.1. Compute the square root of the binary number 110001, decimal 49.

The computation is illustrated in Figure 11.10 and proceeds as follows:

1. Number 110001 is marked into 2 bit groups starting from the binary point. Since there are three 2 bit groups to the left of the binary point, the square root of this number will have 3 integer bits.
2. Number 01 is subtracted from first 2 bit group and a 1 is written on left side, becoming the MSB of the square root of 110001.

* The material in this section originally appeared by the author in *Computer Design*, August 1972, pp. 53–57.

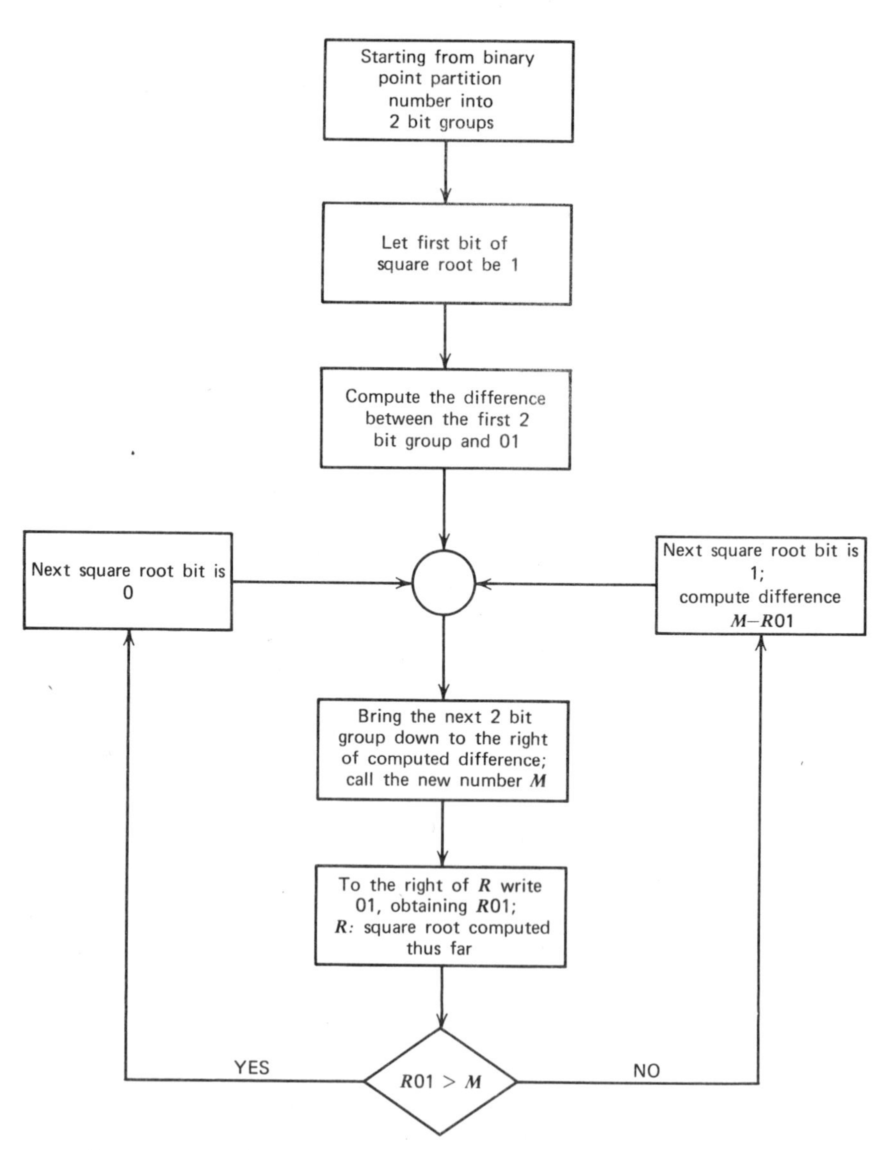

Figure 11.9. Algorithm for the computation of the square root of binary numbers. (Courtesy of *Computer Design.*)

3. The second 2 bit group is now brought down to line 3 next to the difference of lines 1 and 2. On line 4 under the second 2 bit group write 01, and to the left of 01 write the square root obtained thus far—in this case, only bit 1. Now subtract line 4 from line 3. If the difference is negative, a 0 is inserted in the root space of that operation and the number on line 3 is copied onto line 5. If the difference is positive or 0, a 1 is inserted in the root space of that operation and the difference is written on line 5. In this case, the difference is positive and is written on line 5 with a 1 inserted in the root space on the left.
4. The third 2 bit group is now brought down to line 5 next to the difference of lines 3 and 4. On line 6 under the third 2 bit group write 01, and to the left of 01 write the square root obtained thus far—in this case, 11. Now subtract line 6 from line 5. Since the difference is not negative, it is written on line 7, and a 1 is written in root space on left. Since line 7 is 0 and no other 2 bit group exists, computation ends and the square root of 110001 is found to be 111. Indeed, the square root of 49 (110001 in binary) is 7 which is 111 in binary. ▮

Example 11.2. Compute the square root of the binary number 1.101. The computation is illustrated in Figure 11.11.

The square root of 1.101 to six fractional binary places is 1.010001, leaving a remainder of 0.000001011111. The validity of the square root can be verified by squaring the square root and adding the remainder to it.

```
            1.010001      Square root
          × 1.010001
            --------
            1010001
       1010 001
    1 01000 1
  -------------------
    1.10011 0100001       (Square root)²
  + 0.00000 1011111       Remainder
  -------------------
    1.10100 0000000       Given number
```
▮

It is important to note that this method of computing the square root of a binary number requires only amplitude comparisons and subtractions. As a result, inexpensive minicomputers can now have a square root feature using the above algorithm.

Implementation of the Square Root Algorithm

To digitally implement the square root algorithm, the mechanics of the algorithm and their time interrelation must be identified. The process consists

Root	Operations	Line
	11 00 01	1
1 (MSB)	01	2
	10 00	3
1	1 01	4
	11 01	5
1	11 01	6
	00 00	7

Figure 11.10. Step-by-step extraction of the square root of the binary number 110001 of Example 11.1. (Courtesy of *Computer Design*).

of amplitude comparisons and of subtractions. The results of the comparisons are the square root bits, and the result of the last subtraction is the remainder of the operation. There are two methods to this implementation.

Method One—A Sequential Design

If the minuend is held stationary and the subtrahend is properly shifted to the right in agreement with the algorithm, the minuend weights and the subtrahend bits will have the relationship shown in Figure 11.12. The subtrahend can be considered as being of two parts—an expanding one and a fixed one. The expanding part consists of the square root bits obtained until that comparison, and is shifted to the right 1 bit per comparison. The fixed part is 01 and is shifted to the right 2 bits per comparison.

A sequential implementation of the square root algorithm appears in Figure 11.13, where the minuend is held stationary and the subtrahend is shifted to the right 1 bit position per comparison. This implementation has been designed for the computation of the square root of binary numbers having up to 12 bits. The number may or may not have a fractional part, but when partitioned into 2 bit groups, the number of such groups should not exceed six.

N, the number the square root of which is to be computed, is entered into a 12 bit storage register. This register holds the minuend of the subtractions that are performed in the process of the algorithm. The output of the storage register serves as the minuend of the 12 bit binary subtracter. The subtrahend to this subtracter is the output of the square root shift register. This shift register is initially cleared. During each comparison, 1 bit of the square root is computed and entered into the proper flip-flop of the square root shift register.

Root	Operations	Line
	01. 10 10 00 00 00	1
1 (MSB)	01	2
.	00 10	3
0	1 01	4
	00 10 10	5
1	10 01	6
	00 01 00	7
0	1 01 01	8
	00 01 00 00	9
0	10 10 01	10
	00 01 00 00 00	11
0	1 01 00 01	12
	00 01 00 00 00 00	13
1	10 10 00 01	14
	00 00 01 01 11 11	

Figure 11.11. Step-by-step extraction of the square root of binary number 1.101 of Example 11.2. (Courtesy of *Computer Design*.)

Order of comparison	Computed square root bit	Significance of subtrahend bits											
		Minuend's binary weights											
		2048	1024	512	256	128	64	32	16	8	4	2	1
First	R_{12}	0	1										
Second	R_{16}		R_{32}	0	1								
Third	R_{8}			R_{32}	R_{16}	0	1						
Fourth	R_{1}				R_{32}	R_{16}	R_{8}	0	1				
Fifth	R_{2}					R_{32}	R_{16}	R_{8}	R_{4}	0	1		
Sixth	R_{2}						R_{32}	R_{16}	R_{8}	R_{4}	R_{2}	0	1
Square root								R_{32}	R_{16}	R_{8}	R_{4}	R_{2}	R_{1}

Figure 11.12. Significance of the subtrahend bits and their relation to the minuend weights. (Courtesy of *Computer Design*.)

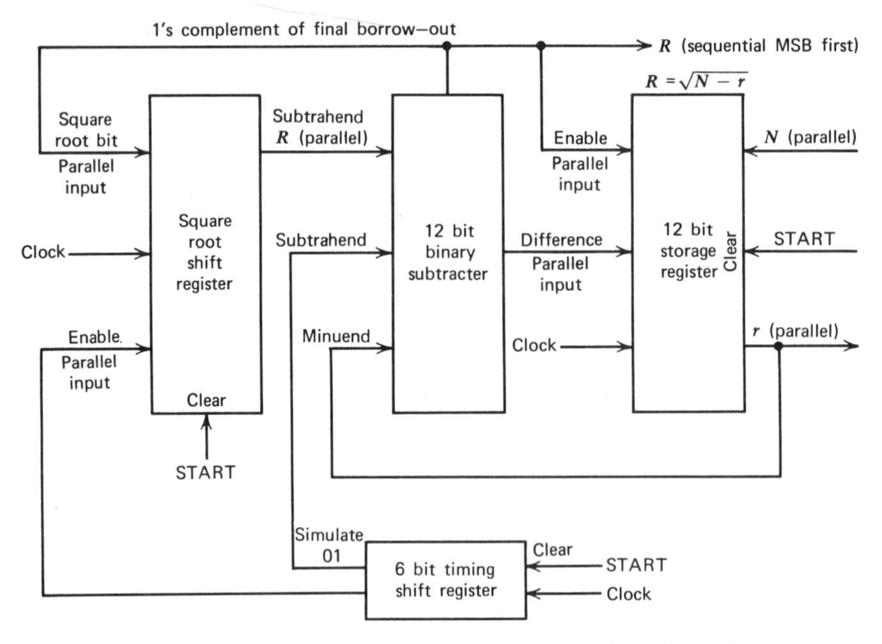

Figure 11.13. Sequential hardware implementation of the binary square root algorithm. (Courtesy of *Computer Design*.)

The most significant borrow of the binary subtracter indicates the magnitude interrelation that exists between the minuend and the subtrahend. The most significant borrow is 0 when the minuend is greater than or equal to the subtrahend. In this case, the square root bit resulting from this comparison is 1 and it is stored in the square root shift register; also, the minuend is updated, becoming the output of the binary subtracter. The most significant borrow is 1 when the minuend is less than the subtrahend. In this case, the square root bit resulting from this comparison is 0 and it is stored in the square root shift register. Since no subtraction was possible, the minuend is not updated.

Timing for the square root extraction process is provided by a 7 bit shift register whose outputs enable the entry of the square root bits into the square root shift register and simulate the 01 fixed part of the subtrahend. The operation lasts seven clock pulses after the activation of the START pulse which clears the square root shift register and sets the initial conditions of the timing register. At the end of the operation, the square root of the number, which was initially in the minuend register, is in the square root shift register. The minuend register now holds the remainder of the operation which is the difference between the given number and the square of the square root computed for that number. The square root is also available in sequential form, with the MSB generated first. A detailed logic diagram of this method is illustrated in Figure 11.14.

Method Two—A Parallel Design

As shown in Figure 11.15, the square root algorithm can be digitally implemented in a more straightforward way by using a separate subtracter for each subtraction. The logical network shown in a direct hardware implementation of the developed computation process, and it has been designed for the computation of the square root of binary numbers with up to 12 bits. Operation of this network is as follows:

The first subtracter performs subtraction $N_{2048}N_{1024} - 01$. If $N_{2048}N_{1024}$ is less than 01, the square root bit R_{32} is 0. In this case, the result of the subtraction is inhibited and the minuend of the first subtraction becomes part of the minuend of the second subtracter. If $N_{2048}N_{1024}$ is equal to or greater than 01, the square root bit R_{32} is 1. In this case, it is the resulting difference $N_{2048}N_{1024} - 01$ that becomes part of the minuend of the second subtracter.

The second subtracter performs the subtraction

$$[(\text{result of first subtraction})\ N_{512}N_{256}] - R_{32}\,01$$

If a borrow is generated, R_{16} is 0; otherwise R_{16} is 1. In the first case, the minuend of the second subtracter becomes part of the minuend of the third subtracter, while in the second case, the difference obtained becomes part of the minuend of the third subtracter. The output of the last subtracter is the remainder r of the square root computation, and its relation to the input number N is $r = N - R^2$.

The advantage of the parallel design over the sequential one is speed, while the disadvantage is that the square root is not available in sequential form. A detailed logic diagram of this method is illustrated in Figure 11.16.

In both designs the indicated subtractions can be performed by means of an adder. In this case what used to be the minuend is added to the 2's complement of what used to be the subtrahend. The 2's complement of a binary number can be obtained by adding an LSB of 1 to the 1's complement of that number—that is,

$$\begin{aligned}
\text{Difference} &= \text{Minuend} - \text{subtrahend} \\
&= \text{Minuend} + (-\text{subtrahend}) \\
&= \text{Minuend} + (\text{2's complement of subtrahend}) \\
&= \text{Minuend} + (\text{1's complement of subtrahend} + 1\ \text{LSB}) \\
&= \text{Minuend} + \overline{\text{subtrahend}} + 1\ \text{LSB}
\end{aligned}$$

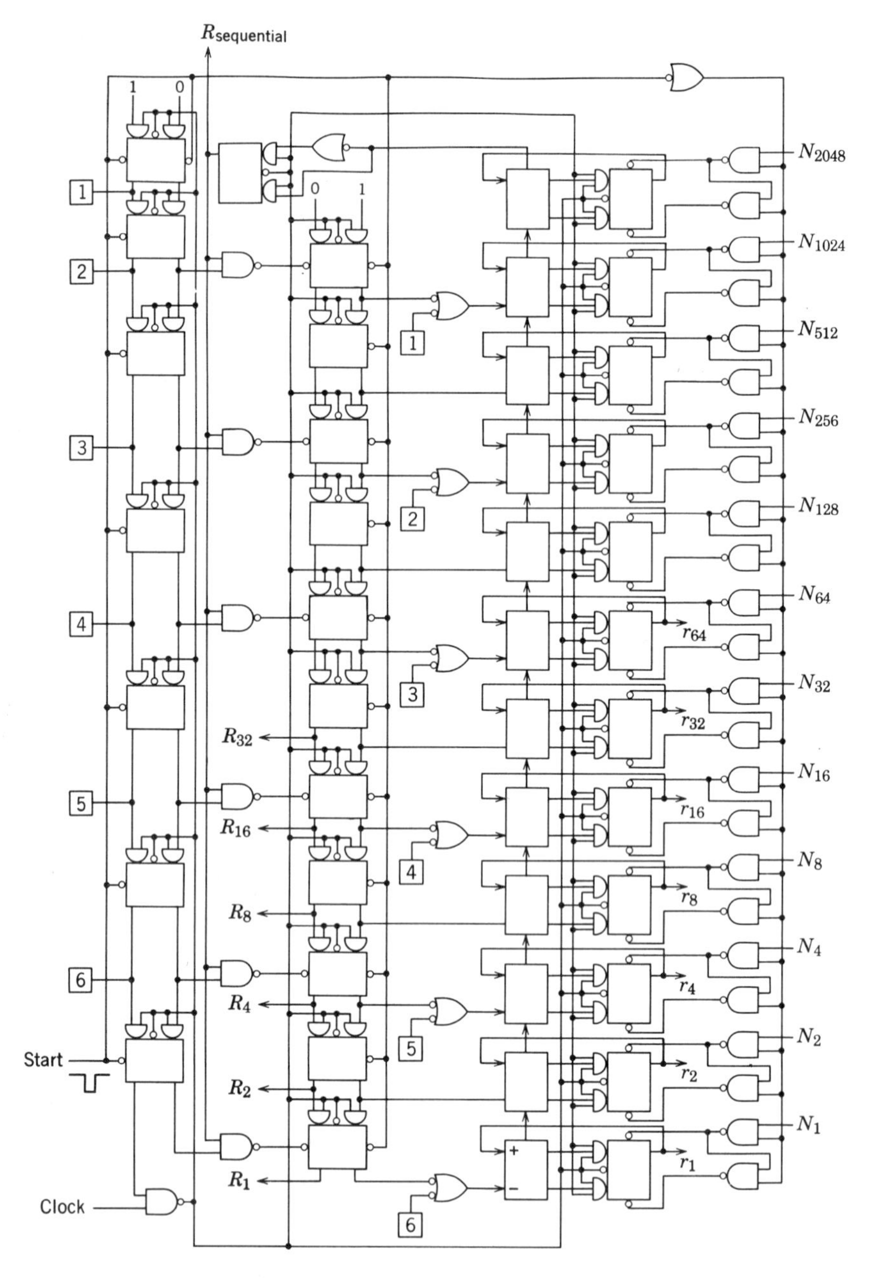

Figure 11.14. Sequential design of the digital implementation of the square root algorithm. (Courtesy of *Computer Design.*)

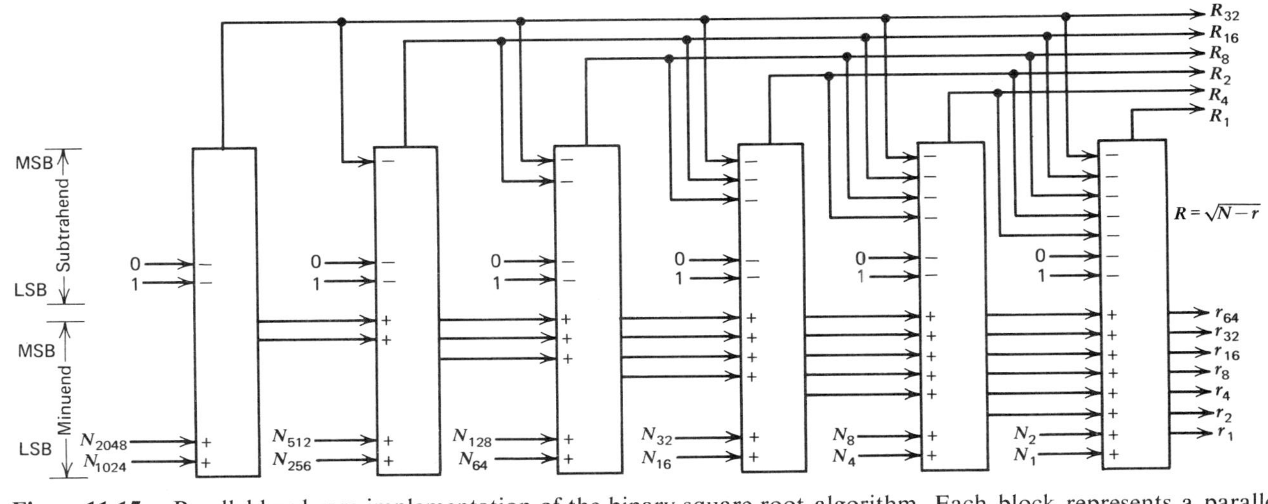

Figure 11.15. Parallel hardware implementation of the binary square root algorithm. Each block represents a parallel binary subtracter network with gated output: horizontal output = (difference) $\overline{\text{(borrow-out)}}$ + (Minuend) (borrow-out); vertical output = borrow-out. (Courtesy of *Computer Design.*)

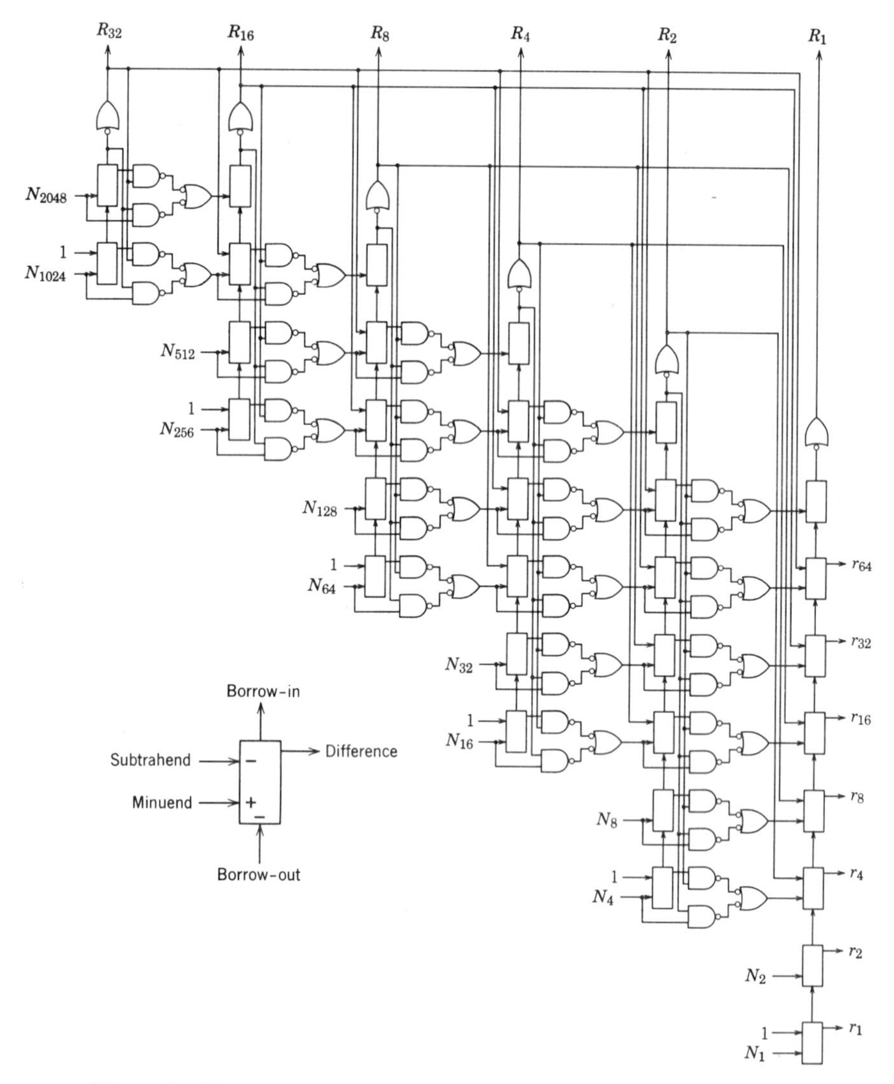

Figure 11.16. Parallel design of the digital implementation of the square root algorithm where $R = \sqrt{N - r}$. (Courtesy of *Computer Design*.)

The use of adders will facilitate the implementation, since adders are readily available. It should be noted that the most significant "carry," which will represent the minuend/subtrahend magnitude comparison, is the complement of the true "borrow"—that is, in the subtraction by addition method when the minuend is less than the subtrahend, the final borrow is 0, and when the minuend is greater than or equal to the subtrahend, the final borrow is 1. The simulated subtracter provides the true value of the square root bits.

The proposed algorithm represents a simple means of taking the square root of a binary number. It is very accurate and lends itself readily to either software or hardware implementation. The only arithmetic operation required to allow programming of the algorithm is subtraction, and its hardware implementation can be either the serial of parallel approach illustrated above.

11.5 CUBE ROOT OF BINARY NUMBERS

The technique for the computation of the cube root of binary numbers presented here has been derived from the general algorithm for the computation of the cube root, and offers a novel approach to the computation of the cube root of binary numbers.

The derivation of this technique is illustrated in Appendix A. It is a simple approach and can be easily performed by hand or by a software routine. As was the case of the square root algorithm, no correction factors are required, and the binary numbers may have fractional parts as well. The algorithm for the computation of the cube root of binary numbers is

Step 1. Starting from the binary point, partition number into 3 bit groups.

Step 2. Let the first bit of the cube root be 1.

Step 3. Subtract 001 from the first 3 bit group of the number.

Step 4. Write the next 3 bit group of the number to the right of the computed difference; call the grouping M.

Step 5. Compute expression $[6R(2R + 1) + 1]$ where R is the cube root computed thus far, call the expression S.

Step 6. If $S > M$, the cube root bit resulting from this iteration is 0. Return to Step 4 and compute the next bit of the cube root.

Step 7. If $S \geq M$, the cube root bit resulting from this iteration is 1. Compute the difference of $M - S$ and return to Step 4 to compute the next bit of the cube root.

Figure 11.17 illustrates the above algorithm in flow diagram form. The simplicity of the algorithm can be best illustrated by the following examples:

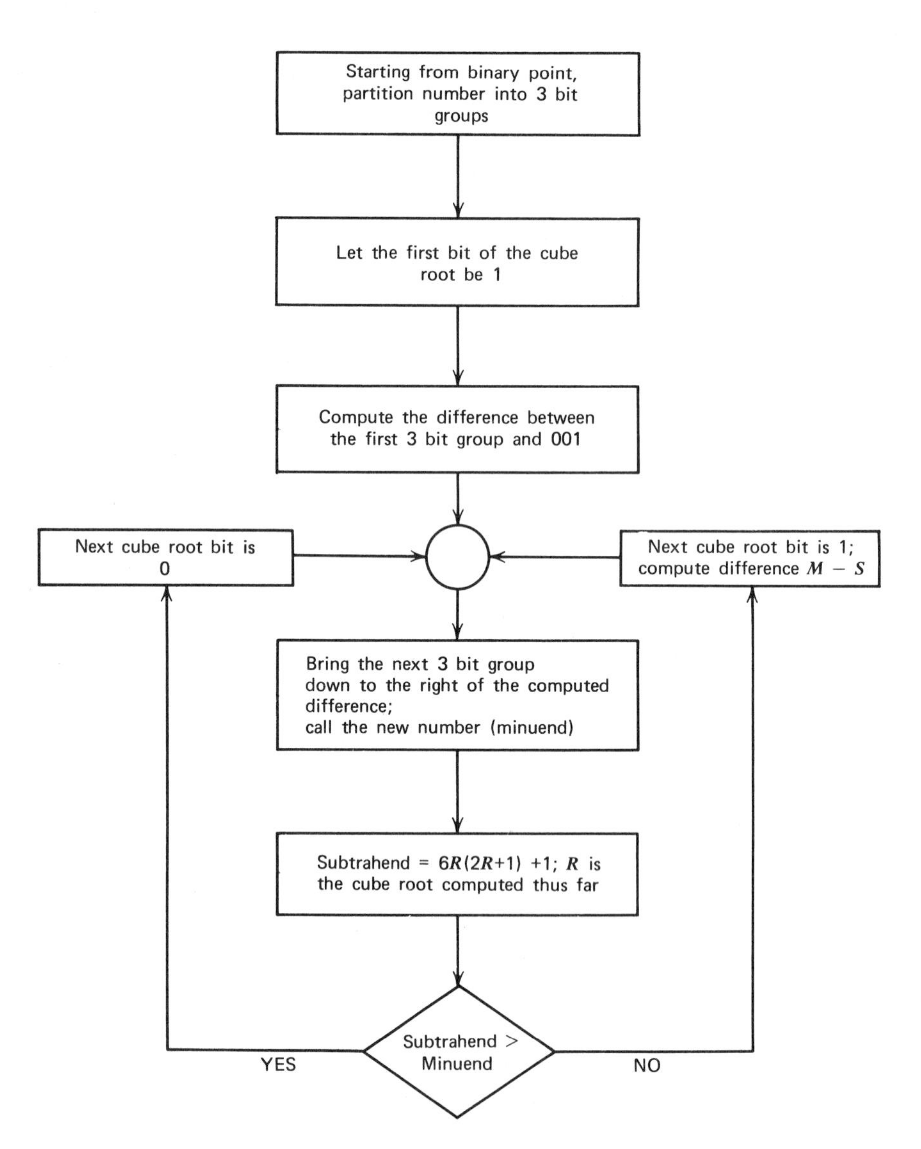

Figure 11.17. Algorithm for the computation of the cube root of binary numbers.

Example 11.3. Compute the cube root of binary number 1111101 (decimal 125).

The computation is illustrated in Figure 11.18, and proceeds as follows:

1. Number 1111101 is marked into 3 bit groups starting from the binary point. Since there are three 3 bit groups to the left of binary point, the cube root of this number will have 3 integer bits.
2. Number 001 is subtracted from first 3 bit group and a 1 is written on the left side, becoming the MSB of the cube of number 1111101.
3. The second 3 bit group is now brought down to line 3 next to the difference of lines 1 and 2. On line 4 write in binary the result of expression $[6R(2R + 1) + 1]$ where $R = 1$, which is 10011 or decimal 19. Now subtract line 4 from line 3. If the difference is negative, a 0 is inserted in the root space of that operation, and line 3 is copied on line 5. In this example this is the case. If the difference were positive, or 0, a 1 would have been inserted in root space of that operation, and difference would have been written on line 5.
4. The third 3 bit group is now brought down to line 5 next to the number copied from line 3. On line 6 write in binary the result of expression $[6R(2R + 1) + 1]$ where $R = 10$ (decimal 2). This is 111101 or

Root	Operations	Line
1	0 0 1 1 1 1 1 0 1	1
	0 0 1	2
0	0 0 0 1 1 1	3
	1 0 0 1 1	4
1	1 1 1 1 0 1	5
	1 1 1 1 0 1	6
	0 0 0 0 0 0	7

line 4 $= 6R(2R + 1) + 1$ (R = 1)
$= 6 \cdot 1(2 \cdot 1 + 1) + 1$
$= 19$
$= 1\ 0\ 0\ 1\ 1$ (binary)

line 6 $= 6R(2R + 1) + 1$ ($R = 10$ binary $= 2$ decimal)
$= 6 \cdot 2(2 \cdot 2 + 1) + 1$
$= 61$
$= 1\ 1\ 1\ 1\ 0\ 1$ (binary)

Figure 11.18. Step-by-step extraction of the cube root of the binary 1111101.

decimal 61. Now subtract line 6 from line 5. Since the difference is not negative, it is written on line 7, and a 1 is written in root space on left.

Since line 7 is 0 and no other 3 bit group exists, computation ends and the cube root of 1111101 is found to be 101. Indeed, the cube root of 125 (1111101 in binary) is 5, which is 101 in binary. ∎

Example 11.4. Compute the cube root of the binary number 10000110101.100111.

The computation is illustrated in Figure 11.19. The cube root of 10000110101.100111 to two fractional binary places is 1010.01, leaving a remainder of 0.101110. The validity of the cube root can be verified by cubing the cube root and adding the remainder to it, as follows:

```
Cube root                             1 0 1 0.0 1
Cube root                             1 0 1 0.0 1   ×
                ---------------------------------
                                      1 0 1 0 0 1
                                1 0 1 0 0 1
                            1 0 1 0 0 1
                ---------------------------------
(Cube root)^2               1 1 0 1 0 0 1 0 0 0 1
Cube root                             1 0 1 0 0 1   ×
                ---------------------------------
                            1 1 0 1 0 0 1 0 0 0 1
                      1 1 0 1 0 0 1 0 0 0 1
                  1 1 0 1 0 0 1 0 0 0 1
                ---------------------------------
(Cube root)^3   1 0 0 0 0 1 1 0 1 0 0.1 1 1 0 0 1
Remainder                            .1 0 1 1 1 0   +
                ---------------------------------
Given number    1 0 0 0 0 1 1 0 1 0 1.1 0 0 1 1 1
```
∎

Digital Implementation of Cube Root Algorithm

To digitally implement the cube root algorithm the mechanics of the algorithm must be first identified. The process is similar to that of the square root computation and consists of amplitude comparisons, subtractions, storage of difference, and subtrahend computations. The results of the comparisons are the cube root bits and the result of the last subtraction is the remainder of the operation.

Figure 11.20 illustrates the block diagram of a cube root extraction network which proceeds as follows: N, the number the square root of which is to be computed is entered into the minuend storage register. This register holds the minuend of the subtractions that are performed in the process of the algorithm.

Root	Operations		Line
1	010 000 110 101.100 111		1
	S = 001	(R = 0)	2
0	001 000		3
	S = 10 011	(R = 1)	4
1	001 000 110		5
	S = 111 101	(R = 10)	6
0	000 001 001 101		7
	S = 101 001 011	(R = 101)	8
0	001 001 101100		9
	S = 10 011 101101	(R = 1010)	10
1	001 001 101100 111		11
	S = 001 001 100111 001	(R = 10100)	12
	000 000 000101 110		13
	$S = 6R(2R + 1) + 1$		

Figure 11.19. Step-by-step extraction of the cube root of the binary number 10000110101.100111.

The output of the minuend storage register serves as the minuend of the binary subtracter. The subtrahend to this subtracter is the output of the subtrahend computation unit, the input to which is the output of the cube root shift register. This shift register is initially cleared. During each comparison, 1 bit of the cube root is computed and entered into the shift register at the serial input.

The most significant borrow of the binary subtracter indicates the magnitude interrelation that exists between the minuend and the subtrahend. The most significant borrow is 0 when the minuend is greater than or equal to the subtrahend. In this case, the cube root is 1, and it is stored in the cube root shift register; also, the minuend is updated, becoming the output of the binary subtracter.

The most significant borrow is 1 when the minuend is less than the subtrahend. In this case, the cube root bit resulting from this comparison is 0, and it is stored in the cube root shift register. Since no subtraction was possible, the minuend is not updated.

Figure 11.21 illustrates the block diagram of the subtrahend computation unit, where the subtrahend expression

$$S = 6R(2R + 1) + 1$$

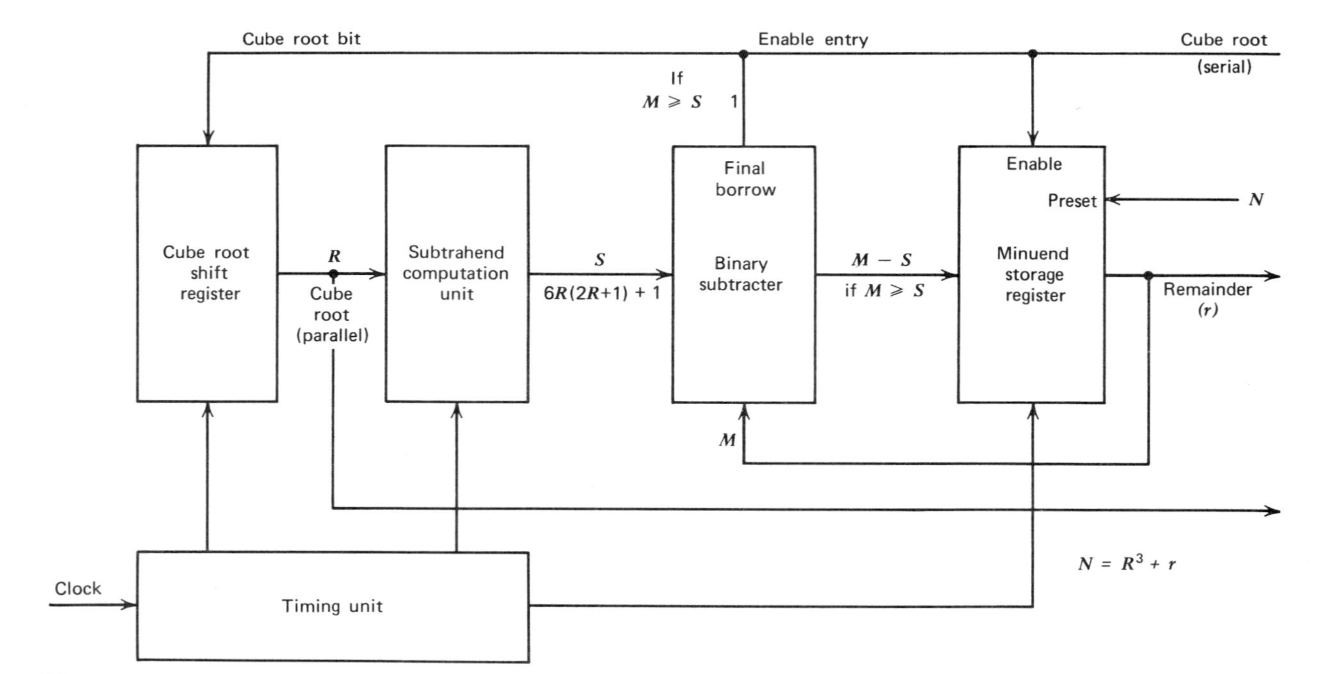

Figure 11.20. Block diagram of the cube root extracting network. (*Note:* At the beginning of the cube root calculation a start pulse clears the root shift register, presets the timing unit, and enters N into the minuend storage register via the dc set/reset inputs.)

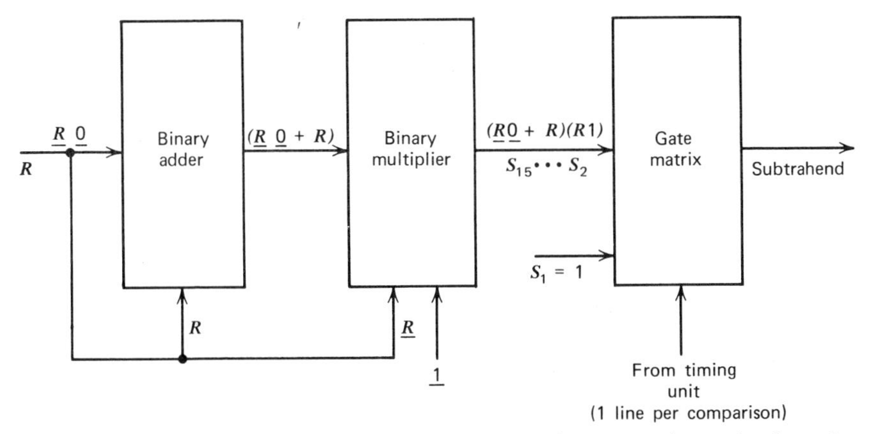

Figure 11.21. Block diagram of a subtrahend computation unit where the subtrahend $= 6R(2R + 1) + 1 = (\underline{R}\ \underline{0} + R)(\underline{R}\ \underline{1})\underline{1}$.

is implemented. For optimum implementation the subtrahend expression is converted to a binary form

$$\begin{aligned} S &= 6R(2R + 1) + 1 && \textbf{Decimal} \\ &= 3R(2R + 1)2 + 1 && \textbf{Decimal} \\ &= (11)R[(10)R + 1](10) + 1 && \textbf{Binary} \\ &= [(10)R + R][10R + 1](10) + 1 && \textbf{Binary} \\ &= [\underline{R}\ \underline{0} + R][\underline{R}\ \underline{0} + 1]\underline{0} + 1 \end{aligned}$$

Underlining means placing numbers one next to the other in the order indicated. Multiplying by 2 is the same as placing a 0 to the right of a binary number. Also, adding 1 to a number the LSB of which is 0 is the same as making that bit a 1. Continuing with the conversion we have

$$S = [\underline{R}\ \underline{0} + R][\underline{R}\ \underline{1}]\underline{1}$$

The above expression can be easily implemented as it is shown in Figure 11.21.

The output of the binary multiplier and the LSB of 1 enter a gate matrix that places the results of the subtrahend computation in the appropriate position with reference to the minuend which is fixed. Figure 11.22 illustrates the performance requirements of the gate matrix.

11.6 *n*TH INTEGER ROOT OF BINARY NUMBERS

The square root method described in Section 11.4, when repeatedly applied, can be also used for the computation of the fourth, eighth, sixteenth, or any

Clock period	MSB					Gate Matrix output									LSB
	G_{15}	G_{14}	G_{13}	G_{12}	G_{11}	G_{10}	G_9	G_8	G_7	G_6	G_5	G_4	G_3	G_2	G_1
1	S_3	S_2	S_1												
2	S_6	S_5	S_4	S_3	S_2	S_1									
3	S_9	S_8	S_7	S_6	S_5	S_4	S_3	S_2	S_1						
4	S_{12}	S_{11}	S_{10}	S_9	S_8	S_7	S_6	S_5	S_4	S_3	S_2	S_1			
5	S_{15}	S_{14}	S_{13}	S_{12}	S_{11}	S_{10}	S_9	S_8	S_7	S_6	S_5	S_4	S_3	S_2	S_1

Figure 11.22. Gate matrix performance requirements.

2^k root, where k is an integer. This is possible because the exponent $\frac{1}{2^k}$ can be represented as a product of k $\frac{1}{2}$ exponents, where each exponent indicates square root computation. For example, the sixteenth root of N can be written

$$\begin{aligned}\sqrt[16]{N} &= N^{1/16}\\ &= N^{1/2\ 1/2\ 1/2\ 1/2}\\ &= [[[N^{1/2}]^{1/2}]^{1/2}]^{1/2}\\ &= \sqrt{\sqrt{\sqrt{\sqrt{N}}}}\end{aligned}$$

Therefore, by taking the square root four times we obtain the sixteenth root.

Similarly, the cube root method described in Section 11.5, by being repeatedly applied, can be also used for the computation of any 3^k root where k is an integer. From the general root algorithm, derived in Appendix A, detailed algorithms can be developed for any integer root. Should their implementation turn out to be impractical, other indirect methods must be attempted.

An indirect method to compute the nth integer root of a binary number K is to compare K to the nth power of a known number. Figure 11.23 illustrates the implementation of this approach.

The operation of this network proceeds as follows: The binary counter, which is initially cleared, counts up, and its output enters the exponentiating

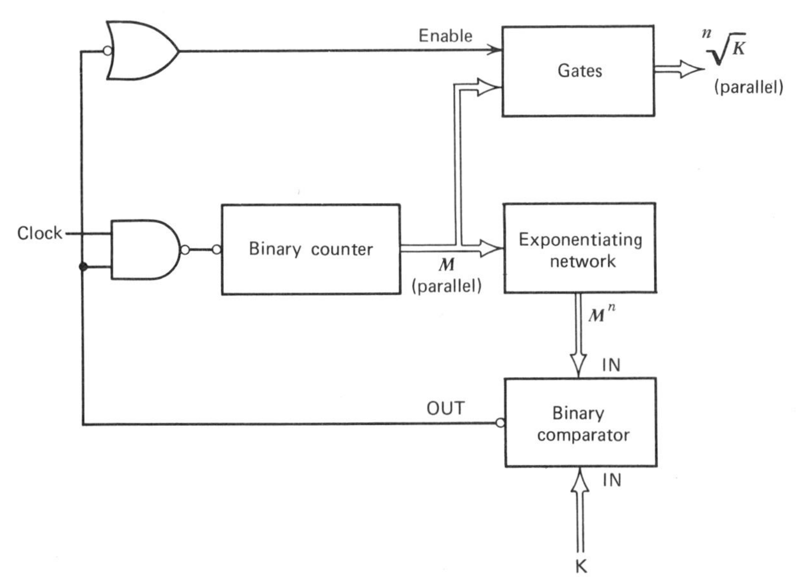

Figure 11.23. Computation of the nth integer root of a binary number.

circuit where it is raised to the nth power. The output of the exponentiating circuit is compared to the binary number K; K is the number the nth root of which is needed.

When the output of the exponentiating circuit reaches or exceeds K, the binary comparator produces a signal which stops the clock from triggering the binary counter. This signal also enables the output gates, making the output of the overall network the number at which the binary counter stopped. This number is the nth root of the output of the exponentiating circuit and the closest upper approximation of the nth root of the input binary number K.

11.7 NONINTEGER POWERS OF BINARY NUMBERS

We have thus far covered digital networks that raise binary numbers to integer exponents or compute integer roots. In many applications, however, the exponent is neither an integer nor the reciprocal of an integer, but it is a number consisting of an integral as well as a fractional part, or of a fractional part only. Numbers 1.75, 0.75, 0.3, 2.9, for example, fall in this category.

In these cases the exponent should be first represented by a fraction of two integers. If the fraction is not exactly equal to the exponent, its value must be within the tolerance dictated by the design requirements. The numerator and denominator integers must be of as small a magnitude as possible, with the denominator preferably being a power of 2. Thus, a given power should be expressed as N^m where $m = n/2^k$ within an acceptable tolerance.

In a fractional exponent the numerator indicates the number of times the base should be multiplied to itself, while the denominator indicates the order of the root that should be extracted. For example,

$$\begin{aligned} N^{0.6} &= N^{3/5} \\ &= (N^3)^{1/5} \\ &= \sqrt[5]{N^3} \end{aligned}$$

When the denominator is a power of 2, 2^p, the square root algorithm can be applied p times, thus computing the 2^p root of the number. For example,

$$\begin{aligned} N^{1.75} &= N^{7/4} \\ &= (N^7)^{1/4} \\ &= [(N^7)^{1/2}]^{1/2} \\ &= \sqrt{\sqrt{N^7}} \end{aligned}$$

which is the square root taken twice.

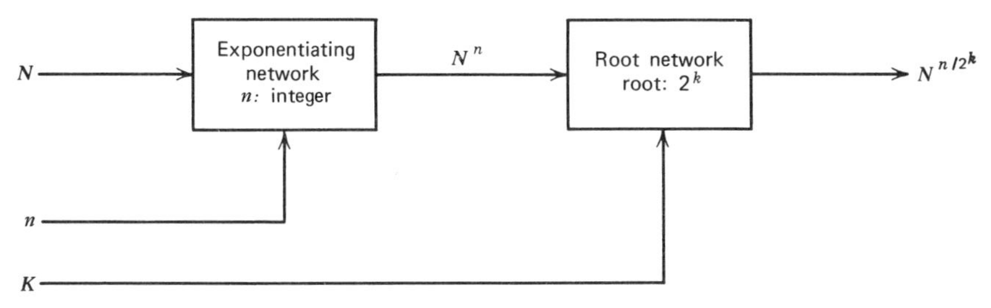

Figure 11.24. Computation of the noninteger nth power of a binary number.

Figure 11.24 shows the basic block diagram of a network that computes binary power expressions where the exponent is a noninteger number. Since the extraction of the cube root of binary numbers has a practical implementation, the root network could also be a cube root computing unit.

PROBLEMS

1. Design a squaring network for a 6 bit binary parallel number.
2. Design a squaring network for a 6 bit binary serial number.
3. Design an exponentiating network for the computation of A^5 where A is a 5 bit parallel binary number. Use the method of Figure 11.8.
4. Compute the square root of the following binary numbers:
   (*a*) 1011101.1 (*b*) 1100110.0
   (*c*) 11011.110 (*d*) 10110.101
5. Compute the cube root of the following binary numbers:
   (*a*) 1101101.0 (*b*) 10110.110
   (*c*) 11100110.101 (*d*) 1011100.0
6. Develop the algorithm for the computation of the fifth root of binary numbers. Consult Appendix *A*.
7. Design a network for the computation of the sixth root of binary numbers.
8. Design a network for the computation of the twelfth root of binary numbers.
9. Design a network for the computation of $A^{0.7}$ where A is a 7 bit binary parallel number. The accuracy of the output should be within 5%.
10. Design a network for the computation of $A^{0.44}$ where A is a 6 bit binary parallel number. The accuracy of the output should be within 3%.

CHAPTER TWELVE

Binary-Coded Decimal System

12.1 INTRODUCTION

The techniques and methods discussed in the preceding chapters were based on the pure binary numerical system. Use of the pure binary system in the digital implementation of functions always results in the least amount of hardware requirements.

Numerical displays of digital systems, however, are very inconvenient when the binary system is used. People are used to working with the decimal numerical system, and all numerical presentations should therefore be in decimal in order to be easily read and understood.

Since digital functions are usually performed in binary, a binary-to-decimal conversion would be required any time a decimal presentation is needed. A direct conversion from one code to another is always undesirable, because of its high hardware requirement, and a compromise between binary code and decimal code has been developed called binary-coded decimal, BCD.

In this code the bits appear in groups of four where each group stands for 1 digit of the decimal form of the represented number. The first group of 4 bits expresses, in binary, the number of units of the decimal number; the second group of 4 bits expresses the number of decades of the number; the third group expresses the number of hundreds of the represented number; and so on.

For example, when the number 369 is expressed in BCD, it appears as 0011 0110 1001. Similarly, the BCD number 0101 1001 0010 corresponds to decimal 592. It should be noted that each BCD digit is always expressed by 4 bits and it never exceeds 1001, which is decimal 9.

The amount of hardware required for the conversion of a BCD number to its corresponding decimal number is significantly less than that required for a binary-to-decimal conversion. Because of that, digital designs employing BCD code are widely used.

For long-hand operations the BCD system is impractical and the only purpose of studying it will be to enable us to design logic networks which will perform arithmetic operations and provide a BCD output which can be easily converted to decimal for numerical display. In this chapter the BCD arithmetic operations and their implementation are discussed.

12.2 ADDITION OF BINARY-CODED DECIMAL NUMBERS

Addition of BCD numbers is performed on digit by digit basis. The digits consist of 4 bits each, and their addition is performed as shown in the following example:

Example 12.1. Add BCD numbers A = 1001 0111 1000 and B = 1001 0101 0001.

First, we write the two numbers as

$$\begin{array}{ll} A & 1001\ 0111\ 1000 \\ B & 1001\ 0101\ 0001 \end{array}$$

Next, we start the addition going from right to left. First, the two 4 bit digits on the right are added in binary fashion

$$\begin{array}{rrl} & 1000\ A & \\ & +\ 0001\ B & \textbf{Digit of units} \\ \hline 0 & 1001 & \end{array}$$

BCD carry

Since the sum is not greater than 9, we write 0 as the carry to the next digit. This carry has a numerical value of 10 and is included in the addition of the next digit, where it is added to the LSB of that digit. The addition takes the following form:

```
      0111  A
      0101  B      Digit of tens
 +       0
 ---------
      1100
```

The numbers are initially added in binary and since the sum is greater than 9, we subtract 10 from it and write 1 as the carry to the next digit.

```
          1100 sum
        – 1010 ten     Digit of tens
        ----------
   1      0010
BCD carry
```

This carry is included in the addition of the next digit and it is added to the LSB of that digit

```
      1001  A
      1001  B        Digit of hundreds
 +       1  carry
 ----------------
    1 0011
```

The above sum which is the result of a binary addition exceeds 10, and gives a carry of 16. Here, also, we write 1 as the carry to the next digit and we subtract 10 from the sum. The final sum for this digit will be the difference of the binary sum, 10011, obtained above and 10, 1010—that is,

```
        1 0011
        – 1010       Digit of hundreds
        ------
   1      1001       Sum
BCD carry
```

To summarize, the addition of the above two BCD numbers was performed as follows:

```
                1001 0111 1000   (decimal 978)
                1001 0101 0001   (decimal 951)    Addition
BCD carry    1     1    0
-------------------------------------------------
              1 0011 1100 1001
                1010 1010                         Subtraction
-------------------------------------------------
Final      0001 1001 0010 1001   (decimal 1929)
BCD sum
```

∎

In BCD addition, instead of subtracting 10 any time the sum exceeds 9, we can add 6 and have the same result. This can be done because -10 and $+6$

in binary is the same when we are dealing with 4 bit numbers. For example, the subtraction

$$\begin{array}{r} 1101 \\ -\ 1010 \\ \hline 0011 \end{array}$$

and the addition

$$\begin{array}{r} 1101 \\ +\ 0110 \\ \hline 0011 \end{array}$$

give the same result, when only the last 4 bits are considered. The advantage of adding 6 rather than subtracting 10 is that throughout the BCD addition, only additions of 4 bit binary numbers are performed.

To perform BCD addition, we directly implement the operation of BCD addition as explained above—that is, we add the two numbers in binary, digit by digit, we detect the carry, and if it is 1, we add 6 to the digit's binary sum.

Parallel Addition

Figure 12.1 shows a single-digit parallel BCD adder. The adder chain on the left, which may have carry-look-ahead, is a 4 bit binary adder. The gating circuit appearing in the upper part of the figure produces a signal anytime the binary sum of the 4 bit binary adder exceeds 9. The signal produced by these gates is the BCD carry, and it is added in the addition of the digit next in significance. Figure 12.2 shows a cascade of single-digit BCD adders forming a multidigit BCD adder for the addition of parallel multidigit BCD numbers.

Serial Addition

The BCD numbers are often available in serial form rather than parallel form. Addition of serial BCD numbers can be performed using the serial BCD adder of Figure 12.3. In this circuit, the two BCD numbers to be added enter the binary full-adder appearing on the left in Figure 12.3.

The two numbers enter with the least significant bit of the least significant digit first, with the rest of the digit-bits following in ascending order of significance. That is, if the two numbers have 2 digits each, and the digits are units and 10s, the sequence of entry into the sequential BCD adder will be

$$1,\quad 2,\quad 4,\quad 8,\quad 10,\quad 20,\quad 40,\quad 80$$

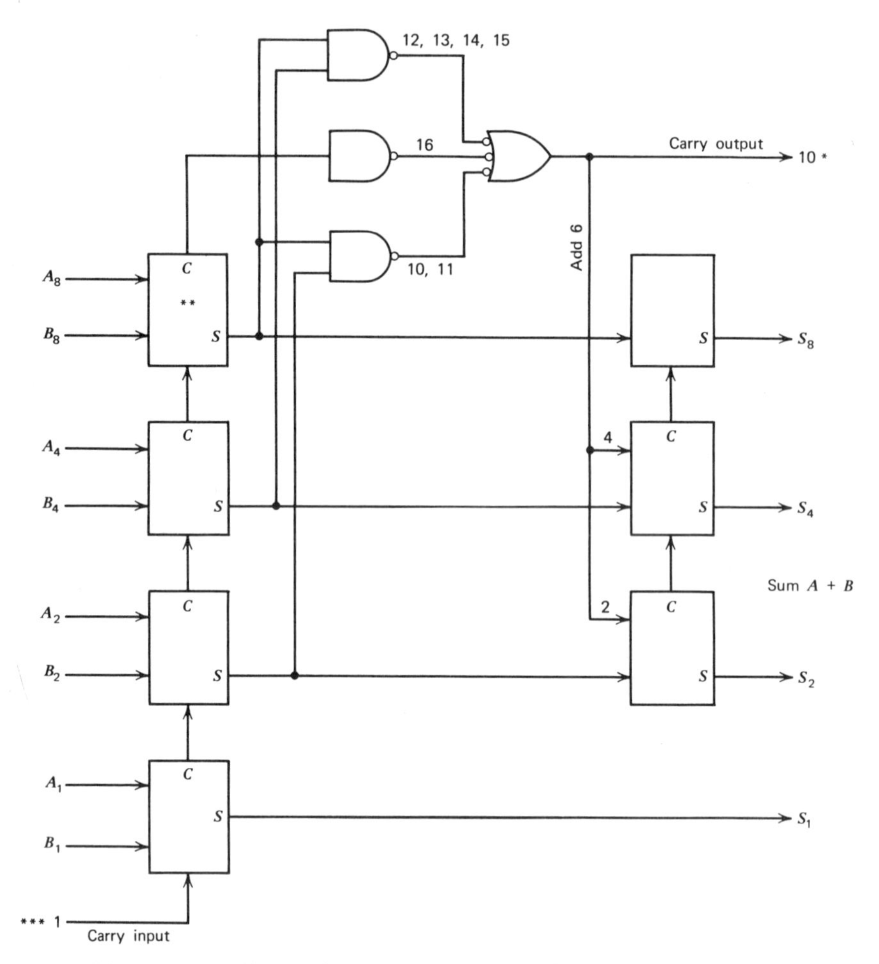

Figure 12.1. Single-digit parallel BCD adder. (*Note:* *To be added with A_{10} and B_{10}. **Each block represents a binary full-adder. ***Produced from the addition of the preceding digit.)

The serial BCD adder of Figure 12.3 works as follows: Before the start of the sequential addition a START pulse presets all bistable elements to the desired initial states. The closed-loop shift register is preset to 1000, and the remaining flip-flops in the serial BCD adder which operate as shift registers are cleared. Full-adder FA_1 and shift register SR_1 form a binary serial adder. In this adder the two BCD numbers A and B are added as if they were binary with the exception that the "carry-in" input is partially controlled by the output of SR_6 of the shift register.

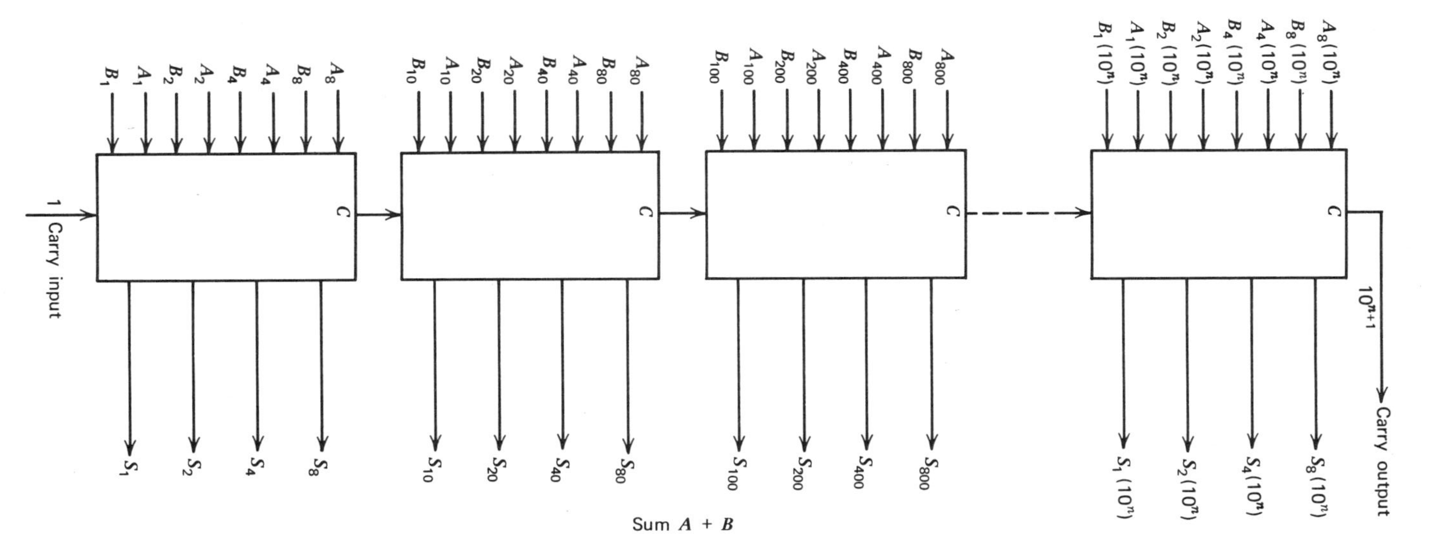

Figure 12.2. Multidigit parallel BCD adder. (*Note:* Each block represents a single-digit parallel BCD adder.)

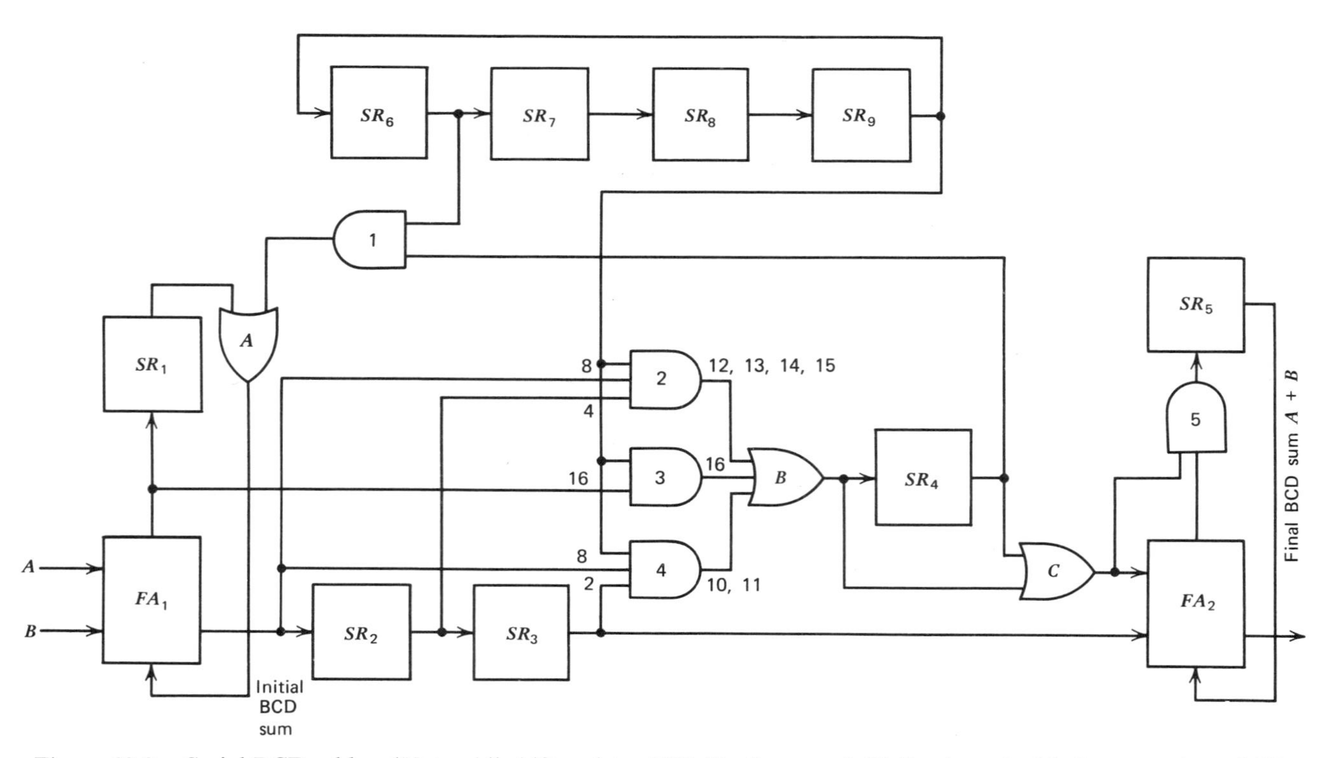

Figure 12.3. Serial BCD adder. (*Note:* All shift register (*SR*) flip-flops are initially cleared with the exception of SR_6, which is set to 1; the synchronization clock of inputs *A* and *B* feeds all flip-flops.)

The sum output of the sequential binary adder formed by SR_1 and FA_1 feeds a 2 bit shift register. The shift register serves as a sequential to parallel converter for 2 bits of the sum obtained from the serial binary adder.

When the sum output of FA_1 has the weight of 8, its carry output has the weight of 16, the SR_2 output has the weight of 4, and the SR_3 output has the weight of 2. With these four outputs we may determine if the sum obtained is greater than 9. In accordance with the rules of BCD addition, when the initial sum exceeds 9, we add 6 to the sum ignoring any produced final carry (weight 16), and we add 1 to the digit next in significance. This is implemented by decoding 12, 16, and 10 by means of AND gates 2, 3, and 4, respectively.

The three gate outputs feed OR gate B, the output of which constitutes the BCD carry output of the digit being computed, the 8×10^n bit of which appears at the input of FA_1. The output of gate B indicates that 6 must be added to the produced initial digit sum. At that time the SR_3 output is 2×10^n and it is added in FA_2 with the output of OR gate C. The present output of gate C is that of gate B which indicates presence of the BCD carry. Thus, gate C output is 1 only when the BCD carry is present.

During the next clock period the SR_4 output will be what gate B output was a period ago. Thus, if there was a BCD carry, the B output would have been 1 and now the SR_4 output would be 1, keeping the output of gate C on 1. The C gate output will be 1 for two periods, and it will be added to the output of SR_3, which is 2×10^n and 4×10^n in the respective two periods. This way when the BCD carry is 1, 6×10^n is added to the initial sum provided by FA_1.

A period later the inputs to the FA_1 adder will have the weight of $1 \times 10^{n+1}$, and the BCD carry obtained from the preceding digit will be added to the $1 \times 10^{n+1}$ A and B inputs. This is provided by the output of SR_4, which feeds the FA_1 carry-in input through gates 1 and A. Gate 1 is ANDed with the output of SR_6 Q, so that the BCD carry will feedback to FA_1 only when the FA_1 input has the weight of 10^{n+1}. This way the BCD carry is added to the LSB of the digit that follows.

When 2 and 4 are added to the FA_1 serial sum at FA_2, only the carries produced during this addition are fed back into the adder. All others are inhibited at gate 5. The sum output of FA_2 constitutes the sum of the two input BCD numbers A and B and is provided in BCD form. The time delay between application of input and availability of the output is two clock periods. This delay is caused by SR_2 and SR_3.

It should be noted, that the hardware requirements of the serial BCD adder are independent of the size of the BCD numbers that are to be added. This is an advantage of serial over parallel BCD addition, where the amount of hardware is directly proportional to the size of the BCD numbers involved.

12.3 SUBTRACTION OF BINARY-CODED DECIMAL NUMBERS

Similarly to the addition, subtraction of BCD numbers is performed on a digit-by-digit basis where each digit consists of 4 bits.

Example 12.2. Perform the BCD subtraction $A - B$ where

$$A = 1001\ 0011\ 1000 \qquad \text{(decimal 938)}$$

and

$$B = 0101\ 0101\ 0001 \qquad \text{(decimal 551)}$$

First, we subtract the digit of units of B from that of A

	1001	(8)	A	**Digit of units**
	− 0001	(1)	B	
Borrow 0	0111	(7)		**Difference**

The subtraction is performed in binary and since the borrow is 0, the difference remains unchanged. If the borrow were 1, we would have added 10 to the difference. The same result is obtained if 6 is subtracted rather than adding 10. The next digit difference will be obtained as follows:

	0011	(3)	A	**Digit of tens**
	− 0101	(5)	B	
1	1110			
Borrow				

Here, the borrow is 1 and we add 10 to the obtained difference, ignoring any resulting new borrow.

1110		1110	**Digit of tens**
+ 1010	or	− 0110	
1000		1000	**Difference**

Next, we subtract the digit of hundreds of B from that of A, including the borrow obtained in the subtraction of the 10s. Here, the borrow is first added to the B digit and the obtained sum is subtracted from the A digit.

0101	***B* digit**
+ 1	**Borrow**
0110	**New subtrahend**

Digit of hundreds

1001	(9)	***A* digit**
− 0110	(6)	**New subtrahend**
0011	(3)	**Difference digit**

To summarize, the subtraction $A - B$ was performed as follows:

(938)	1001	0011	1000		} **Subtraction**
(551)	0101	0101	0001	} **Addition**	
	1	0			
	0011	1110	0111	(if the digit borrow is 1, add 1010 and ignore produced carry)	
		1010			
(387)	0011	1000	0111		

Subtraction of BCD numbers can be also performed by addition, where the minuend is added to the 2's complement of the subtrahend and 1010 is added when the digit carry is a 0. The method takes the following form:

	$A_{800}\ A_{400}\ A_{200}\ A_{100}$	$A_{80}\ A_{40}\ A_{20}\ A_{10}$	$A_8\ A_4\ A_2\ A_1$
1's complement	$B_{800}\ B_{400}\ B_{200}\ B_{100}$	$B_{80}\ B_{40}\ B_{20}\ B_{10}$	$B_8\ B_4\ B_2\ B_1$
			1
Carry	⊗	⊗	
(when carry bit is 0 add 1010, ignoring any produced new carries)	Initial difference 1 0 1 0	⊗ Initial difference 1 0 1 0	⊗ Final difference
	Final difference	**Final difference**	**Final difference**

The above subtraction 938 − 551 performed by means of this method is as follows:

938	1 0 0 1	0 0 1 1	1 0 0 0	
− 551	1 0 1 0	1 0 1 0	1 1 1 0	**Addition**
			1	
	0	1		
	1 0 0 1 1	0 1 1 1 0	1 0 1 1 1	(1010 is added when carry is 0)
		1 0 1 0		
387	0 0 1 1	1 0 0 0	0 1 1 1	

The above method is widely used, because its implementation requires only additions.

Parallel Subtraction

Figure 12.4 shows a single-digit parallel BCD subtracter. This network consists of two groups of full-adders, where the group on the left performs a

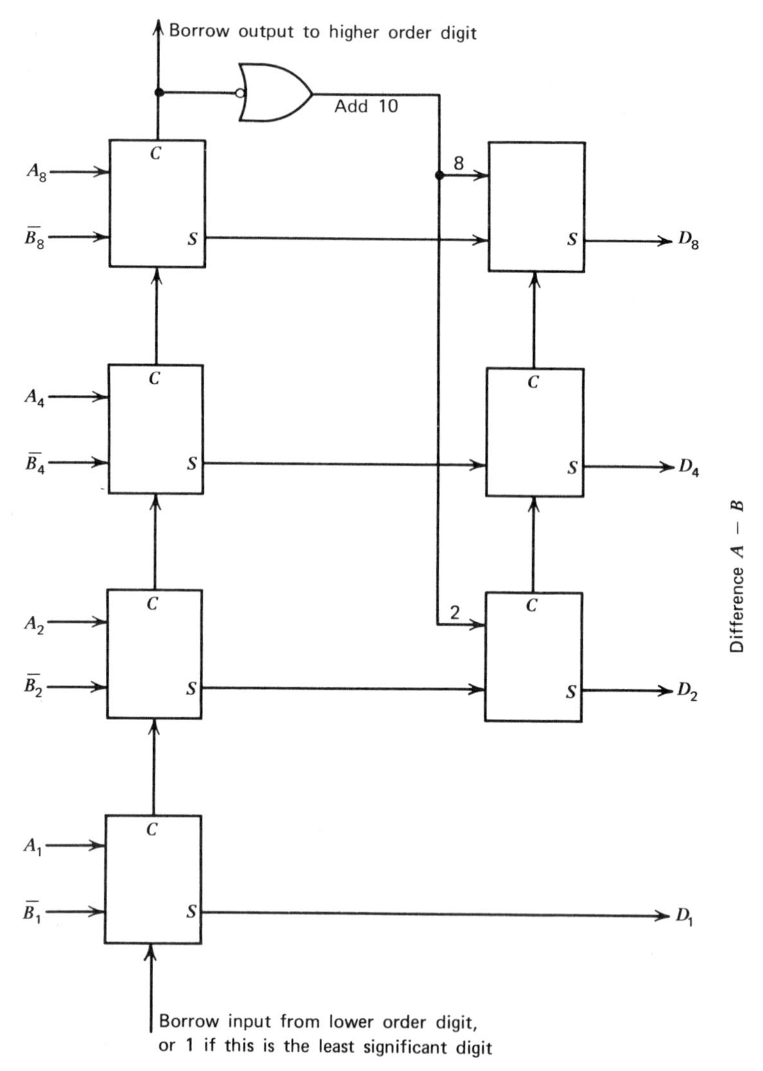

Figure 12.4. Single-digit BCD subtracter. (*Note:* Each block represents a binary full-adder.)

typical binary addition, the group on the right adds 10 to the sum obtained from the group on the left; 10 is added any time the final carry of the addition performed by the group on the left is 0.

The output of the single-digit BCD subtracter of Figure 12.4 consists of the three outputs of the adder group on the right, representing weights of 2, 4, and 8; that of the least significant full-adder of the group on the left representing the weight of 1; and of the final carry of the adder group on the left representing borrow 10. Figure 12.5 shows a chain of single-digit BCD subtracters forming a multibit BCD subtracter to be used for the subtraction of parallel multidigit BCD numbers.

Serial Subtraction

Subtraction of serial BCD numbers can be performed using the serial BCD subtracter configuration of Figure 12.6, which works as follows: A "start subtraction" pulse presets all flip-flops of the circuits, establishing the initial conditions of the subtracter. In that state all flip-flops are cleared with the exception of SR_1 and SR_7. Then, SR_1 is preset to 1 in order to provide an LSB of 1 for serial subtraction employing adders; SR_7 is also preset to 1 providing the timing signal that will circulate in the closed-loop shift register, the output of which indicates the period during which the 8×10^n bit enters the subtracter.

During the period of this bit, the borrow digit is determined, and if it is a 0, number 1010 is added to the 4 bits of the computed initial difference of the 2 digits A and B that produced the borrow. If the digit borrow is a 1, no addition is necessary.

At the arrival of the first clock pulse the least significant bit of the least significant digit of serial numbers A and B appears at the input of the subtracter. The operation performed in this period is the addition

$$A_1 + \bar{B}_1 + 1$$

The 1 is provided by SR_1 which has been preset by the start subtraction pulse. The sum produced will be applied to the input of flip-flop SR_2, and the carry to the input of SR_1.

When the second pulse arrives, the computed sum and carry bits will enter their respective flip-flops, and the input to the subtracter will be A_2 and $\bar{B}_2$. The operation performed in FA_1 will be

$$A_2 + \bar{B}_2 + SR_1 \text{ output}$$

The sum produced will be applied to SR_2 and the carry to SR_1. During this period the least significant bit of the least significant digit of the difference $A - B$ is available at the output of the serial subtracter. Similarly, after the arrival of the third pulse, the addition

$$A_3 + \bar{B}_3 + SR_1 \text{ output}$$

is performed. After the fourth clock pulse triggers the subtracter, the last bits of the first digit appear and operation

$$A_4 + \bar{B}_4 + SR_1 \text{ output}$$

is performed.

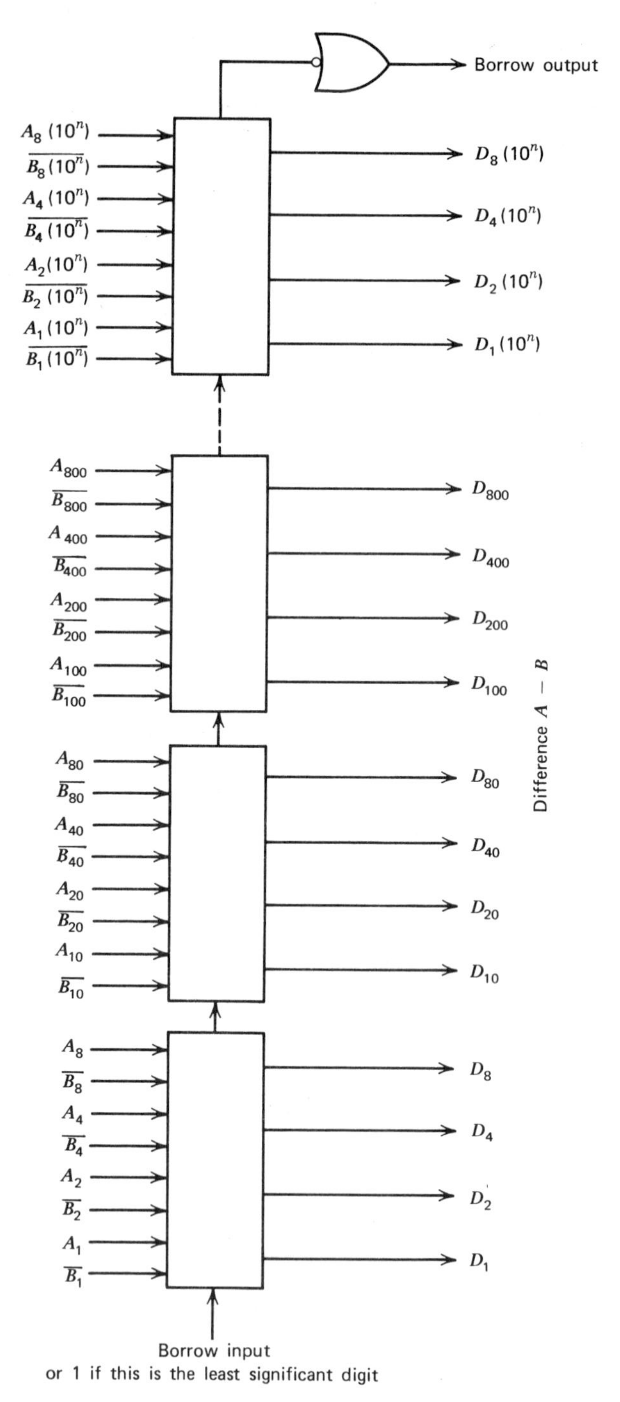

Figure 12.5. Multidigit parallel BCD subtracter. (*Note:* Each block represents a single-digit BCD subtracter.)

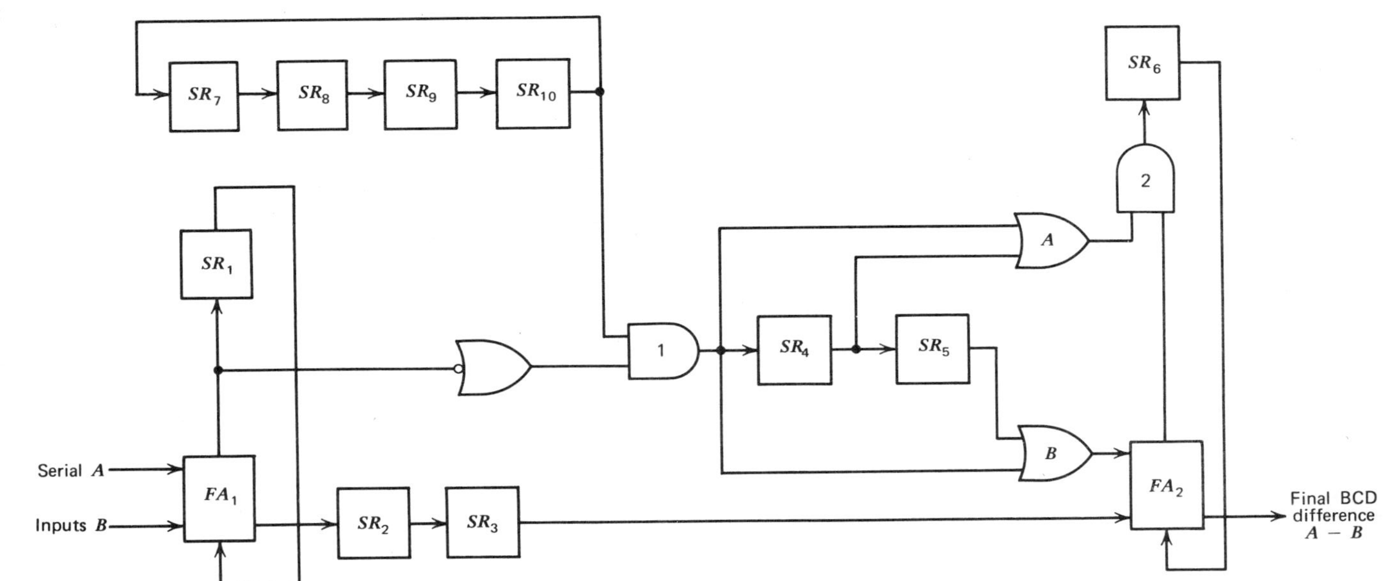

Figure 12.6. Serial BCD subtracter. (*Note:* All shift register (SR) flip-flops are initially cleared with the exception of SR_1 and SR_7, which are set to 1; the synchronization clock of inputs A and B feeds all flip-flops.)

During this period the carry output of FA_1 indicates whether there is borrow from the subtraction of the least significant digits. If this output is 0, there is a borrow and 10 must be added to the computed difference of the two input digits. Otherwise, the output of FA_1 is the correct difference. At that time SR_2 holds bit D_4 and SR_3 holds D_2 of the initial difference produced by FA_1. Also, SR_{10} is 1, enabling gate 1.

If there is no carry at FA_1, indicating that there is a borrow, gate 1 output is 1 and through gate B it is applied to FA_2 where it is added to D_2, which is the output of SR_3. Thus, a bit of weight 2 is added to the serial sum output of FA_1. The output of gate 1 through gates A and 2 also enables the carry of FA_2 to be stored in SR_6, while the sum output of FA_2 provides the second bit of the computed difference.

In the fifth period the output of gate B is 0 and FA_2 provides the third bit of the computed difference. Any generated carry in FA_2 will be stored in SR_6 since SR_4 is 1, propagating through gate A to gate 2.

In the sixth period FA_2 provides the fourth bit of the computed difference. This bit is the sum of the following three terms: the output of SR_3, which is D_8; the output of SR_5 through gate B, which is 1, and has a weight of 8; and the output of SR_6, which is the carry that was produced at the immediately preceding addition in FA_2. Thus, should the borrow digit be 1 or the carry digit 0, a total of 10 (2 + 8) is added to the initially produced difference.

Interesting to note is that the serial BCD subtracter configuration of Figure 12.6 is independent of the size of the input operands, and its speed is limited only by the speed of the individual flip-flops used, also, that the generated difference follows the input operands by two clock periods.

12.4 MULTIPLICATION OF BINARY-CODED DECIMAL NUMBERS

Multiplication of BCD numbers is similar to that of binary numbers with the difference being that special considerations must be exercised regarding the decimal weight of the bits.

Example 12.3. Multiply 9 × 3.

		1	0	0	1	(9)	
	×	0	0	1	1	(3)	
		1	0	0	1		
	1	0	0	1			
	1	1	0	1	1	(27)	**Binary product**
Decimal weight	(16)	(8)	(4)	(2)	(1)		

Up to this point, the multiplication is performed as if the operands were binary numbers. The product, however, must be converted to BCD. Thus, the bit having weight of 16 is broken down to 10 and 6; 10 becomes part of the 10s digit of the product and 6 is added to the units digit of the product as shown below:

```
(10)  (8)  (4)  (2)  (1)
 1     1    0    1    1   (21)
    +  0    1    1    0    (6)
----------------------------------
 1  1  0    0    0    1   (27)
(10)(16)
Carry
```

Now, the carry, which is 16 is broken down to 10 and 6. The 10 bit is added to the 10 bit previously obtained, and the 6 bit is added to the units digit obtained in the above addition.

```
(20)  (10)    (8)  (4)  (2)  (1)
        1      0    0    0    1
        1      0    1    1    0
----------------------------------------
  1     0      0    1    1    1     (27)     BCD product
```

Example 12.4. Multiply 9 × 9.

```
          1  0  0  1   (9)
          1  0  0  1   (9)
------------------------
          1  0  0  1
 1  0  0  1
------------------------
 1  0  1  0  0  0  1  (81)     Binary product
```

Now the product is partitioned into two groups where one group has the first 3 bits and the other the last 4,

```
(64)  (32)  (16)    (8)  (4)  (2)  (1)
  1     0     1      0    0    0    1
```

Bit 16 is relabeled 10, and 6 is added to the group of 4 bits. Bit 64 broken down into 40 and 20 and the remaining 4 is added to the group of 4 bits. Thus, the computation of the product takes the following form:

```
(40)  (20)  (10)    (8)  (4)  (2)  (1)
  1     1     1      0    0    0    1
                     0    1    1    0   (6)
                     0    1    0    0   (4)
---------------------------------------------
  1     1     1      1    0    1    1
```

Since the number in the 4 bit group is greater than 9 (1001), 10 is subtracted from it and 1 is added to the column of 10s, giving

(80)	(40)	(20)	(10)						
0	1	1	1	1	0	1	1		
		+	1	− 1	0	1	0		
1	0	0	0	0	0	0	1	(81)	**Product in BCD** ▮

Multidigit BCD multiplication can be performed with single-digit BCD multiplications, adding the individual products to form the final product—that is, to multiply

$$(A_{80}A_{40}A_{20}A_{10}A_8A_4A_2A_1) \times (B_{80}B_{40}B_{20}B_{10}B_8B_4B_2B_1)$$

we first form partial products which are subsequently added. These are

$$(A_{80}A_{40}A_{20}A_{10}) \times (B_8B_4B_2B_1) + (A_{80}A_{40}A_{20}A_{10}) \times (B_{80}B_{40}B_{20}B_{10}) + (A_8A_4A_2A_1) \times (B_8B_4B_2B_1) + (A_8A_4A_2A_1) \times (B_{80}B_{40}B_{20}B_{10})$$

It should be noted that in the addition of the products, the weight of the bits must be carefully considered so that each bit can be added to bits of its own weight.

Parallel-Parallel BCD Multiplication

Multiplication of BCD numbers is performed by means of 4 × 4 binary multiplier modules. Figure 12.7 illustrates a 4 × 4 digit BCD multiplier handling operands available in parallel form. Each of the blocks of the matrix represents a single-digit BCD multiplier.

A single-digit BCD multiplier consists of a 4 × 4 bit binary multiplier and a binary to BCD converter as shown in Figure 12.8. The binary multiplier produces a product that ranges from 0 to 81 which is 8 bits. This 8 bit binary number enters a binary to BCD converter which expresses the number in two BCD digits, each consisting of 4 bits.

The partial 2 digit products generated by each of the 16 multiplier modules of Figure 12.7 enter a parallel multidigit BCD adder where they are added together to form the final BCD product of the multiplication. This approach offers high speed at the expense of hardware which increases exponentially with operand digit size.

Parallel-Serial BCD Multiplication

Considerable hardware savings result if BCD multiplication is performed in a parallel-serial fashion, where one operand is in parallel form and the other

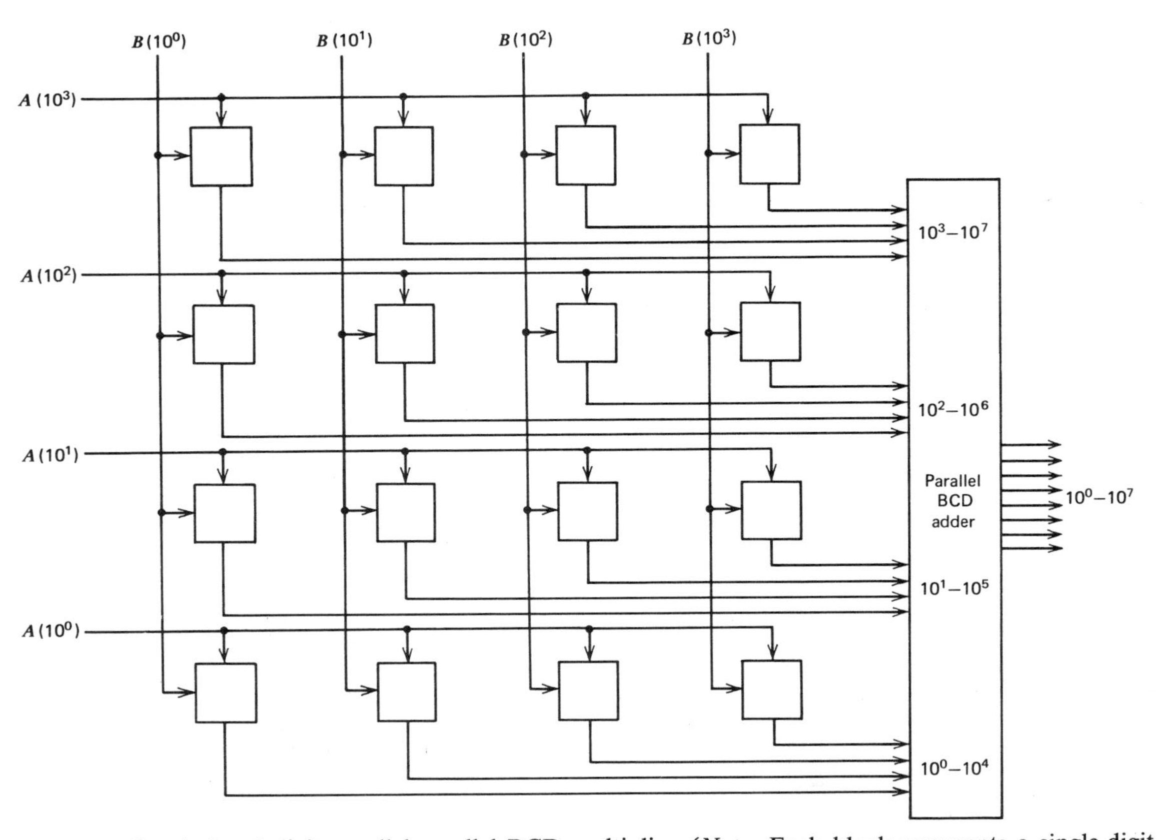

Figure 12.7. A 4 × 4 digit parallel-parallel BCD multiplier. (*Note:* Each block represents a single-digit BCD multiplier and each line four parallel bits.)

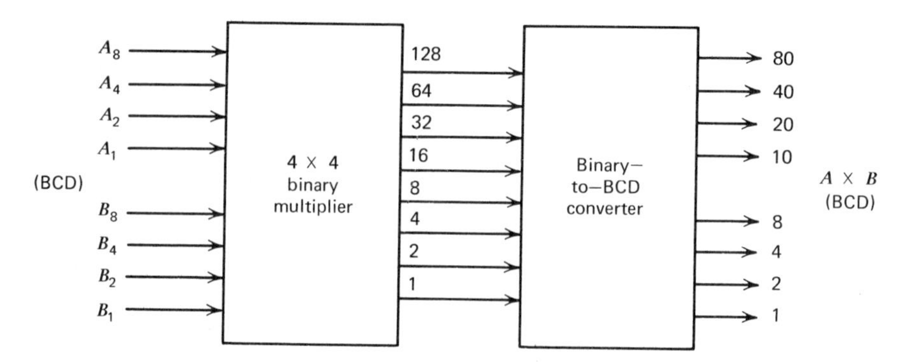

Figure 12.8. Single-digit BCD multiplier (the binary-to-BCD converter is covered in Chapter Fourteen).

in digit-serial form. That is, each digit consisting of 4 bits is available serially by means of four lines.

Figure 12.9 illustrates the general form of a parallel-serial BCD multiplier. In this circuit input A, which is an $(n + 1)$ – digit BCD number available in parallel form, is present for the entire duration of the multiplication, and input B, which is also a BCD number, provides the multiplier with 1 digit at a time.

During the first clock period of the multiplication the digit at the B input is

$$B_8 \; B_4 \; B_2 \; B_1$$

This is multiplied by each of the digits of input A, resulting in $n + 1$ partial 2 digit products. The least significant digit that results from the performed computation is a digit of the final product which is

$$P_8 \; P_4 \; P_2 \; P_1 \tag{12.1}$$

All higher order partial products enter the parallel BCD adder where they are added together and are then stored to be added with the partial products of the next partial multiplication.

During the second clock period of the multiplication the digit at the B input is

$$B_{80} \; B_{40} \; B_{20} \; B_{10}$$

It is multiplied by each digit of input A, generating the product

$$A(B_{80} \; B_{40} \; B_{20} \; B_{10}) \tag{12.2}$$

which is available at the input of the parallel BCD adder. In the adder this product (12.2) is added to the one computed in the preceding clock period

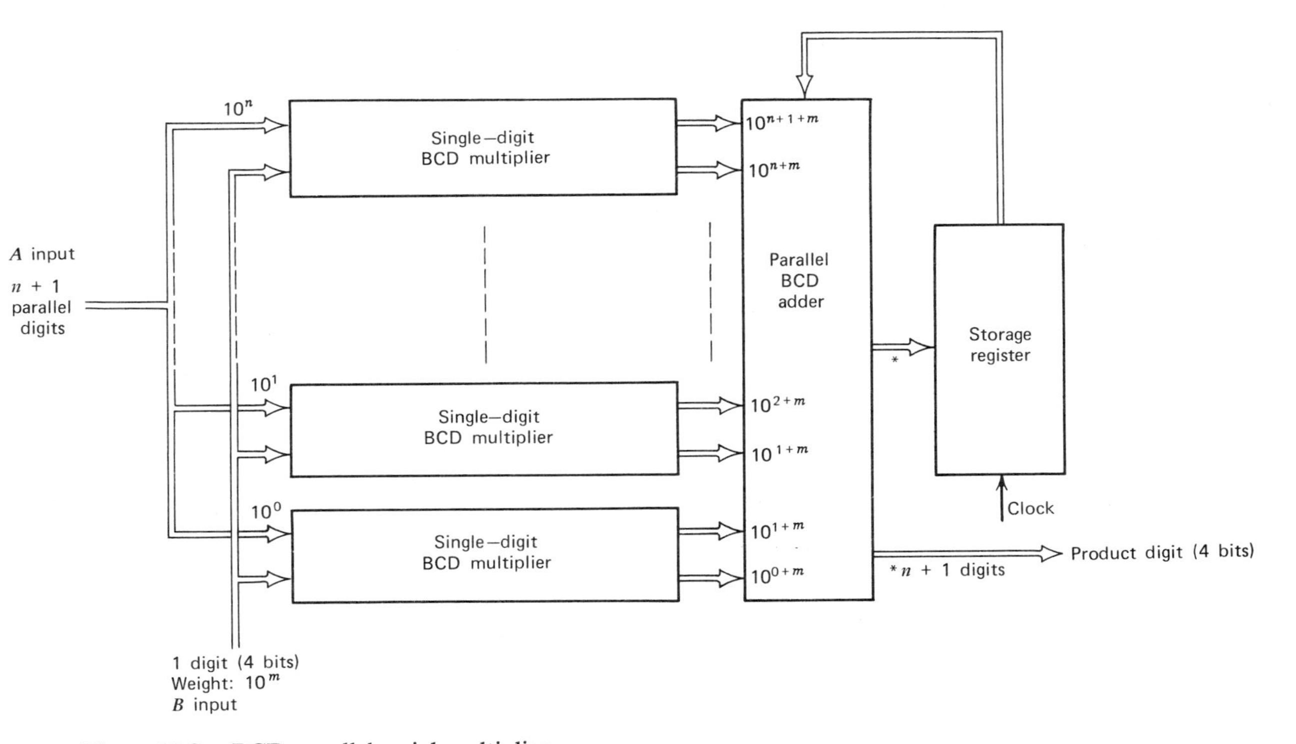

Figure 12.9. BCD parallel-serial multiplier.

(12.1) resulting in a partial product for which the least significant digit is the second digit of the final BCD product,

$$P_{80}\ P_{40}\ P_{20}\ P_{10}$$

Similarly, at the third clock period the generated product digit will be

$$P_{800}\ P_{400}\ P_{200}\ P_{100}$$

This process must continue for $n \cdot m$ clock periods where n is the number of digits of operand A and m is that of B. After each entry in the storage register, which is a shift register, the stored BCD number is shifted down in significance by 1 digit so that it lines up with the next partial products that will be produced by the single-digit BCD multipliers. The storage elements used and the concept employed is the same as that at the binary parallel-serial multiplier of Figure 9.12.

12.5 DIVISION OF BINARY-CODED DECIMAL NUMBERS

Division of BCD numbers is performed through repeated subtractions where the initial minuend is the dividend and the subtrahend is BCD multiples of the divisor. The basic block diagram of a BCD divider is illustrated in Figure 12.10.

In the first subtraction the 8 times multiple of the divisor $8d$ is subtracted from the dividend D. The final borrow produced from the subtraction is

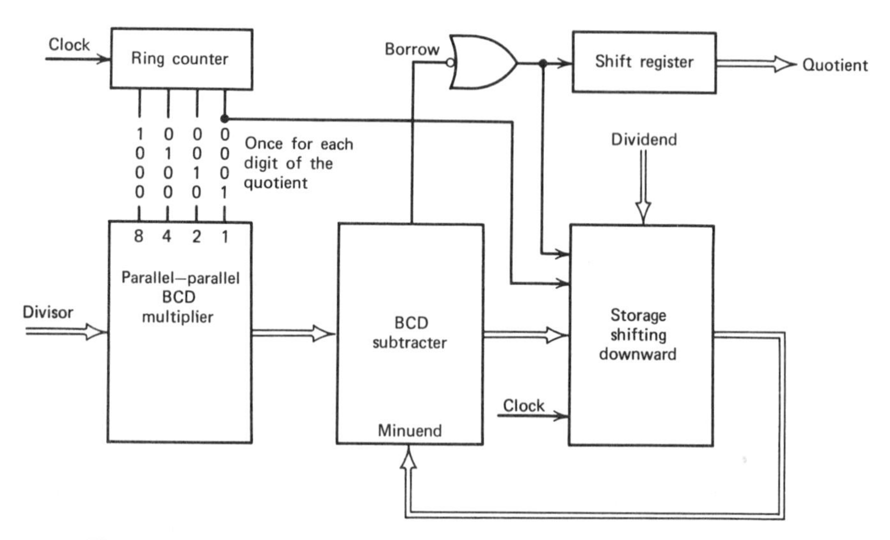

Figure 12.10. BCD divider.

the inverse of the weight-8 bit of the most significant digit of the quotient.

If the final borrow is 0, the computed difference serves as the minuend for the subtraction of the computation of the next bit weight-4 of the quotient. Otherwise, the dividend remains as the minuend for the next subtraction.

In the second subtraction the 4 times multiple of the divisor $4d$ is subtracted from the minuend that was determined in the first subtraction. The final borrow produced from the subtraction is the inverse of the weight-4 bit of the most significant digit of the quotient.

Similarly, if the final borrow of the subtraction is 0, the difference serves as the minuend of the subtraction for the computation of the next bit, weight-2, of the quotient. Otherwise, the minuend of this subtraction remains as the minuend for the next subtraction which will determine the weight-2 bit of the most significant digit of the quotient.

In the third subtraction the 2 times multiple of the divisor is subtracted from the minuend determined in the preceding subtraction, and the weight-2 bit of the quotient's most significant digit is computed.

Similarly, in the fourth subtraction the weight-1 bit of the currently computed digit of the quotient is determined. In the next four subtractions the digit second in most significance of the quotient is computed. In these subtractions the most significant digit of the subtrahend, which is the multiple of the divisor is lined up with the position of the digit second in most significance of the dividend which was the initial minuend.

This process may continue until the desired number of quotient digits has been determined. Figure 12.11 is an example of BCD division illustrating this method.

Division of BCD numbers can be also achieved by repeated subtraction, where the divisor is repeatedly subtracted from the dividend until the difference becomes less than the divisor. The number of subtractions having taken place is the quotient and the last difference obtained is the remainder of the division. For example, to perform the division

$$0110\ 0111\ 0110 \div 0011\ 0001$$

we repeatedly subtract 0011 0001 from 1001 0111 0110 as follows on p. 354.

From this process, we see that number 0011 0001 can be subtracted 21 times from 0110 0111 0110, leaving a remainder of 0010 0101. Therefore, in the above division the quotient is 21 and the remainder is 25 or 0010 0001 and 0010 0101, respectively. A BCD divider employing this method is illustrated in Figure 12.12. It consists of a BCD subtracter, a storage register, and of a BCD counter. The division's dividend is initially preset in the storage register, the output of which constitutes the minuend input of the BCD subtracter. While the divisor is the subtrahend input.

First quotient digit	$Q_8 = 0$	$M_1 = D$	M_1		0 1 0 1	1 0 0 0	0 1 0 1	. 0 0 1 1
		$S_1 = 8d$	S_1	0 0 0 1	0 1 0 0	0 1 0 0		
		$M_1 < S_1$						
	$Q_4 = 0$	$M_2 = M_1$	M_2		0 1 0 1	1 0 0 0	0 1 0 1	. 0 0 1 1
		$S_2 = 4d$	S_2		0 1 1 1	0 0 1 0		
		$M_2 < S_2$						
	$Q_2 = 1$	$M_3 = M_2$	M_3		0 1 0 1	1 0 0 0	0 1 0 1	. 0 0 1 1
		$S_3 = 2d$	S_3		0 0 1 1	0 1 1 0		
		$M_3 > S_3$	**Difference**		0 0 1 0	0 0 1 0	0 1 0 1	0 0 1 1
	$Q_1 = 1$	$M_4 =$ Difference	M_4		0 0 1 0	0 0 1 0	0 1 0 1	. 0 0 1 1
		$S_4 = d$	S_4		0 0 0 1	1 0 0 0		
		$M_4 > S_4$	**Difference**		0 0 0 0	0 1 0 0	0 1 0 1	0 0 1 1
Second quotient digit	$Q_8 = 0$	$M_5 =$ Difference	M_5			0 1 0 0	0 1 0 1	. 0 0 1 1
		$S_5 = 8d$	S_5		0 0 0 1	0 1 0 0	0 1 0 0	
		$M_5 < S_5$						
	$Q_4 = 0$	$M_6 = M_5$	M_6			0 1 0 0	0 1 0 1	. 0 0 1 1
		$S_6 = 4d$	S_6			0 1 1 1	0 0 1 0	
		$M_6 < S_6$						
	$Q_2 = 1$	$M_7 = M_6$	M_7			0 1 0 0	0 1 0 1	. 0 0 1 1
		$S_7 = 2d$	S_7			0 0 1 1	0 1 1 0	
		$M_7 > S_7$	**Difference**			0 0 0 0	1 0 0 1	0 0 1 1
	$Q_1 = 0$	$M_8 =$ Difference	M_8				1 0 0 1	. 0 0 1 1
		$S_8 = d$	S_8			0 0 0 1	1 0 0 0	
		$M_8 < S_8$						

Digit	Quotient bit	Relations	Operand	Value
Third quotient digit	$Q_8 = 0$	$M_9 = M_8$	M_9	1 0 0 1 . 0 0 1 1
		$S_9 = 8d$	S_9	0 0 0 1 0 1 0 0 0 1 0 0
		$M_9 < S_9$		
	$Q_4 = 1$	$M_{10} = M_9$	M_{10}	1 0 0 1 . 0 0 1 1
		$S_{10} = 4d$	S_{10}	0 1 1 1 0 0 1 0
		$M_{10} > S_{10}$	**Difference**	0 0 1 0 0 0 0 1
	$Q_2 = 0$	$M_{11} =$ Difference		
		$S_{11} = 2d$	M_{11}	0 0 1 0 . 0 0 0 1
		$M_{11} < S_{11}$	S_{11}	0 0 1 1 0 1 1 0
	$Q_1 = 1$	$M_{12} = M_{11}$	M_{12}	0 0 1 0 . 0 0 0 1
		$S_{12} = d$	S_{12}	0 0 0 1 1 0 0 0
		$M_{12} > S_{12}$	**Difference**	0 0 0 0 0 0 1 1
Fourth quotient digit	$Q_8 = 0$	$M_{13} =$ Difference	M_{13}	. 0 0 1 1 0 0 0 0
		$S_{13} = 8d$	S_{13}	0 0 0 1 0 1 0 0 0 1 0 0
		$M_{13} < S_{13}$		
	$Q_4 = 0$	$M_{14} = M_{13}$	M_{14}	0 0 1 1 0 0 0 0
		$S_{14} = 4d$	S_{14}	0 1 1 1 0 0 1 0
		$M_{14} < S_{14}$		
	$Q_2 = 0$	$M_{15} = M_{14}$	M_{15}	0 0 1 1 0 0 0 0
		$S_{15} = 2d$	S_{15}	0 0 1 1 0 1 1 0
		$M_{15} < S_{15}$		
	$Q_1 = 1$	$M_{16} = M_{15}$	M_{16}	0 0 1 1 0 0 0 0
		$S_{16} = d$	S_{16}	0 0 0 1 1 0 0 0
		$M_{16} > S_{16}$	**Difference**	0 0 0 1 0 0 1 0 **Remainder**

Figure 12.11. Example of BCD division. (*Note:* D = dividend; d = divisor; M = minuend; S = subtrahend; Q = quotient; and $D = 585.3_{10}$ and $d = 18_{10}$.)

		BCD	Decimal	Step
Dividend		0110 0111 0110	676	
Divisor	−	0011 0001	31	1
		0110 0100 0101	645	
	−	0011 0001	31	2
		0110 0001 0100	614	
	−	0011 0001	31	3
		0101 1000 0011	583	
	−	0011 0001	31	4
		1010 0101 0010	552	
		⋮ ⋮ ⋮	⋮	
		0000 0101 0110	56	
	−	0011 0001	31	21
Remainder		0010 0101	25	

When the clock is applied to the network, the difference obtained is stored in the register and it is fed back becoming the new minuend of the subtraction. As long as the obtained difference (minuend) is greater than the divisor (subtrahend), the subtracter's borrow output is 0, and its inverse is used to enable the clock input of a BCD counter.

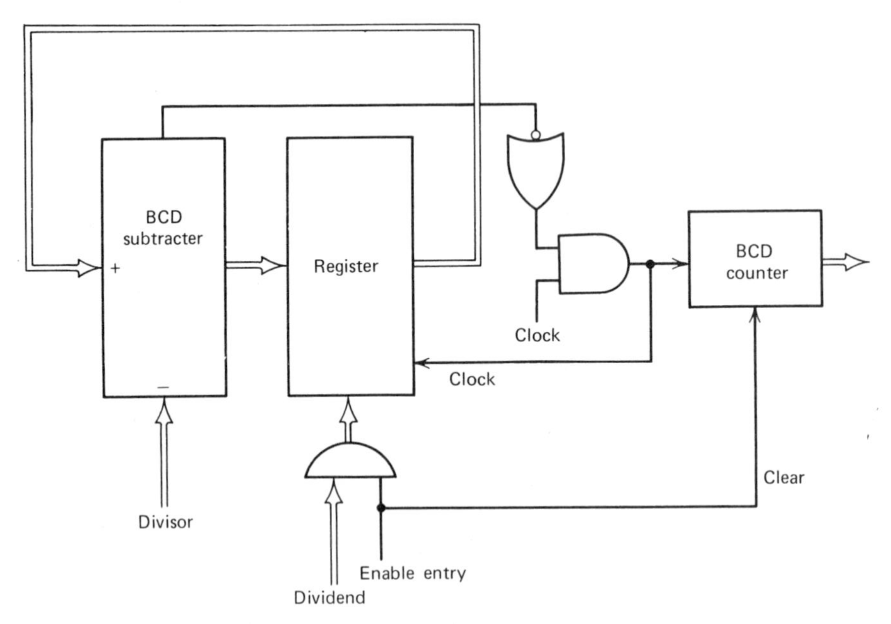

Figure 12.12. BCD divider employing repeated subtraction.

The clock pulses enter the counter from the beginning of the operation until the borrow of the subtracter becomes 1 or until the division's remainder is less than the divisor.

The final number in the BCD counter will be the number of times the divisor was subtracted from the dividend before the borrow became 1. Thus, at the end of the division the quotient is in the BCD counter and the remainder is in the storage register.

12.6 COMPLEMENTS OF BINARY-CODED DECIMAL NUMBERS

There are two BCD complements that are mostly used—the 9's complement and 10's complement.

The 9's complement of a BCD number is the difference between that number and 9. For example, the 9's complement of 0101 is 0100, and it is obtained from the subtraction

$$\begin{array}{rr} 1001 & (9) \\ -\ 0101 & (5) \\ \hline 0100 & (4) \end{array}$$

Similarly, the 10's complement of a BCD number is the difference between that number and 10. For example, the 10's complement of 0100 is 0110, and it is obtained from the subtraction

$$\begin{array}{rr} 1010 & (10) \\ -\ 0100 & (4) \\ \hline 0110 & (6) \end{array}$$

To design a logic circuit for the computation of a 9's complement, we first prepare a table showing the desired conversion. Shown below is the conversion table for the 9's complement. The sum of the BCD number and its complement is always 9 (1001).

In the table on p. 356 we see that the LSB of the 9's complement is the inverse of that of the BCD number—that is, to obtain the LSB of the 9's complement, an inverter is connected to the LSB of the BCD number. The output of that inverter will be the LSB of the 9's complement. We also see that the bit 2 of the 9's complement is the same as that of the BCD number. Therefore, no computation is required for bit 2 of the 9's complement. Bit 4 of the 9's complement is 1 anytime bit 4 and bit 2 of the BCD number are not the same. We can also see that bit 8 of the 9's complement is 1 anytime bit 8 4, and 2 of the BCD number are all 0.

BCD number	9's complement
0000	1001
0001	1000
0010	0111
0011	0110
0100	0101
0101	0100
0110	0011
0111	0010
1000	0001
1001	0000

The above information, which was derived from the conversion table, can be used to design a 9's complement generator. Such circuit is shown on Figure 12.13. The BCD 10's complement is similarly computed. Shown below is the conversion table for the 10's complement. Here, the sum of the BCD number and its complement always equals 10 (ten).

Conversion

BCD number	10's complement
0000	1 0000
0001	0 1001
0010	0 1000
0011	0 0111
0100	0 0110
0101	0 0101
0110	0 0100
0111	0 0011
1000	0 0010
1001	0 0001

In the above table we see that the LSB of the 10's complement is always the same as that of the BCD number and no computation is required for the generation of the LSB of the 10's complement.

Bit 2 of the 10's complement is 1 anytime bits 1 and 2 of the BCD number are the same, except when all bits of the BCD number are 0. Bit 4 of the 10's complement is 1 when bit 8 of the BCD number is 0 and bits 4 and 2 of the same number are not both 0. Bit 8 of the 10's complement is 1 when the BCD number is 0001 or 0010.

From the above information, the 10's complement generator was designed and is shown on Figure 12.14.

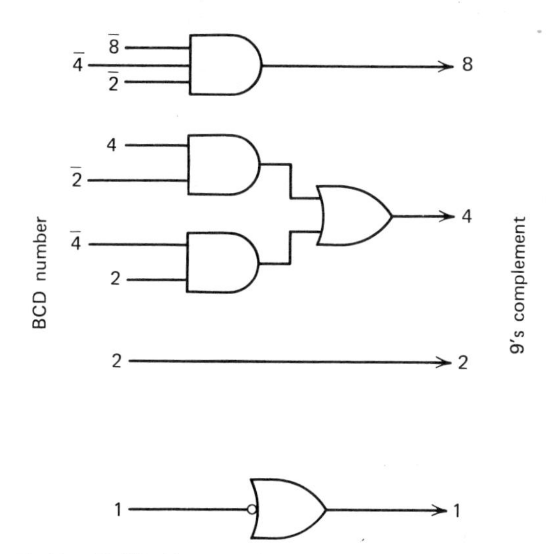

Figure 12.13. BCD 9's complement generator.

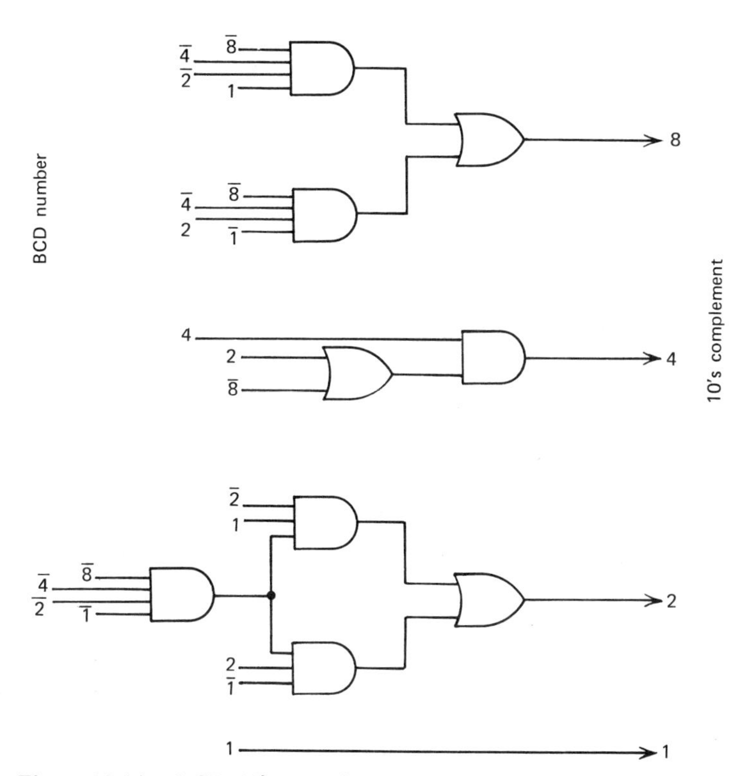

Figure 12.14. BCD 10's complement generator.

12.7 NEGATIVE BINARY-CODED DECIMAL NUMBERS

Negative BCD numbers can be expressed in two forms. One is as a positive number with an additional bit indicating the polarity and the other is as the complement of some greater BCD number. For example, −58 in BCD can be expressed as

0 0101 1000

where 0 will indicate that the BCD number following it is negative.

The other expression is the complement of 58 with respect to a number which is considered to be the maximum BCD number in the system. If 100 (one hundred) is that maximum number, then −58 will be 42—that is,

−0101 1000 0100 0010

The rule to be followed in the conversion is that the least significant digit of the positive number expressing negative quantity is the 10's complement of the same digit of the negative number, and the rest of the digits of the positive number are 9's complements of those of the negative number.

Example 12.5. Express −367 by means of a complement.

−(367)	0011	0110	0111	
	9's	9's	10's	**Complement**
(633)	0110	0011	0011	

∎

Example 12.6. Express −367 in complement form in a 4 digit BCD space. Number −367 in 4 BCD digits is

−0000 0011 0110 0111 −367

Taking the 10's complement of the least significant digit and the 9's complement of the remaining digits we obtain

1001 0110 0011 0011 +9633

which is the number that when added to +367, gives 0 in the 4 digit BCD space. ∎

12.8 SERIES BINARY-CODED DECIMAL COUNTERS

The BCD counter is a counter that counts in binary code either from zero to nine and returns to zero, or from nine to zero and returns to nine.

Series-Up BCD Counter

Figure 12.15 shows a series BCD counter that counts up with its timing wave-forms. In that circuit, the first flip-flop acts as a divide-by-2 device and the other three flip-flops act together as a divide-by-5 circuit. All four flip-flops together make a divide by 10 (2×5) circuit.

The circuit of Figure 12.15 is a series-up BCD counter and works as a typical binary counter until it reaches number 9 which is 1001. At that state, flip-flops 8 and 1 are in the 1 state, and flip-flops 4 and 2 are in the 0 state.

At 1001, the $\bar{Q}$ output of flip-flop 8 inhibits flip-flop 2 from changing to 1. The input pulse following this state will change flip-flop 1 from its present 1 state to the 0 state. This change will have no effect on flip-flop 2, because both inputs S_s and C_s of the second flip-flop will be inhibited; S_s is inhibited by $\bar{Q}$ of flip-flop 8 and C_s is inhibited by Q of flip-flop 2; thus flip-flop 2 remains in the 0 state.

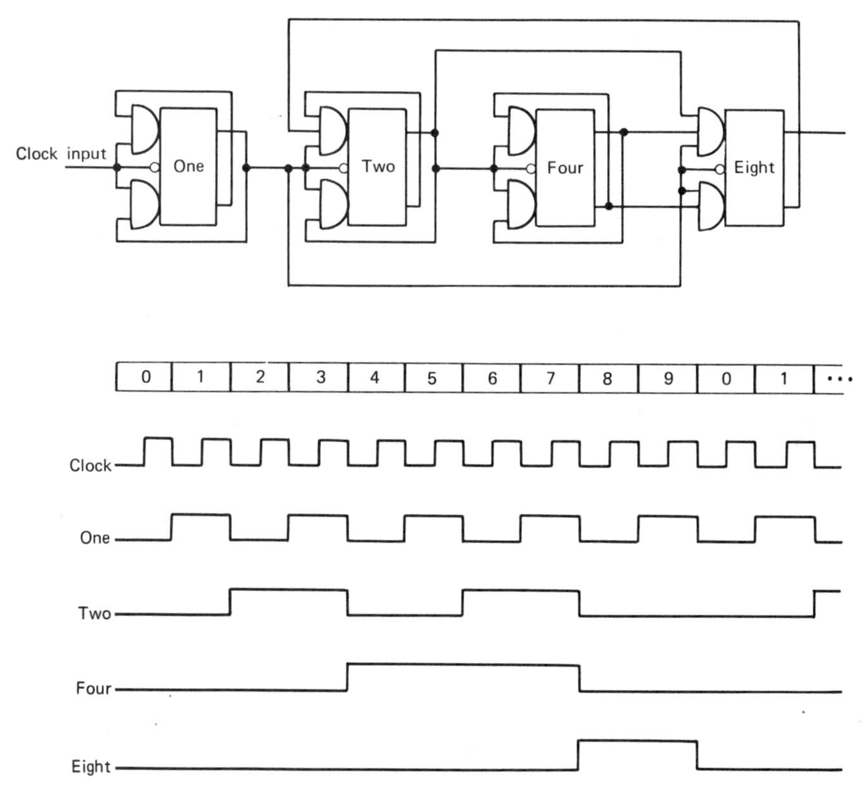

Figure 12.15. Single-digit serial BCD counter and timing diagram where the counter counts up.

When the output of flip-flop 1 changes from 1 to 0, which in this case is from HIGH to LOW, flip-flop 8 "transfers" the information existing at its input to its output. This happens because the output of flip-flop 1 feeds the *CP* input of flip-flop 8. The information transferred to the output of flip-flop 8 is the information in flip-flop 4, which is 0—that is, after the transfer flip-flop 8, which was 1 before, changes to 0. Therefore, the clock pulse that triggers the counter after the counter has reached nine (1001), brings the counter to zero (0000). The next clock pulse will make the counter 0001, and it will start another cycle. When it reaches 1001, it will again change to 0000, and so on.

The ten states of a BCD counter are as follows:

0000
0001
0010
0011
0100
0110
0111
1000
1001 Next state 0000

To decode each of the ten states of a BCD counter, the following expressions may be used:

Zero $= \bar{1}\cdot\bar{2}\cdot\bar{4}\cdot\bar{8}$
One $= 1\cdot\bar{2}\cdot\bar{4}\cdot\bar{8}$
Two $= \bar{1}\cdot 2\cdot\bar{4}$
Three $= 1\cdot 2\cdot\bar{4}$
Four $= \bar{1}\cdot\bar{2}\cdot 4$
Five $= 1\cdot\bar{2}\cdot 4$
Six $= \bar{1}\cdot 2\cdot 4$
Seven $= 1\cdot 2\cdot 4$
Eight $= \bar{1}\cdot 8$
Nine $= 1\cdot 8$

where 1 is the output of the first flip-flop, 2 is the output of the second flip-flop, 4 is the output of the third flip-flop and 8 is the output of the fourth flip-flop of the counter in Figure 12.15.

The circuit of Figure 12.15 is good for counting from zero to nine—that is, it is good for 1 digit. If more digits are needed, circuits like the one of Figure 12.15 may be connected in series as shown in Figure 12.16. This circuit provides the counted number in BCD form. From this form it may be easily converted to decimal to facilitate readout.

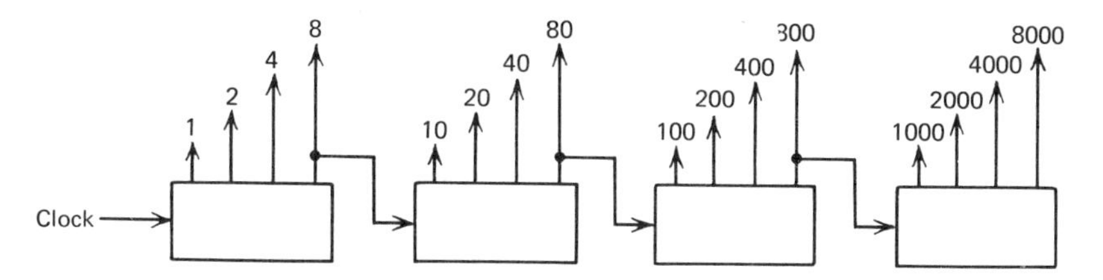

Figure 12.16. A 4 digit serial-up BCD counter. (*Note:* Each block represents a single-digit counter module.)

Series-Down BCD Counter

Figure 12.17 shows a series BCD counter that counts down. This counter is good for 1 digit—that is, it counts from nine to zero, and consists of four master-slave flip-flops.

The first three flip-flops have cross-feedback and operate as trigger flip-flops, while the fourth one does not have cross-feedback and operates as a shift register. The output of the first three flip-flops changes state any time their clock input changes from HIGH to LOW. In the fourth flip-flop, information enters when the clock input is HIGH and is transferred to the output when the clock input changes to LOW.

To achieve a count-down, the first three flip-flops of the counter are connected so that the $\bar{Q}$ output of the first flip-flop feeds the clock input of the second and the $\bar{Q}$ output of the second flip-flop feeds the clock input of the third flip-flop. The counter counts down from nine, 1001, to zero, 0000. At 0000 all flip-flop outputs are 0, and the $\bar{Q}$ outputs of all four flip-flops feed a NAND gate, the output of which prepares the counter so that at the arrival of the next pulse it will change to nine, 1001. This is achieved by setting flip-flop 8 to 1, and by inhibiting flip-flop 2 from changing state, thus remaining at 0. After the counter is set to nine, it starts counting down again until it reaches zero. After that it jumps to nine, and continues the process.

12.9 SYNCHRONOUS BINARY-CODED DECIMAL COUNTERS

There is a synchronous BCD counter similar to the binary synchronous counter. In this counter, all stages (flip-flops) are simultaneously triggered by the clock input.

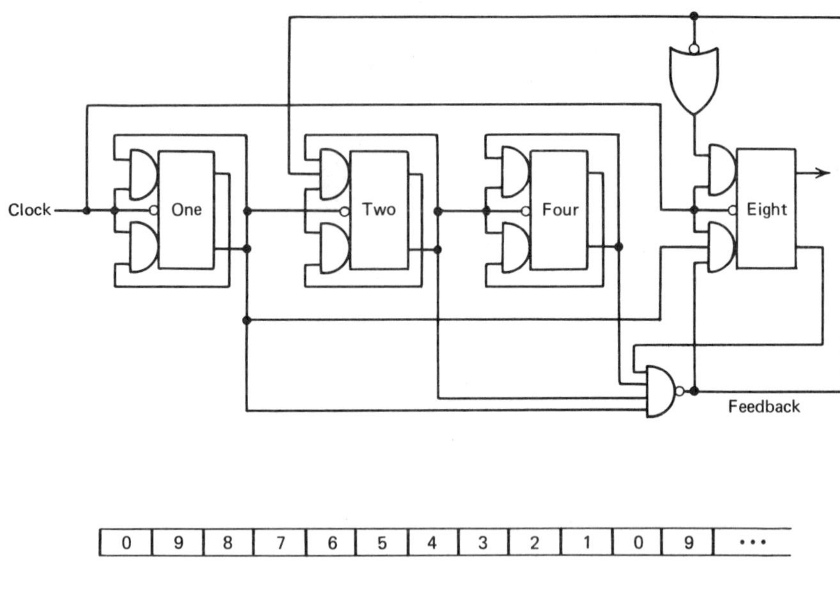

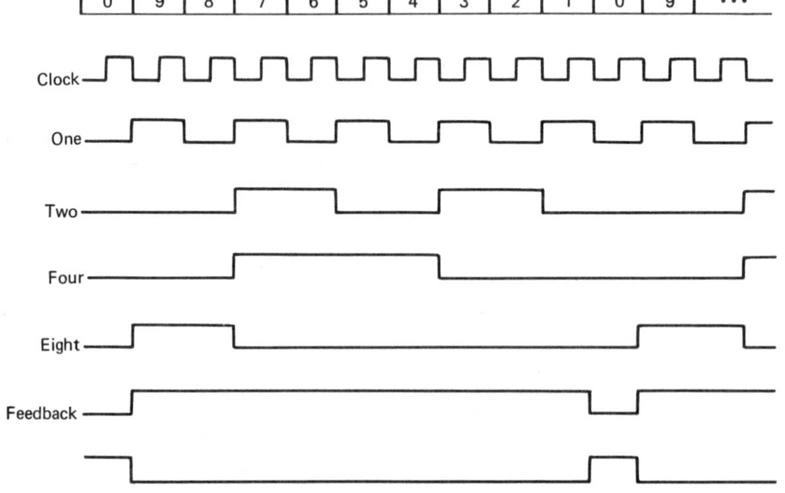

Figure 12.17. Single-digit serial-down BCD counter and its waveforms of operation.

Synchronous-Up BCD Counter

Figure 12.18(*a*) shows a counting up synchronous BCD counter and its timing waveforms. In this circuit, there is no cumulative delay in the generation of the outputs because a stage does not have to wait for the preceding stages to change. In this counter, the state of each flip-flop for time $n + 1$ is "prepared" during time n.

Information enters each flip-flop at the flip-flop's synchronous inputs S_s and C_s. This information is then transferred to the flip-flop output when the clock input changes from HIGH to LOW. This way, all flip-flops change into their new state simultaneously at the arrival of the clock pulse.

The advantage of this operation is that the counter output is ambiguous only during the time information transfers into the flip-flops. This ambiguity starts right after the clock input changes to LOW and lasts until all flip-flop outputs have stabilized to their new state. In a typical master-slave flip-flop this transfer time is approximately 15nsec, depending on the type of circuit. From Figure 12.18(*a*) it can be seen that to achieve simultaneous flip-flop changes and higher speed of operation the synchronous BCD counter uses more hardware than the series BCD counter. Therefore, use of the synchronous type should be justified by the need for higher speeds. High speed is usually needed when the counter outputs feed other circuits that are sensitive to delay variation existing among the outputs.

Synchronous operation or high speed, however, is not needed when the counter output is solely going to a display, because a few nanoseconds, or even a second, of delay will not be noticeable to the viewer.

Figure 12.18(*a*) shows the general form of a 1 digit BCD counter, and Figure 12.18(*b*) shows the same network simplified for use as the least significant digit in a multidigit BCD counter. Both circuits operate as synchronous binary counters until they reach nine, which is 1001; after that, the second flip-flop is inhibited from entering new information. At the arrival of the next clock pulse, which is the tenth pulse, the first flip-flop changes state, becoming 0, the second and third flip-flops remain the same (in the 0 state), and the fourth flip-flop, which works as a shift register, changes from 1 to 0. The last flip-flop becomes 0 after the tenth pulse because before its arrival both clear synchronous inputs C_s were HIGH, while one of the set synchronous inputs S_s was LOW.

Thus, at the arrival of the tenth pulse, starting since the counter was at 0000, the counter changes from nine (1001) to zero (0000) and is ready to start again its cycle of ten counts, which are as follows:

Zero	0000	
One	0001	
Two	0010	
Three	0011	
Four	0100	
Five	0101	
Six	0110	
Seven	0111	
Eight	1000	
Nine	1001	Next count 0000

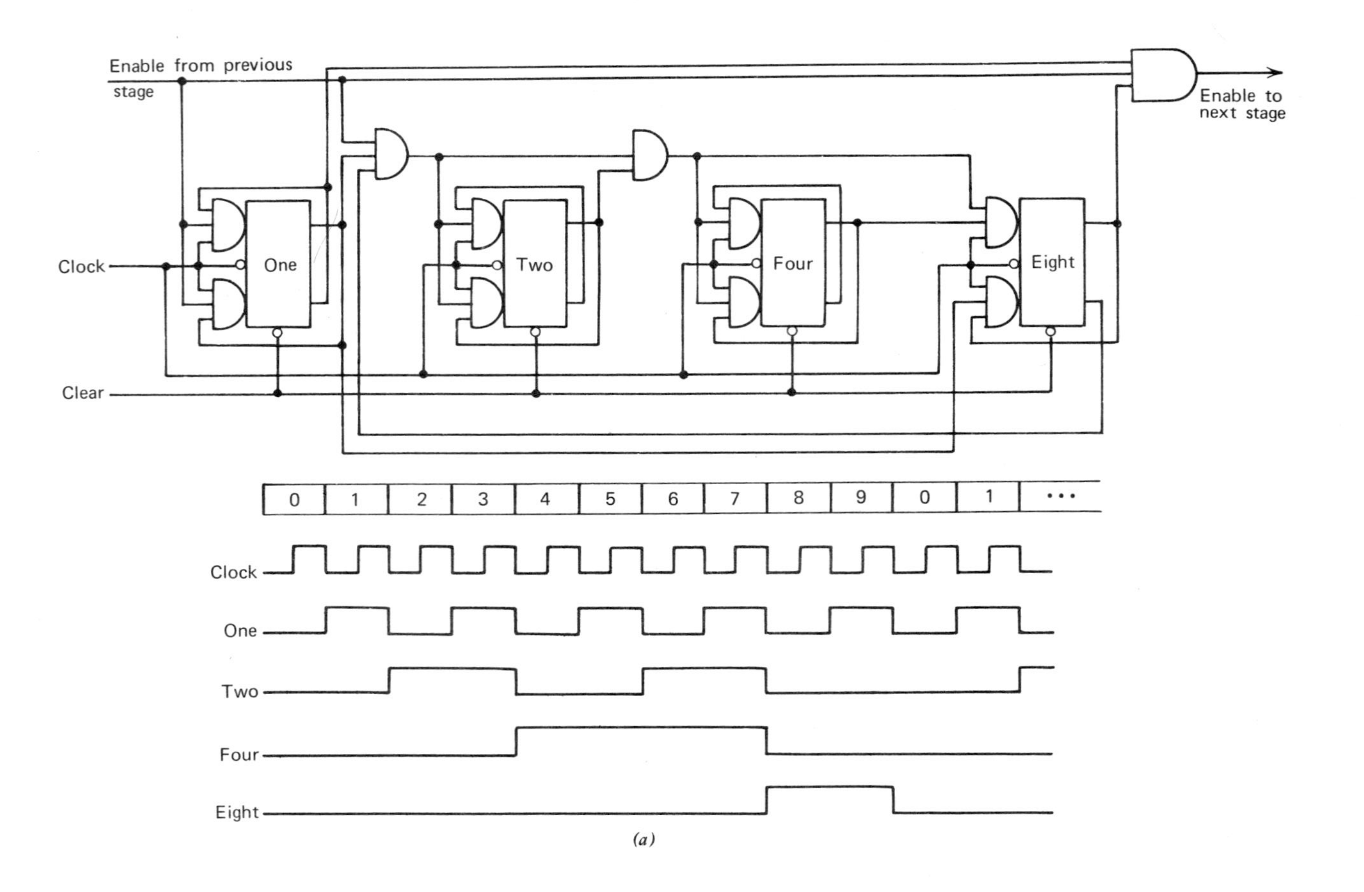

Enable from previous
stage
Enable to
next stage
Clock
Clear
One
Two
Four
Eight
0 1 2 3 4 5 6 7 8 9 0 1 • • •
Clock
One
Two
Four
Eight

(a)

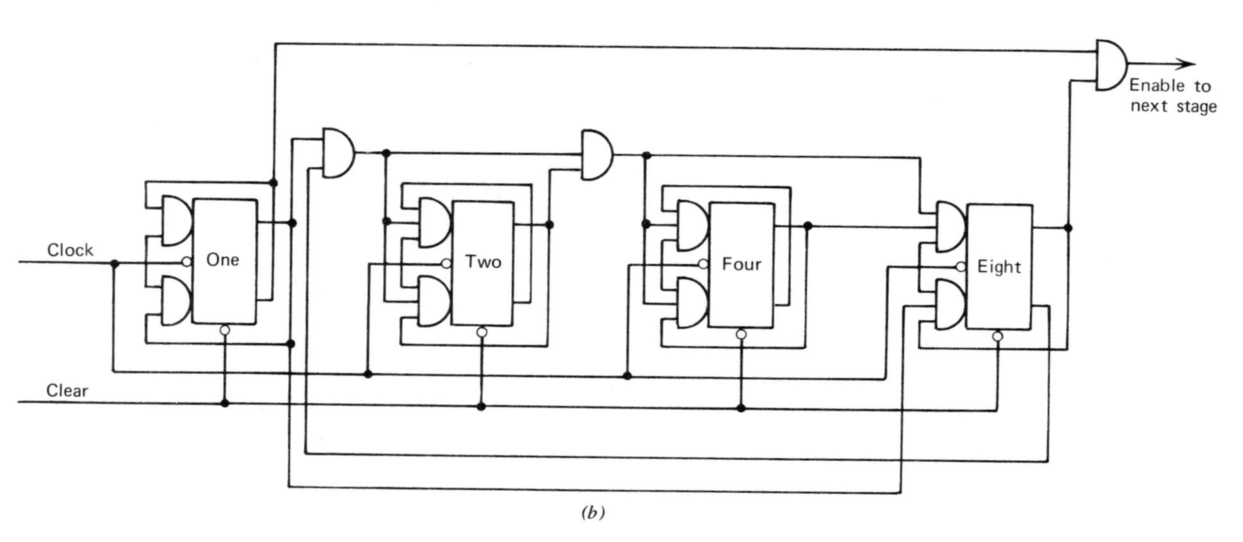

(b)

Figure 12.18. Single-digit synchronous-up BCD counter: (*a*) general configuration; (*b*) to be used as a least significant digit.

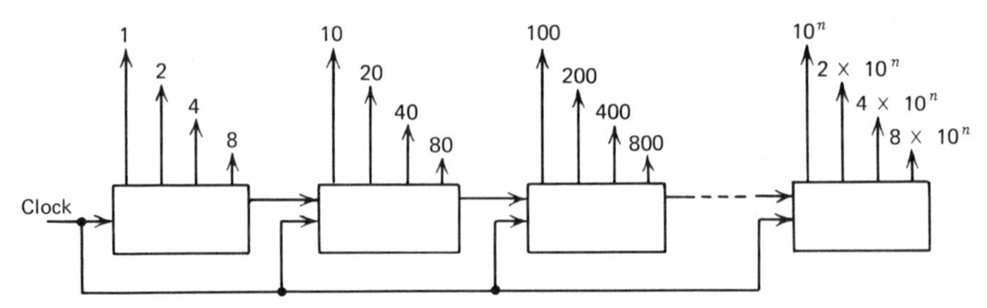

Figure 12.19. Multidigit synchronous BCD counter counting up. (*Note:* Each block represents a single-digit synchronous-up BCD counter similar to that of Figure 12.18.)

To count to numbers beyond nine, circuits like the one of Figure 12.18(*a*) can be connected in cascade as shown in Figure 12.19. The output of this configuration is in BCD form, which can be very easily converted to decimal, and drive decimal displays.

Synchronous-Down BCD Counter

The configuration of a synchronous BCD counter that counts downwards is shown in Figure 12.20. This circuit in principle works like the series-down BCD counter discussed in the preceding section—that is, it counts downwards from nine, 1001, to zero, 0000, and then jumps to nine, 1001, starting all over again.

To achieve that, when the counter reaches zero, 0000, proper gating inhibits the two middle flip-flops from entering new information and provides the fourth flip-flop with a 1 at its synchronous set S_s and a 0 at its synchronous reset C_s.

This way, at the arrival of the next clock pulse after the counter is at 0000, the first flip-flop, which is the LSB, changes state, becoming 1; the two middle flip-flops remain what they were, which is 0, and the fourth flip-flop, which is the MSB, changes state to 1.

Circuits like the one of Figure 12.20 can be connected in cascade to accommodate numbers greater than nine. Such a configuration is shown in Figure 12.21. The output of this chain of individual single-digit circuits may be very easily converted to decimal for convenience in readout.

Synchronous Reversible BCD Counter

In the design of digital systems we often encounter the need for a reversible BCD counter. Figure 12.22 shows the logic diagram of such a counter using master-slave flip-flops.

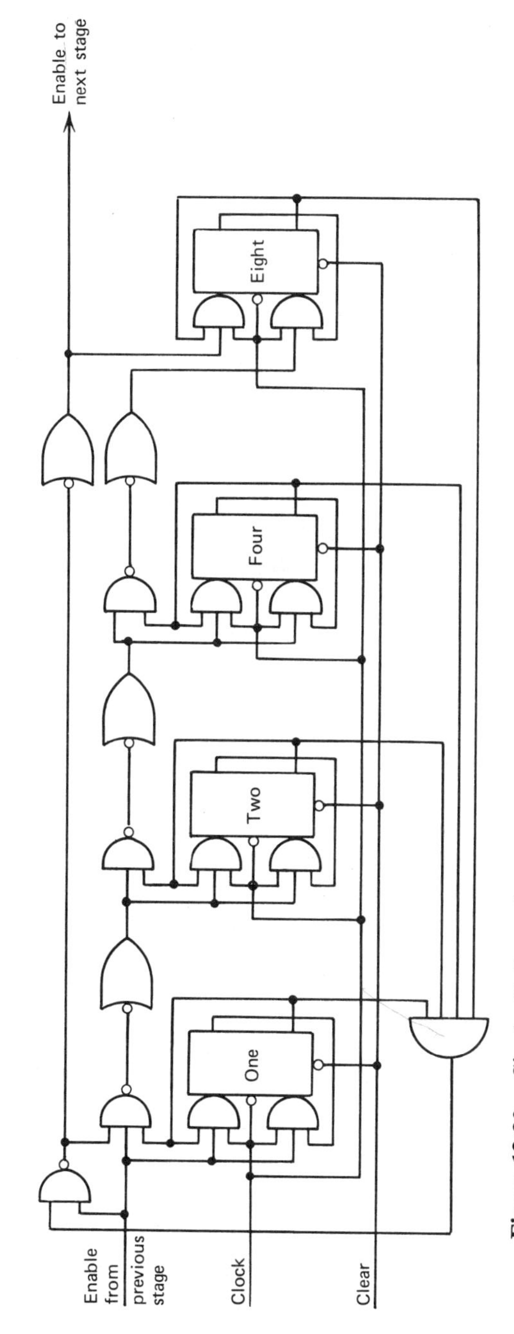

Figure 12.20. Single-digit synchronous-down BCD counter.

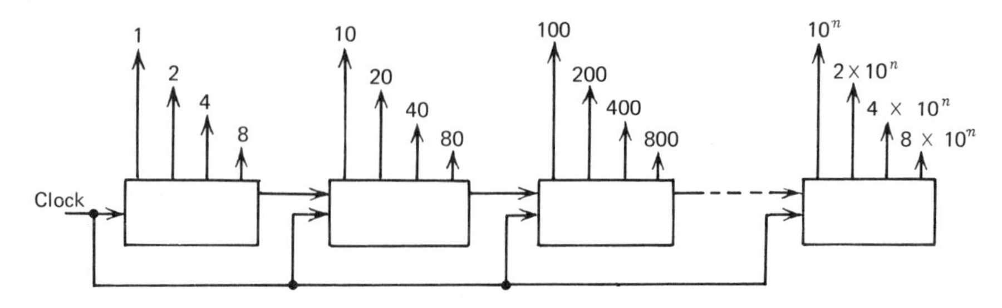

Figure 12.21. Multidigit synchronous BCD counter counting down. (*Note:* Each block represents a single-digit synchronous-down BCD counter similar to that of Figure 12.20.)

The thinking in designing this circuit is as follows: The functions to be implemented are first defined for each of the two modes of operation, the count-up mode and the count-down mode. In the count-up mode we want the counter to count up from 0000 to 1001, change back to 0000, and count up again.

If the counter were a pure binary one, the count after 1001 would have been 1010. To make the counter a BCD one, we must somehow force it to become 0000 after it reaches 1001.

Looking at the desired count of 0000 and the binary count of 1010, that would normally follow count 1001, we see that the LSB of 0000 and 1010 is the same which tells us that the flip-flop representing the LSB should remain unchanged. We also see that the second LSB in the binary count changes from 0 to 1, while in the BCD we want it to remain 0. To achieve that, when the counter is in the count of 1001, we decode this count and inhibit the second flip-flop from entering new information. This way after count 1001 the second flip-flop will remain in the 0 state. The third LSB is in both cases, binary and BCD, 0 and therefore should remain unchanged.

The MSB, however, should change from 1 to 0. To achieve that we treat the fourth flip-flop, which represents the MSB, as a separate master-slave flip-flop as shown in the circuit of Figure 12.22. To change the MSB from 1 to 0, we decode count 1001 and with this signal we enable the synchronous clear of the fourth flip-flop so that at the arrival of the clock pulse this flip-flop will change to 0.

Since we treat the fourth flip-flop as a separate flip-flop, we must also provide information for its synchronous set. Since the synchronous set is the one that sets the flip-flop to 1, and we want this to happen after count 0111, we decode that count, and with this signal we enable the synchronous clear of the fourth flip-flop so that, at the arrival of the clock pulse following

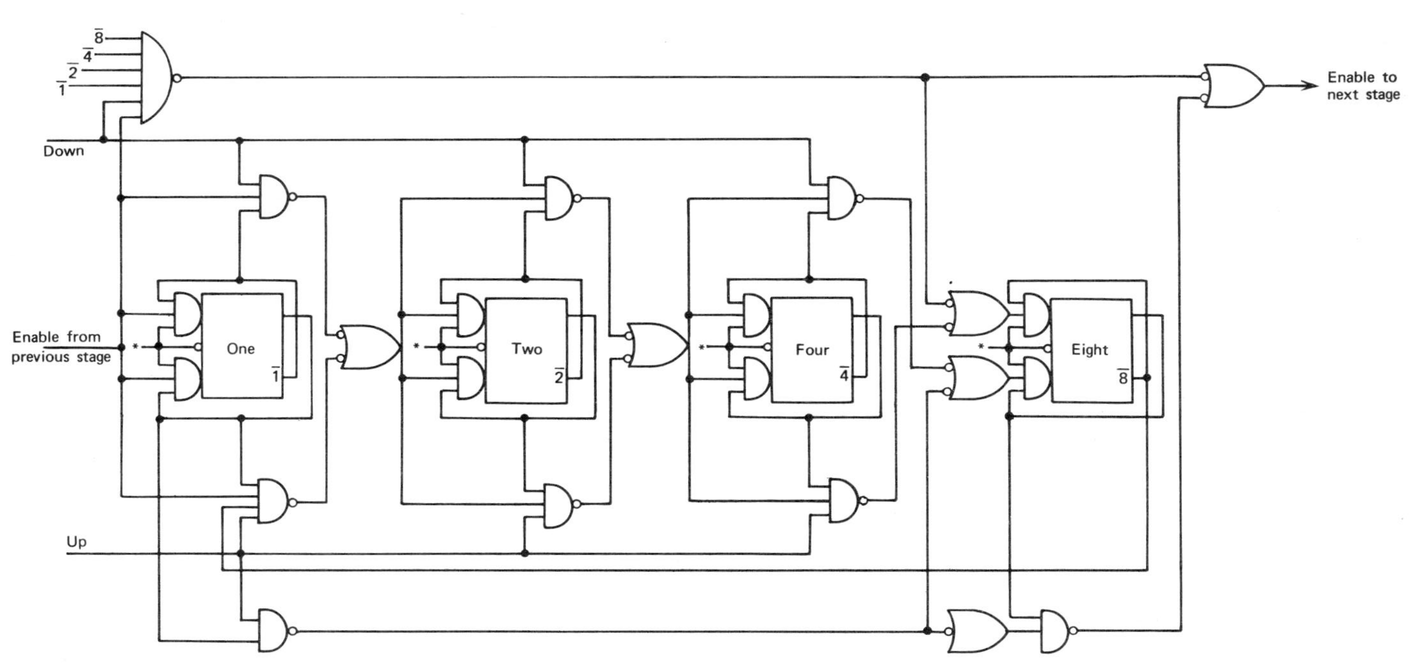

Figure 12.22. Single-digit synchronous reversible BCD counter. (*Note:* Clock connection is not shown.)

1000, this flip-flop will change to 1. For the count-up mode the following sequence and operations take place:

	0000	
	0001	
	0010	
	0011	
	0100	
	0101	
	0110	
	0111	(Enable synchronous set of fourth flip-flop)
	1000	
(Enable next stage)	1001	(Enable synchronous clear of fourth flip-flop and
	0000	inhibit second flip-flop from entering new
	0001	information)
	etc.	

In the count-down mode we want to achieve the opposite of what we did in the count-up mode—that is, we want to count down and when we reach 0000 we want to change to 1001 rather than to 1111 which would be the normal binary count.

A philosophy similar to that applied for the design of the count-up circuitry used in the design of the count down circuitry will result in the following sequence of events:

	1001	
	1000	(Enable synchronous clear of fourth flip-flop)
	0111	
	0110	
	0101	
	0100	
	0011	
	0010	
	0001	
(Enable next stage)	0000	(Enable synchronous set of fourth flip-flop, and
	1001	inhibit second flip-flop from entering new
	1000	information)
	etc.	

Circuits like the one of Figure 12.22 can be connected in cascade to form a multidigit reversible BCD counter. When the circuit of Figure 12.22 is used as the least significant digit, the input level to the "enable input" should be HIGH—that is, it should either be left open-circuited or connected to the V_{CC} of the circuit.

12.10. DECIMAL RING COUNTERS

The decimal ring counter is a ten-stage shift register, the sequential output of which feeds its sequential input directly. That is, the output of the tenth stage of the shift register feeds the input of the first flip-flop stage of the ten-stage shift register. This way, a closed-loop shift register is formed as shown in Figure 12.23, where the clock triggers all stages simultaneously and shifts the information contained in each stage to the stage that follows it.

Thus, if the first stage is preset to 1 and the remaining nine to 0, 1 will move from stage to stage, making one move per clock pulse as shown on Figure 12.23. If the first flip-flop of the circuit of Figure 12.23(*a*) is labeled decimal 0, the second decimal 1, the third decimal 2, and so on, the location of 1 in the ring counter will determine the number of pulses having "hit" the ring counter since the counter was at 1000000000. The decimal ring counter does not generate new information as do the BCD counters discussed in the previous sections of this chapter, but it simply circulates the information present in it.

The ten outputs of a decimal ring counter can be used to directly control the display of decimal displays. This way, information goes directly from the counting circuit to the display without the need of any decoding networks.

The advantage of the decimal ring counter over the BCD counter is simplicity, when full decoding is needed, while the disadvantage is that the decimal ring counter does not provide BCD code which is useful for the performance of arithmetic operations. Figure 12.23(*a*) illustrates a conventional decimal ring counter. Circuit (*b*) is a modified decimal ring counter. In this circuit the tenth flip-flop is replaced by a multi-input NAND gate and an inverter, forming a noninverting AND gate. This gate produces an output of 1 whenever all inverted outputs of the flip-flops are 1. This output is the same as that of the tenth flip-flop of the conventional decimal ring counter. The information generated by this noninverting AND gate feeds the first flip-flop in the same way that the output of the tenth flip-flop feeds the first flip-flop in the conventional decimal ring counter.

The advantage of the modified decimal ring counter over the conventional type is that the former is self-starting—that is, even if the counter is not preset, it will go into its normal sequence within a maximum of ten pulses.

12.11 DECIMAL SHIFT COUNTERS

The decimal shift counter like the decimal ring counter is a shift register with feedback. In the case of the decimal shift counter, however, the feedback is crossed—that is, output Q of the last flip-flop feeds the synchronous clear

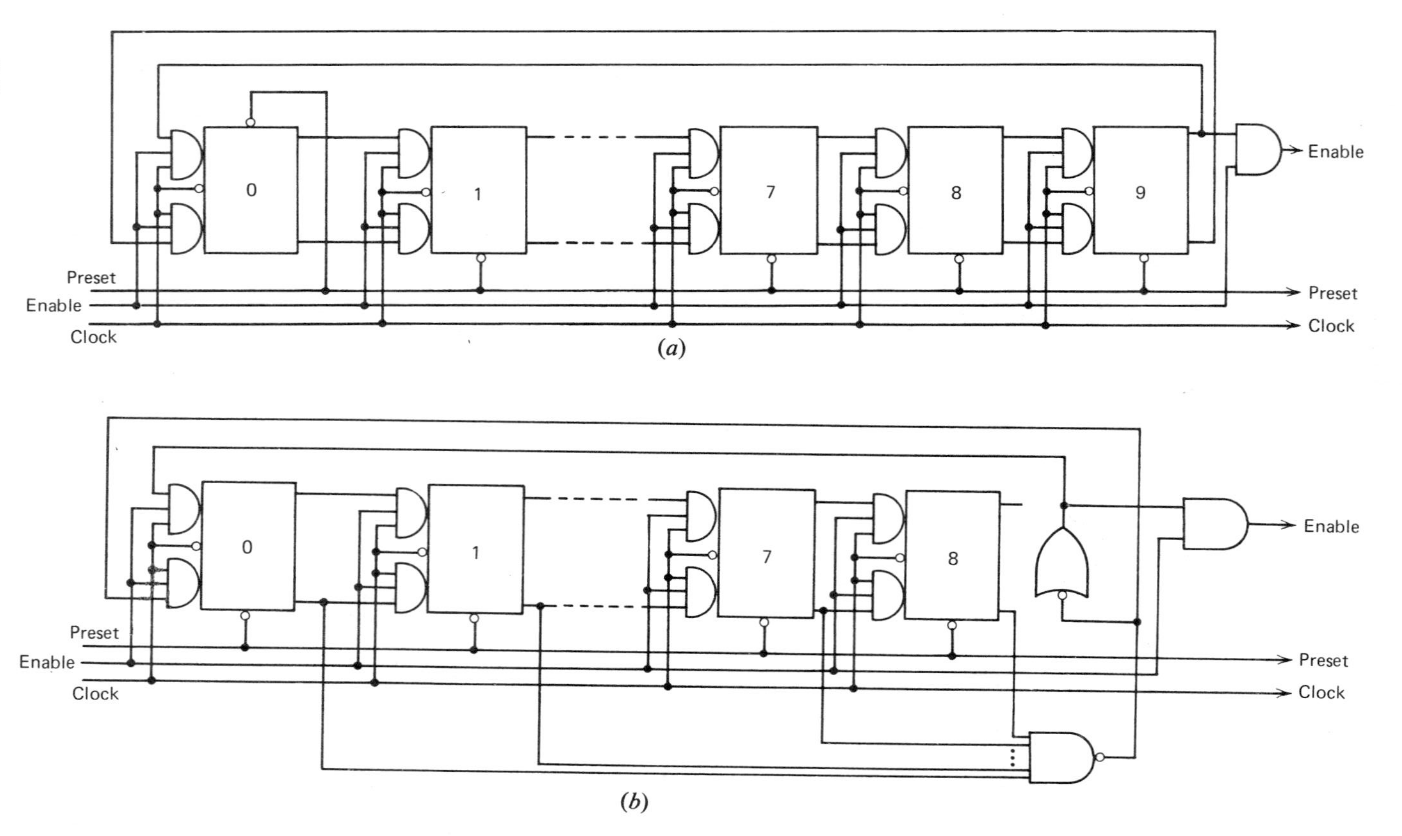
0
1
7
8
9
Enable
Preset
Enable
Clock
Preset
Clock
0
1
7
8
Enable
Preset
Enable
Clock
Preset
Clock

(a)

(b)

	Ring counter outputs									
	0	1	2	3	4	5	6	7	8	9
Preset	1	0	0	0	0	0	0	0	0	0
Status of outputs after clock pulse										
1	0	1	0	0	0	0	0	0	0	0
2	0	0	1	0	0	0	0	0	0	0
3	0	0	0	1	0	0	0	0	0	0
4	0	0	0	0	1	0	0	0	0	0
5	0	0	0	0	0	1	0	0	0	0
6	0	0	0	0	0	0	1	0	0	0
7	0	0	0	0	0	0	0	1	0	0
8	0	0	0	0	0	0	0	0	1	0
9	0	0	0	0	0	0	0	0	0	1
10 (start of new cycle)	1	0	0	0	0	0	0	0	0	0

(*c*)

Figure 12.23. (*a*) Conventional decimal ring counter. (*b*) Modified decimal ring counter. (*c*) Truth table.

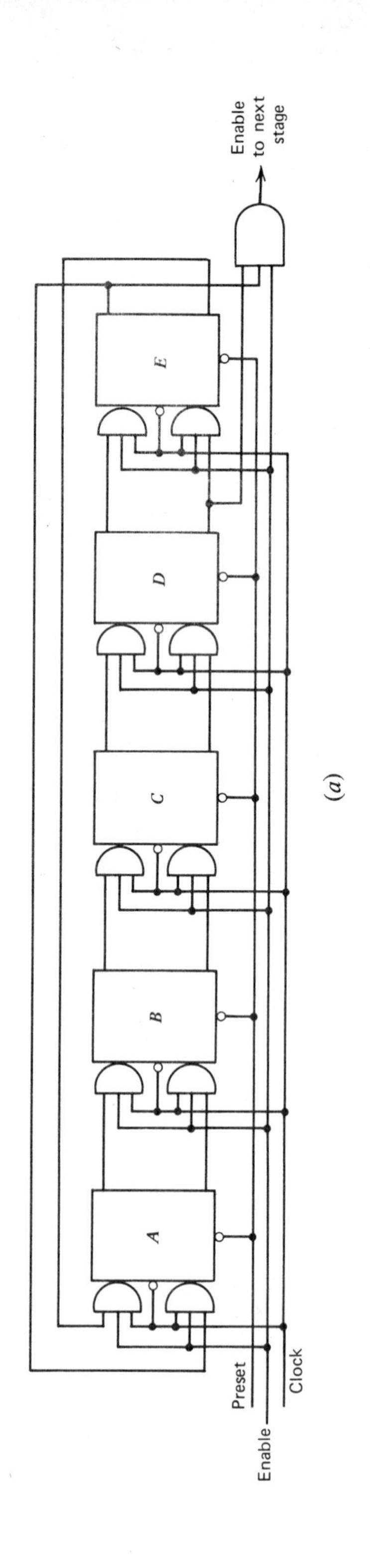

(a)

Decimal	Shift counter output					Code
	A	B	C	D	E	
0	0	0	0	0	0	$\bar{A}\bar{E}$
1	1	0	0	0	0	$A\bar{B}$
2	1	1	0	0	0	$B\bar{C}$
3	1	1	1	0	0	$C\bar{D}$
4	1	1	1	1	0	$D\bar{E}$
5	1	1	1	1	1	AE
6	0	1	1	1	1	$\bar{A}B$
7	0	0	1	1	1	$\bar{B}C$
8	0	0	0	1	1	$\bar{C}D$
9	0	0	0	0	1	$\bar{D}E$

(*c*)

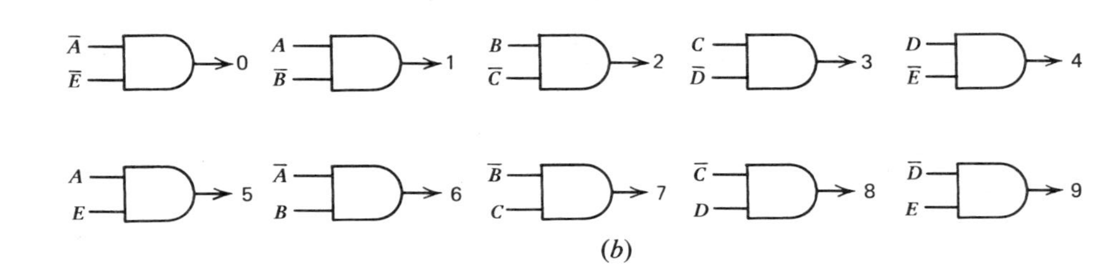

(*b*)

Figure 12.24. Decimal shift counter: (*a*) logic diagram; (*b*) truth table; (*c*) decoding gates.

C_s of the first flip-flop and the inverse output $\bar{Q}$ of the last flip-flop feeds the synchronous set S_s of the first flip-flop.

A decimal shift counter and its truth table are shown on Figure 12.24. Also shown are the decoding gates for decimal readout. The decimal shift counter requires presetting where all flip-flops must be initially preset to the same state, either all to the 0 state or all to the 1 state.

To decode a specific state of a decimal counter, the uniqueness of that state must be recognized and detected. The third column of Figure 12.24(*b*) shows the uniqueness of each of the ten counts of the decimal shift counter when the counter is initially preset so that all Q outputs are 0. If the number five, for example, is to be decoded, the Q outputs of the first and the last flip-flops should be ANDed. These two outputs are both 1 only in the fifth count, counting from the time the counter was preset—or the counter was in the count of preset which in this case is 00000. The decimal shift counter provides simplicity and speed and can be used where plain decimal counting is required, and no BCD code is needed. With the use of decoding gates the decimal shift counter can provide decimal code.

Similarly to the other decimal-oriented counters, the decimal shift counter can be used for the configuration of a multidigit decimal counter, where each digit stage uses the network of Figure 12.24.

12.12 COMPARISON OF DECIMAL-ORIENTED COUNTERS

The advantages and disadvantages of the decimal-oriented counters are summarized in Figure 12.25.

Type of counter	Advantages	Disadvantages
Series BCD counter	Simplicity BCD code	Non synchronous operation
Synchronous BCD counter	Speed BCD code	Required additional gates
Decimal ring counter	Speed Simplicity Decimal code	No BCD code
Decimal shift counter	Speed Simplicity Less hardware	No BCD code No decimal code

Figure 12.25. Advantages and disadvantages of decimal-oriented counters.

12.13 MEDIUM-SCALE INTEGRATION FOR BINARY-CODED DECIMAL OPERATIONS

The BCD system finds extensive use in the field of instrumentation and numerical calculations. To meet these needs, several types of BCD networks are available. The most common are BCD-to-decimal converters used to drive decimal displays and BCD counters.

Some BCD counters also include a storage register that enables operation of the counter while a previously obtained output has been saved. This feature considerably facilitates operations where intermediate results need be temporarily stored before they are used for either display or further processing.

Figure 12.26 illustrates typical BCD related logical networks that are available in MSI form.

PROBLEMS

1. Convert the following decimal numbers to BCD:

   375, 201, 37, 11

2. Convert the following binary numbers to BCD:

   1001, 11101, 101, 000011

3. Convert the following BCD numbers to decimal:

   (*a*) 0010 0011 0101
   (*b*) 1001 0111 0110
   (*c*) 0111 0010 0101

4. Perform the following BCD operations:

   (*a*) 0101 1001 + 1000 1001
   (*b*) 0111 1001 + 0010 0001
   (*c*) 0101 0111 − 0100 1000
   (*d*) 0010 0011 − 0001 0100
   (*e*) 1000 × 0010
   (*f*) 0111 × 0101

5. Compute the 9's and 10's complements of the following BCD numbers:

   1001, 0000, 0101, 0010

6. Draw the timing waveforms of a sequential BCD adder that adds two 3 digit BCD numbers.

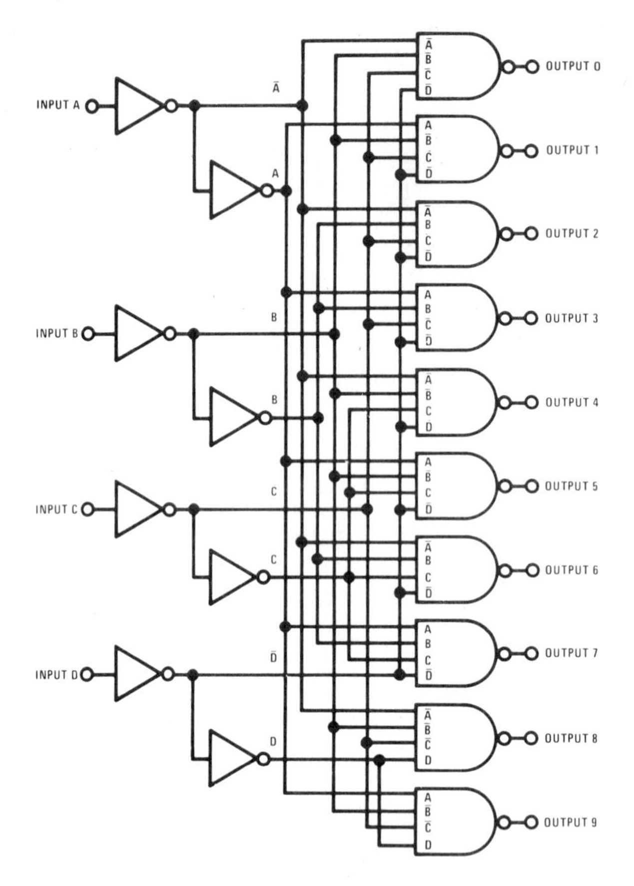

TRUTH TABLE

INPUTS				OUTPUTS									
D	C	B	A	0	1	2	3	4	5	6	7	8	9
0	0	0	0	0	1	1	1	1	1	1	1	1	1
0	0	0	1	1	0	1	1	1	1	1	1	1	1
0	0	1	0	1	1	0	1	1	1	1	1	1	1
0	0	1	1	1	1	1	0	1	1	1	1	1	1
0	1	0	0	1	1	1	1	0	1	1	1	1	1
0	1	0	1	1	1	1	1	1	0	1	1	1	1
0	1	1	0	1	1	1	1	1	1	0	1	1	1
0	1	1	1	1	1	1	1	1	1	1	0	1	1
1	0	0	0	1	1	1	1	1	1	1	1	0	1
1	0	0	1	1	1	1	1	1	1	1	1	1	0
1	0	1	0	1	1	1	1	1	1	1	1	1	1
1	0	1	1	1	1	1	1	1	1	1	1	1	1
1	1	0	0	1	1	1	1	1	1	1	1	1	1
1	1	0	1	1	1	1	1	1	1	1	1	1	1
1	1	1	0	1	1	1	1	1	1	1	1	1	1
1	1	1	1	1	1	1	1	1	1	1	1	1	1

(a)

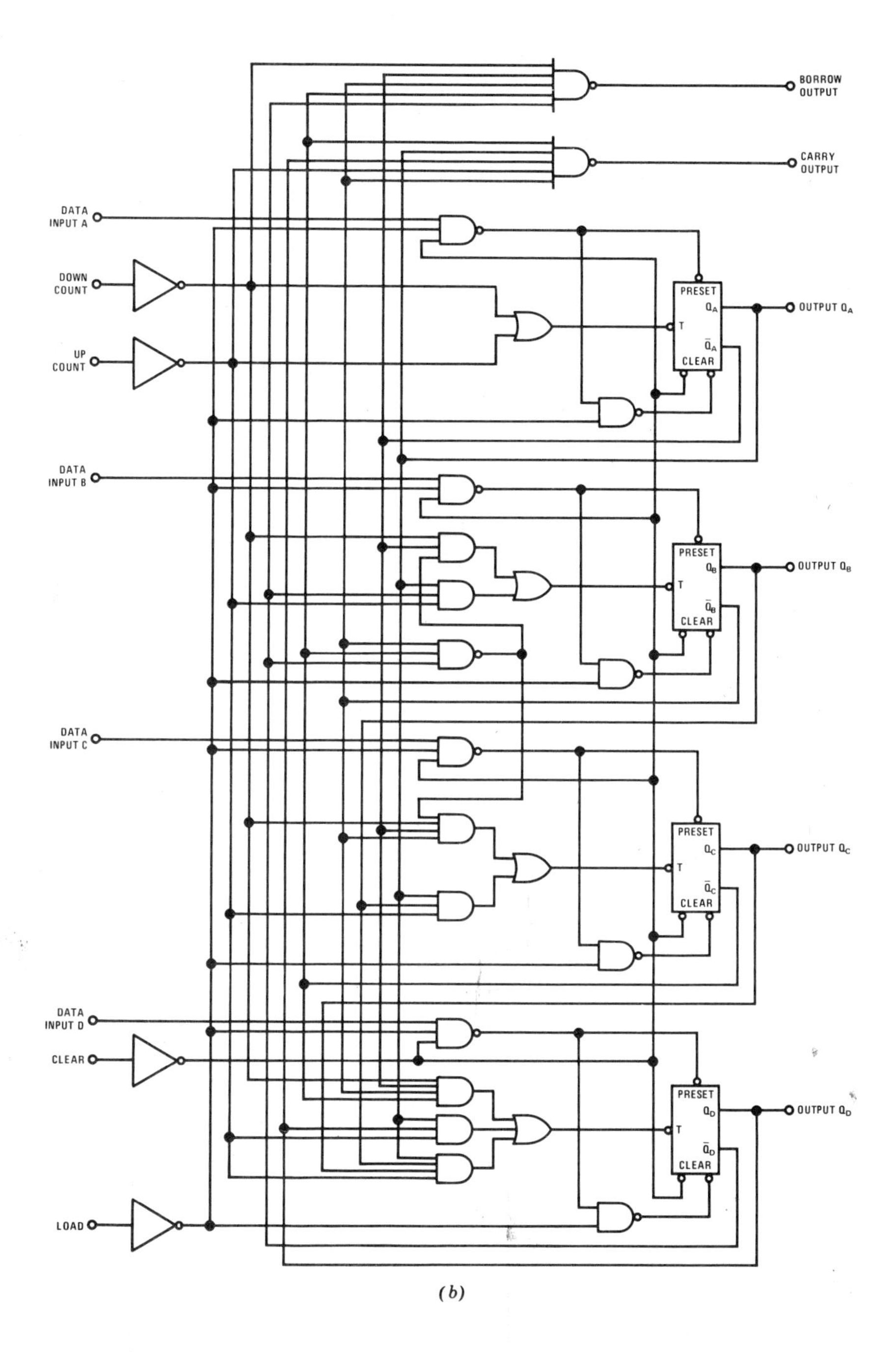
BORROW OUTPUT
CARRY OUTPUT
DATA INPUT A
DOWN COUNT
UP COUNT
PRESET
Q_A
T
$\bar{Q}_A$
CLEAR
OUTPUT Q_A
DATA INPUT B
PRESET
Q_B
T
$\bar{Q}_B$
CLEAR
OUTPUT Q_B
DATA INPUT C
PRESET
Q_C
T
$\bar{Q}_C$
CLEAR
OUTPUT Q_C
DATA INPUT D
CLEAR
PRESET
Q_D
T
$\bar{Q}_D$
CLEAR
OUTPUT Q_D
LOAD

(*b*)

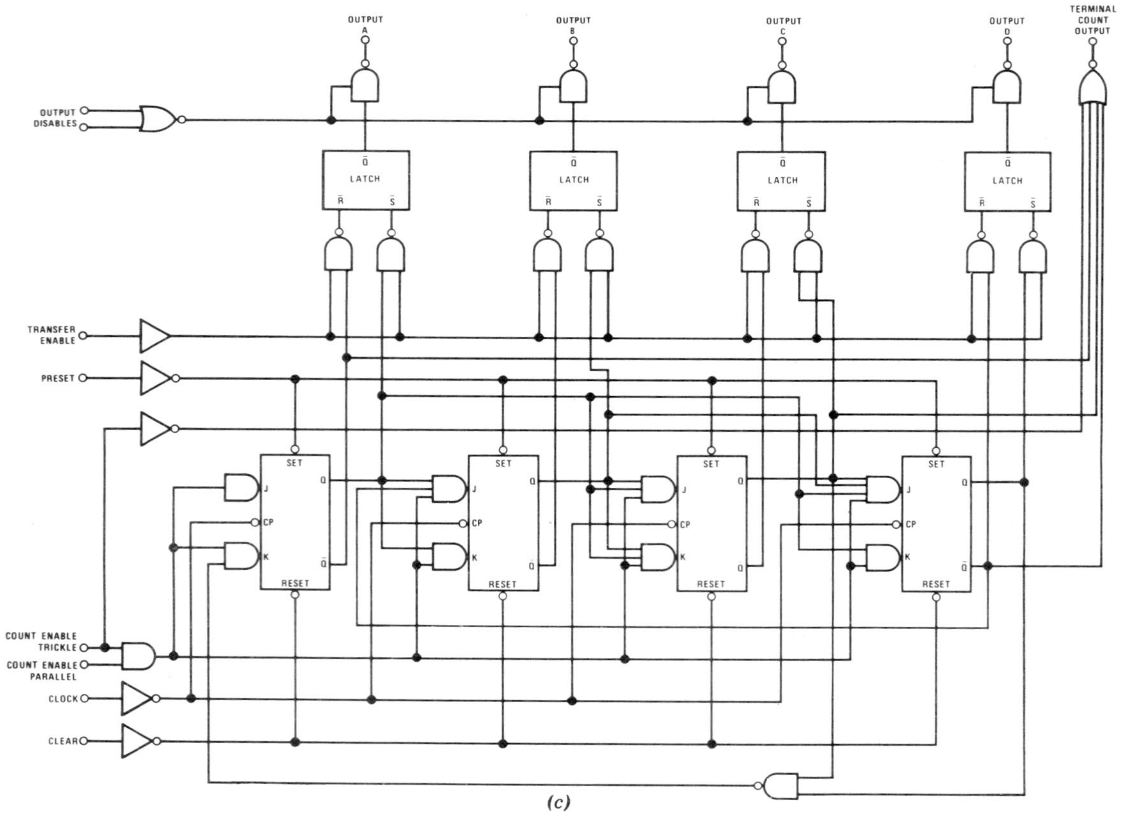

(c)

Figure 12.26. Typical BCD logic networks available in MSI form. (Courtesy of National Semiconductor): (*a*) BCD-to-decimal decoder and drivers; (*b*) up-down decade counter; (*c*) decade counter with output latch flip-flops.

7. Design a 2 digit sequential BCD multiplier and draw the timing diagram for the following multiplication:

0111	1001	1000	0010	(sequential in digits)
$\times$		0101	0011	(parallel)

8. Design a BCD divider that handles dividends having up to 7 digits and divisors having up to 6 digits.
9. Design a 10's complement generator for sequential BCD numbers.
10. A 3 digit decimal counter is needed, the output of which will be used only for numerical display. Select the counter that will require the least amount of hardware and draw its diagram.
11. Design a counting network to be used in a parking lot which will tell the attendant the number of available parking spaces. The lot has three entry ports, one exit port, and a capacity of 375 cars. Determine the necessary car sensor requirements. Base your design on currently available hardware.

CHAPTER THIRTEEN

Logical Symmetry*

13.1 INTRODUCTION

Logical symmetry is a property that exists in certain logical functions, making them subject to special implementation techniques. Such functions are called symmetric functions.

Symmetric functions are of special significance in logical design and form an important class of logical functions. Special design techniques exist for symmetric functions, and it is therefore very important to know whether a given function is symmetric. If these techniques are not used, the implementation of symmetric functions is difficult because there is no simple conventional method of logical simplification which leads to their minimizations.

This chapter covers the definitions and properties of symmetric functions and presents methods for their detection and identification. Also included are methods of implementation of symmetric functions.

* This chapter is an excerpt from the author's doctoral dissertation presented at the Arizona State University.

13.2 DESCRIPTION OF LOGICAL SYMMETRY

A logical function of n variables $X_1, X_2, \ldots, X_n$ is said to be *symmetric* if an interchange in these variables leaves the function identically the same. If this property extends to all variables, the function is said to be *totally symmetric*. If it extends only to a subset of the n variables, the function is said to be *partially symmetric*. And, if symmetry exists only in a subset of the function, the function is said to be *independently symmetric*.

Interchangeability in variables may exist between nominal values (i.e., X_i and X_j), inverted values (i.e., $\overline{X}_i$ and $\overline{X}_j$), or mixed values (i.e., $\overline{X}_i$ and X_j or X_i and $\overline{X}_j$).

Figure 13.1 illustrates examples of the above three types of logical symmetry where

$$f(A, B, C, D) = \sum(5, 9, 12, 15)$$

variables A, B, $\overline{C}$, and D are interchangeable. If variables A and B are interchanged, we have

$$\begin{aligned} f(A, B, C, D) &= B\overline{A}\overline{C}D + \overline{B}A\overline{C}D + BA\overline{C}\overline{D} + BACD \\ &= \sum(9, 5, 12, 15) \\ &= \sum(5, 9, 12, 15) \end{aligned}$$

that is, the interchange in these variables left the function unchanged. If variables $\overline{C}$ and D are interchanged, we have

$$\begin{aligned} f(A, B, C, D) &= \overline{A}B\overline{D}C + A\overline{B}\overline{D}C + ABDC + AB\overline{D}\overline{C} \\ &= \sum(5, 9, 15, 12) \end{aligned}$$

that is, interchanges in variables A, B, $\overline{C}$, D do not affect the function. Variables $\overline{A}$, $\overline{B}$, C, $\overline{D}$, also have the same properties. If variables $\overline{A}$ and C are interchanged, we have

$$\begin{aligned} f(A, B, C, D) &= CBAD + C\overline{B}\overline{A}D + CB\overline{A}\overline{D} + \overline{C}B\overline{A}D \\ &= \sum(15, 9, 12, 5) \end{aligned}$$

that is, symmetric functions have two sets of variables of symmetry. In this case they are A, B, $\overline{C}$, and D, and $\overline{A}$, $\overline{B}$, C, and D. These two sets also represent the symmetry's two centers; $AB\overline{C}D = 13$ and $\overline{A}\overline{B}C\overline{D} = 2$.

Symmetric functions are designated by means of the following notation:

$$\text{Symmetric function} = S_i^n(X_1, X_2, \ldots, X_n)$$

where S indicates symmetry, n is the number of variables in the considered logical space, $X_1, X_2, \ldots, X_n$ are the variables of symmetry, and i is the number of variables that are the same between the function's terms and the set of variables of symmetry.

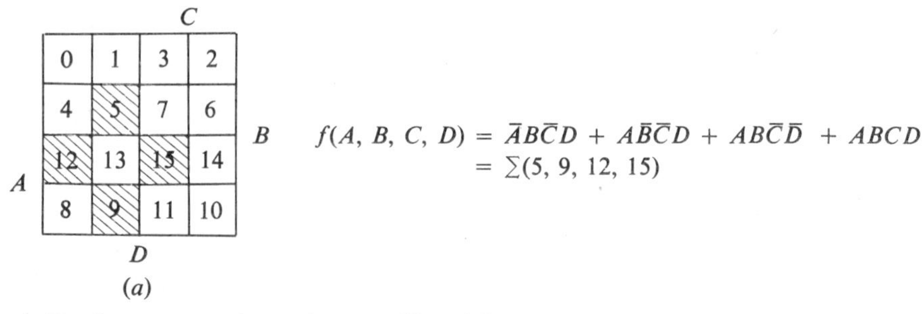

(*a*)

a) Total symmetry about elements 13 and 2.
Interchangeability exists between elements A, B, $\bar{C}$, and D.

0	1	3	2
4	5	7	6
12	13	15	14
8	9	11	10

$$f(A, B, C, D) = A\bar{B}\bar{C}D + AB\bar{C}\bar{D} + ABCD = \sum(9, 12, 15)$$

(*b*)

b) Partial symmetry. Interchangeability exists only between variables B, $\bar{C}$, and D, and symmetry is about elements 2, 5, 13, and 10.

0	1	3	2
4	5	7	6
12	13	15	14
8	9	11	10

$E = 0$

16	17	19	18
20	21	23	22
28	29	31	30
24	25	27	26

$E = 1$

(*c*)

$$\begin{aligned} f(A, B, C, D, E) &= Eg(A, B, C, D) + \bar{E}h(A, B, C, D) \\ &= \bar{E}(\bar{A}\bar{B}\bar{C}D + \bar{A}B\bar{C}\bar{D} + \bar{A}BCD + AB\bar{C}D) \\ &\quad + E(\bar{A}\bar{B}C\bar{D} + \bar{A}B\bar{C}\bar{D} + \bar{A}BCD + ABC\bar{D}) \\ &= \sum(1, 4, 7, 13, 18, 20, 23, 30) \end{aligned}$$

c) Independent symmetry. In the $E = 0$ subset, symmetry exists about elements 5 and 10, and the interchangeability is among $\bar{A}$, B, $\bar{C}$, and D; while in the $E = 1$ domain the symmetry is about elements 22 and 25, and the interchangeability is among $\bar{A}$, B, C, and $\bar{D}$.

Figure 13.1. Examples of logical symmetry.

If a logical function consists of more than one symmetric subset, where all subsets have the same set of variables of symmetry, the function is written as

$$S^n_{i,j,\ldots}(X_1, X_2, \ldots, X_7)$$

which is the same as

$$S^n_i(X_1, X_2, \ldots, X_7) + S^n_j(X_1, X_2, \ldots, X_7) + \cdots$$

Symmetric functions having the same set of variables of symmetry can be ANDed, ORed, or inverted, as if they were single variables. These properties are defined by the following theorems.

Theorem 13.1. *The sum of two given symmetric functions of the same set of variables of symmetry is a symmetric function of these variables. The subscripts of this function are all the subscripts appearing in either or both of the two given symmetric functions.*

For example, if

$$f(A, B, C, D, E) = S^5_{1,2,3}(A\bar{B}C\bar{D}E) + S^5_{2,4}(A\bar{B}C\bar{D}E)$$

it can be written as

$$f(A, B, C, D, E) = S^5_{1,2,3,4}(A\bar{B}C\bar{D}E)$$

Theorem 13.2. *The product of two given symmetric functions of the same set of variables of symmetry is a symmetric function of these variables. The subscripts of this function are all the subscripts common to both of the given symmetric functions.*

For example, if

$$f(A, B, C, D, E) = S^5_{1,2,3}(A\bar{B}C\bar{D}E) \cdot S^5_{2,4}(A\bar{B}C\bar{D}E)$$

it can be written as

$$f(A, B, C, D, E) = S^5_{2}(A\bar{B}C\bar{D}E)$$

Theorem 13.3. *The inverse of a symmetric function is also a symmetric function. The subscripts of this function are those which are not included in the given function.*

For example, if

$$f(A, B, C, D, E) = \overline{S^5_{1,2,3}}(A\bar{B}C\bar{D}E)$$

it can be written as

$$f(A, B, C, D, E) = S^5_{0,4,5}(A\bar{B}C\bar{D}E)$$

Theorem 13.4. *A symmetric function of* n *variables is equal to the symmetric function in which each of the original variables is complemented and each subscript* j *of the original function is replaced by the* n − j.

For example, if

$$f(A, B, C, D, E) = S^5_{1,2,3}(A\bar{B}C\bar{D}E)$$

it can also be written as

$$\begin{aligned} f(A, B, C, D, E) &= S^5_{5-1,5-2,5-3}(\bar{A}B\bar{C}D\bar{E}) \\ &= S^5_{4,3,2}(\bar{A}B\bar{C}D\bar{E}) \end{aligned}$$

Theorem 13.5. *The number of* n*-variable terms of which a single-subscript symmetric function* $S^n_j(X_1, X_2, \ldots, X_7)$ *consists of is*

$$\binom{n}{j} \quad \text{or} \quad \frac{n!}{(n-j)!\,j!}$$

For a function to be symmetric it must contain all such terms.

If a symmetric function is plotted onto a logical map a symmetric pattern is formed. The exhibited symmetry is equidistant to two certain points which are defined as the centers of the function. For example, if the symmetric function

$$f(A, B, C, D) = \bar{A}B\bar{C}D + A\bar{B}\bar{C}D + AB\bar{C}\bar{D} + ABCD$$

is plotted onto a four variable Veitch diagram, Figure 13.1(*a*), the two centers can be visually identified. They are elements 13 and 2.

All four terms of this symmetric function are one element away from center 13, and three elements away from center 2. A four variable logical space with centers 13 and 2 can be also thought of as being the surface of a sphere where equidistant groups are latitude zones. Such representation is shown in Figure 13.2.

In addition to the logical and geometric properties described above, symmetric functions also possess arithmetic properties.

When the variables of symmetry of a symmetric function are used as inputs to a binary adder, the outputs of the adder represent symmetric functions corresponding to that set of variables of symmetry. For example, if the variables of symmetry are A, B, $\bar{C}$, and D, the outputs of the binary adder that they input will be as shown in Figure 13.3(*a*).

To obtain all elementary (single-subscript) symmetric functions with variables of symmetry $AB\bar{C}D$, the appropriate outputs of the binary adder are logically combined as shown in Figure 13.3(*b*).

Thus, implementation of logical symmetric functions need not go through boolean simplifications at all, but binary adders should be used instead. This

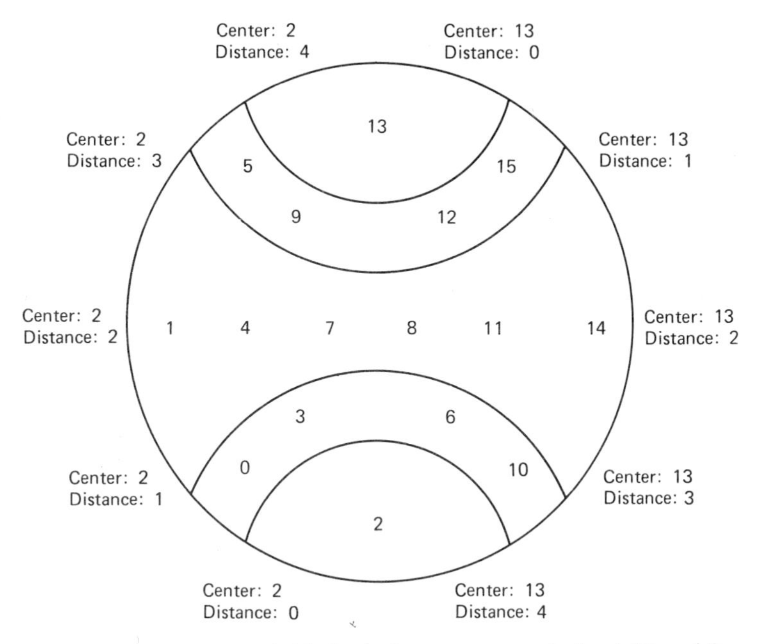

Figure 13.2. Four-variable logical space centered about 13 and 2.

is the most important and desirable feature of symmetric functions. To be able, however, to take advantage of the arithmetic properties of symmetric functions, the presence of logical symmetry must be first detected and identified.

13.3 DETECTION OF TOTAL SYMMETRY

Method One

The method presented here was originally developed by E. R. Robbins of the Arizona State University and has been further developed by the author. This is the simplest available method, where all required computations are in the decimal system. The method is as follows:

Step 1. Using the following equations, identify the two centers of symmetry.

$$\text{Nearest center} = \frac{1}{2}\left((2^n - 1) - \frac{m(2^n - 1) - 2\sum_{i=1}^{i=m} k_i}{|m - 2e|}\right)$$

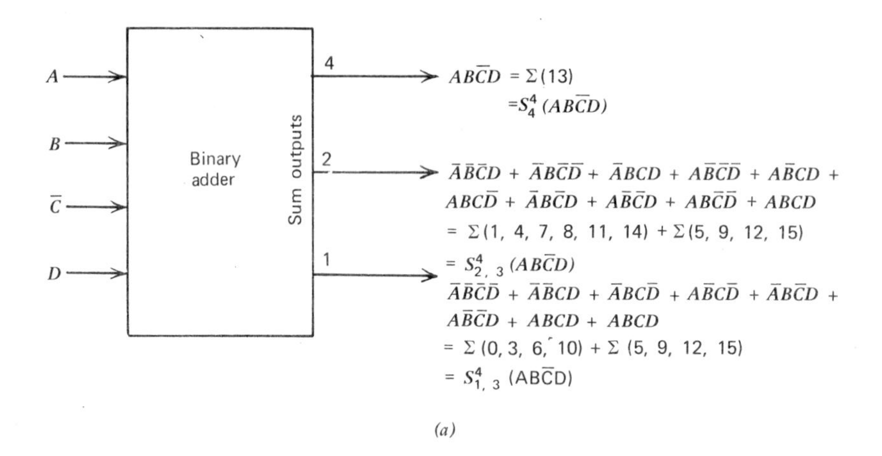

(a)

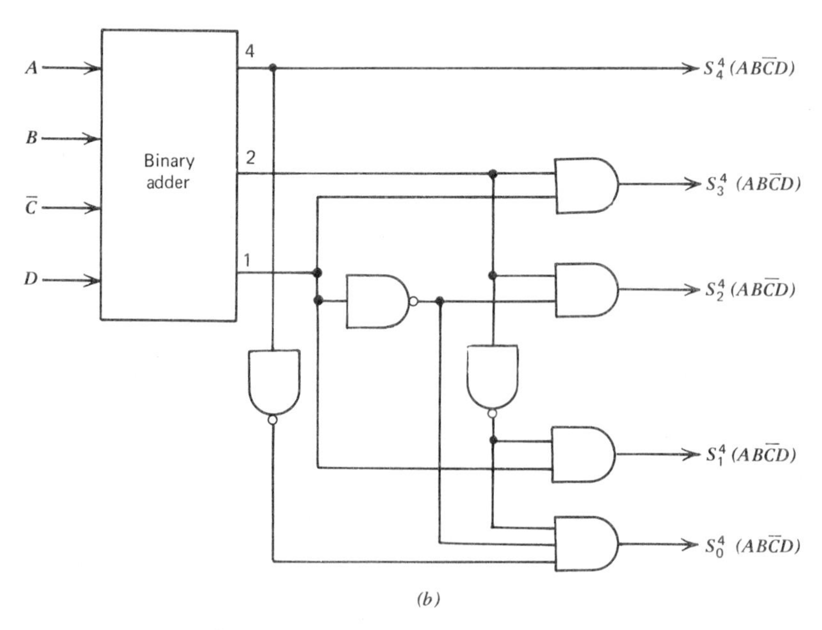

(b)

Figure 13.3. (*a*) Addition of variables of symmetry. (*b*) Elementary symmetric functions obtained from the addition of the variables of symmetry.

where n is the number of variables, m is the number of canonical terms, and e is the number of even canonical terms. The formula for the other center is

$$\text{Furthest center} = (2^n - 1) - \text{Nearest center}$$

Step 2. Convert all terms and center to sums of powers of 2. Compare all powers of 2 in each of the terms to those of center, and determine number of different powers. Compute number of terms having same number of different powers with center. These terms are equidistant to center.

Step 3. Compute the number of equidistant terms that the function should have. They should be $n!/(n - d)!\, d!$, where n is the number of variables and d is the distance; if they are not, the function is not totally symmetric.

Example 13.1. Determine if the following logical function is totally symmetric. If it is, identify the parameters of symmetry. (Centers, variables of symmetry, distances)

$$f(A, B, C, D) = \sum (7, 11, 13, 14, 15)$$

Using the above method, we have

Step 1.

$$\text{Nearest center} = \frac{1}{2}\left[(2^n - 1) - \frac{m(2^n - 1) - 2\sum_{i=1}^{i=m} P_i}{|m - 2^e|}\right]$$

$$= \frac{1}{2}\left[(2^4 - 1) - \frac{5(2^4 - 1) - 2(7, 11, 13, 14, 15()}{5 - 2(1)}\right]$$

$$= \frac{1}{2}\left[15 - \frac{75 - 2(60)}{3}\right]$$

$$= 15$$

Step 2.

$$\text{center: } 15 = 2^3 + 2^2 + 2^1 + 2^0$$

$$\text{terms}\begin{cases} 7 = 2^2 + 2^1 + 2^0 & \text{(one different powers)} \\ 11 = 2^3 + 2^1 + 2^0 & \text{(one different powers)} \\ 13 = 2^3 + 2^2 + 2^0 & \text{(one different powers)} \\ 14 = 2^3 + 2^2 + 2^1 & \text{(one different powers)} \\ 15 = 2^3 + 2^2 + 2^1 + 2^0 & \text{(zero different powers)} \end{cases}$$

Step 3. From Step 2 we see that there are four terms at distance 1, and one term at distance 0. Computing the correct number of terms with these characteristics we have

$$\text{distance: } 1 \quad \frac{n!}{(n-d)!\,d!} = \frac{4!}{(4-1)!\,1!} = \frac{1\cdot 2\cdot 3\cdot 4}{(1\cdot 2\cdot 3)\cdot 1} = 4 \text{ terms}$$

$$\text{distance: } 0 \quad \frac{n!}{(n-d)!\,d!} = \frac{4!}{(4-0)!\,0!} = \frac{4!}{4!\,0!} = 1 \text{ term}$$

Both figures agree with the findings of Step 2 indicating that the given logical function is totally symmetric.

The parameters of symmetry are

Nearest center $= 15$
Furthest center $= (2^n - 1) - 15 = 0$
Variables of symmetry $= \overline{A}\overline{B}\overline{C}\overline{D}$ and $ABCD$

distances $= 0, 1$
distances $= 4, 3$

Thus, the given expression can be written as

$$\begin{aligned} f(A, B, C, D) &= S^4_{0,1}\,(\overline{A}\overline{B}\overline{C}\overline{D}) \\ &= S^4_{4,3}\,(ABCD) \end{aligned}$$

Example 13.2. Determine if the following logical function is totally symmetric:

$$f(A, B, C, D) = \sum (5, 11, 13, 14, 15)$$

Using the expression for the computation of the nearest center, we have

$$\begin{aligned} \text{Nearest center} &= \frac{1}{2}\left[(2^n - 1) - \frac{m(2^n - 1) - 2\sum_{i=1}^{i=m} P_i}{|m - 2e|}\right] \\ &= \frac{1}{2}\left[(2^4 - 1) - \frac{5(2^4 - 1) - 2\sum (5, 11, 13, 14, 15)}{5 - 2(1)}\right] \\ &= \frac{17}{3} \end{aligned}$$

Since the number for the nearest center is not even an integer, the given function cannot be totally symmetric.

Method Two

To determine whether a logical function is totally symmetric, the function is first expanded into its canonical terms. With the canonical terms defined, a table is formed consisting of the binary form of these terms. The general form of such table of m terms is

$$\begin{matrix} X_{11} & X_{12} \ldots X_{1n} \\ X_{21} & X_{22} \ldots X_{2n} \\ \vdots & \vdots \qquad \vdots \\ X_{m1} & X_{m2} \ldots X_{mn} \end{matrix}$$

where the first subscript designates the term and the second subscript designates the variable. The values of the variables are 1 or 0. In this method there are two procedures to determine if a function is totally symmetric and its two centers.

Procedure **13.1.** Unbalanced symmetry: The majority of terms are closer to one of the two centers.

Step 1. Form a table consisting of all terms of the function in binary form. Compute the vertical sums of 1s and 0s. If more than two different numbers are obtained, the function is not totally symmetric. If all sums are identical, go to Procedure 13.2.

Step 2. Replace the larger number in vertical sums by 1 and the smaller number by 0. The binary number formed from the vertical sums of the 1s will represent the nearest center. Similarly, the binary number formed from the vertical sums of the 0s will represent the furthest center.

Step 3. Use one of the centers as a transformation, apply it to the table of terms and compute the horizontal sums of 1s of the new table. Test for sufficiency of occurrences.

If sufficiency is found, the function is symmetric. The two centers of symmetry are, then, those obtained in Step 2, and the variables of symmetry for each center are indicated by the 1s complement of the respective center. ▮

Procedure 13.2. Balanced symmetry: The function is equidistant to the two centers.

Step 1. (Entry from Step 1 of Procedure 13.1) Partition the terms into two groups. The first group should have all the terms where a selected variable is 1. Any variable can be considered for this selection. The second group will have the remaining terms where the value of the selected variable is 0.

Step 2. Considering either of the above two groups, compute the vertical sums of 1s and 0s of the selected group. If more than two different sums are obtained, the function is not symmetric. If all sums are equal, then go to Step 1 of this procedure.

Step 3. Replace the larger number in vertical sums with 1 and smaller with 0.

The binary number formed by the vertical sums of the 1s will be $n - 1$ bits of the nearest center. The missing bit will be 0, if the first group (Step 1) is considered, and 1 if the second group is considered. If the number of terms in the group considered is either n or $n + 1$, the reverse holds regarding the value of the missing bit. The furthest center will be the 1s complement of the nearest center just computed.

Step 4. Use one of the centers as a transformation and apply it to the table of terms obtained in Step 1 of Procedure 13.1. Then compute the horizontal sums of the 1s of the new table.

Step 5. Test for sufficiency of occurrences. If sufficiency is found, the function is symmetric and the two centers of the function are those obtained in Step 3. The variables of symmetry for each center are indicated by the 1's complement of the respective center.

Example 13.3. Determine if the following function is totally symmetric. If it is, identify the parameters of symmetry (centers, variables of symmetry, distances.)

$$f(X_1, X_2, X_3, X_4, X_5, X_6, X_7) = \sum (13, 50, 69, 73, 76, 77, 79, 93, 109)$$

To analyze the given function the first procedure is applied.

Procedure 13.1.

Step 1. Compute partial sums.

	X_1	X_2	X_3	X_4	X_5	X_6	X_7
	0	0	0	1	1	0	1
	0	1	1	0	0	1	0
	1	0	0	0	1	0	1
	1	0	0	1	0	0	1
	1	0	0	1	1	0	0
	1	0	0	1	1	0	1
	1	0	0	1	1	1	1
	1	0	1	1	1	0	1
	1	1	0	1	1	0	1
Vertical sum of 1s	7	2	2	7	7	2	7
Vertical sum of 0s	2	7	7	2	2	7	2

Step 2. Identify possible centers.

Nearest center	1	0	0	1	1	0	1
Furthest center	0	1	1	0	0	1	0

Step 3. Apply transformation (invert if center bit is 0), obtain horizontal sums.

X_1	X_2	X_3	X_4	X_5	X_6	X_7	
0	1	1	1	1	1	1	6
0	0	0	0	0	0	0	0
1	1	1	0	1	1	1	6
1	1	1	1	0	1	1	6
1	1	1	1	1	1	0	6
1	1	1	1	1	1	1	7
1	1	1	1	1	0	1	6
1	1	0	1	1	1	1	6
1	0	1	1	1	1	1	6

Step 4. Check for sufficiency of occurrences. The occurrences are

seven	6
one	0
one	7

For sufficiency, the occurrences should be

$$\text{Sums of } 6 = \binom{n}{d} = \binom{7}{6} = \frac{7!}{(7-6)!\,6!} = 7$$

$$\text{Sums of } 0 = \binom{n}{d} = \binom{7}{0} = \frac{7!}{(7-0)!\,0!} = 1$$

$$\text{Sums of } 7 = \binom{n}{d} = \binom{7}{7} = \frac{}{(7-7)!\,7!} = 1$$

All occurrences are correct; therefore, the given function is totally symmetric and the parameters are

Centers	Variables of symmetry	Distances
1 0 0 1 1 0 1	$\bar{X}_1\ X_2\ X_3\ \bar{X}_4\ \bar{X}_5\ X_6\ \bar{X}_7$	0, 1, 7
0 1 1 0 0 1 0	$X_1\ \bar{X}_2\ \bar{X}_3\ X_4\ X_5\ \bar{X}_6\ X_7$	0, 6, 7

The given function can be then written as

$$\begin{aligned} f(X_1\ X_2\ X_3\ X_4\ X_5\ X_6\ X_7) &= S^7_{0,1,7}(\bar{X}_1\ X_2\ X_3\ \bar{X}_4\ \bar{X}_5\ X_6\ \bar{X}_7) \\ &= S^7_{0,6,7}(X_1\ \bar{X}_2\ \bar{X}_3\ X_4\ X_5\ \bar{X}_6\ X_7) \end{aligned}$$ ∎

Example 13.4. Determine if the following function is totally symmetric. If it is, identify the parameters of symmetry.

$$f(X_1\ X_2\ X_3\ X_4) = \sum(1, 4, 7, 8, 11, 14)$$

Procedure 13.1

Step 1. Compute partial sums.

	X_1	X_2	X_3	X_4
	0	0	0	1
	0	1	0	0
	0	1	1	1
	1	0	0	0
	1	0	1	1
	1	1	1	0
Vertical sums of 1s	3	3	3	3
Vertical sums of 0s	3	3	3	3

The function is symmetrically balanced. Therefore, Procedure 13.2 of symmetry identification should be used.

Procedure 13.2

Step 1. $X_1 = 1$. Selected half:

X_2	X_3	X_4
0	0	0
0	1	1
1	1	0

Step 2. Determine vertical sums.

Vertical sums of 1s	1	2	1
Vertical sums of 0s	2	1	2
Nearest subgroup center	0	1	0

Step 3. Determine centers.

The number of terms in the subgroup is three, which is different than $n + 1$ or n which are 5 and 4, respectively. Therefore, in determining the center of the function nearest to this subgroup, the complement of the value of the selected variable X_1 is taken. That center is

0 0 1 0 **Center of X_1 subgroup**

Since the two centers are complementary, the second center will be

1 1 0 1

Step 4. The sets of variables of symmetry corresponding to the above two centers are, respectively,

$$X_1 \; X_2 \; \overline{X}_3 \; X_4$$

and

$$\overline{X}_1 \; \overline{X}_2 \; X_3 \; \overline{X}_4$$

The validity of the centers is now verified by using either set of variables of symmetry as a transformation, as shown below:

$\overline{X}_1$	$\overline{X}_2$	X_3	$\overline{X}_4$	
1	1	0	0	2
1	0	0	1	2
1	0	1	0	2
0	1	0	1	2
0	1	1	0	2
0	0	1	1	2

Step 5. Check for sufficiency of occurrences. The occurrences are six 2s. For sufficiency the occurrences should be

$$\text{Sums of } 2 = \binom{n}{d} = \binom{4}{2} = \frac{4!}{(4-2)!\,2!} = 6$$

The number of occurrences is correct. Therefore, the given function is symmetric and its parameters of symmetry are

Centers	Variables of Symmetry	Distance
0 0 1 0	$X_1 \; X_2 \; \overline{X}_3 \; X_4$	2
1 1 0 1	$\overline{X}_1 \; \overline{X}_2 \; X_3 \; \overline{X}_4$	2

The function can be written as

$$\begin{aligned} f(X_1, X_2, X_3, X_4) &= S_2^4(X_1, \overline{X}_2, \overline{X}_3, \overline{X}_4) \\ &= S_2^4(X_1, X_2, \overline{X}_3, X_4) \end{aligned}$$

▮

Example 13.5. Determine if the following function is totally symmetric. If it is, identify the parameters of symmetry.

$$f(X_1, X_2, X_3, X_4) = \sum (3, 5, 10, 12, 13)$$

To analyze the given function, Procedure 13.1 is applied.

Procedure 13.1

Step 1. Compute partial sums.

	X_1	X_2	X_3	X_4	**Decimal equivalent**
	0	0	1	1	3
	0	1	0	1	5
	1	0	1	0	10
	1	1	0	0	12
	1	1	0	1	13
Vertical sums of 1s	3	3	2	3	
Vertical sums of 0s	2	2	3	2	

Step 2. Identify possible centers.

Nearest center	1	1	0	1
Furthest center	0	0	1	0
Respective variables of	$\overline{X}_1$	$\overline{X}_2$	X_3	$\overline{X}_4$
symmetry	X_1	X_2	$\overline{X}_3$	X_4

Step 3. The given function is transformed under one of the two sets of variables of symmetry and the distances are identified.

X_1	X_2	$\overline{X}_3$	X_4	
0	0	0	1	1
0	1	1	1	3
1	0	0	0	1
1	1	1	0	3
1	1	1	1	4

Step 4. Check for sufficiency of occurrences. The occurrences are

two 1
two 3
one 4

For sufficiency the occurrences should be

$$\text{Occurrences of } 1 = \binom{n}{d} = \binom{4}{1} = \frac{4!}{(4-1)!\,1!} = 4$$

$$\text{Occurrences of } 3 = \binom{n}{d} = \binom{4}{3} = \frac{4!}{(4-3)!\,3!} = 4$$

$$\text{Occurrences of } 4 = \binom{n}{d} = \binom{4}{4} = \frac{4!}{(4-4)!\,4!} = 1$$

Two of the obtained numbers of occurrences are incorrect. Therefore, the given function is not totally symmetric. ∎

13.4 DETECTION OF PARTIAL SYMMETRY

After a given function $f(X_i, i = 1 \text{ to } n)$, n: number of variables, is found to be non-totally symmetric, a test can be performed to determine if partial symmetry exists in the function.

Procedure 13.3

Step 1. Perform parts (*a*) and (*b*) of this step

$$\binom{n}{n-1} = n \text{ times}$$

where n is the number of variables in the given function. Each time let j be a different variable out of the n variables of the function.

(*a*) Partition the given function into groups

$$\bar{X}_j \cdot g(X_i, i = 1 \text{ to } n, i \neq j)$$
$$X_j \cdot h(X_i, i = 1 \text{ to } n, i \neq j)$$

(*b*) Test functions g and h for total symmetry.

If functions g and h are both totally symmetric for the same arrangement of variables, then the given function

$$f(X_i, i = 1 \text{ to } n)$$

is said to be partially symmetric. The variables of partial symmetry are those variables for which functions g and h are totally symmetric.

If functions g and h are not totally symmetric both for the same arrangement of variables, then the given function

$$f(X_i, i = 1 \text{ to } n)$$

is not partially symmetric in variables

$$(X_i, i = 1 \text{ to } n, i \neq j)$$

If all n tests, performed in this step, indicate that no partial symmetry exists in $n - 1$ variables, then proceed to Step 2.

Step 2. Perform parts (*a*) and (*b*) of this step

$$\binom{n}{n-2} = \frac{(n-1)n}{2} \text{ times}$$

Each time let j and k be a different two-variable selection out of the n variables of the function.

(*a*) Partition the given function into groups

$$\begin{aligned}&\overline{X}_i\,\overline{X}_k\,p\,(X_i,\,i = 1 \text{ to } n,\, i \neq j \neq k)\\&\overline{X}_j\,X_k\,q\,(X_i,\,i = 1 \text{ to } n,\, i \neq j \neq k)\\&X_j\,\overline{X}_k\,r\,(X_i,\,i = 1 \text{ to } n,\, i \neq j \neq k)\\&X_j\,X_k\,s\,(X_i,\,i = 1 \text{ to } n,\, i \neq j \neq k)\end{aligned}$$

(*b*) Test functions p, q, r, and s for total symmetry.

If all four functions are totally symmetric for the same arrangement of variables, then the given function

$$f(x_i,\, i = 1 \text{ to } n)$$

is said to be partially symmetric. The variables of partial symmetry are those variables for which all four functions p, q, r, and s are totally symmetric.

If functions p, q, r, and s are not all totally symmetric for the same arrangement of variables, then the given function

$$f(x_i,\, i = 1 \text{ to } n)$$

is not partially symmetric in variables

$$(X_i,\, i = 1 \text{ to } n,\, i \neq j \neq k)$$

If all $(n - 1)n/2$ tests performed in this step indicate that no partial symmetry exists in $n - 2$ variables, then proceed to Step 3.

Step 3. Perform parts (*a*) and (*b*) of this step

$$\binom{n}{n-3} = \frac{(n-2)(n-1)n}{6} \text{ times}$$

In each time let j, k, and l be a different three-variable selection out of the n variables of the given function.

(*a*) Partition the given function into groups

$$\begin{aligned}&\overline{X}_j\overline{X}_k\overline{X}_l\,a\,(X_i,\,i = 1 \text{ to } n,\, i = j \neq k \neq l)\\&\overline{X}_j\overline{X}_kX_l\,b\,(X_i,\,i = 1 \text{ to } n,\, i = j \neq k \neq l)\\&\overline{X}_jX_k\overline{X}_l\,c\,(X_i,\,i = 1 \text{ to } n,\, i = j \neq k \neq l)\\&\overline{X}_jX_kX_l\,d\,(X_i,\,i = 1 \text{ to } n,\, i = j \neq k \neq l)\\&X_j\overline{X}_k\overline{X}_l\,e\,(X_i,\,i = 1 \text{ to } n,\, i = j \neq k \neq l)\\&X_j\overline{X}_kX_l\,f\,(X_i,\,i = 1 \text{ to } n,\, i = j \neq k \neq l)\\&X_jX_k\overline{X}_l\,g\,(X_i,\,i = 1 \text{ to } n,\, i = j \neq k \neq l)\\&X_jX_kX_l\,h\,(X_i,\,i = 1 \text{ to } n,\, i = j \neq k \neq l)\end{aligned}$$

(*b*) Test functions *a*, *b*, *c*, *d*, *e*, *f*, *g*, and *h* for total symmetry.

If all eight functions are totally symmetric for the same arrangement of variables, then the given function

$$f(X_i, i = 1 \text{ to } n)$$

is said to be partially symmetric. The variables of partial symmetry are those variables for which all eight functions *a*, *b*, *c*, *d*, *e*, *f*, *g*, and *h* are totally symmetric.

If functions *a*, *b*, *c*, *d*, *e*, *f*, *g*, and *h* are not all totally symmetric for the same arrangement of variables, then the given function

$$f(X_i, i = 1 \text{ to } n)$$

is not partially symmetric in variables

$$(X_i, i = 1 \text{ to } n, i \neq j \neq k \neq l)$$

If all $(n - 2)(n - 1)n/6$ tests performed in this step indicate that no partial symmetry exists in $n - 3$ variables, then proceed to Step 4.

Step 4. As in Step 3, partition the given function into 16 groups and test for total symmetry in these groups.

The process of partitioning and testing continues until partial symmetry is found, or until the number of variables in the functions, that resulted from the partitions, decrease to two. ∎

Example 13.6. Determine if the following function is symmetric. If it is, identify the parameters of symmetry—that is, the centers, the variables of symmetry, and the distances.

$$f(X_1\, X_2\, X_3\, X_4) = \sum (1, 2, 7, 8, 9, 13, 14)$$

We first test the given function to determine if it exhibits total symmetry. We start by computing the vertical sums of the function

	X_1	X_2	X_3	X_4	**Decimal equivalent**
	0	0	0	1	1
	0	0	1	0	2
	0	1	1	1	7
	1	0	0	0	8
	1	0	0	1	9
	1	1	0	1	13
	1	1	1	0	14
Vertical sums of 1s	4	3	3	4	
Vertical sums of 0s	3	4	4	3	

Possible centers				
nearest	1	0	0	1
furthest	0	1	1	0
Respective variables	$\bar{X}_1$	X_2	X_3	$\bar{X}_4$
of symmetry	X_1	$\bar{X}_2$	$\bar{X}_3$	X_4

The function is transformed under one of the two sets of variables of symmetry.

$\bar{X}_1$	X_2	X_3	$\bar{X}_4$	
1	0	0	0	1
1	0	1	1	3
1	1	1	0	3
0	0	0	1	1
0	0	0	0	0
0	1	0	0	1
0	1	1	1	3

For four variables the number of occurrences of 1 is four; since we only have three, the given function is not totally symmetric.

We now test the function to determine the possible existence of partial symmetry.

Procedure 13.3

Step 1. Determine if the function is partially symmetric in $n - 1$ (3) variables. This step is performed n (4) times. This is because n variables can be taken $n - 1$ at a time n times.

$$f(X_1\ X_2\ X_3\ X_4) = \bar{X}_1\, g(X_2\ X_3\ X_4) + X_1\, h(X_2\ X_3\ X_4)$$

(Variables considered: $X_2\ X_3\ X_4$)

	$X_1\, h(X_2$	X_3	$X_4)$
	0	0	0
	0	0	1
	1	0	1
	1	1	0
Vertical sums of 1s	2	1	2
Vertical sums of 0s	2	3	2

The presence of more than two different sums immediately excludes the possibility of total symmetry existing in $h(X_2\ X_3\ X_4)$. At the same time it excludes the possibility that

$$f(X_1\ X_2\ X_3\ X_4)$$

be partially symmetric in variables $X_2\ X_3\ X_4$.

Step 1 is now repeated, testing the given function for partial symmetry in variables $X_1\ X_3\ X_4$.

$$f(X_1\ X_2\ X_3\ X_4) = \bar{X}_2\, g(X_1\ X_3\ X_4) + X_2\, h(X_1\ X_3\ X_4)$$
$(X_2$ excluded)

	$\bar{X}_2\, g(X_1$	X_3	$X_4)$
	0	0	1
	0	1	0
	1	0	0
	1	0	1
Vertical sums of 1s	2	1	2
Vertical sums of 0s	2	3	2

The presence of more than two different sums immediately excludes the possibility of $g(X_1\ X_3\ X_4)$ being totally symmetric. At the same time it excludes the possibility of $f(X_1\ X_2\ X_3\ X_4)$ being partially symmetric in variables $X_1\ X_3\ X_4$.

Step 1 is now repeated, testing the given function for partial symmetry in variables $X_1\ X_2\ X_4$.

$$f(X_1\ X_2\ X_3\ X_4) = \bar{X}_3\, g(X_1\ X_2\ X_4) + X_3\, h(X_1\ X_2\ X_4)$$
$(X_3$ excluded)

		$\bar{X}_3\, g(X_1$	X_2	$X_4)$
		0	0	1
		1	0	0
		1	0	1
		1	1	1
Vertical sums of 1s		3	1	3
Vertical sums of 0s		1	3	1
Possible centers	**nearest**	1	0	1
	furthest	0	1	0
Respective variables of symmetry		$\bar{X}_1$	X_2	$\bar{X}_4$
		X_1	$\bar{X}_2$	X_4

The function $g(X_1\ X_2\ X_4)$ is now transformed under one of the two sets of variables of symmetry, so that the distances of $g(X_1\ X_2\ X_4)$ to the center corresponding to the selected set of variables of symmetry are determined.

X_1	$\bar{X}_2$	X_4	
0	1	1	2
1	1	0	2
1	1	1	3
1	0	1	2

The number of occurrences of distances 2 and 3 are correct, since

$$\binom{3}{2} = 3 \qquad \text{and} \qquad \binom{3}{3} = 1$$

Therefore, function $g(X_1\ X_2\ X_4) = S^3_{2,3}\ (X_1\ \bar{X}_2\ X_4) = S^3_{0,1}\ (\bar{X}_1\ X_2\ \bar{X}_4)$.

	$X_3\,h(X_1$	X_2	$X_4)$
	0	0	0
	0	1	1
	1	1	0
Vertical sums of 1s	1	2	1
Vertical sums of 0s	2	1	2
Possible centers nearest	0	1	0
Possible centers furthest	1	0	1
Respective variables of symmetry	X_1	$\bar{X}_2$	X_4
	$\bar{X}_1$	X_2	$\bar{X}_4$

The function $h(X_1\ X_2\ X_4)$ is now transformed under one of the two sets of variables of symmetry, so that the distances of $h(X_1\ X_2\ X_4)$ to the center, corresponding to the selected set of variables of symmetry, are determined.

X_1	$\bar{X}_2$	X_4	
0	1	0	1
0	0	1	1
1	0	0	1

The number of occurrences of distance 1 is correct, since

$$\binom{3}{1} = 3$$

Therefore, function $h(X_1\ X_2\ X_4) = S^3_1\ (X_1\ \bar{X}_2\ X_4) = S^3_2\ (\bar{X}_1\ X_2\ \bar{X}_4)$.

Functions $g(X_1\ X_2\ X_4)$ and $h(X_1\ X_2\ X_4)$ are totally symmetric to the same set of variables of symmetry, namely,

$$X_1\bar{X}_2X_4 \qquad \text{or} \qquad \bar{X}_1X_2\bar{X}_4$$

Consequently, the given function is partially symmetric in variables

$$X_1\bar{X}_2X_4 \qquad \text{or} \qquad \bar{X}_1X_2\bar{X}_4$$

Since

$$f(X_1X_2X_3X_4) = \bar{X}_3g(X_1X_2X_4) + X_3h(X_1X_2X_4)$$

the given function can be written as

$$\begin{aligned} f(X_1X_2X_3X_4) &= \bar{X}_3S^3_{2,3}(X_1\bar{X}_2X_4) + X_3S^3_2(X_1\bar{X}_2X_4) \\ &= \bar{X}_3S^3_{0,1}(\bar{X}_1X_2\bar{X}_4) + X_3S^3_1(\bar{X}_1X_2\bar{X}_4) \end{aligned}$$

Step 1 is applied for the last time, testing the given function for partial symmetry in variables X_1 X_2 X_3.

$$f(X_1X_2X_3X_4) = \bar{X}_4g(X_1X_2X_3) + X_4h(X_1X_2X_3)$$

(X_4 excluded)

	$X_4h(X_1$	X_2	$X_3)$
	0	0	0
	0	1	1
	1	0	0
	1	1	0
Vertical sums of 1s	2	2	1
Vertical sums of 0s	2	2	3

The presence of more than two different sums immediately excludes the possibility of $h(X_1X_2X_3)$ being totally symmetric. At the same time it excludes the possibility that $f(X_1X_2X_3X_4)$ be partially symmetric in variables $X_1X_2X_3$. Therefore, the given function

$$f(X_1X_2X_3X_4) = \sum(1, 2, 7, 8, 9, 13, 14)$$

is partially symmetric in variables

$$X_1\bar{X}_2X_4 \quad \text{or} \quad \bar{X}_1X_2\bar{X}_4$$ ∎

13.5 IMPLEMENTATION OF LOGICAL SYMMETRY

The availability of binary adders in integrated circuit form makes the implementation of symmetric functions a very simple matter. With the use of binary adders the problem of synthesizing symmetric functions leaves the domain of boolean algebra and finds its solution in binary arithmetic.

Theoretically, symmetric functions can be realized by means of any logical synthesis method. However, some methods are more practical than others. The following examples illustrate the simplicity in the implementation of logical symmetric functions.

Example 13.7. Design a network which realizes the symmetric function

$$f(A, B, C, D, E, F) = S^6_5(ABCDEF)$$

This function can be written as

$$\begin{aligned} f(A, B, C, D, E, F) &= ABCDE\bar{F} + ABCD\bar{E}F + ABC\bar{D}EF + AB\bar{C}DEF \\ &\quad + A\bar{B}CDEF + \bar{A}BCDEF \\ &= (ABCDE + ABCDF + ABCEF + ABDEF \\ &\quad + ACDEF + BCDEF)\overline{ABCDEF} \\ &= [BC(ADE + ADF) + BE(ACF + ADF) \\ &\quad + EF(ACD + BCD)]\overline{ABCDEF} \end{aligned}$$

Figure 13.4 shows the implementation of this expression which amounts to six packages. The given function can be also written as

$$\begin{aligned} f(A, B, C, D, E, F) &= S_5^6(ABCDEF) \\ &= S_{4,5,6}^6(ABCDEF) \cdot S_{1,3,5}^6(ABCDEF) \end{aligned}$$

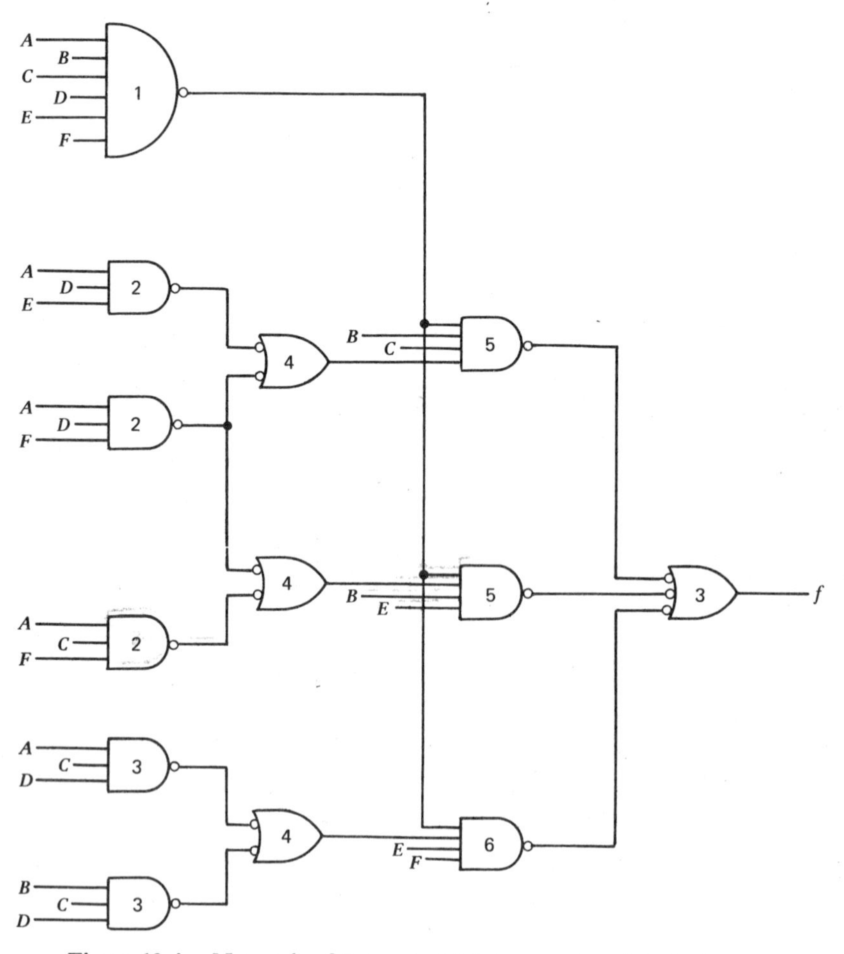

Figure 13.4. Network of Example 13.7 (conventional design).

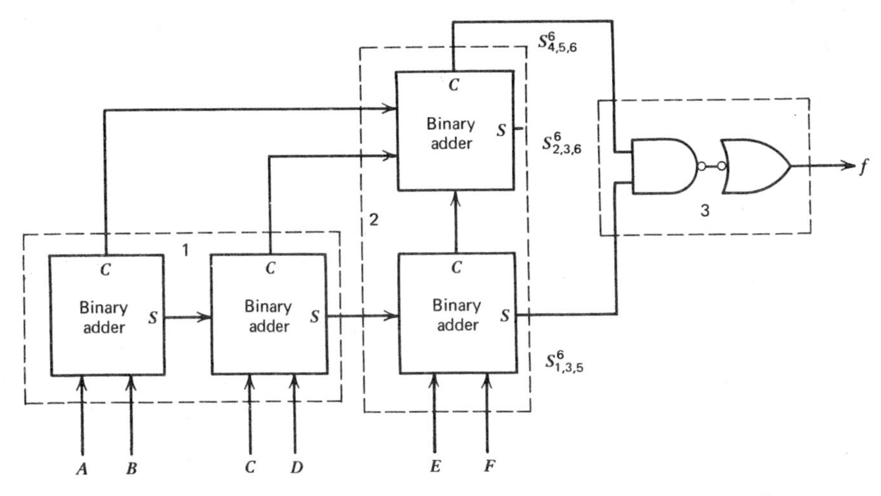

Figure 13.5. Network of Example 13.7 (binary addition method).

Figure 13.5 shows the implementation of this expression which uses binary adders and amounts to three packages. ∎

Example 13.8. Design a network which realizes the symmetric function

$$f(ABCDE) = S^5_3(\bar{A}B\bar{C}D\bar{E})$$

center: 21
distance: 3

With the aid of the Veitch diagram of this function its canonical terms are identified. The given expression can now be written as

$$\begin{aligned} f(A, B, C, D, E) = {} & \bar{A}\bar{B}\bar{C}\bar{D}\bar{E} + \bar{A}\bar{B}\bar{C}DE + \bar{A}\bar{B}CD\bar{E} + \bar{A}B\bar{C}\bar{D}E \\ & + \bar{A}BC\bar{D}\bar{E} + \bar{A}BCDE + A\bar{B}\bar{C}D\bar{E} + AB\bar{C}\bar{D}\bar{E} \\ & + AB\bar{C}DE + ABCD\bar{E} \end{aligned}$$

which can be simplified to

$$\begin{aligned} f(A, B, C, D, E) = {} & \bar{A}\bar{B}(\bar{C}\bar{D}\bar{E} + \bar{C}DE + CD\bar{E}) \\ & + AB(\bar{C}\bar{D}\bar{E} + \bar{C}DE + CD\bar{E}) \\ & + \bar{A}B(\bar{C}\bar{D}E + C\bar{D}\bar{E} + CDE) \\ & + A\bar{B}\bar{C}D\bar{E} \end{aligned}$$

Figure 13.6 shows the conventional implementation of this expression which amounts to six packages. The given function can be also written as

$$\begin{aligned} f(A, B, C, D, E) & = S^5_3(\bar{A}B\bar{C}D\bar{E}) \\ & = S^5_{2,3}(\bar{A}B\bar{C}D\bar{E}) \cdot S^5_{1,3,5}(\bar{A}B\bar{C}D\bar{E}) \end{aligned}$$

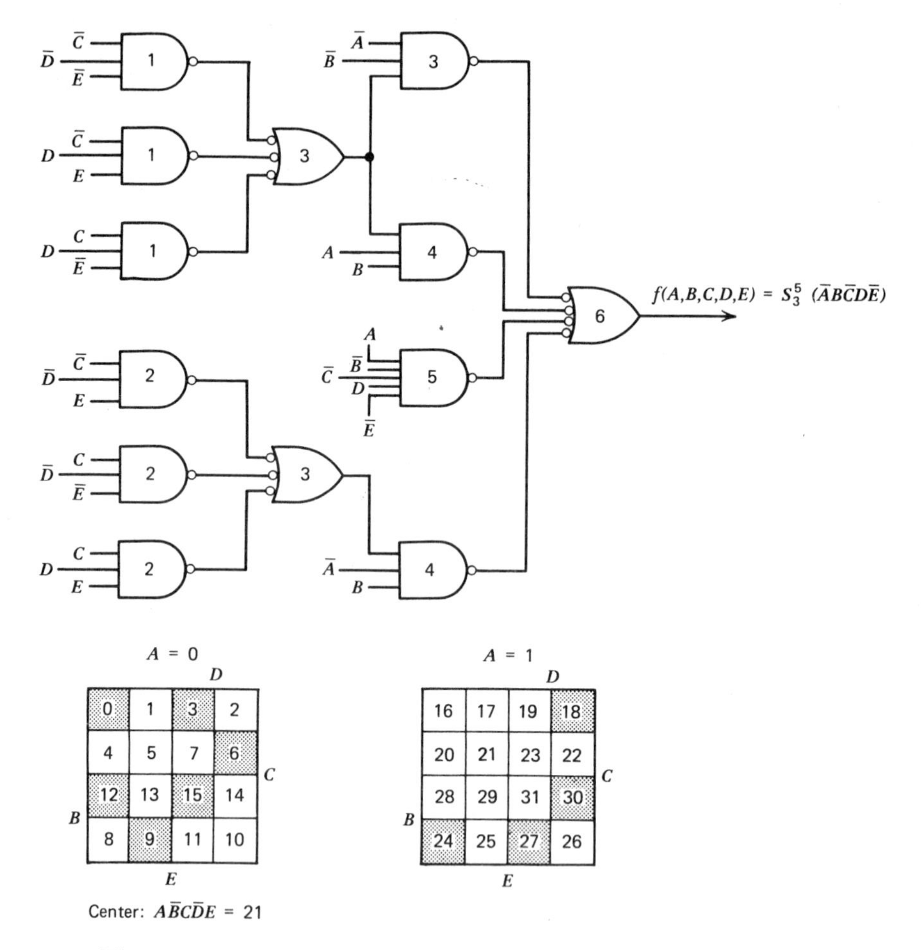

Figure 13.6. Network of Example 13.8 (conventional design).

Figure 13.7 shows the implementation of this expression which uses binary adders and amounts to two packages. ▮

Example 13.9. Design a network which realizes the symmetric function

$$f(A, B, C, D) = ABCD + A\bar{B}\bar{C}\bar{D} + \bar{A}B\bar{C}\bar{D} + \bar{A}\bar{B}C\bar{D} + \bar{A}\bar{B}\bar{C}D$$

By inspection it can be seen that this is

$$f(A, B, C, D) = S^4_{1,4}(ABCD)$$

The designed network must produce an output anytime the binary values

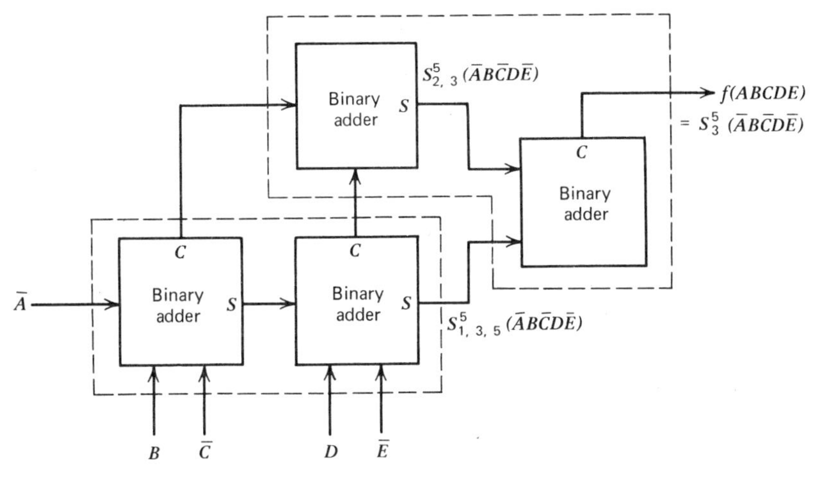

Figure 13.7. Network of Example 13.8 (binary addition method).

of A, B, C, and D add up to either 1 or 4. Figure 13.8 shows the necessary network which is implemented by means of binary adders.

The hardware requirements for this design are three packages in small-scale design (two double-adders and one quad-gate), or $2\frac{1}{3}$ packages in large scale design ($1\frac{1}{2}$ double-adders, $\frac{1}{2}$ quad-gate and $\frac{1}{3}$ hexinverters). Figure 13.9 shows the conventional implementation of the same function, where the package count is five. ∎

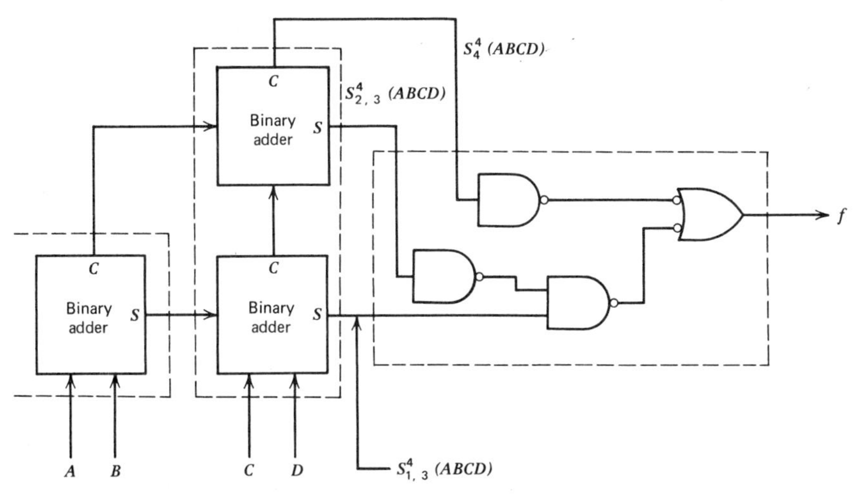

Figure 13.8. Network of Example 13.9 (binary addition method).

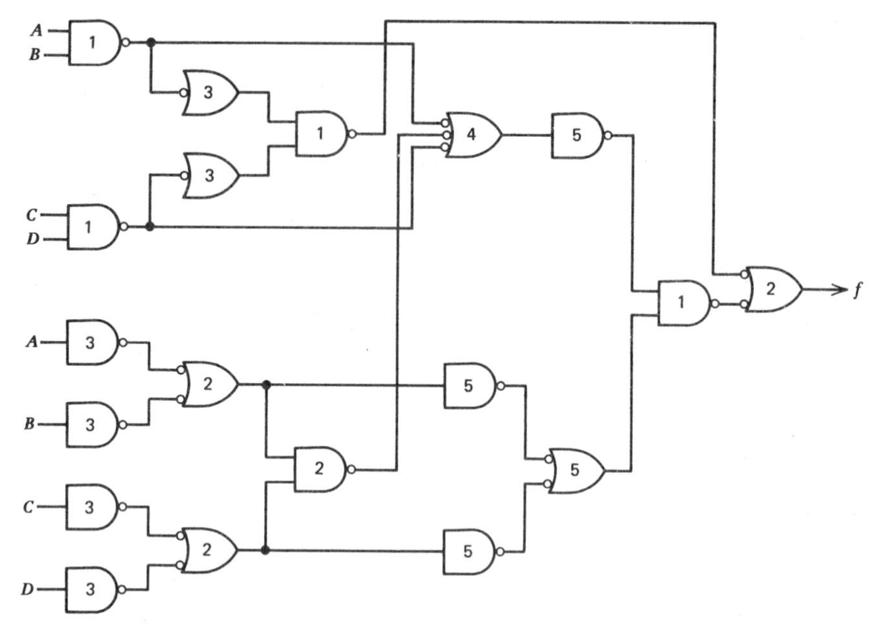

Figure 13.9. Network of Example 13.9 (conventional design).

13.6 INDEPENDENT SYMMETRY

There is an additional type of function which, although it exhibits logical symmetry, cannot be classified in the categories of total and partial symmetry. This is the case where the logical subsets of the given function are symmetric to different centers.

In the detection of partial symmetry described earlier the given function was partitioned into logical subsets and the criterion for partial symmetry was that all logical subsets be totally symmetric in the same set of variables of symmetry. The definition of independent symmetry resembles that of partial symmetry and can be stated as follows:

> A logical function is said to be *independently symmetric* if all of its logical subdivisions exhibit total symmetry in different sets of variables of symmetry.

The most usual objective in analyzing a logical function is to extract useful information regarding the function's properties. This information will hopefully lead into the conception of a simple hardware implementation for the analyzed function.

From this implementation viewpoint, the property of independent symmetry is as important as that of total or partial symmetry. This is because detection and identification of independent symmetry provides information which leads to simple hardware designs. The following example illustrates an independently symmetric function and its realization.

Example 13.10. Determine a network that realizes the following logical function:

$$f(A, B, C, D, E) = \sum (3, 5, 6, 15, 16, 19, 22, 25, 28, 31)$$

The given function is first examined for total symmetry. Using the modified version of Robbins' method (Method One), we have

Step 1.

Nearest center

$$= \frac{1}{2}\left[(2^5-1) - \frac{11(2^5-1)-2\sum(3+5+6+15+16+19+22+25+28+31)}{11-2(5)}\right]$$

$= 15$ (binary 01111)

Step 2. Center: $15 = 2^3 + 2^2 + 2^1 + 2^0$.

		Number of different powers between terms and center
Function's terms	$3 = 2^1 + 2^0$	Two
	$5 = 2^2 + 2^0$	Two
	$6 = 2^2 + 2^1$	Two
	$15 = 2^3 + 2^2 + 2^1 + 2^0$	Zero
	$16 = 2^4$	Five
	$19 = 2^4 + 2^1 + 2^0$	Three
	$22 = 2^4 + 2^2 + 2^1$	Three
	$25 = 2^4 + 2^3 + 2^0$	Three
	$28 = 2^4 + 2^3 + 2^2$	Three
	$31 = 2^4 + 2^3 + 2^2 + 2^1 + 2^0$	One

The number of ones should have been

$$\binom{5}{1} = 5$$

The number of twos should have been

$$\binom{5}{2} = 10$$

The number of threes should have been

$$\binom{5}{3} = 10$$

Since the numbers of powers of 2 are not correct, the given function is not totally symmetric. The function is now tested for partial symmetry. For this test Method One can be again applied. The logical subdivision where $A = 0$ is

$$\textstyle\sum_{A=0}(3, 5, 6, 15)$$

Step 1. The center is

$$\frac{1}{2}\left[(2^4 - 1) - \frac{4(2^4 - 1) - 2(3 + 5 + 6 + 15)}{|\,4 - 2(1)\,|}\right] = 7$$

The variables of symmetry corresponding to this center are $BCDE$. The distances between this center and each of the terms of the $A = 0$ logical subdivision is determined as follows:

Step 2. Center: $7 = 2^2 + 2^1 + 2^0$

	Number of different powers between terms and center
$3 = 2^1 + 2^0$	One
$5 = 2^2 + 2^0$	One
$6 = 2^2 + 2^1$	One
$15 = 2^3 + 2^2 + 2^1 + 2^0$	One

The number of ones must be

$$\binom{4}{1} = 4$$

Since the number of powers of 2 is correct, the $A = 0$ subdivision is totally symmetric in variables B, C, D, and E in the arrangement $B\overline{C}\overline{D}\overline{E}$. The terms of this subdivision can be expressed as

$$\overline{A}S_1^4\,(B\overline{C}\overline{D}\overline{E})$$

The logical subdivision where $A = 1$ is

$$\textstyle\sum_{A=1}(16, 19, 22, 25, 28, 31)$$

Normalizing the terms to four variables by subtracting 16 from each term, we have

$$\textstyle\sum_{A=1}(0, 3, 6, 9, 12, 15) = \sum_{A=1\ B=0}(0, 3, 6) + \sum_{A=1\ B=1}(9, 12, 15)$$

The number of even terms equals that of the odd. Therefore, only one-half of the terms need be examined. Let us consider the logical space in subdivision $A = 1$ where $B = 0$.

Step 1. The center is

$$\frac{1}{2}\left[(2^3 - 1) - \frac{3(2^3 - 1) - 2(0 + 3 + 6)}{|\,3 - 2(2)\,|}\right] = 2 = \text{Binary } 010$$

Therefore, one of the centers of logical subdivision $A = 1$ is 1010, and the corresponding variables of symmetry are $\bar{B}C\bar{D}E$. The distance between that center and each of the terms of the $A = 1$ logical subdivision is determined in the following step:

Step 2. Center: $10 = 2^3 + 2^1$

	Number of different powers between terms and center
0	Two
$3 = 2^1 + 2^0$	Two
$6 = 2^2 + 2^1$	Two
$9 = 2^3 + 2^0$	Two
$12 = 2^3 + 2^2$	Two
$15 = 2^3 + 2^2 + 2^1 + 2^0$	Two

The number of twos must be

$$\binom{4}{2} = 6$$

Since the number of occurrences of powers of 2 is correct, the $A = 1$ subdivision is totally symmetric in variables B, C, D, and E in the arrangement $\bar{B}C\bar{D}E$. The terms of this subdivision can be expressed as

$$AS_2^4(\bar{B}C\bar{D}E)$$

Therefore, the given expression can be written as

$$f(A, B, C, D, E) = \bar{A}S_1^4(B\bar{C}\bar{D}\bar{E}) + AS_2^4(\bar{B}C\bar{D}E)$$

Logical subdivisions $B, \bar{B}$; $C, \bar{C}$; $D, \bar{D}$; and $E, \bar{E}$ were also tested and exhibited no symmetry in four variables. Since the two subdivisions $\bar{A}\phi_1(B, C, D, E)$ and $A\phi_2(B, C, D, E)$ are not totally symmetric for the same arrangement of variables B, C, D, and E, the given function cannot be classified as partially symmetric.

The given function is independently symmetric and can be realized by means of the techniques applicable to totally or partially symmetric functions.

Figure 13.10 shows the Veitch diagram of this function. From this diagram the independent symmetry can be seen. Figure 13.11 illustrates the implementation of the given function

$$f(A, B, C, D, E) = \sum (3, 5, 6, 15, 16, 19, 22, 25, 28, 31)$$

using the method of binary addition, and Figure 13.12 shows the conventional design of this function.

The network of the conventional design not only needs more hardware, but it also requires a considerably greater number of interconnections, an item which greatly reduces system reliability.

The concept of logical symmetry can be used in an unlimited number of applications. Among them is error detection.

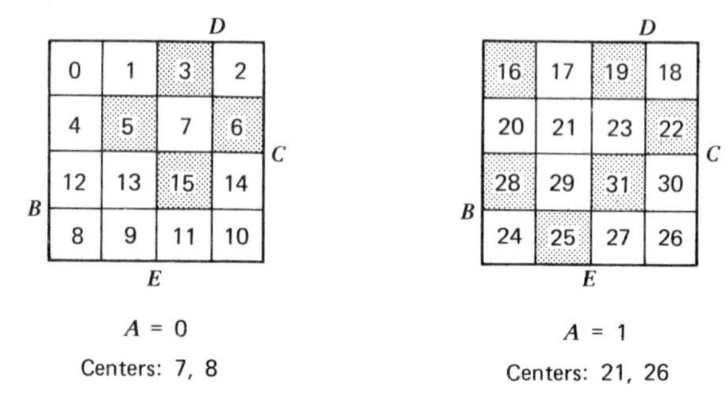

Figure 13.10. Veitch diagram of $f(A, B, C, D, E) = \sum (3, 5, 6, 15, 16, 19, 22, 25, 28, 31)$.

Transmission of data words always requires some verification method at the point of reception. To simplify the data verification procedure, acceptable words can be selected out of a specific symmetric function and tested against this function at their arrival at the receiver. For example, if a data set consists of 35 different data words, they may all be out of symmetric function $S_3^7(ABCDEFG)$. This function consists of

$$\binom{7}{3} \text{ terms}$$

which when expanded is

$$\binom{7}{3} = \frac{7!}{(7-3)!\,3!}$$
$$= 35$$

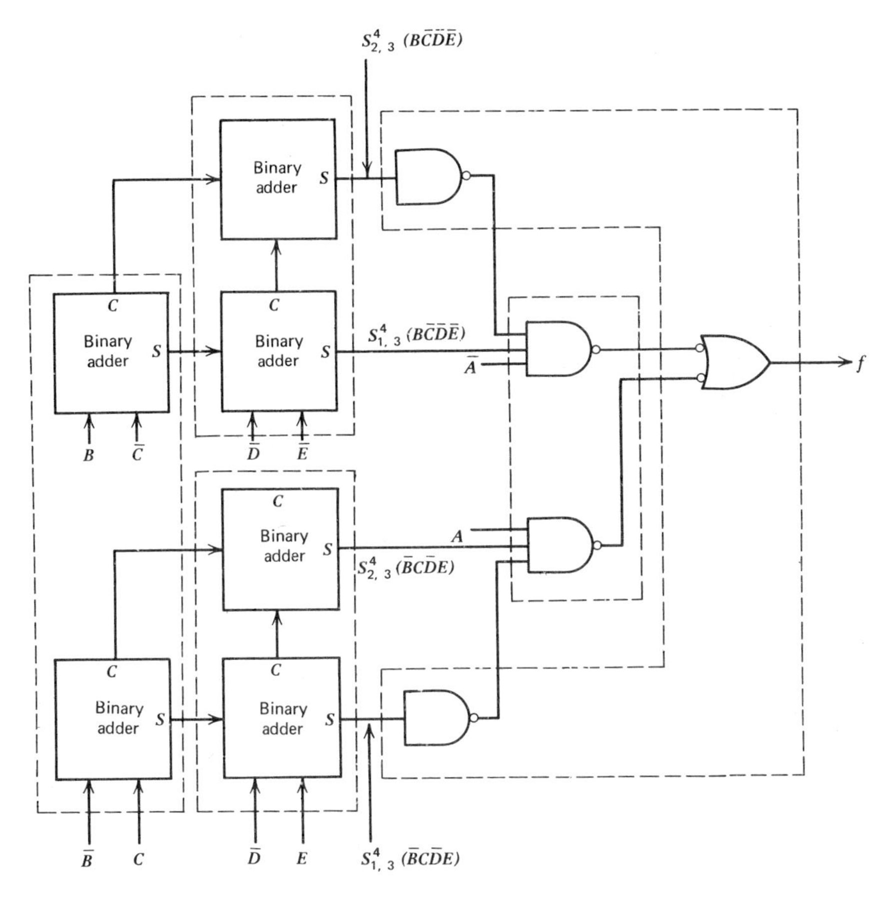

Figure 13.11. Network for the implementation of

$$f(A, B, C, D, E) = \sum (3, 5, 6, 15, 16, 19, 22, 25, 28, 31)$$

with five packages and 25 interconnections (binary addition method).

The characteristic of this function is that the binary sum of all terms always gives a sum of 3. Thus, if this sum is tested at the receiver, the presence of errors will be always detected with the exception of cases where errors cancel each other like when a 0 changes to 1 and a 1 to a 0 in the same data word. Figure 13.13 shows the necessary network that passes the terms of the function $S^7_3(ABCDEFG)$ and blocks all others.

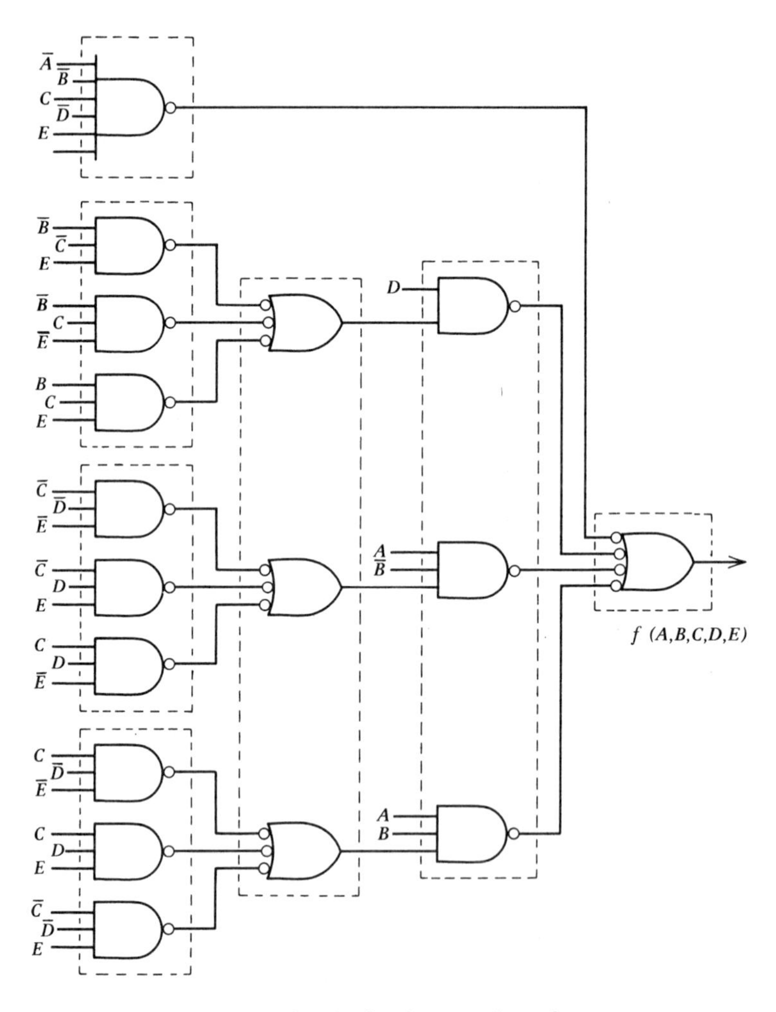

Figure 13.12. Network for the implementation of

$$\begin{aligned} f(A, B, C, D, E) &= \sum (3, 5, 6, 15, 19, 22, 25, 28, 31) \\ &= D(\bar{B}\bar{C}E + \bar{B}C\bar{E} + BCE) \\ &\quad + A\bar{B}(\bar{C}\bar{D}\bar{E} + \bar{C}DE + CD\bar{E}) \\ &\quad + \bar{A}\bar{B}C\bar{D}E + AB(C\bar{D}\bar{E} + CDE + \bar{C}\bar{D}E) \end{aligned}$$

with seven packages and 60 interconnections (conventional design).

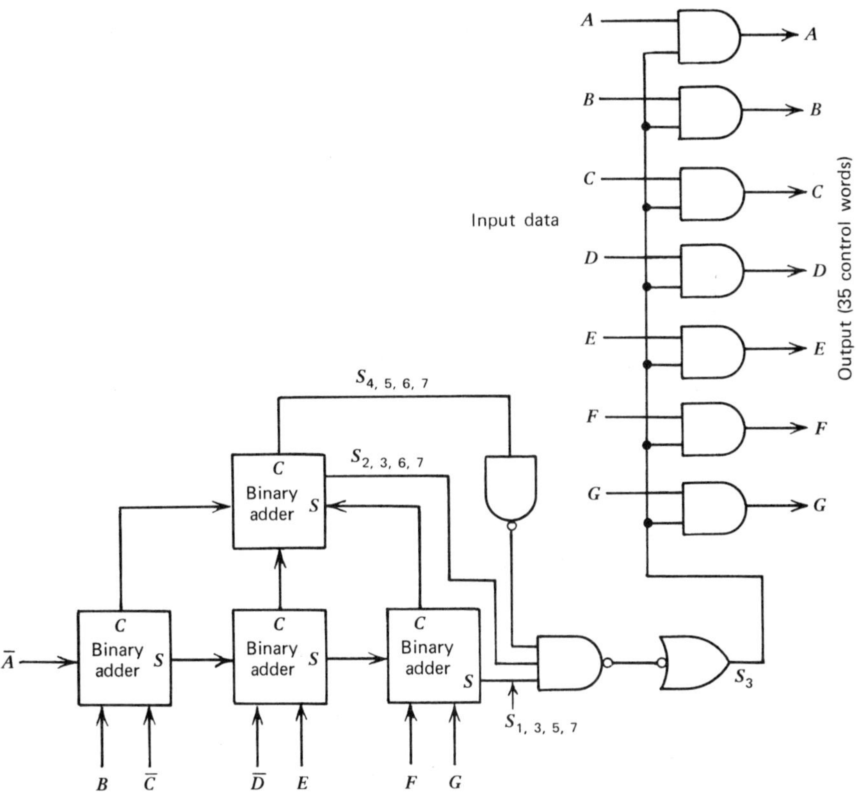

Figure 13.13. Logical band-pass network passing terms of the expression $S_3^7(\bar{A}B\bar{C}\bar{D}EFG)$ only.

PROBLEMS

1. State the difference between total and partial symmetry.
2. Determine whether the following logical function is symmetric; if it is, identify the type and parameters of symmetry:

$$f(A, B, C, D, E) = \sum (1, 4, 7, 13, 21)$$

3. Using the Robbins' method determine the parameters of symmetry of the following totally symmetric functions:

   (*a*) $f(A, B, C, D) = \sum (1, 2, 4, 11, 13, 14)$
   (*b*) $f(A, B, C, D) = \sum (0, 3, 5, 10, 12, 15)$

4. State the difference between partial and independent symmetry.

5. Determine whether the following logical function is symmetric; if it is, identify the type and parameters of symmetry:

$$f(A, B, C, D) = \sum (5, 6, 9, 12, 15)$$

6. Determine the parameters of symmetry of the following partially symmetric functions:

   (*a*) $f(A, B, C, D, E) = \sum (5, 9, 12, 15, 21, 25, 28, 31)$
   (*b*) $f(A, B, C, D) = \sum (1, 4, 7, 10)$

7. Design the digital implementation of the logical expression

$$f(A, B, C, D, E) = \sum (2, 5, 8, 10, 11, 14, 17, 20, 21, 23, 26, 29)$$

8. Using both approaches (conventional logical design and binary addition) implement the following partially symmetric function:

$$f(A, B, C, D) = \sum (1, 2, 4, 11, 14)$$

9. Design the digital implementation of the following logical expression:

$$f(A, B, C, D, E) = \sum (1, 4, 7, 13, 18, 20, 23, 30)$$

10. Using binary adders, implement the following logical expression:

$$f(A, B, C, D, E) = [S^5_{0,1}(A\bar{B}\bar{C}DE) + S^5_{0,1}(\bar{A}BC\bar{D}\bar{E})]S^5_{0,2,3}(A\bar{B}\bar{C}DE)$$

PART THREE

GENERAL TOPICS IN DIGITAL ENGINEERING

CHAPTER FOURTEEN

Conversions

14.1 INTRODUCTION

In the design of digital systems it is very often required to convert the format of a given set of data for a new format. The new format normally facilitates processing of these data in devices where the old format is not acceptable.

For example, the computerized evaluation of a patient's health may require that all medical parameters enter the computer in a certain digital format. This may be serial or parallel form where in both cases the analog signals provided by the sensors will need to be converted to the appropriate digital form. In this case analog-to-digital converters will be used.

Even inside digital systems format conversions may be required. A calculator, for example, may accept and present numbers in decimal form while its arithmetic unit is binary. In this case a decimal-to-binary converter will provide the interface between the keyboard and the arithmetic unit, and a binary-to-decimal converter will convert the computed binary result to decimal form that will drive the calculator's display.

There are numerous types of converters since there are many forms in which data may be found or processed. In this chapter converter design approaches are described which will provide the reader with the necessary background for evaluating currently available converters or building his own.

14.2 BINARY-TO-GRAY CONVERSION

The numerical system described in the preceding chapters was the pure binary system where the general rules of arithmetic are directly applicable. Other nonbinary digital codes exist that offer special features which should be known to the digital design engineer. The binary coded decimal code, known as BCD, and the Gray code are the most commonly used nonbinary codes. (BCD was covered in Chapter Twelve.)

The special characteristic of the Gray code is that any two successive numbers differ only by 1 bit. In binary code two successive numbers differ by one or more bits as is illustrated in Figure 14.1, where a comparison of decimal, binary, and Gray codes appears. From this figure it can be seen that when counting from number 15 to 16 in the binary code, all 5 bits of the number change while in the Gray code only 1 bit changes. The Gray code is merely a counting scheme and not a numerical system since its bits carry no numerical significance of their own but all together define a number. The logic relation that links Gray to the binary code is defined by the following expressions:

$$G_n = B_n \oplus B_{n+1} \tag{14.1}$$

$$= B_n\bar{B}_{n+1} + \bar{B}_nB_{n+1} \tag{14.2}$$

where G is the Gray code, B is the Binary code, and n is the order of bit—that is, a Gray bit having the weight of 2^0 equals the sum of the binary bits

Decimal	Binary	Gray
0	0000	0000
1	0001	0001
2	0010	0011
3	0011	0010
4	0100	0110
5	0101	0111
6	0110	0101
7	0111	0100
8	1000	1100
9	1001	1101
10	1010	1111
11	1011	1110
12	1100	1010
13	1101	1011
14	1110	1001
15	1111	1000

Figure 14.1. Comparison of the decimal, binary, and Gray systems.

having the weight of 2^0 and 2^1. For example, to convert binary number 0110 to the Gray code, we will proceed as follows:

$$\begin{aligned} G_8 &= B_8 + B_{16} = 0 + 0 = 0 \\ G_4 &= B_4 + B_8 \ \ = 1 + 0 = 1 \\ G_2 &= B_2 + B_4 \ \ = 1 + 1 = 0 \\ G_1 &= B_1 + B_2 \ \ = 0 + 1 = 1 \end{aligned}$$

Therefore, the Gray code equivalent of binary number 0110 is 0101. In the above additions only the sum output is considered. The carry output is of no significance.

Figure 14.2 shows a parallel binary-to-Gray converter. This network is used for the conversion of parallel binary numbers to parallel Gray coded numbers and implements (14.1). If the binary number is available in sequential

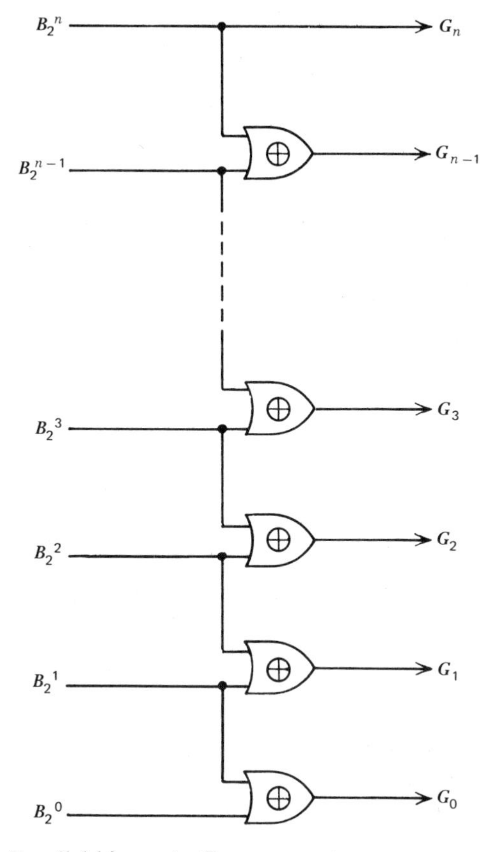

Figure 14.2. Parallel binary-to-Gray converter.

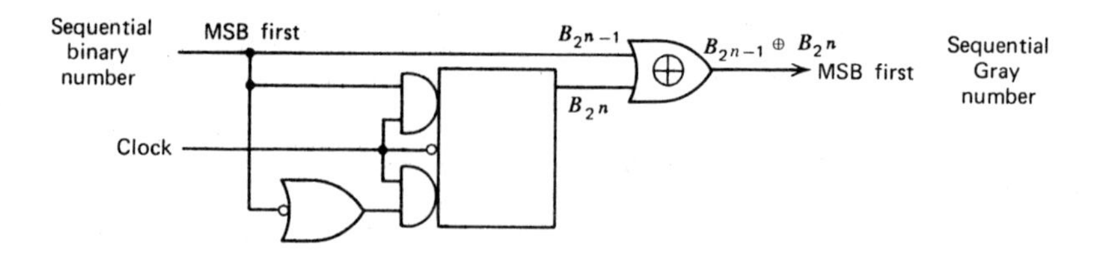

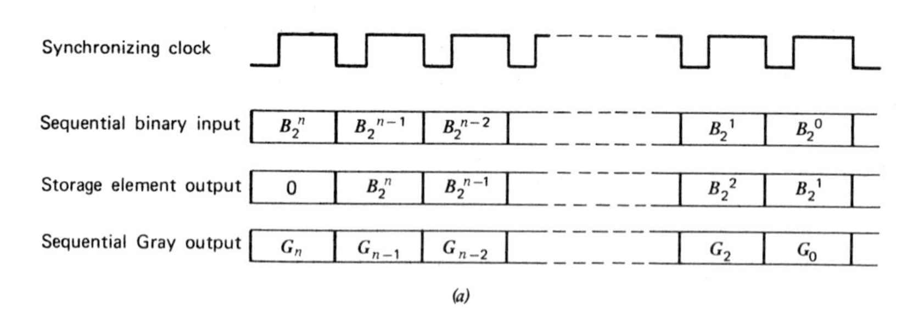

(*a*)

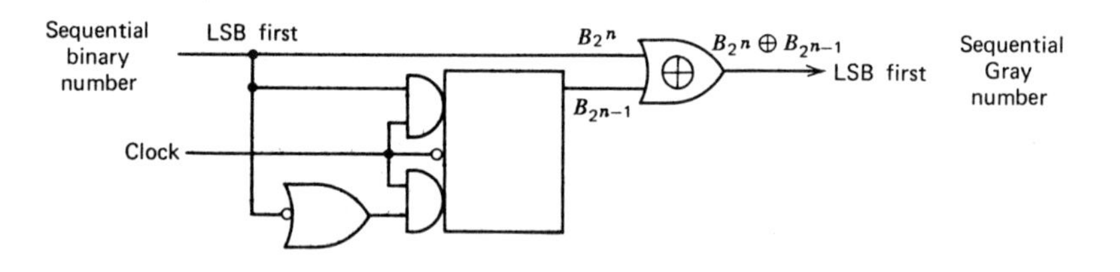

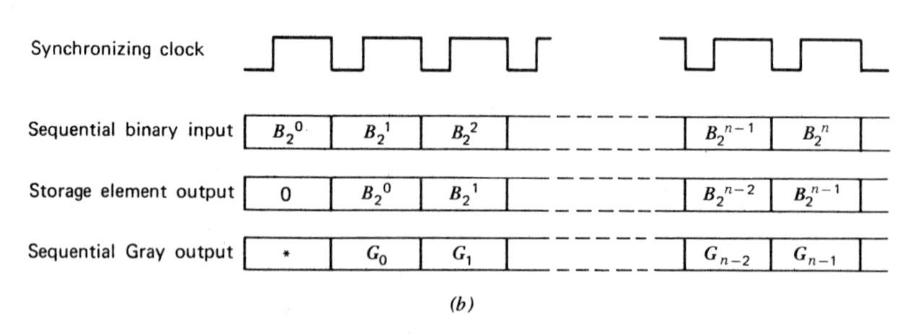

(*b*)

Figure 14.3. Sequential binary-to-Gray converter: (*a*) MSB enters first; (*b*) LSB enters first. (*Note:* Storage element is initially cleared. *Output is of no meaning during this period.)

form, one of the two circuits shown in Figure 14.3 can be used. Circuit (*a*) should be used when the sequential binary number is available with its MSB first, while circuit (*b*) should be used when the seqential binary number to be converted is available with its LSB first. Also shown in Figure 14.3 are the timing diagrams of the respective sequential operations.

14.3 GRAY-TO-BINARY CONVERSION

To convert a number from Gray to binary code, we first solve (14.1) for B to derive the Gray-to-binary relation.

$$G_n = B_n + B_{n+1} \tag{14.3}$$

Solving for B_n, we obtain

$$B_n = G_n - B_{n+1} \tag{14.4}$$

where B_n equals the difference bit of $G_n - B_{n+1}$, where the resulting borrow is not considered. Equation (14.4) can also be written as

$$B_n = G_n + B_{n+1} \tag{14.5}$$

Thus, since the sum and difference of any 2 bits is always the same, the binary bit having weight 2^0 equals the sum of the Gray bit having weight 2^0 and the binary bit having weight 2^1. For example, to convert the Gray number 1010 to binary, we proceed as follows:

$$\begin{aligned}
B_{16} &= G_{16} = 0 \\
B_8 &= G_8 + B_{16} = 1 + 0 = 1 \\
B_4 &= G_4 + B_8 = 0 + 1 = 1 \\
B_2 &= G_2 + B_4 = 1 + 1 = 0 \\
B_1 &= G_1 + B_2 = 0 + 0 = 0
\end{aligned}$$

where B_{16} is 0 because the Gray number has less than 5 bits. Therefore, the binary equivalent of Gray number 1010 is 1100.

Figure 14.4 shows a parallel Gray-to-binary converter where eq. (14.5) is implemented. This network can be used for the conversion of parallel Gray coded numbers to parallel binary coded numbers. If the Gray number is available in sequential form, the sequential Gray-to-binary converter of Figure 14.5(*a*) can be used.

In the network of Figure 14.5(*a*), the Gray coded number must enter with its MSB first. The converter is simply a single flip-flop used as a divide-by-2 device. Here, advantage is taken of the fact that in a binary number the bits change from 1 to 0 or from 0 to 1 anytime the Gray number equivalent bit changes from 0 to 1.

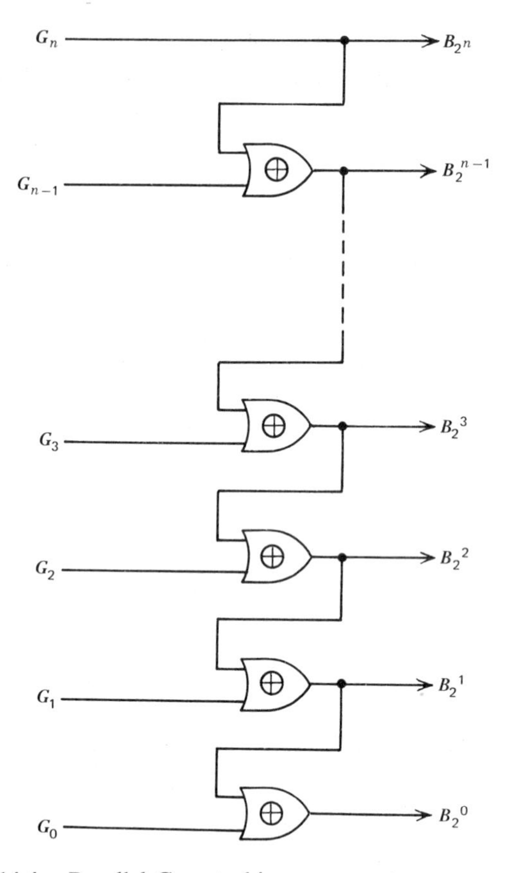

Figure 14.4. Parallel Gray-to-binary converter.

For example, consider the Gray number 1001 and its binary counterpart 1110. If the two numbers are represented by the pulse trains shown in Figure 14.5(*b*), we can easily see that if the Gray number is the input to a divide-by-2 device, the binary equivalent will be the output of that device. If 1 for the Gray number is HIGH, the divide-by-2 circuit should trigger at the positive-going edge.

14.4 DIGITAL-TO-ANALOG CONVERSION

Digital-to-analog conversion is accomplished by the use of summing and ladder networks.

Figure 14.6(*a*) shows a resistive current-summing network, the resistors

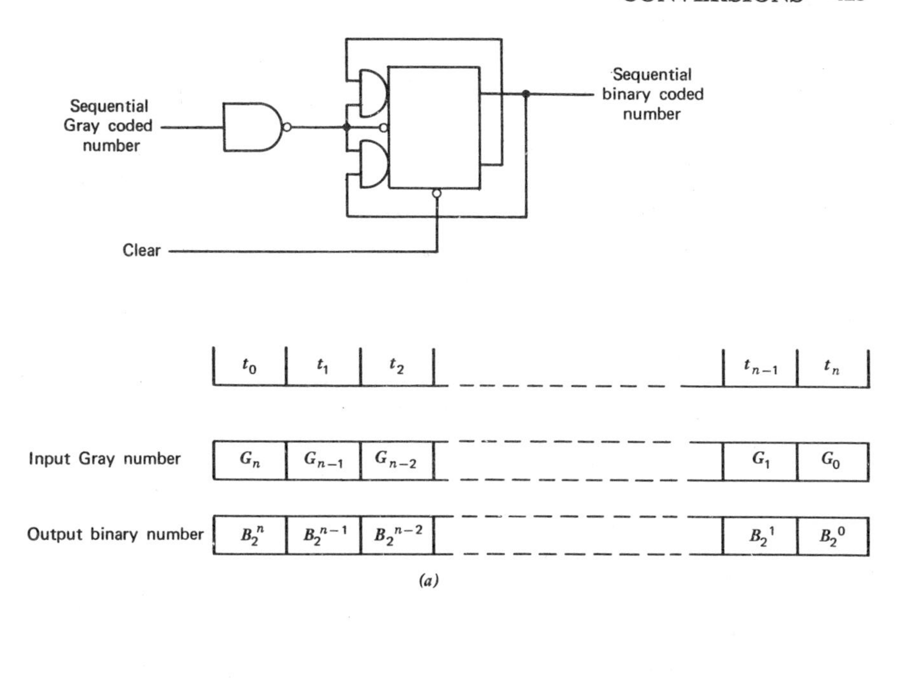

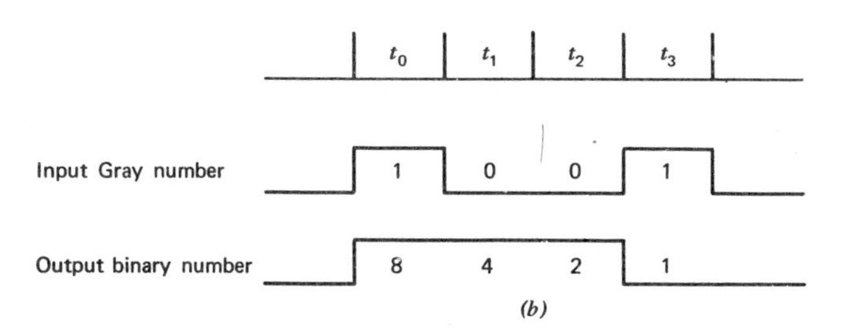

Figure 14.5. Sequential Gray-to-binary converter where the MSB enters and exits first.

of which are binary-weighted. The amplitude of the output voltage of this circuit is directly proportional to the input binary number. This output has a high impedance and it should feed a high-impedance amplifier before it is used. The input to the summing network must switch from a voltage representing 1 to ground or an open circuit representing 0. This circuit requires a different resistor value for each bit of the digital input. The per-bit load provided by the summing network is directly proportional to the corresponding resistor of the network.

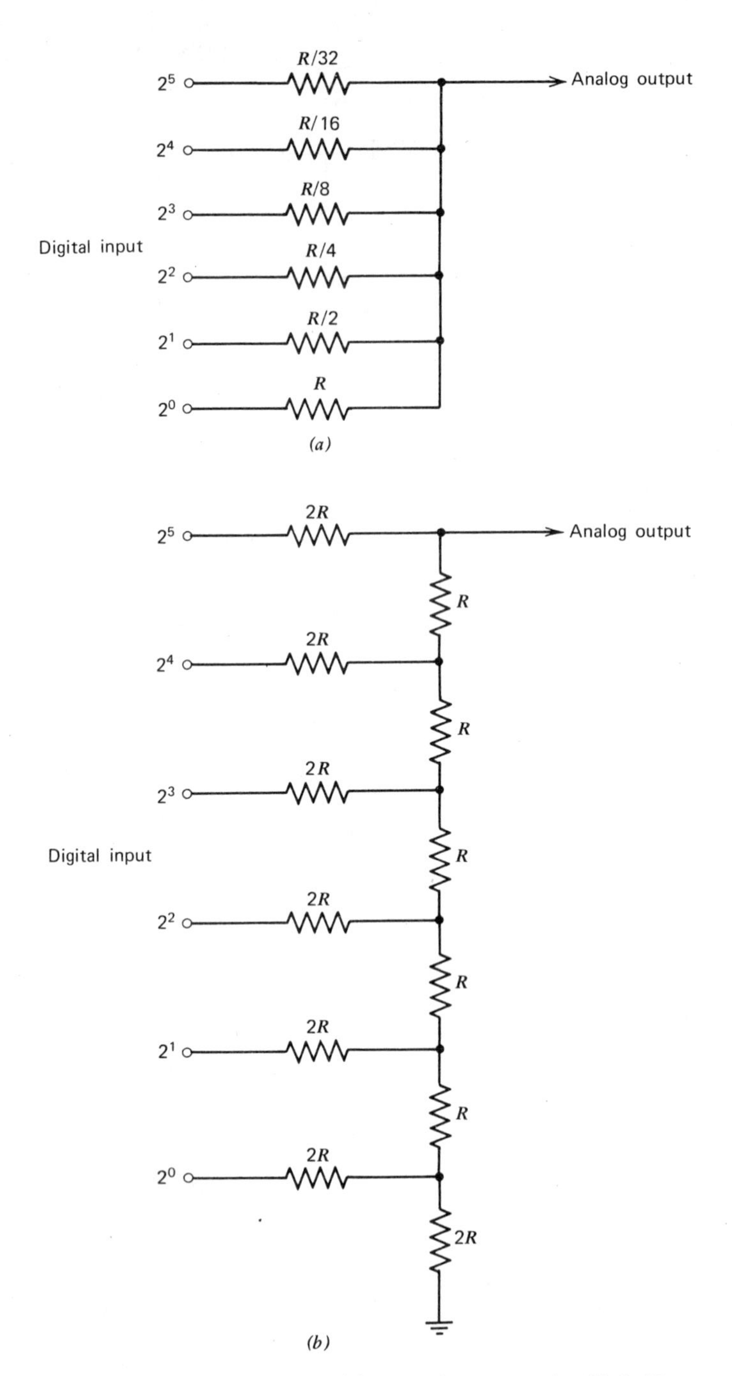

Figure 14.6. D/A converters: (*a*) summing network; (*b*) ladder network.

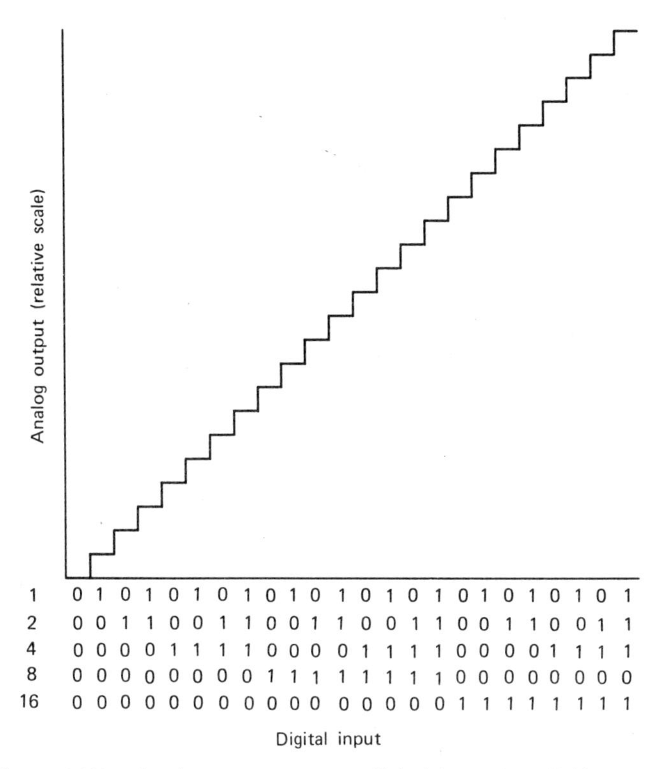

Figure 14.7. Analog output versus digital input in a D/A converter.

Figure 14.6(*b*) shows a ladder network which is used for D/A conversion. The input to the ladder network must switch from a voltage representing 1 to ground representing 0. It should never be left open-circuited. The advantage of this circuit is that it requires only two types of resistors regardless of the number of bits of the digital input. Another advantage is that the load the resistors provide to the digital input is almost the same for all bits of the digital input.

Figure 14.7 shows the analog output of a D/A converter versus the digital input. If the steps are very small, the staircase can be considered to be a straight line.

14.5 ANALOG-TO-DIGITAL CONVERSION

In digital systems, inputs are very often in analog form. To digitally process the information they carry, these inputs must be first converted from their analog form to digital. There are various methods of converting analog information to digital. Described below are the three major configurations.

Conversions Using Counters

Figure 14.8 shows a ramp-counter A/D converter. This circuit consists of a binary counter, a D/A converter, and an analog comparator. The output of the binary counter feeds the D/A converter and the output of the D/A converter feeds one of the two inputs of the analog comparator. The analog input feeds the other input. The output of the analog comparator feeds the clear input of the counter thus completing the loop.

The circuit operates as follows: The binary counter starts from 0. . .00 and counts up, making the output of the D/A converter start from 0 V and increase with time, forming a ramp. The analog input is continuously compared to this ramp and when the ramp becomes greater than the analog

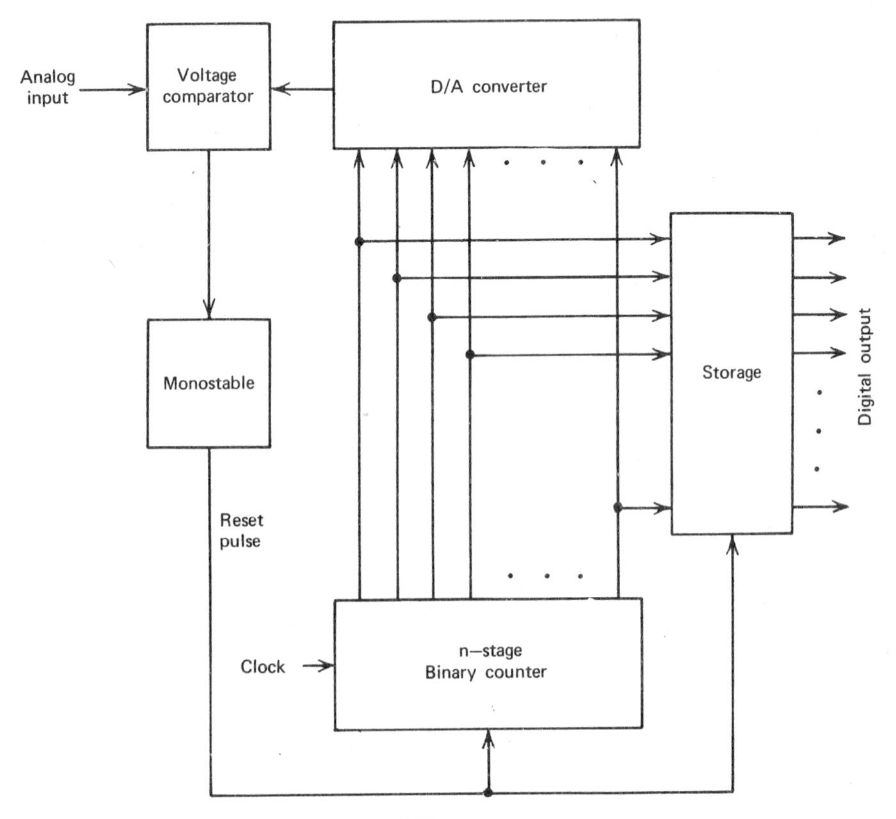

Figure 14.8. Ramp-counter A/D converter.

input, the comparator output which is a digital output switches state. The monostable which is driven by the comparator output senses that change in voltage and produces a pulse which clears the binary counter, bringing it to 0...00. The output of the binary counter, before it is cleared, is the digital equivalent of the analog input appearing at the comparator at that time.

The digital output of this A/D converter can be stored in a register where the storage entry will be controlled by the output of the analog comparator.

The A/D converter of Figure 14.8 may require up to $2^n - 1$ clock periods before it produces the digital equivalent of the analog input. For applications that require faster A/D conversion, the network of Figure 14.9 can be used.

This circuit consists of an up-down counter, and D/A converter, and an analog comparator. In this A/D converter, the analog comparator output controls the mode of counting of the up-down counter, where the counter actually "tracks" the input analog voltage. The output of the up-down counter is the digital equivalent of the analog input.

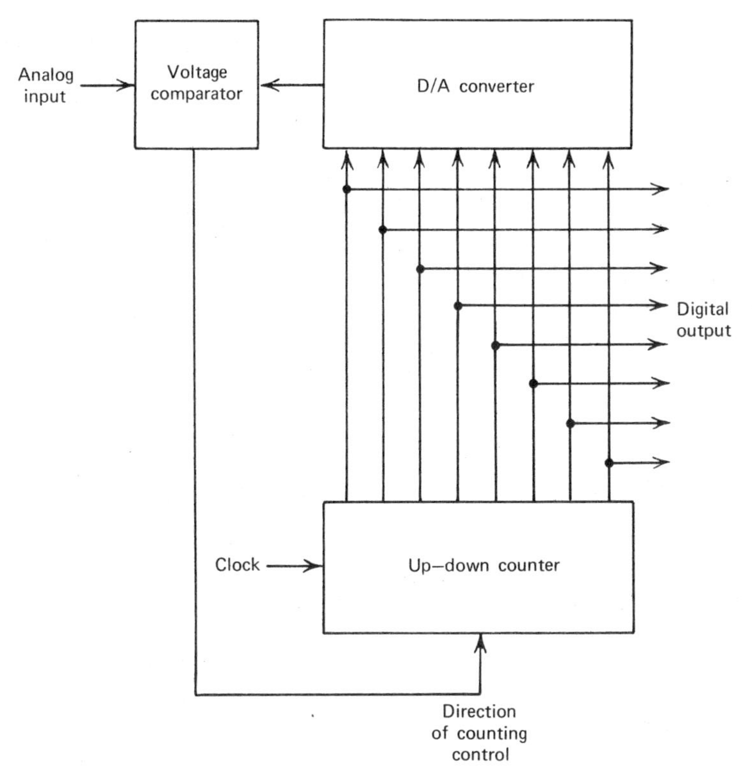

Figure 14.9. Up-down counter A/D converter.

Conversion by Successive Approximation

Another method of converting information from analog to digital is by successively approximating the digital equivalent until it is determined. The method is called A/D conversion by successive approximation. The logic configuration of this method is shown on Figure 14.10.

This converter consists of a shift register, a storage register, a D/A converter, and an analog comparator. The shift register provides the timing for the conversion operation and it is preset so that all stages except one are 0.

The storage register is initially cleared and its outputs drive the D/A converter, the output of which is an input to the analog comparator. The inputs to the storage register are modified by the output of the analog comparator at time intervals determined by the timing register. The comparator output feeds all stages of the conversion register at the synchronous clear input, where it is ANDed with a timing signal. If the input analog signal is less than that provided by the D/A converter, the content of the storage register is decremented.

This A/D converter operates as follows: Initially, the two registers are preset so that all stages of the two registers are 0 with the exception of the first stage of the timing register.

At the arrival of the first pulse, the output of the 2^n stage of the conversion register becomes 1. At that period the input to the D/A converter is 1000. . .0, and its analog equivalent is compared to the input analog signal. If the input signal is greater, the comparator output is LOW, otherwise it is HIGH. If HIGH at the arrival of the second clock pulse, the 2^n stage of the conversion counter will change from 1 to 0, otherwise it will remain unchanged. At the arrival of the second clock pulse, regardless of the outcome in stage 2^n of the conversion register, the output of stage 2^{n-1} of the same register will become 1. Now the output of the D/A converter will correspond to either 110. . .0 if the input signal was greater than 100. . .0 or to 010. . .0 if the input signal was less than 100. . .0. If the input signal is greater than the output of the D/A converter, its output will be LOW which will maintain 1 at the 2^{n-1} stage of the conversion register; otherwise at the arrival of the third clock pulse, the 2^{n-1} stage will change from 1 to 0.

Thus, at the arrival of the third clock pulse, the 2^{n-2} stage of the conversion register will become 1 and the analog equivalent of the new content of the conversion register will be compared to the input signal. From this comparison the correct state of the 2^{n-2} stage will be determined.

This process of successively approximating the digital equivalent of the analog input signals continues until the desired number of bits has been obtained. The number of clock periods required for the conversion equals n, the number of the desired bits, while in the preceding method it could go as

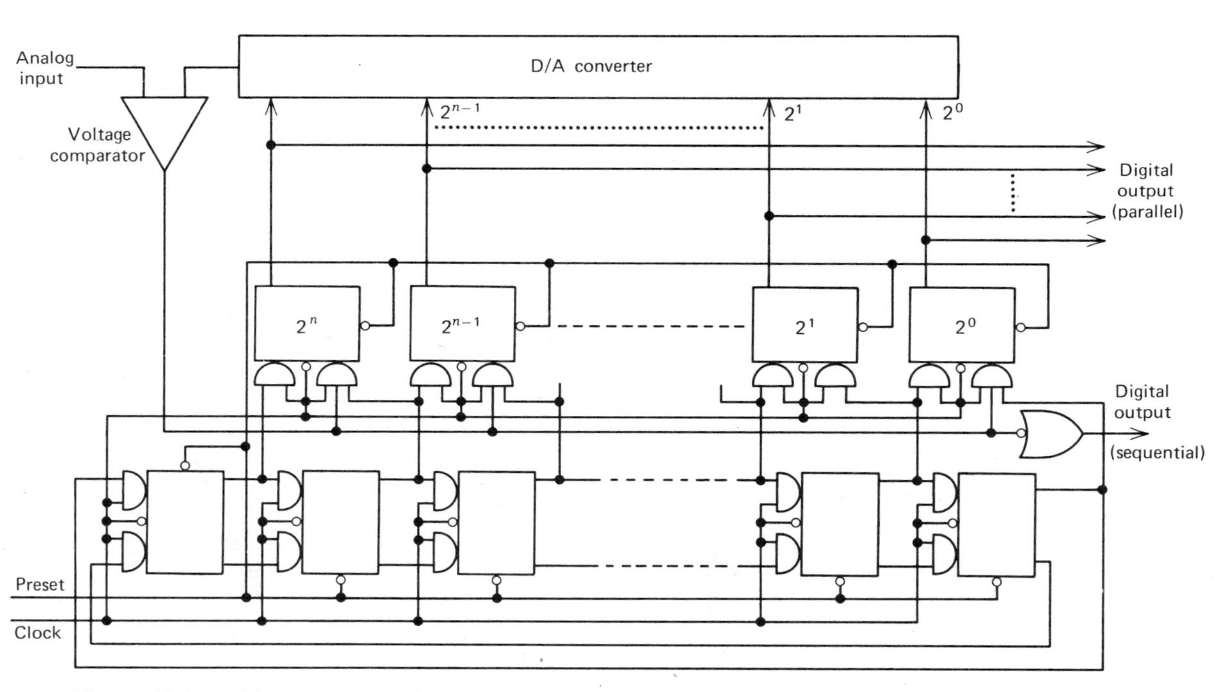

Figure 14.10. A/D conversion by successive approximation.

high as $2^n - 1$ clock periods. An additional advantage of this method is that the digital equivalent of the analog input signal is provided in parallel as well as in sequential form.

Asynchronous Conversion

The A/D converters discussed thus far are almost totally digital; they all require a clock to drive them and they all have some kind of a closed loop. The A/D converter of Figure 14.11 is almost totally analog, requires no clock, has no closed loop, and, in addition, is very fast, being limited only by the propagation delay of its components.

This converter works as follows: The analog input is initially compared to a reference voltage corresponding to the MSB of the digital output. If the analog input is found to be greater than the reference voltage, the comparator output is HIGH, indicating 1 for that bit. The reference voltage is then subtracted from the input analog signal and the obtained difference is compared to a new reference corresponding to the bit next in significance.

This procedure repeats until the input analog signal has passed through the cascade of amplifiers and comparators. In this A/D configuration the conversion time is limited only by the propagation delay of the selected components.

14.6 VOLTAGE-TO-FREQUENCY CONVERSION

Conversion of voltage to frequency can be achieved with the use of a modified analog-to-digital converter. Figure 14.12 shows a voltage-to-frequency converter. It consists of a down-counter, a D/A converter, an analog voltage comparator, and a monostable multivibrator.

The counter starts at 1...11 and counts down until its analog equivalent becomes less than the input voltage. Then, the comparator output changes state, initiating the generation of a pulse at the output of the monostable device. This pulse presets the counter to 1...11 and the cycle starts again. The analog equivalent of the counter output when the counter is 1...11 must be always greater than the maximum input voltage. The time it takes the counter to count down and have its output coincide with the input voltage depends on the amplitude of that voltage—the higher the input voltage, the shorter the time for the counter to reach the level of the input voltage and initiate a comparison pulse.

The time interval between comparisons is therefore inversely proportional to the input voltage, or the repetition rate of comparison pulse is proportional to the input voltage. Thus as the input voltage increases, the output frequency increases also.

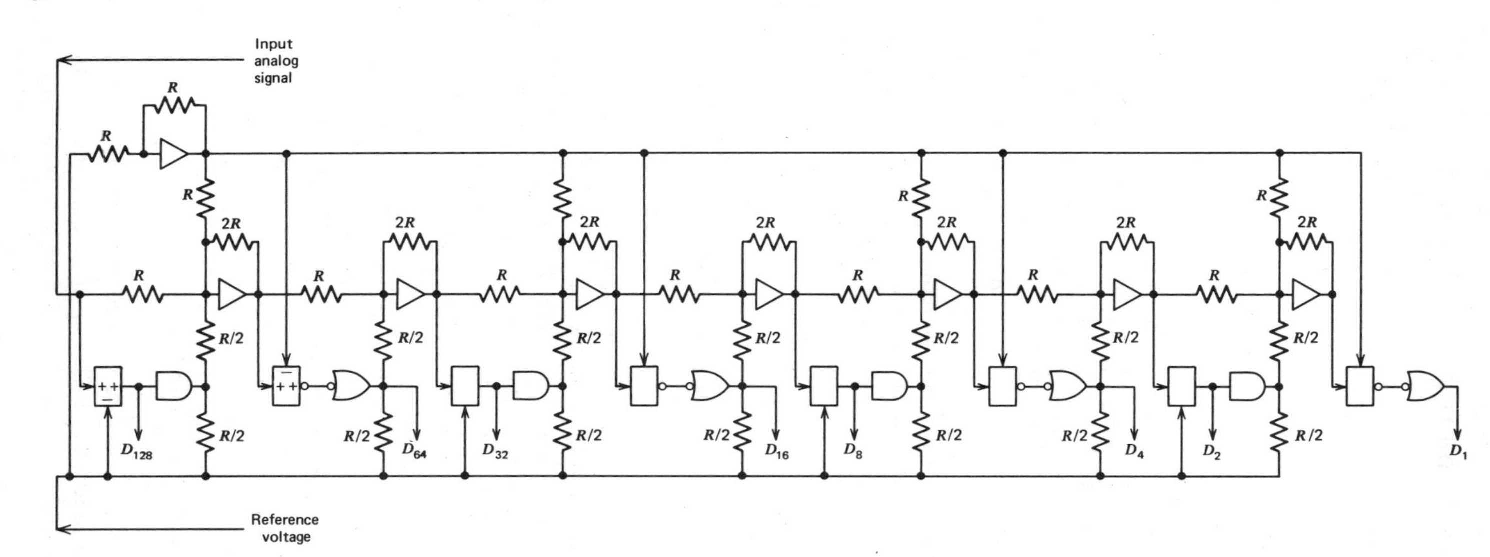

Figure 14.11. A/D converter (asynchronous conversion). (*Note:* Squares represent voltage comparators; the digital output is $D_{128}\ D_{64}\ D_{32}\ D_{16}\ D_8\ D_4\ D_2\ D_1$.)

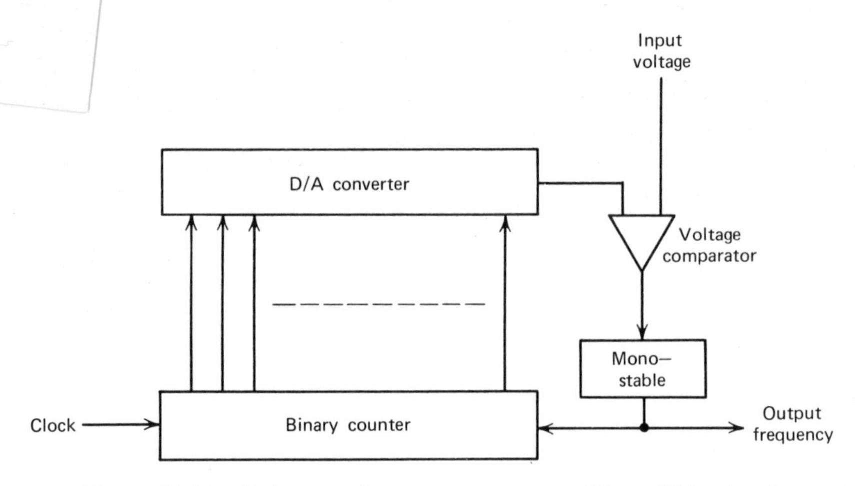

Figure 14.12. Voltage-to-frequency converter. (*Note:* This signal presets the counter to 1...11; the counter counts down.)

14.7 FREQUENCY-TO-VOLTAGE CONVERSION

A frequency can be converted to voltage by means of a counter, a storage register, and a D/A converter. A frequency-to-voltage converter is shown in Figure 14.13.

Here the input frequency feeds a counter which is cleared by a signal of known repetition rate. Between the clearing pulses, the counter counts up at a rate depending on the input frequency. Upon the arrival of the clear pulse, the number contained in the counter enters the storage register. This number represents the input frequency and is directly proportional to it.

The output of the storage register feeds a D/A converter which provides a voltage directly proportional to the binary content of the storage register. This binary value is directly proportional to the input frequency, and so becomes the output of the D/A converter. This circuit can be also used as a digital frequency meter where the storage register output will digitally represent the input frequency.

14.8 BINARY-TO-BCD CONVERSION

Conversion of binary numbers into BCD is accomplished by expressing each bit of the binary number in terms of BCD numerical quantities and then add-

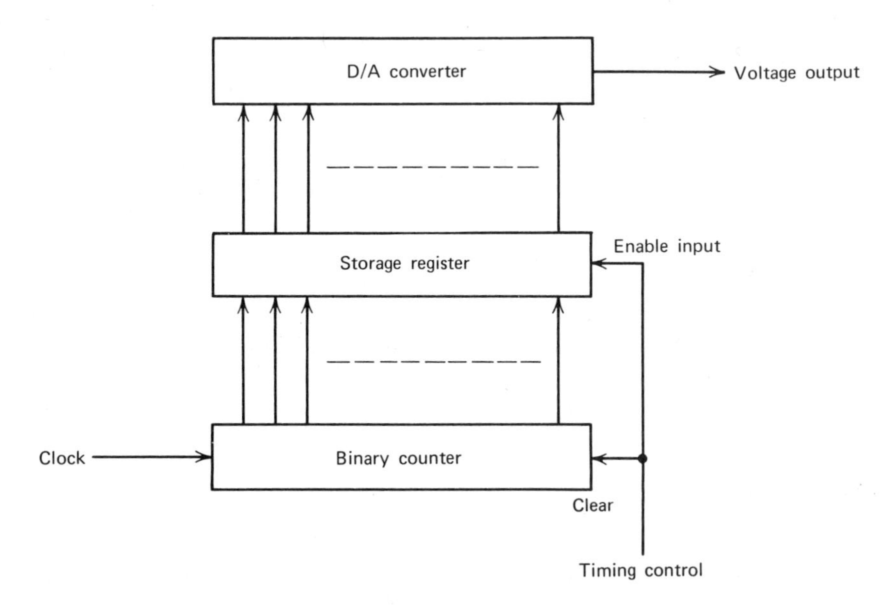

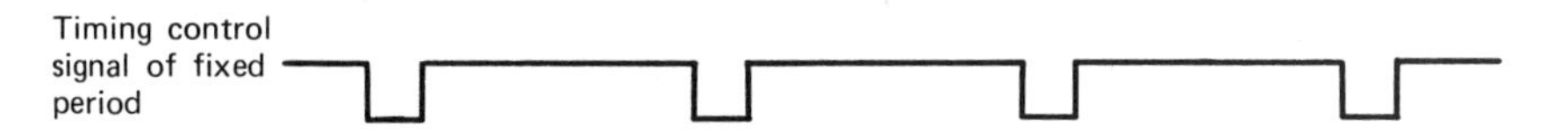

Figure 14.13. Frequency-to-voltage converter.

ing these quantities in a BCD adder. For example, the binary bit that represents 128 is expressed as BCD number

0001 0010 1000

Figure 14.14 shows an 8 bit binary-to-BCD converter.

14.9 BCD-TO-BINARY CONVERSION

Conversion from BCD to binary code employs the same basic approach used in the binary-to-BCD conversion—that is, the numerical quantities of the code to be converted are expressed in terms of numerical quantities of the new code which are added together in using an adder of the new code. Figure 14.15 illustrates a 2 digit BCD-to-binary converter employing this principle.

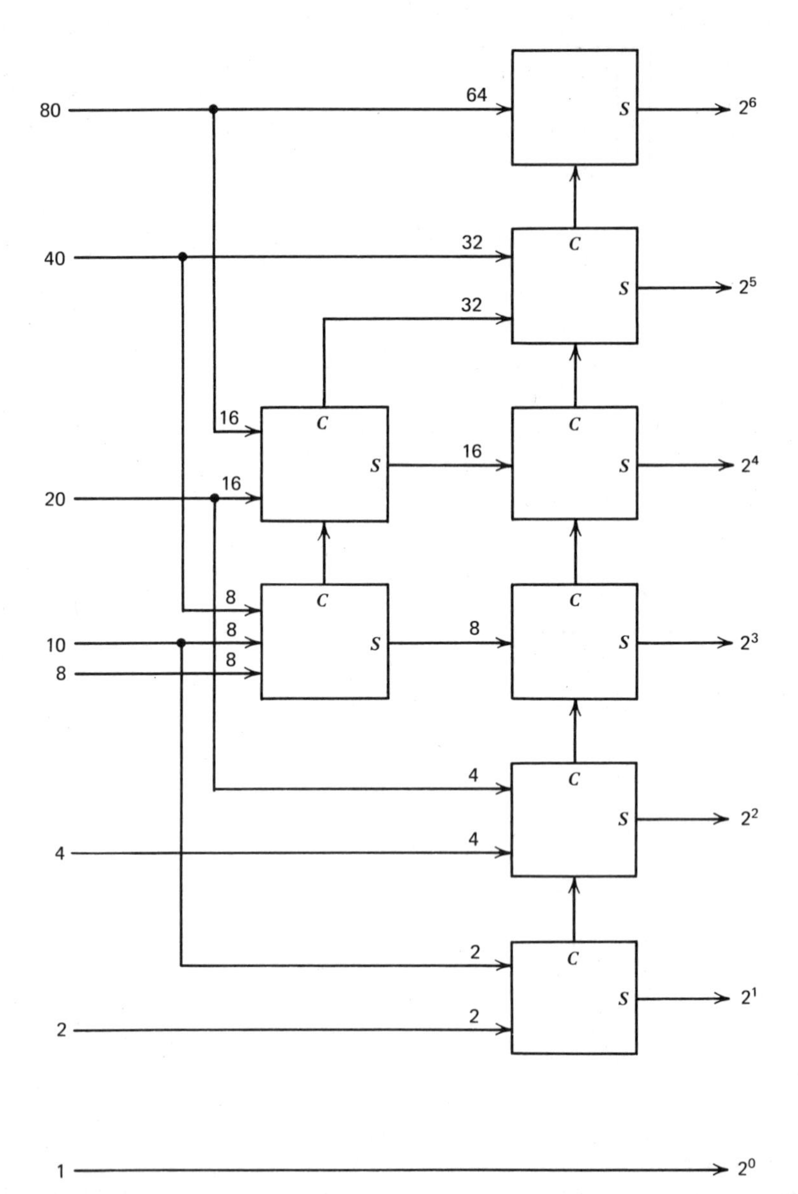

Figure 14.14. Binary-to-BCD converter.

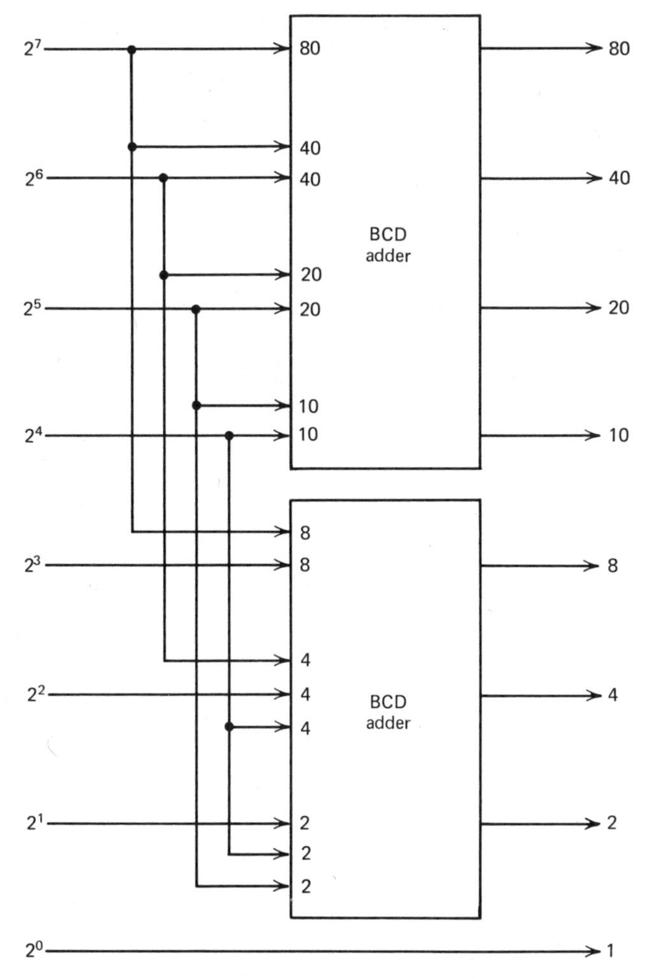

Figure 14.15. BCD-to-binary converter. (*Note:* Each block represents a binary adder.)

PROBLEMS

1. Draw the timing diagram of a sequential Gray-to-binary converter showing the conversion of Gray number 10110110. Assume that the number is available with its MSB leading the sequence.
2. When speed is of most importance, what method of analog-to-digital conversion should be used?

3. When reduction in hardware is of most importance, what method of analog-to-digital conversion should be used?
4. State the advantages and disadvantages of the ladder type of D/A converter over the summing network type.
5. Design a binary-to-BCD converter for the conversion of an 8 bit parallel binary number.
6. Design a BCD-to-binary converter for the conversion of a 3 digit parallel BCD number.
7. Design a parallel single-digit decimal-to-BCD converter.
8. Design a parallel single-digit BCD-to-decimal converter.
9. Design a 6 bit A/D converter employing the method of successive approximation and draw its timing diagram.
10. Make a study on the currently available hardware performing analog-to-digital conversions.

CHAPTER FIFTEEN

Miscellaneous Operations

15.1 INTRODUCTION

In the preceding chapters, network and functions falling into definite categories were discussed. In digital system design, however, very often the need arises for special logical networks which may appear only once in the system.

In this chapter various special circuits are presented and discussed. Some of the circuits consist of a single logic component, while others may be very complex. The complexity of these circuits, however, does not indicate the complexity of the function they perform. Many times implementation of simple functions requires complex circuitry, while complex functions may be implemented easily. The reason for this occasional peculiarity is the relationship these functions have to binary mathematics.

15.2 SYNCHRONIZATION OF ASYNCHRONOUS INPUTS

To synchronize a given digital input signal, means to make that input change state only during the activating edge of a synchronizing clock. This can be achieved by passing the signal through a clocked flip-flop as shown in Figure 15.1. Also shown are the waveforms of the input, the output, and that of the clock.

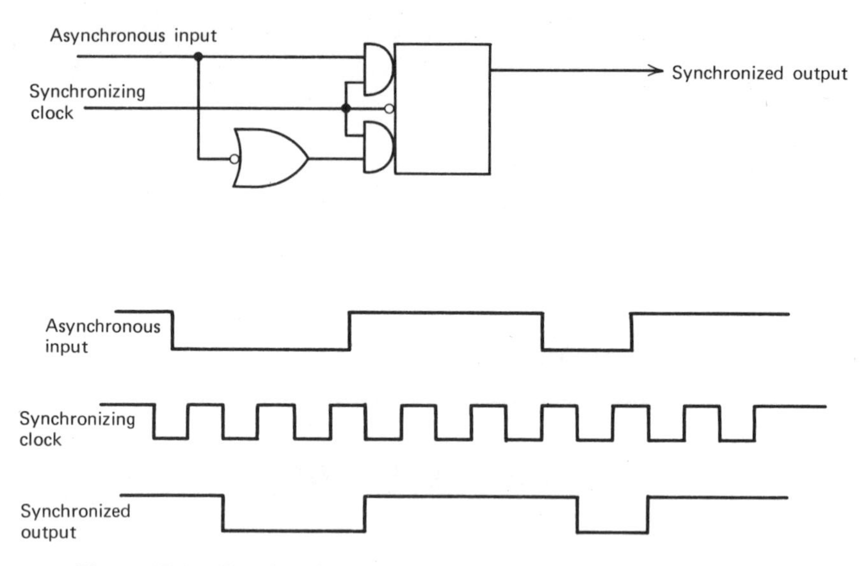

Figure 15.1. Synchronization of asynchronous inputs.

When the clock input is HIGH, the asynchronous input enters the master stage of the master-slave flip-flop and when it changes to LOW, the information is transferred to the slave stage of the flip-flop where it stays until the next clock input change from HIGH to LOW.

After synchronization, the signal pulse width and pulse-spacing become multiples of the clock period of the synchronous system. The maximum pulse width and pulse-spacing error caused by the synchronization equals the period of the clock frequency. This error is computed in section 15.3 where a shift-register digital-delay line is discussed.

15.3 DIGITAL-DELAY LINE

A shift register can be also used as a digital-delay line where the generated time delay is a multiple of the period of the triggering clock. An n stage digital-delay line and its timing waveforms are shown in Figure 15.2. For synchronous inputs, inputs that change state during the activating edge of the clock, the delay provided by a shift register is

$$\text{Delay of synchronous input} = nT \tag{15.1}$$

where n is the number of stages through which the input signal has to travel and T is the period of the clock frequency used by the digital-delay line.

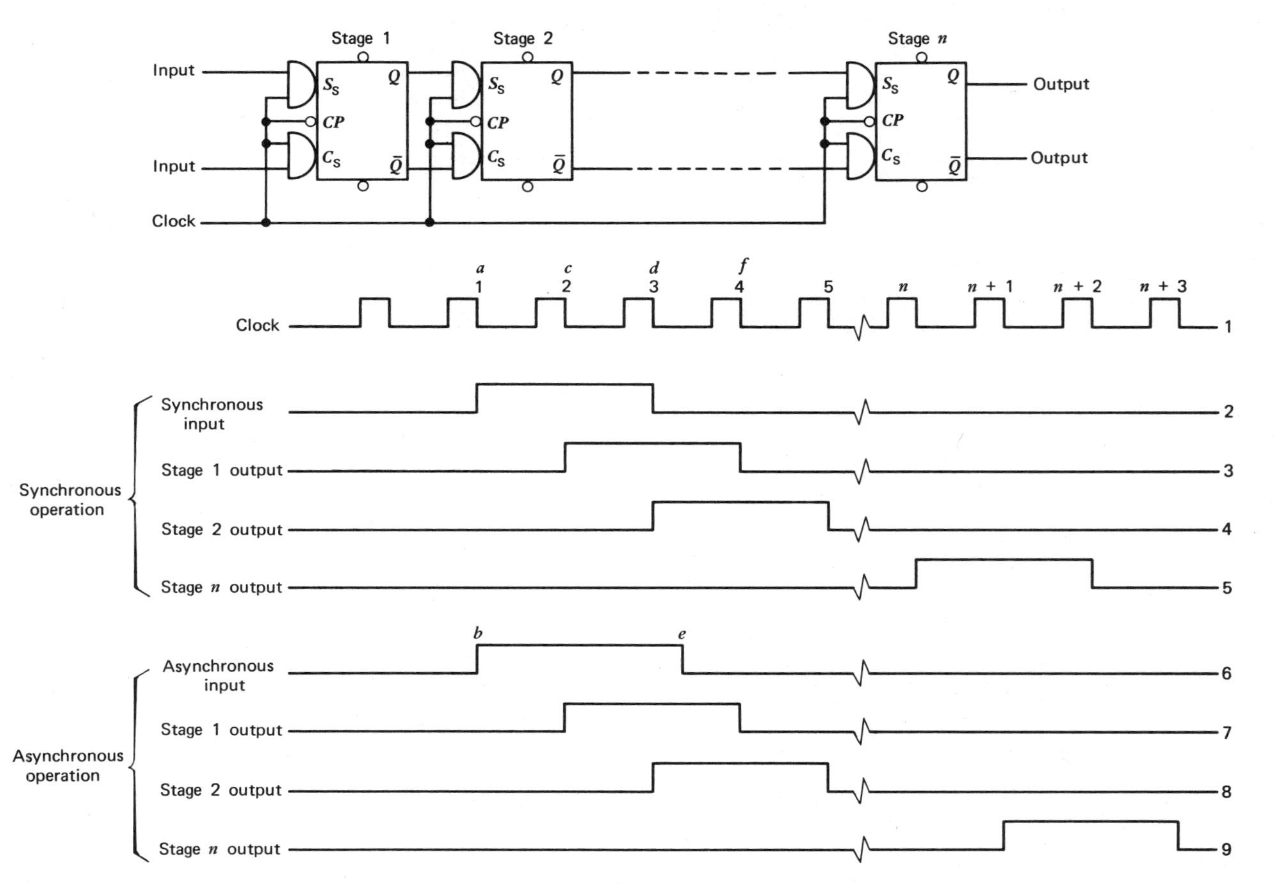

Figure 15.2. Shift-register delay and waveforms of operation.

When synchronous signals are delayed, their pulse width and pulse-spacing remain unchanged. Waveforms 2, 3, 4, and 5 of Figure 15.2 illustrate the operation of an n bit shift-register delay when its input is synchronous. When the input signal is asynchronous with respect to the clock—that is, when the input signal does not change state during the activating edge of the clock—the delay will not be nT, as above, but will slightly vary, depending on the time relationship of the asynchronous input waveform and that of the clock frequency. Waveform 6 of Figure 15.2 shows a pulse which is asynchronous with respect to the clock. The positive-going edge of that pulse may lie any place between time a and time c of the clock waveform without affecting the positive-going edge of the delayed output. The delay provided by the first stage will be bc. Since bc may vary from 0 to T where T is the period of the clock frequency, the delay provided will also vary from 0 to T.

The output of the first stage with respect to the second stage is a synchronous input, and the delay provided by the second stage will always be T. Similarly, the delay provided by the third stage will always be T, and so on. The sum of the delays provided by each of the n stages represents the total delay of the positive-going edge of the asynchronous input shown on Figure 15.2, waveform 6, and is

$$D^+ = bc + (n - 1)T \tag{15.2}$$

Since the time interval bc may vary from 0 to T, (15.2) becomes

$$\begin{aligned} D^+ &= (0 \rightarrow T) + (n - 1)T \\ &= [(n - 1) \rightarrow n]T \end{aligned} \tag{15.3}$$

The negative-going edge of the asynchronous pulse may similarly lie any place between time d and time f without affecting the negative-going edge of the delayed output. A similar time analysis will show that the delay of the negative-going edge of the asynchronous input will be

$$D^- = [(n - 1) \rightarrow n]T \tag{15.4}$$

The pulse width of the delayed asynchronous pulse will then be

$$\text{delayed output} = mT \tag{15.5}$$

where m is the number of "transferring" clock edges (negative-going edges that occurred during the asynchronous pulse) and T is the period of the clock frequency. The difference between the pulse width of the original signal and that of the delayed one equals the difference between the delay of the negative-going edge and the delay of the positive-going edge. Thus, the pulse width error is

$$
\begin{aligned}
PW\,\text{error} &= D^- - D^+ \\
&= (0 \rightarrow T) + (n-1)T - (0 \rightarrow T) - (n-1)T \\
&= (0 \rightarrow T) - (0 \rightarrow T) \\
&= (0 \rightarrow T)
\end{aligned}
\qquad (15.6)
$$

To summarize, when a shift register is used to delay an asynchronous pulse, the maximum error in the delay is T and the pulse width can be expanded or compressed by as much as T. When the pulse width of the asynchronous input is a large multiple of the clock frequency period, these errors can be negligible.

For example, if the clock frequency of a shift-register delay is 10kHz and a 3.75 msec asynchronous pulse is to be delayed, the maximum variation in the delay of the two edges of the asynchronous pulse will be T, which is

$$T = \frac{1}{f} = \frac{1}{10\ \text{kHz}} = 100\ \mu\text{sec}$$

The maximum difference between the original pulse width and the delayed one will also be 100 μsec. The pulse width of the delayed signal will be quantized to a multiple of 100 μsec, becoming either 3.7 msec or 3.8 msec.

The variation in the delay and the quantization of the pulse width of the delayed pulse are synchronization errors and are independent of the pulse width of the asynchronous input. Therefore, if the input signal were 1.5 msec, 10 sec, or any other pulse width, it would still have been subject to the same error in total delay or pulse width.

15.4 MULTIPLEXING

Multiplexing is the use of a single item by many circuits, units, or systems. That item can be a wire connecting two installations, or it can be a computer having many users, in which case it is known as time-sharing.

Figure 15.3(a) shows a digital multiplexing circuit. Here, a timing circuit in System I enables inputs $A, B, C, D, \ldots, N$ in sequence and allows their information to make use of the line joining System I to System II. In System II the same timing signal is generated to enable the corresponding gates, thus reconstructing information $A, B, C, D, \ldots, N$.

During multiplexing only one of the AND gates of System I is enabled at a time, thus allowing one of the signal inputs to feed the OR gate and then proceed to System II. In System II the only gate enabled at that time is the one corresponding to that signal. This approach is often used in transferring digital information through slip-rings where the number of slip-ring contacts must be kept to a minimum.

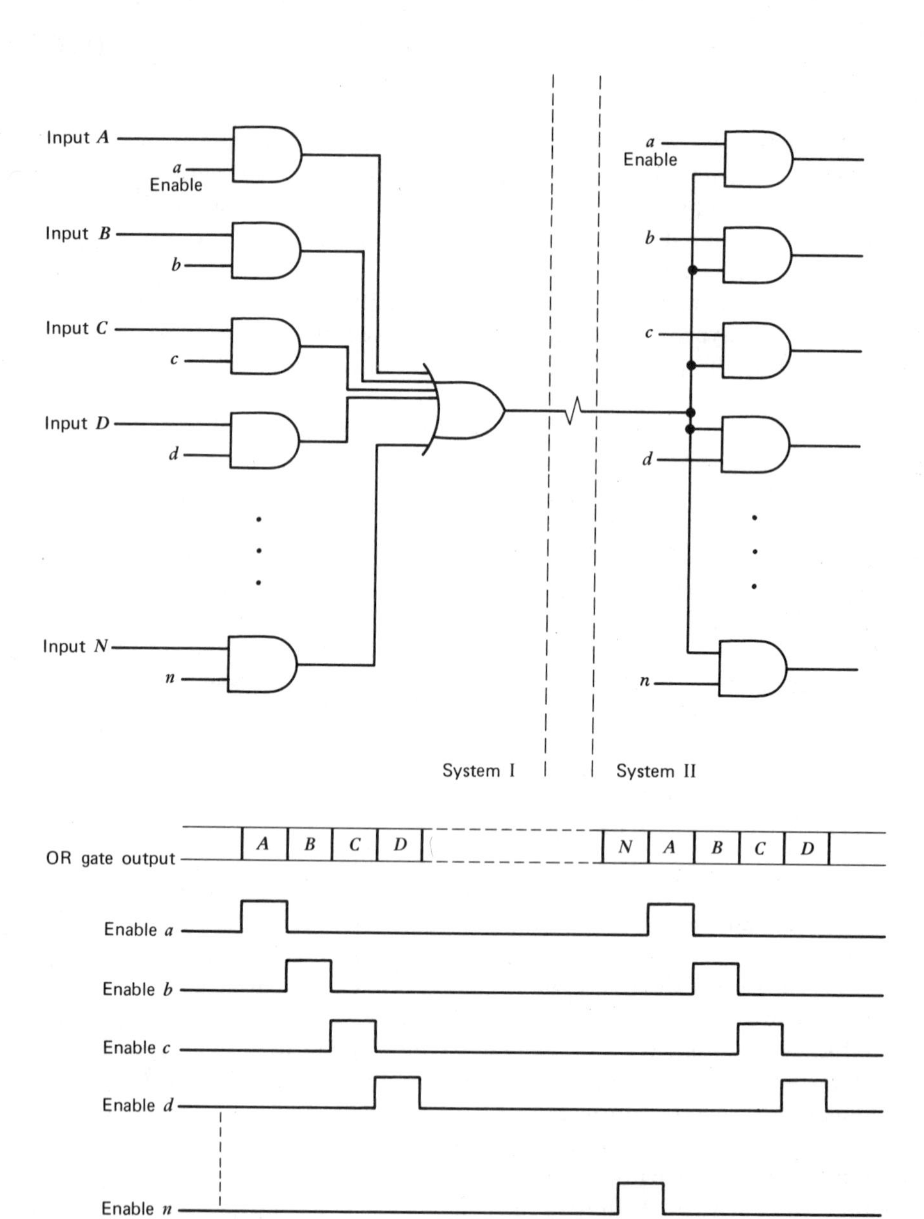
Input A
a
Enable
Input B
b
Input C
c
Input D
d
Input N
n
a
Enable
b
c
d
n
System I
System II
OR gate output
A
B
C
D
N
A
B
C
D
Enable a
Enable b
Enable c
Enable d
Enable n

(a)

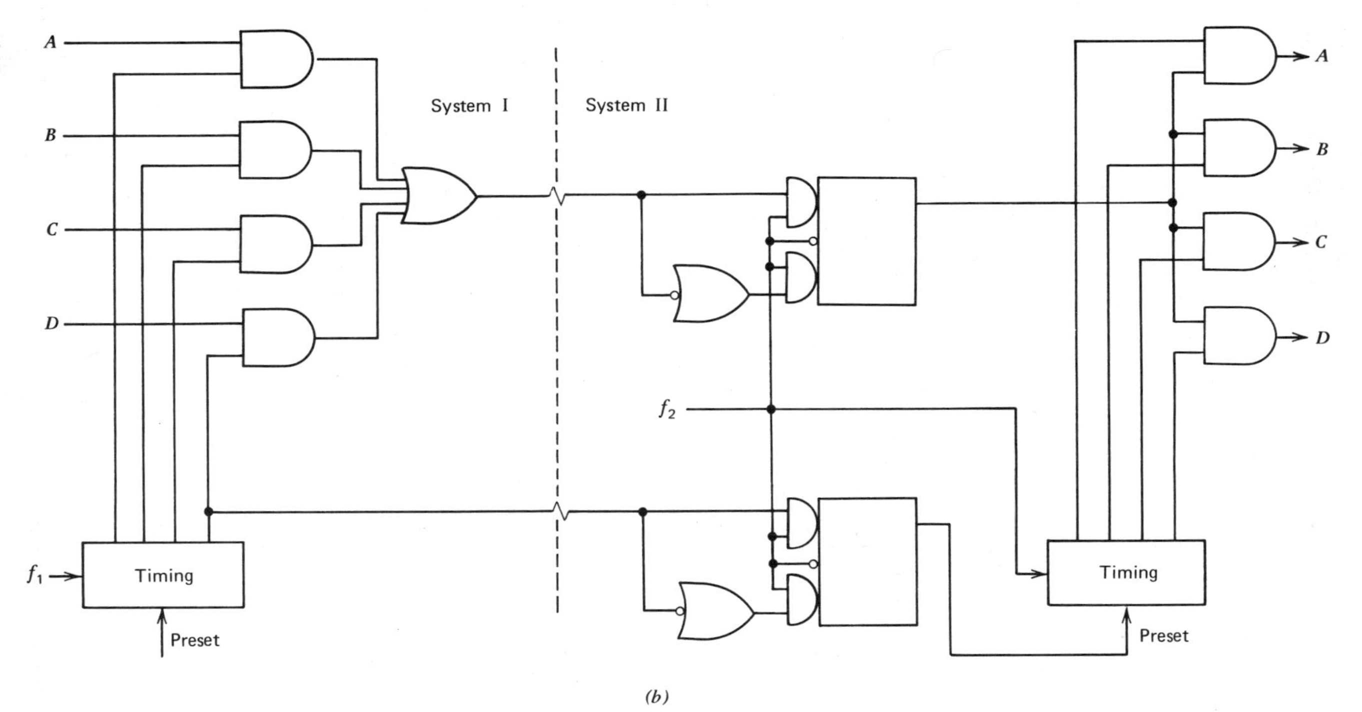

Figure 15.3. (*a*) Digital multiplexing. (*b*) Four-channel digital multiplexer. (*Note:* $f_1 = f_2$ = clock frequency.)

The enable signals of a multiplexing circuit, similar to that of Figure 15.3(*a*), can be outputs of a ring counter. In the ring counter all stages except one should be initially preset to 0. This way, as the clock counts the stages will sequentially take the value of 1, thus enabling the appropriate gate.

For example, if a seven-stage ring counter is used and is initially preset to 1000000, as the clock counts the counter output will change to

0100000
0010000
0001000
0000100
0000010
0000001

it will return to 1000000 and start all over again. If each output enables one of the AND gates of Figure 15.3, only one gate will be ON at a time.

System II should have an identical ring counter to provide the enable lines for the AND gates of that system. A synchronization line, however, should feed System II from System I. That signal will preset the System II ring counter keeping the two systems synchronized.

Preferably, the same clock line should feed both ring counters. If this is not possible, both counters should be clocked by the same frequency and a synchronization circuit similar to that of Figure 15.1 should interface the two systems.

The clock frequency of the two counters is determined by the resolution needed in the transfer of information from System I to System II. For infinite resolution the clock frequency should be infinite. The basic rule, however, in selecting that frequency should be that

$$\frac{f_{\text{clock}}}{n} \geq \frac{1}{PW_{\text{min}}} \tag{15.7}$$

where f_{clock} is the clock frequency that triggers the ring counters, n is the number of multiplexed digital inputs and PW_{min} is the minimum pulse width of the expected digital input signals.

The error involved in this type of multiplexing is in terms of pulse width variation, where the reproduced pulse in System II is longer or shorter than the original that was transmitted from System I. The equation of the error is

$$\text{Error} = \frac{n}{f_{\text{clock}}} \text{ sec} \tag{15.8}$$

where n is the number of multiplexed digital inputs and f_{clock} is the clock frequency used in the multiplexing.

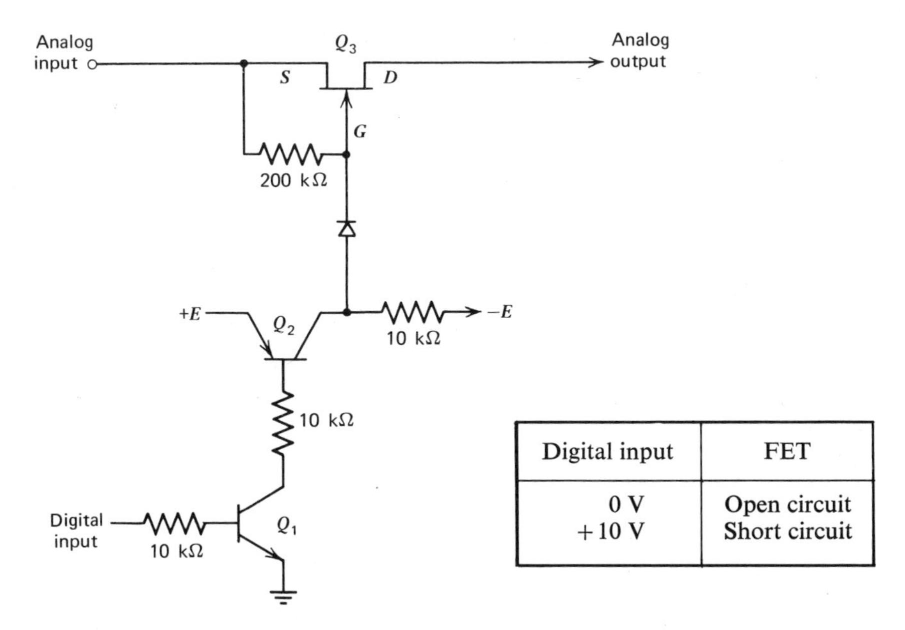

Digital input	FET
0 V	Open circuit
+10 V	Short circuit

Figure 15.4. Digitally controlled analog signal switch.

Figure 15.3(*b*) illustrates a digital multiplexing configuration where each system has a separate clock. In this multiplexing circuit, where there are four digital channels and the clock frequency is 1 MHz, the pulse width error will be

$$\text{Error} = \frac{4}{1\ \text{MHz}} = 4\ \mu\text{sec}$$

Therefore, for minimal signal reproduction the minimum pulse width of the multiplexed inputs should not be less than 4 μsec.

The same concept of multiplexing can be equally extended to the multiplexing of analog signals. In this case the AND gates of Figure 15.3(*a*) that control the selection of the source will be replaced by analog gates, often called transmission gates. Figure 15.4 illustrates a typical configuration of analog gate.

15.5 COMPUTATION OF PARITY OF DIGITAL WORDS

Parity is a term pertaining to the number of 1s in a digital word, where the digital word may or may not be a binary number. The parity of a digital word indicates whether the number of 1s in the word is even or odd, and it is respectively called *even* and *odd parity*. The parity is computed at the point where the digital word is received so that the validity of the transmission be verified.

Sequential Transmission

When the digital word is in sequential form, an initially cleared trigger flip-flop can be used for the computation of the parity, as shown in Figure 15.5. In this circuit, for every 1 of the given digital word, the flip-flop will change state once. If the flip-flop is initially 0 and the number of 1s in the sequential word is even, the flip-flop will change state for an even number of times resulting in 0. If the number of 1s in the sequential word is odd, the flip-flop will change state for an odd number of times resulting in 1.

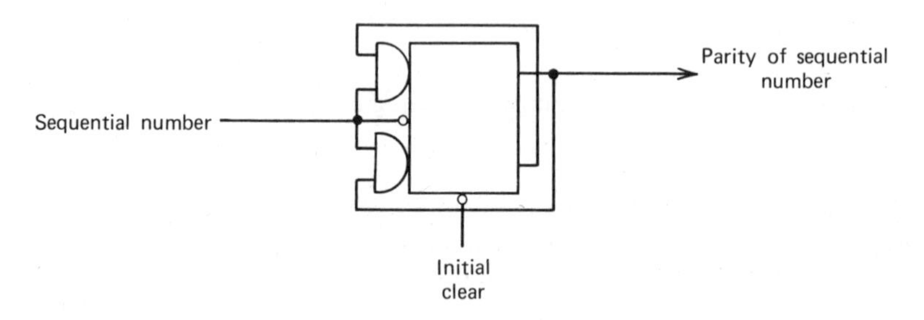

Figure 15.5. Parity generator of sequential numbers.

Parallel Transmission

If the digital word is available in parallel form, the LSB of the sum of the individual bits of the word is the parity of that word. For example, if the word is

$$1011101$$

we compute the sum of all bits, which is

$$1 + 0 + 1 + 1 + 1 + 0 + 1 = 5$$

or, in binary

$$101$$

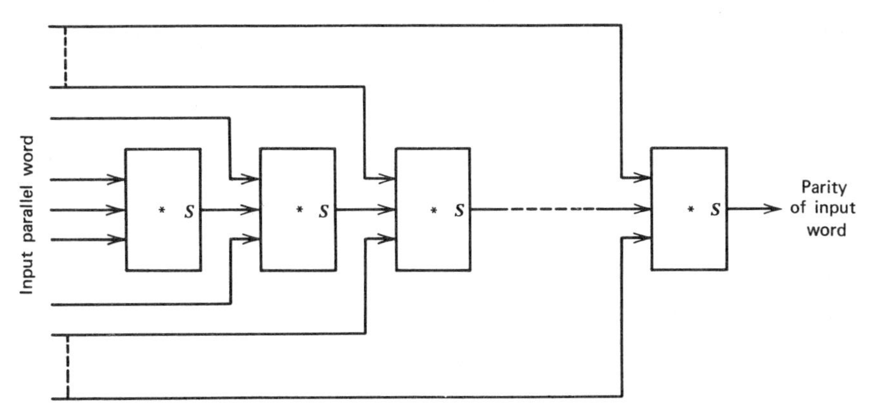

Figure 15.6. Parity generator of parallel numbers. (*Note:* The order of input bits is immaterial. *Binary adder producing only sum output.)

The parity of the given word is the LSB of the sum, which is 1 indicating that the number of 1 bits in the given digital word is an odd number. A parallel parity generator is shown in Figure 15.6. It consists of a chain of binary adders where the order in which the bits are added is immaterial.

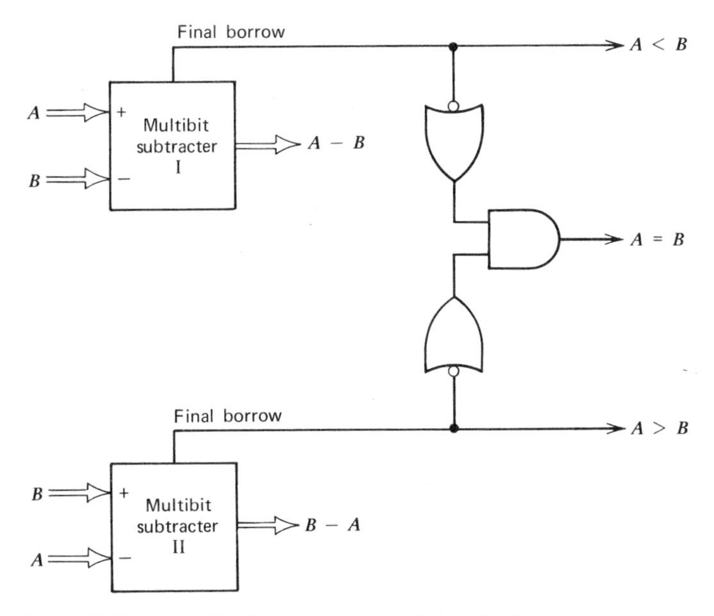

Figure 15.7. Amplitude comparison of parallel binary numbers providing the relationship and actual differences.

15.6 COMPARISON OF PARALLEL BINARY NUMBERS

In the design of digital systems, it is often required to compare two binary numbers and to determine whether they are equal or, if they are not, to determine which of the two is greater. This comparison can be easily made using two multibit subtracters as shown in Figure 15.7.

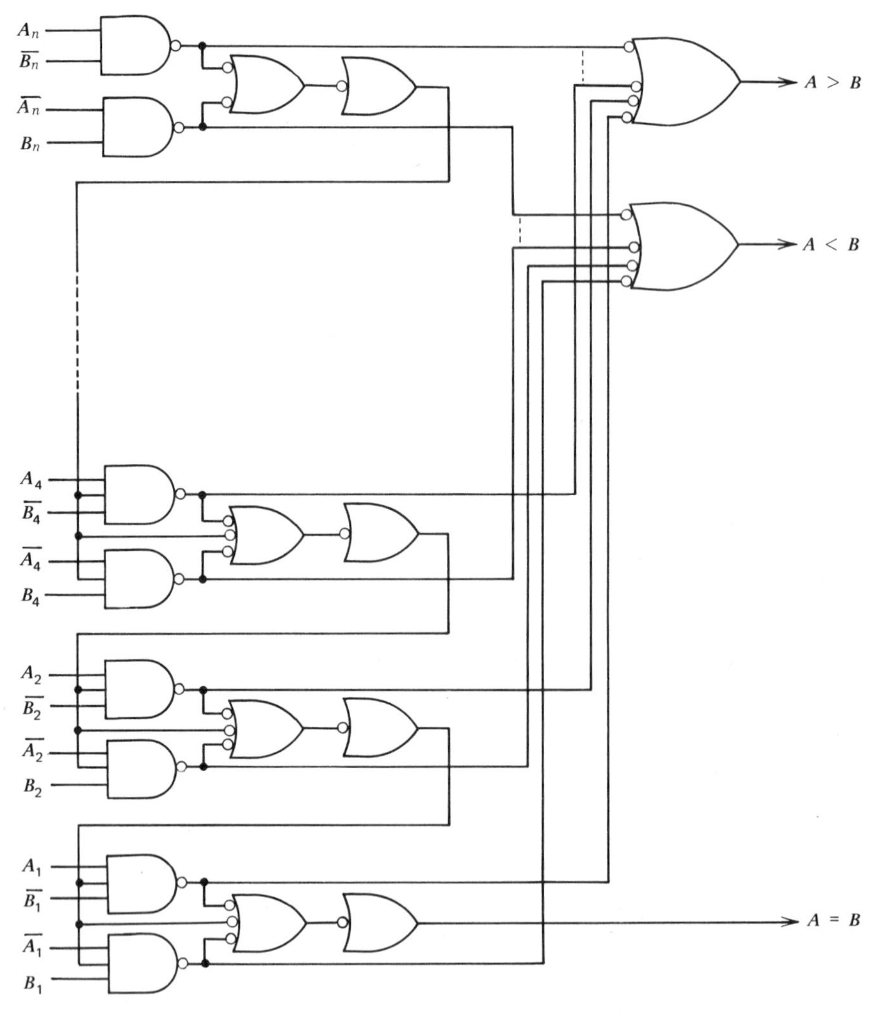

Figure 15.8. Amplitude comparison of parallel binary numbers providing the relationship only.

In a subtracter, a final borrow of 1 indicates that the minuend is less than the subtrahend, and a borrow of 0 indicates that the minuend is greater than or equal to the subtrahend. The two subtracters in Figure 15.7 produce the following information:

Multibit subtracter I $(A - B)$	final borrow $= 1$	$A < B$
	$= 0$	$A \geq B$
Multibit subtracter II $(B - A)$	final borrow $= 1$	$A > B$
	$= 0$	$A \leq B$

From these four relationships we see that the inequality signal can be directly obtained, while the equality signal is obtained by ANDing the inverse of the two inequality signals. The same result can be obtained using the circuit of Figure 15.8. The advantage of this circuit is less hardware, while the disadvantage is that it does not provide the differences $A - B$ and $B - A$. The specific application will determine the selection between these two circuits.

15.7 COMPARISON OF SEQUENTIAL BINARY NUMBERS

When the binary numbers under comparison are in sequential form, the networks of Figures 15.9 or 15.10 can be used.

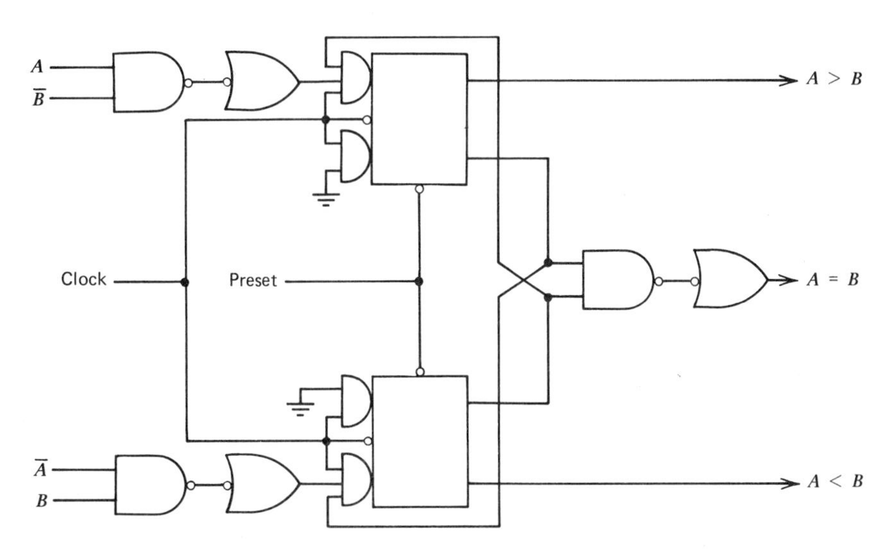

Figure 15.9. Amplitude comparator of sequential binary numbers; MSB enters first.

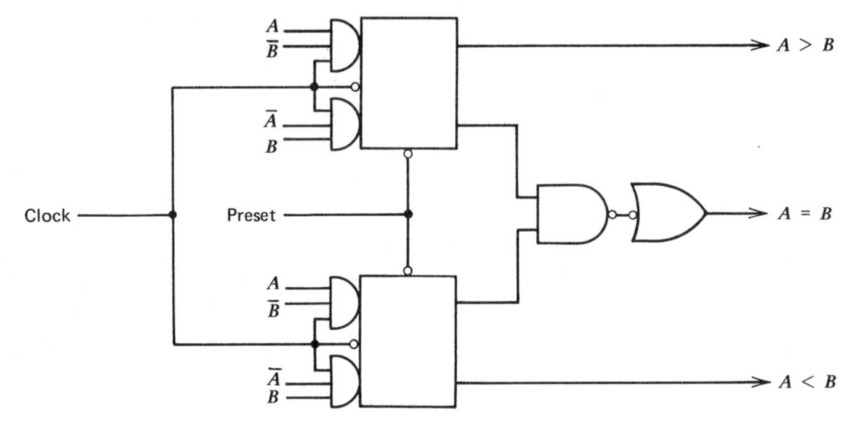

Figure 15.10. Amplitude comparator of sequential binary numbers; LSB enters arst.

Most Significant Bit First

In the network of Figure 15.9 information A, $\bar{A}$, B, and $\bar{B}$ must be available at the input of the circuit sequentially with the MSB appearing first—that is, if A and B are two 6 bit numbers, their bits of information should appear in the following order:

$A_{32}B_{32}$	first
$A_{16}B_{16}$	second
$A_8\ B_8$	third
$A_4\ B_4$	fourth
$A_2\ B_2$	fifth
$A_1\ B_1$	sixth

The network will compare the A and B bits, starting with the MSB continuing down the line.

As soon as A differs from B, one of the two flip-flops will detect the inequality, and upon arrival of the clock that flip-flop will change state, giving out a HIGH output. The inverse of that output will lock the other flip-flop into its initial state.

If the two numbers are equal, the inequality outputs will remain LOW and the equality output will keep its initial HIGH value until all bits have paraded through the inputs of the comparator.

For example, if $A = 000101011$ and $B = 000011011$, the comparator circuit will show an equality until the third bit. After that, the circuit will show that A is greater than B. The $A > B$ output will be HIGH and the other two outputs will be LOW. The outputs will remain in these states until a new start-pulse arrives.

Least Significant Bit First

Figure 15.10 shows a comparator for sequential numbers available with their LSB first. The circuit is initially preset to the state that indicates equality.

As the bits of the two numbers arrive at the proper inputs of the circuit, a bit-by-bit comparison is made. The two outputs $A > B$ and $A < B$ may change states several times before the entire numbers have gone through the comparator circuit, and the magnitude relationship that exists between the two numbers will be determined by the last bit inequality observed by the comparator.

For example, if $A = 00101100$ and $B = 00110010$, during the comparison the comparator output will vary as follows:

	Inputs		Comparator outputs				Clock period
	A	B	$A > B$	$A = B$	$A < B$		
LSB	0	0	LOW	HIGH	LOW		1
	0	1	LOW	HIGH	LOW		2
	1	0	LOW	LOW	HIGH		3
	1	0	HIGH	LOW	LOW		4
	0	1	HIGH	LOW	LOW		5
	1	1	LOW	LOW	HIGH		6
	0	0	LOW	LOW	HIGH		7
	0	0	LOW	LOW	HIGH		8
			LOW	LOW	HIGH	Final	

The outputs are delayed by one clock period because the flip-flops act as shift register units.

15.8 DIGITAL-ANALOG MULTIPLICATION

An analog signal can be multiplied by a digital binary number by means of a D/A converter of variable analog reference. Such a network has the property of converting digital-binary numbers to proportionate analog signals normalized to a given analog voltage level.

If the analog voltage level is the analog operand and the converted

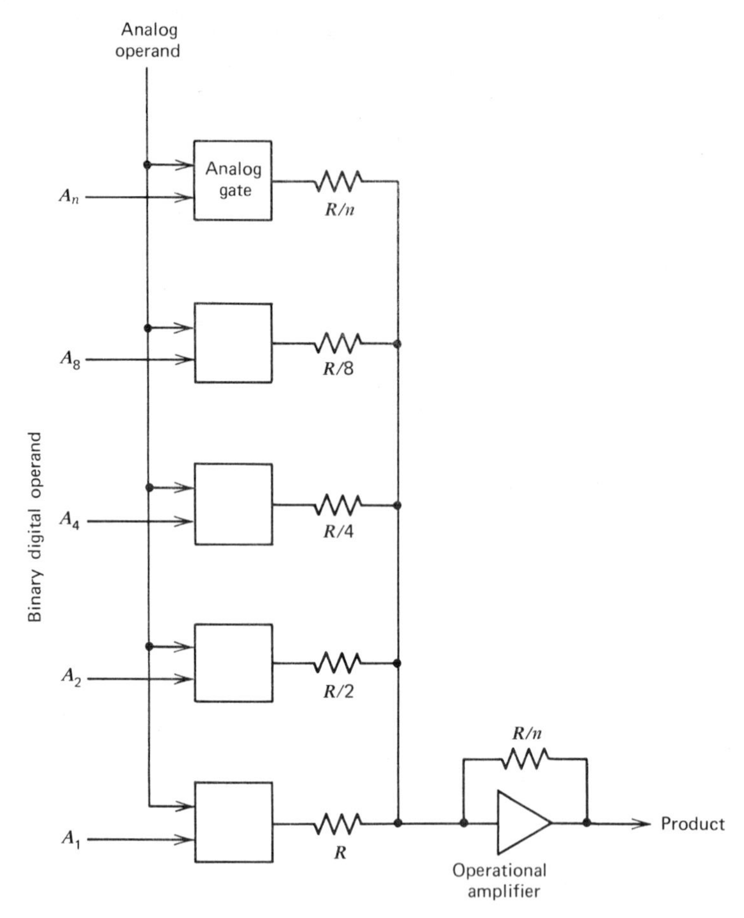

Figure 15.11. Digital-analog multiplier. The analog gate's schematic appears in figure 15.4.

digital-binary number is the digital operand of the multiplication, the output of this D/A converter, which is in analog form, represents the product of the two operands. Figure 15.11 illustrates a D/A multiplier, the operation of which is represented by the following equations.

$$\text{Product} = \text{Analog input} \cdot \text{digital input}$$

$$= \text{Analog input} \cdot \left[\frac{A_n}{R/n} + \cdots + \frac{A_8}{R/8} + \frac{A_4}{R/4} + \frac{A_2}{R/2} + \frac{A_1}{R/1}\right](R/n)$$

$$= \text{Analog input} \cdot \frac{1}{n}(nA_n + \cdots + 8A_8 + 4A_4 + 2A_2 + 1A_1)$$

where the product is the output of the network of Figure 15.11, the analog input is the analog operand of the multiplication, and the binary number $A_n \ldots A_8A_4A_2A_1$ is the digital operand of the multiplication. In this multiplication the digital input indicates the number of analog units the output should consist of while the analog input determines the size of these increments.

15.9 GENERATION OF RANDOM FUNCTIONS

The availability of noise sources makes the design of random digital-function generators a simple task. The noise source may be as plain as the circuit arrangement of Figure 15.12. Using this circuit as a building block, a number of digital generators providing random functions can be designed. Described below are three such generators.

Random Frequency—Random Pulse Width

If the output of the random noise generator of Figure 15.12 is applied to the clock input of a master-slave flip-flop connected in a divide-by-2 configuration, the output of the flip-flop will be a digital waveform of pulses randomly spaced and of random width. Figure 15.13 illustrates this design and the corresponding waveforms of operation.

Random Frequency—Fixed Pulse Width

If the output of the random noise generator is applied to the input of a monostable device, the generated output will have a random repetition rate, while the pulse width will be fixed and a function of the selected monostable device. This configuration is shown in Figure 15.14.

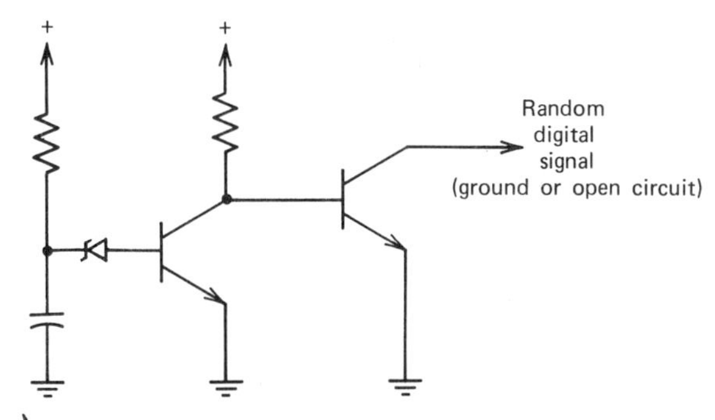

Figure 15.12. Random noise generator.

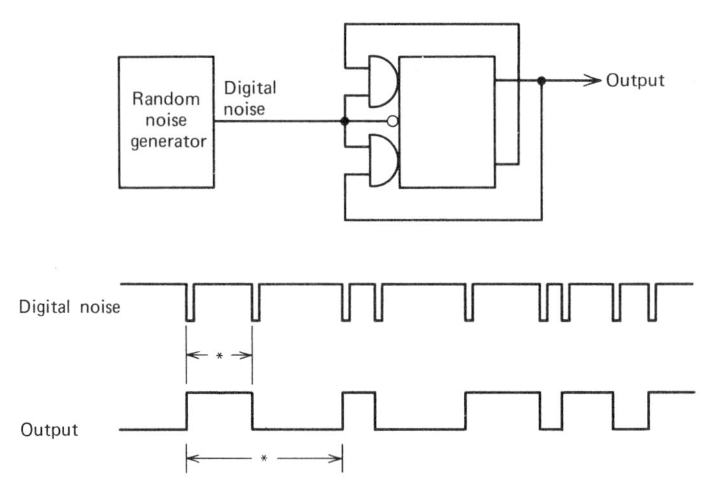

Figure 15.13. Pulse generator of random pulse repetition rate and pulse width. (*Note:* *Random time length.)

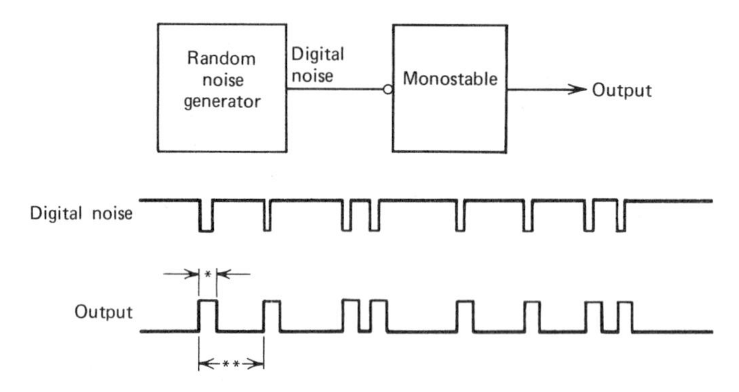

Figure 15.14. Pulse generator of random pulse repetition rate and fixed-pulse width. (*Note:* *Fixed. **Random.)

Constant Frequency—Random Pulse Width

The random noise generator of Figure 15.12 can be also used in the generation of a digital waveform where the pulse repetition rate is fixed and the pulse width is random. This is accomplished by appplying the output of the random noise generator to the clear input of a flip-flop while the set input is connected to the output of a constant frequency generator. Figure 15.15 illustrates this configuration and its waveforms of operation.

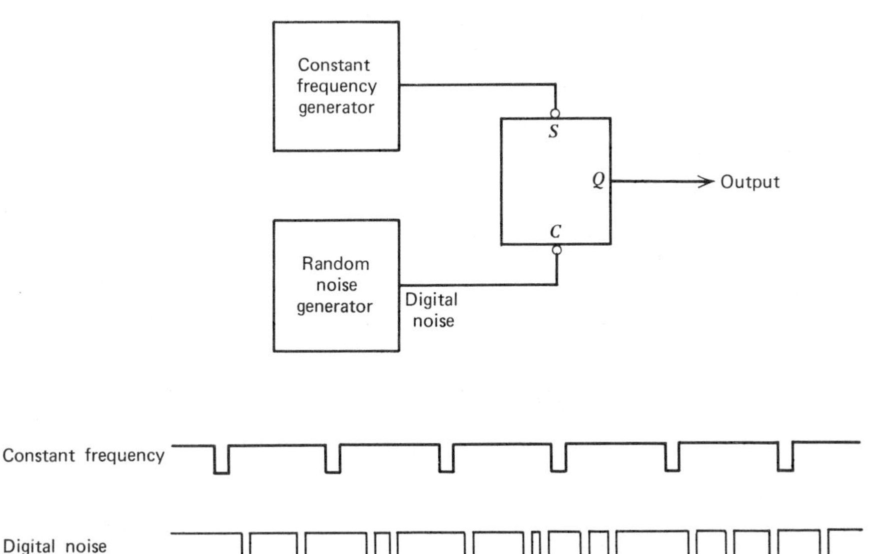

Figure 15.15. Pulse generator of fixed pulse repetition rate and of random pulse width.

15.10 GENERATION OF FREQUENCY-MODULATED SLIDE

The FM slide is a function the frequency of which varies linearly with time. It is a time function that can be generated using two binary counters and a multibit comparator. Figure 15.16 shows the approach used in block diagram form.

A "start operation" signal presets the Y counter to 0...000 and the X counter to 0...001. Upon removal of the start signal, the Y counter starts counting up. At the arrival of the first clock pulse, the number in the Y counter becomes 0...001, it is compared with the number in the X counter which is already 0...001, and a comparison signal is generated by the comparator.

This signal, which is the output of a monostable circuit, clears the Y counter and triggers the X counter. The number in the Y counter is 0...000 while that in the X counter has become 0...010. The next two clock pulses will trigger the Y counter, making its output 0...010. Now both counters will

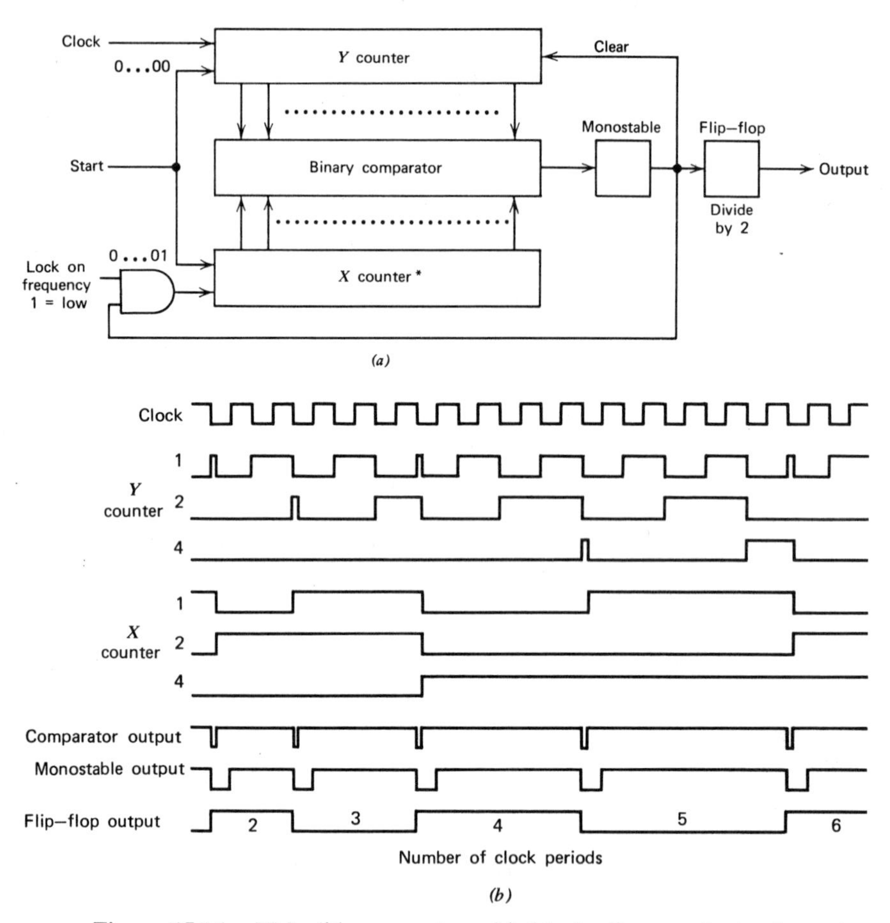

Figure 15.16. FM slide generator: (*a*) block diagram (*note:* *counter counts from 0...01 to 1...11 and returns to 0...01); (*b*) timing diagram for the first 16 clock pulses.

have the same output and the comparator will produce a signal which will clear the Y counter and increment the X counter.

This operation continues as illustrated by the waveforms of Figure 15.16. From these waveforms we can see that the frequency of the signal of the final output is a linear function of time and it is decreasing. The initial frequency of the output signal is

$$f_{\text{OUT}}\,(\text{initial}) = \tfrac{1}{2} f_{\text{clock}}$$

$$= \frac{1}{2\,(\text{number in } X \text{ counter})} f_{\text{clock}} \qquad (15.9)$$

After the first comparison of the content of the two counters the output frequency becomes

$$f_{\mathrm{OUT}} = \frac{1}{2\,(2)} f_{\mathrm{clock}} \tag{15.10}$$

After the second comparison the frequency of the output will be

$$f_{\mathrm{OUT}} = \frac{1}{2\,(3)} f_{\mathrm{clock}} \tag{15.11}$$

And the final frequency of the output will be

$$f_{\mathrm{OUT}}\,(\text{final}) = \frac{f_{\mathrm{clock}}}{2(\text{maximum number in the } X \text{ counter})} \tag{15.12}$$

Expression (15.12) can be simplified as follows:

$$f_{\mathrm{OUT}} = \frac{f_{\mathrm{clock}}}{2(2^n - 1)}$$

$$= \frac{f_{\mathrm{clock}}}{2^{n+1} - 2}$$

where n is the number of stages of the X counter. The number of stages of the Y counter should be the same as that of the X counter. The output waveform of the circuit of Figure 15.16 dwells at each frequency step for only half a period of that frequency. To stay at the various frequency steps longer, a frequency divider should be placed between the output of the monostable and the clock input of the X counter, as shown in Figure 15.17.

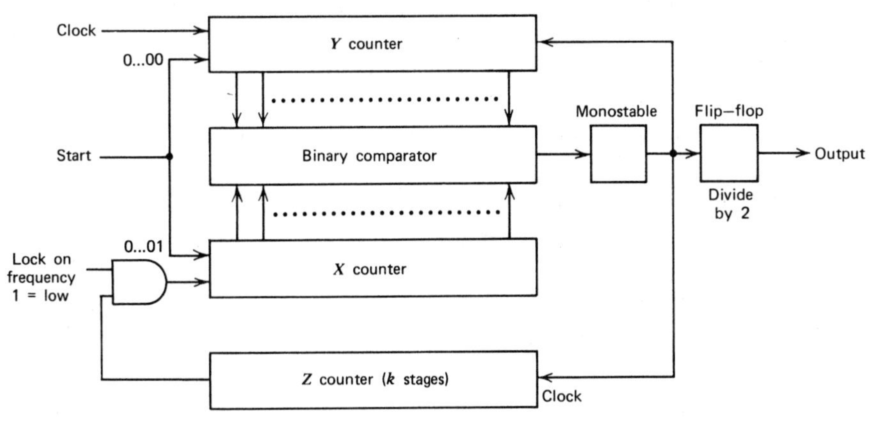

Figure 15.17. FM slide generator dwelling at each frequency for 2^{k-1} clock periods.

Then, the number of periods for which the output will dwell at each frequency step will equal 2^{k-1} where k is the number of flip-flops of the additional counter.

The frequency of the output waveform can be made to increase by using a counting-down X counter which is preset to 111...11. Then the initial frequency will be defined by (15.12) and the final frequency by (15.9).

PROBLEMS

1. Design a 10-channel digital multiplexing circuit. Include control units.
2. Determine the error in a 6-channel digital multiplexer where $f_{\text{clock}} = 3$ MHz.
3. A 20 kHz signal is to be delayed for 47 μsec by means of a shift register. The total pulse width variation of the delayed signal should not exceed 4 microseconds. Draw the circuit in block diagram form and determine the number of stages and the clock frequency required.
4. Design a 6-channel analog multiplexing circuit for the conversion of a single probe oscilloscope to a six-probe oscillosope. Two of the channels should handle digital information and the other four should handle audio frequencies. The design should be at logic component level.
5. Compute the parity of the following digital words:

   (*a*) 101101 (*b*) 100010
   (*c*) 010110 (*d*) 000001

6. Design a 6 bit comparator for parallel binary numbers A and B. The comparator should determine $A > B$, $A < B$, and $A = B$. Show actual logic components, using NAND/NOR gates.
7. Design an 8 bit comparator for parallel numbers A and B. The comparator must have a single output indicating that the absolute difference between A and B is less than 8.
8. Design a digital random number generator updated every 1 msec producing two 8 bit parallel numbers.
9. Draw the timing diagram of operation for an FM slide generator for the first 31 clock periods since it was preset to Y counter = 0...00, X counter = 0...01. Each counter is counting up and has 5 bits.
10. Design an FM slide generator meeting the following characteristics:
    Frequency period range: 0.1 to 2 milliseconds
    Frequency period steps: 0.01 milliseconds
    Dwell at each step: 10 milliseconds

CHAPTER SIXTEEN

Digital Memories

16.1 INTRODUCTION

A very important part in any advanced digital system is the memory. Similar to human memory, digital memory must store information and be able to retrieve it in relatively short periods of time. Digital memory systems can be classified into the following three general categories:

1. Semiconductor memories.
2. Magnetic memories.
3. Mechanical memories.

Figure 16.1(*a*) illustrates the general representation of a memory system. Excluding the power connections, a memory system can be considered as having four ports. These are the input data terminals through which new data enter the memory, the control terminals for the entry of the appropriate read or write command, the output data terminals from where the selected data are available, and the address terminals through which a digital code is applied identifying the location of the retrieved or entering data.

Memory concepts are subdivided into random-access memories and sequential memories.

In a random-access memory, the access time of stored data is fixed and independent of data location. Random-access memories, often called direct-access memories, are word-oriented—that is, retrieval or entry of data is accomplished one word at a time where each transfer command, read or write, is accompanied by a separate address.

In sequential memories, data motion in one form or another takes place. This motion of data is accomplished either by moving the medium in which the data are stored or by moving the data themselves through the medium. Consequently, access time is variable and it depends on the position of the retrieved data relative to the reading network at the moment of the request.

Sequential memories are block-oriented in which case groups of words are entered into or retrieved from consecutive memory locations. The reason for this approach is that their relatively long access time makes sequential memories impractical for single-word data transfers.

16.2 SEMICONDUCTOR MEMORIES

In recent years semiconductor memories have found a wide acceptance in special-purpose digital systems as well as in commercial computers. Semiconductor memories offer speed and simplicity at the expense of cost,

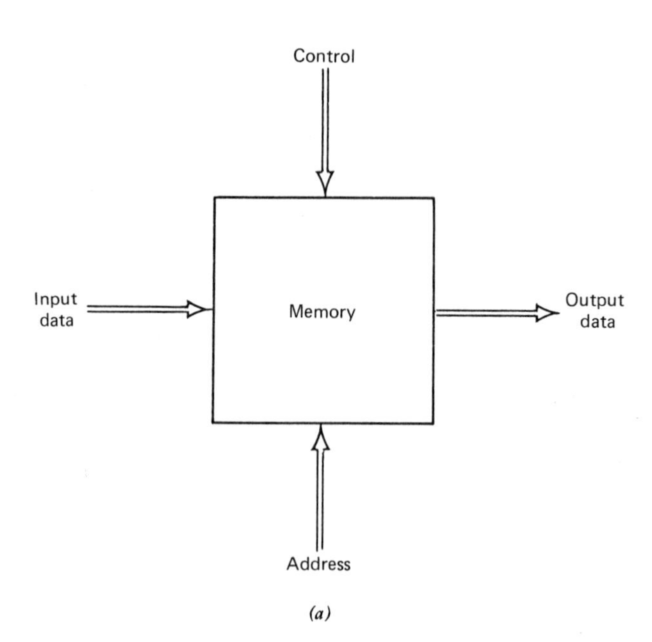

(a)

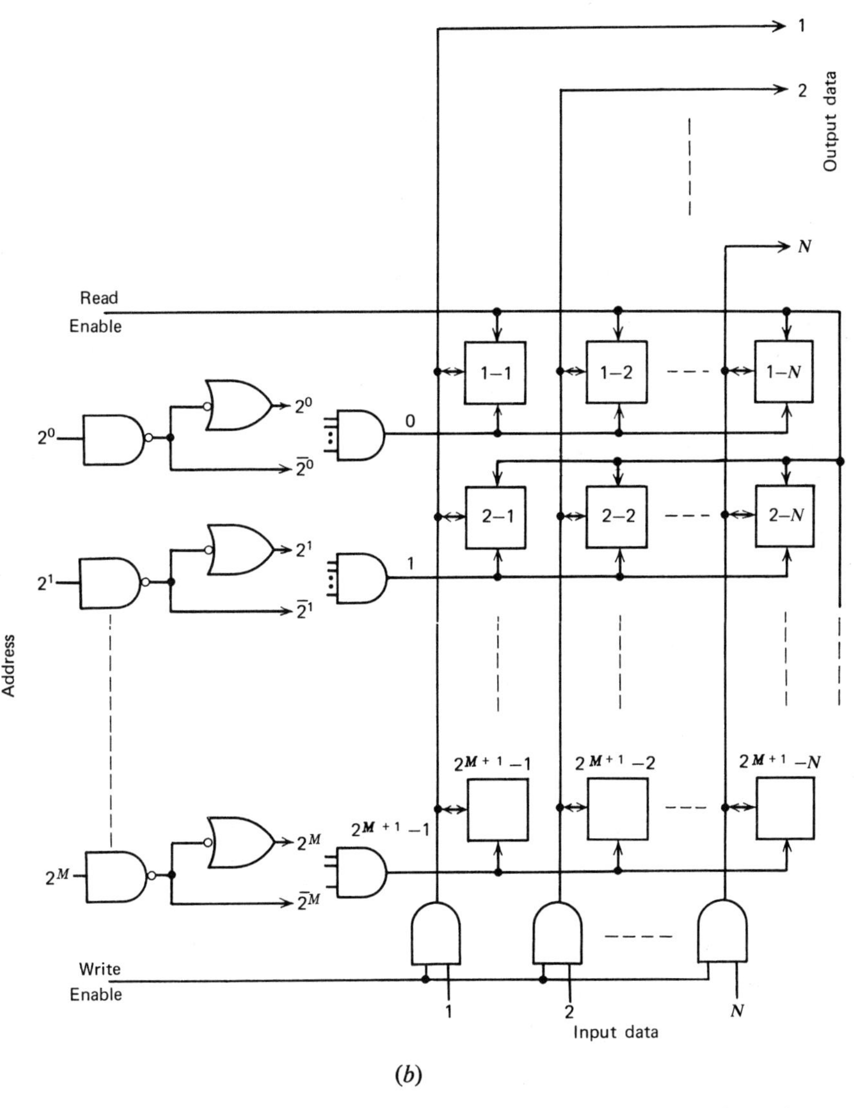

(*b*)

Figure 16.1. Simplified diagrams of (*a*) memory system; and (*b*) random-access memory.

power, and volatility of data. In this section the various types of semiconductor memories are discussed and a comparison of their relative advantages and disadvantages is illustrated at the end of the chapter.

Random-Access Memory (RAM)

The term RAM pertains to semiconductor memories where data can be read out as well as written in and where access time is independent of data location in memory.

The RAMs are available in bipolar or in MOS form. These two terms refer to the fabrication method, where bipolar memories offer high speed and low density, while MOS memories offer low speed and high density.

Bipolar memories interface directly with DTL and TTL logic, but MOS memories need special-level translation logic to interface with DTL or TTL logic networks. Normally, such translations are performed inside the MOS memories and no external circuitry for this purpose is needed.

Figure 16.1(*b*) illustrates the simplified diagram of an $(M + 1) \times N$ RAM. This device consists of 2^{M+1} rows and N columns of memory cells, and it is arranged in 2^{M+1} words of N bits each. The address of the referenced word

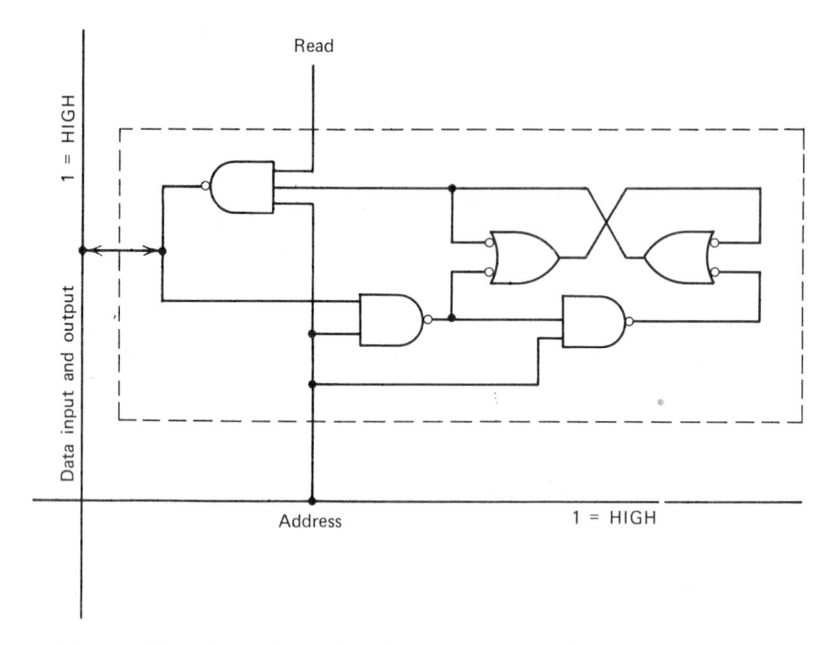

Figure 16.2. Equivalent of an RAM memory cell.

is applied in binary form at the address input of the RAM enabling the data gates of all cells of the corresponding row.

Subsequent application of the write-enable control input will enter the word at the input data into the selected row of cells. Application of the read-enable input will make the content of the selected row available at the output data.

The logical equivalent of an RAM memory cell is illustrated in Figure 16.2. In this circuit when the address line is 1, the memory cell is open to change while at the same time has its output available at the data line.

The MOS RAMs are classified into two categories, static and dynamic. Static MOS RAMs operate in the same manner as bipolar RAMs, as were illustrated in the above discussion. Dynamic MOS RAMs have the same general block structure as the static ones with the exception that their cells maintain the entered logic state, not by means of a flip-flop, but by means of a charge-holding capacitance.

Such a concept necessitates periodic refreshment of the stored charge and the appropriate circuitry to perform this process. Figure 16.3 illustrates a simplified diagram of a 1024 bit dynamic memory, and Figure 16.4 shows a dynamic MOS memory cell.

To prevent the charge representing the logic value to discharge below its usable level, the logic level is read during clock ϕ_1 and written back into the cell during clock ϕ_2. This process must continue indefinitely if the initially entered logical state is to be maintained in the cell.

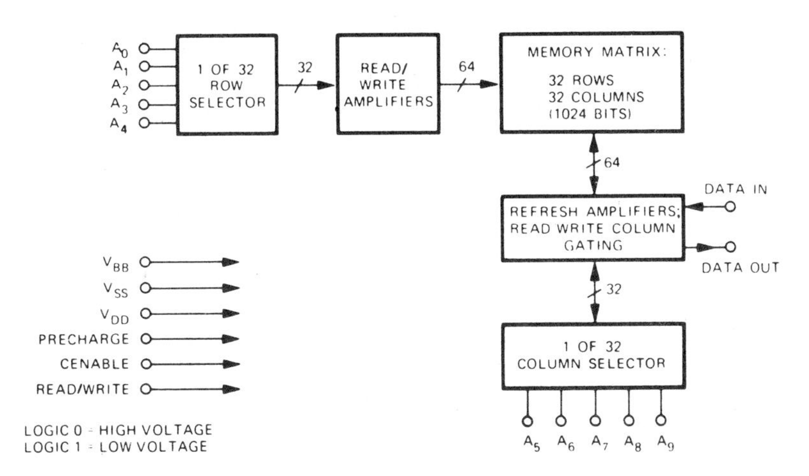

Figure 16.3. Fully decoded random-access 1024 bit dynamic memory. (Courtesy of Intel).

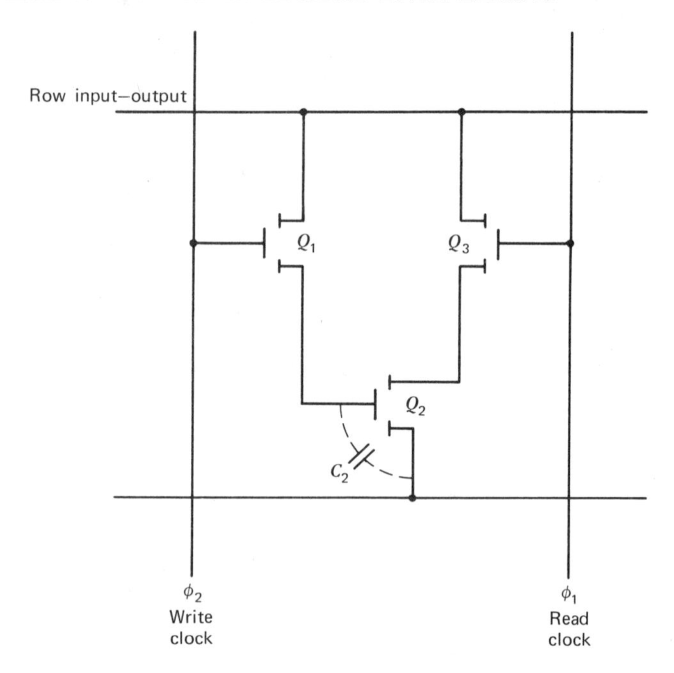

Figure 16.4. Dynamic MOS memory cell.

Dynamic MOS RAMs are simpler than their static counterparts and as a result they always offer higher densities and less power requirements.

Read-Only Memory (ROM)

The ROMs are memory networks that contain information which can be read but not altered. Data are arranged in words and are permanently entered into the memory during fabrication of the device; ROMs find extensive use in applications where a fixed input-to-output relationship exists such as look-up tables.

Figure 16.5 shows the block diagram of a typical ROM. This device contains 512 words of 8 bits each where each word location is defined by a 9 bit address.

Certain types of read-only memories are electrically programmable by the user, and are called programmable ROMs or PROMs. These devices are fabricated with all memory cells in the 0 state where any of the cells may have their state changed to 1 by means of a simple fusing procedure; PROMs find wide use in applications where small quantities of ROMs are required. Figure 16.6 illustrates the block diagram of a field PROM.

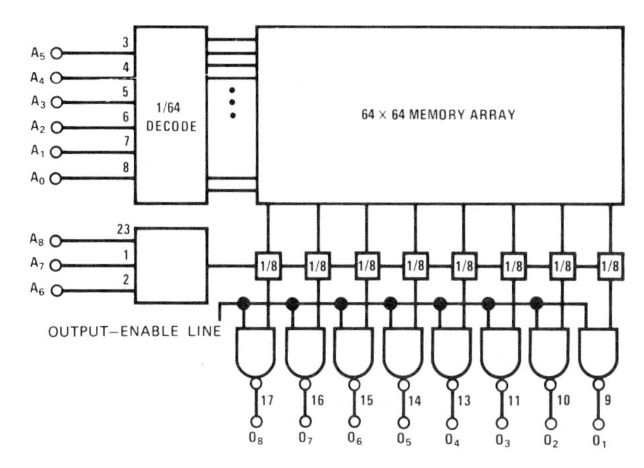

Figure 16.5. A 4096 bit bipolar ROM. (Courtesy of National Semiconductor.)

Some ROMs are even erasable. In such devices data can be preprogrammed and at a later time erased to be preprogrammed again. The packages of these devices have a transparent lid which allows the chip to be exposed to ultraviolet light which erases the preprogrammed bit pattern.

These devices are ideal for experimental work and they have a *pin*-to-*pin* counterpart that is factory-programmable. This way, at the end of the experimentation the user may order ROMs with his final bit arrangement permanently programmed.

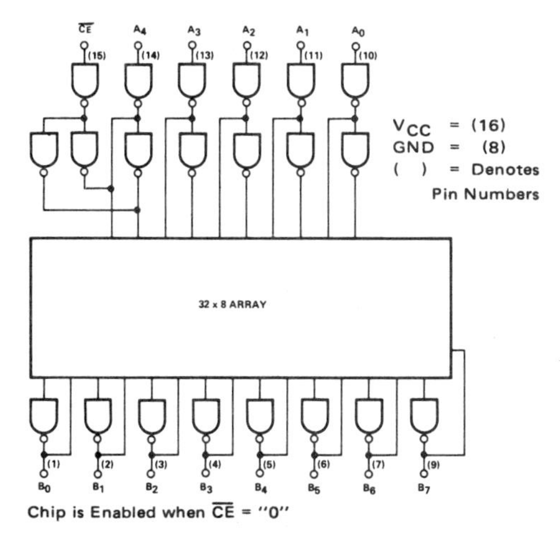

Figure 16.6. A 256 bit bipolar field PROM. (Courtesy of Signetics.)

Content-Addressed Memory (CAM)

Of special importance is a type of memory, the contents of which are addressed directly and not through a memory location address. That is, to determine if a certain word has been stored in memory, the reference word is applied to the data input of the device, and coincidence lines, one per memory location, indicate presence of the reference word in memory. The same device is also a read/write RAM. Figure 16.7 illustrates the logical diagram of a 4(words)-by-4(bits) CAM. This device operates in three modes. These are write, read, and compare.

In the write mode the input enable lines ($E_0 - E_3$) are in the 1 state allowing input data ($D_0 - D_3$) to be applied to the input gates of all memory word-cells. Memory word-cells are vertically arranged. With the write-enable (WE) line in the 1 state, upon arrival of an address signal, the input data will be stored in the word-cell that corresponds to the enabled address line A. The data will remain in this word-cell until another entry is made in that cell.

In the read mode the only necessary action is the application of a 1 to the appropriate address line. The content of the word-cell that corresponds to that address line will then become available at the output ($O_0 - O_3$).

In the compare mode, which is the most significant one, the reference data are applied to the data input ($D_0 - D_3$) and upon application of the enable signals the content of each word-cell is compared to the reference data. Application of the enable signals may be selective. In such case comparison takes place only for the bits with an enable line of 1. Figure 16.8 illustrates the logical diagram of a single cell of a CAM. When the enable signal is 1, the input data become available at the input of gates 3 and 4, and 7 and 8. If the wired-OR output of gates 7 and 8 is HIGH, it will indicate that the stored bit and the reference bit applied at the input are the same.

If the address signal is HIGH, it will enable gate 9 and the storage output from gate 6 will become available at the output data terminal. If, in addition to the address signal, the write-enable is also HIGH, gates 3 and 4 will be enabled allowing the output of gates 1 and 2 to be applied to the storage element that consists of gates 5 and 6.

Content-addressable memories are very useful devices and understanding their operation and proper use may lead to the design of systems that would otherwise require a considerable amount of hardware and time.

Cyclomemory—A Sequentially Accessed Memory*

Advances in MOS technology, especially in the area of dynamic shift registers, have opened up the road to the design of storage equipment similar to the

* The material in this section originally appeared by the author in *Digital Design*, June 1973, pp. 34–37.

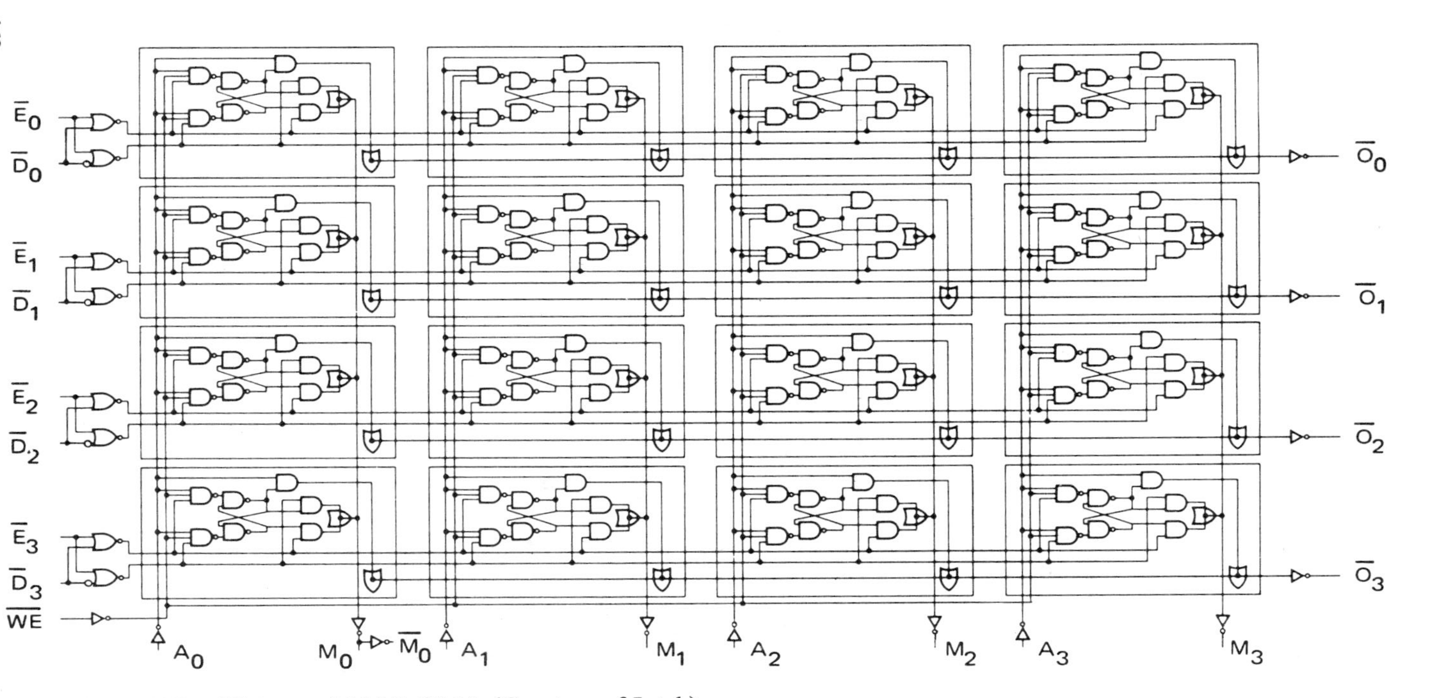

Figure 16.7. High-speed 16 bit CAM. (Courtesy of Intel.)

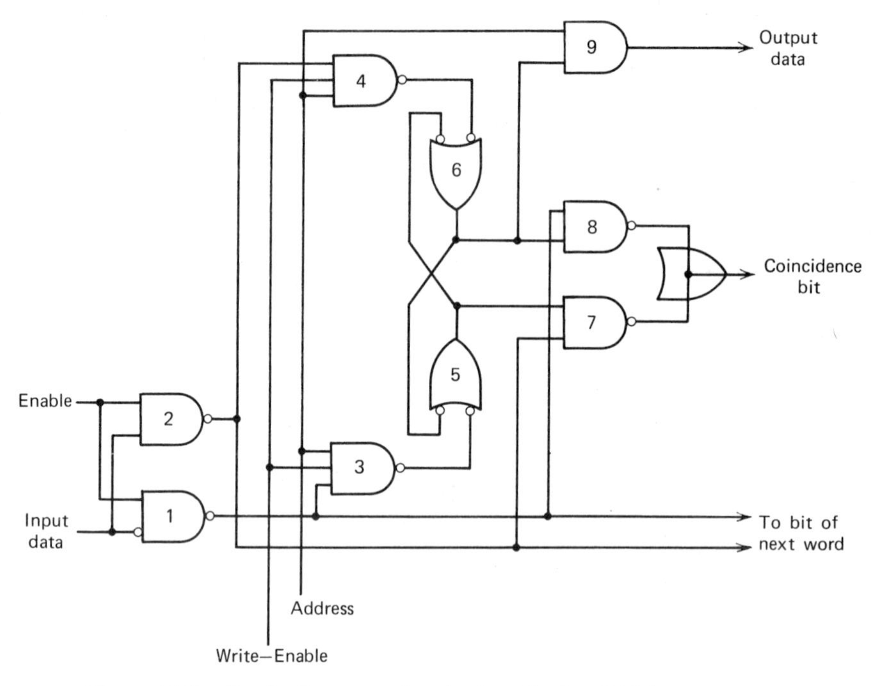

Figure 16.8. Logic diagram of a single cell of a CAM.

magnetic disk but of totally solid-state design. The cyclomemory, considerably faster than its magnetic counterpart, offers features not possible in magnetic systems. Its name comes from its operation in which sorted data move in a rotary fashion.

The cyclomemory consists of a number of closed-loop shift registers with single or multiple I/O ports. The solid-state counterpart of the magnetic

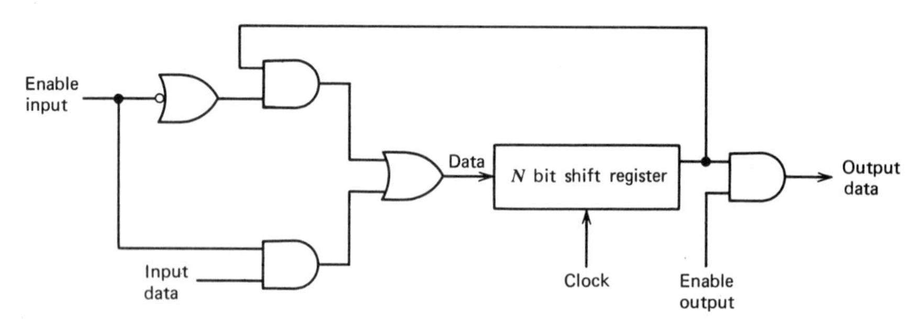

Figure 16.9. Closed-loop shift register with one I/O port simulating a single disk-track and magnetic head. (Courtesy of *Digital Design*.)

track, the register, uses the I/O ports as solid-state counterparts of the magnetic head. Figure 16.9 illustrates a closed-loop shift register having one I/O port. Data present in the shift register circulate in a continuous unidirectional motion at high clock rates, possibly reaching 20 MHz or more. The I/O ports serve as points of entry for new data that should be available at the loop's clock rate. From the I/O ports, data present in the loop can be read off at the loop's clock rate.

The continuous unidirectional circular motion of data in the loop is synchronized by means of a binary counter which explicitly identifies the data locations by a number of $\log_2 N$ bits where N is the bit length of the closed-loop shift register.

A single counter can provide synchronization for any number of parallel loops. In this case, if loops hold individual bits of words, the counter output serves as a memory location address holder, where the words are available in parallel and synchronously rotate in m loops (m is the word size).

Figure 16.10 illustrates a single-port, 16 loop cylcomemory where each closed-loop shift register stores 8192 (8K) bits. This device may very well serve for the storage of 8K 16 bit words where the synchronization counter holds the address of the word currently available at the I/O port.

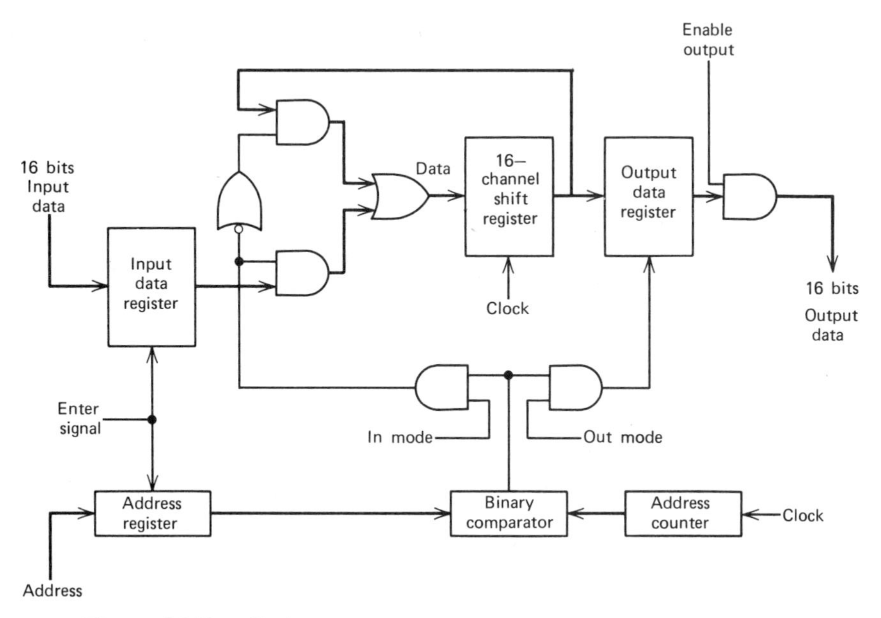

Figure 16.10. Cyclomemory holding 8K 16 bit words; for simplicity, separate data input and output channels are shown. (Courtesy of *Digital Design.*)

The maximum access time of this 8K storage device is the product of the clock frequency period and the loop bit size. If the frequency is 20 MHz, the maximum access time will be

$$AT_{\max} = \frac{1}{20\text{ MHz}}\,8192 = 409.6\ \mu\text{sec}$$

This level of performance surpasses that of the magnetic disk by 4 to 1 and offers single-word retrieval rather than the block retrieval normally provided by disk systems. Being totally solid-state, the cyclomemory exhibits high reliability and unlimited life cycle. In addition, its operation is not restricted by environmental factors that limit the performance of electromechanical devices.

Figure 16.11 illustrates a sixteen-channel cyclomemory and its basic controls, the operation of which is as follows: The operator first initializes the device. This is accomplished by sending to the device an appropriate command which causes N consecutive words of 0 value to enter the closed-loop shift register. This command also initializes the address counter of the device.

Entry of data into the cyclomemory is accomplished by placing the address of the desired memory location into the address channel and by placing the data that should be stored in that memory location into the I/O data channel.

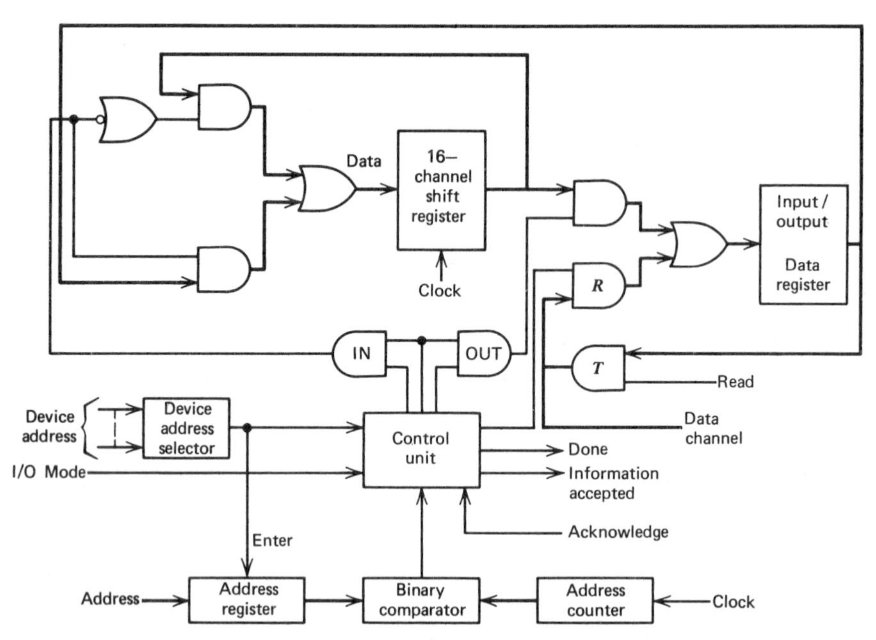

Figure 16.11. Sixteen-channel cyclomemory and its basic controls. (Courtesy of *Digital Design*.)

Accompanying the address and data, a control line commands the device to go into the input mode. This signal enables the bus receivers (R gates) of the cyclomemory and also enters the address from the address bus into the address register. Upon acceptance of data, address, and control commands, the cyclomemory responds with an appropriate signal indicating to the master device that all information has been recorded and that the communication channel need not be occupied any further.

When the address counter reaches the address present in the address register, the binary comparator output becomes 1 and enables the IN gate. The output of this gate allows the data present in the data register to be applied to the input of the N bit shift register. At the arrival of the clock pulse, following the application of the data, the data enter into the stream of the closed-loop shift register. The binary comparator output also sets the DONE flip-flop, the output of which can be sensed under program control, or it may initiate an interrupt to indicate completed data transfer.

Data stored in a cyclomemory can be retrieved by applying the address of the desired memory location into the address channel. Accompanying the address a control line commands the device to go into the output mode. This signal also enables the address to enter into the address register. Upon acceptance of address and control commands, the cyclomemory responds with an appropriate signal indicating to the master device that all necessary information has been recorded or interpreted and that the communication channel need not be occupied anymore.

When the address counter reaches the address in the address register, the binary comparator output becomes 1 and enables the OUT gate. The output of this gate in turn enables the output of the closed-loop shift register to enter into the data register. This signal also sets the "done" flip-flop, indicating to the master device that the word has been retrieved and is ready for application to the data bus. Sensing of the done flip-flop effects a read signal which places the output of the data register onto the data bus by enabling the transmit gates.

Optional Configurations

Block Transfers. The controls of the cyclomemory can be further refined to provide for block transfers. Such transfers can be accomplished by replacing the data register by a shift register which serves as a buffer between the closed-loop shift register and the computer or any other device with which the cyclomemory communicates.

Appropriate changes are needed in the control circuitry for the counting of words and the comparison of addresses. In such a configuration, data must transfer between the closed-loop shift register and the buffer at the shift register's clock rate, while the data transfer between the buffer and external devices are performed at the data rate of the external device. Thus, the cyclomemory can be easily interfaced with devices of either higher or slower data rates than its own.

Multiport Configuration. An alternative to the single long closed-loop shift register encompasses many smaller shift registers interconnected by means of I/O ports, forming a long closed-loop shift register. In this way, more than one external device can have simultaneous access to the data stored in the cyclomemory. Figure 16.12 illustrates an eight-port cyclomemory capable of

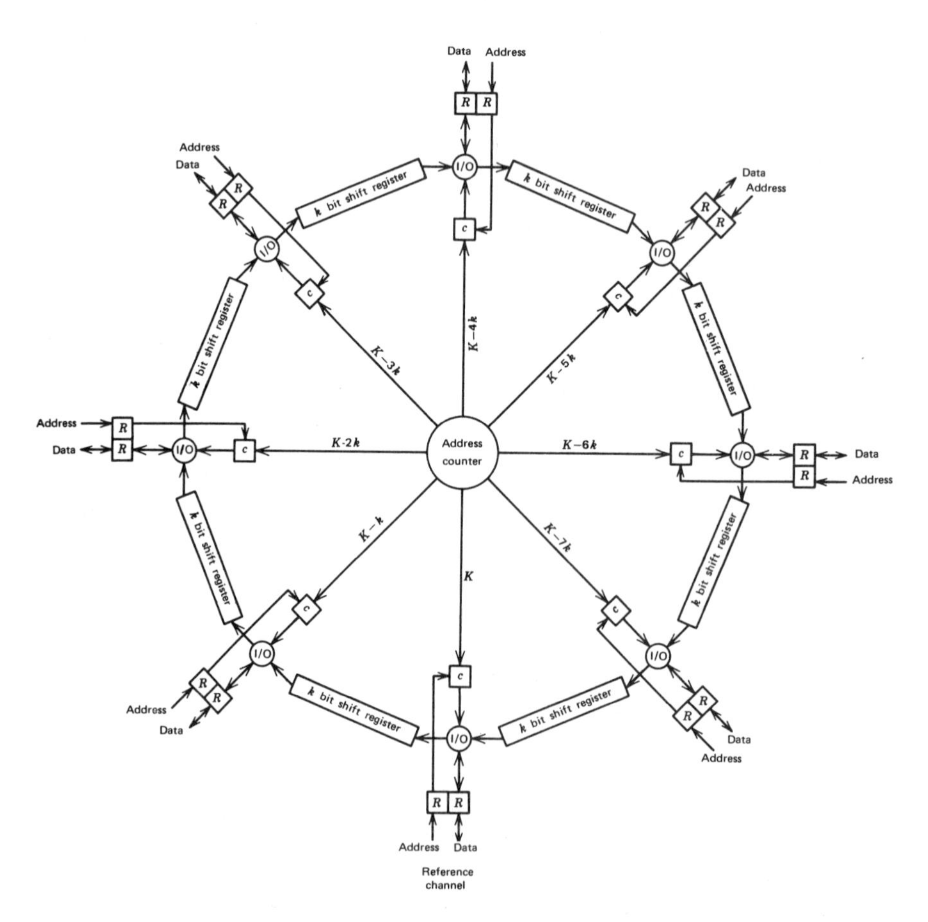

Figure 16.12. Eight-port cyclomemory. (*Note:* C = Binary comparator; R = register; I/O = input/output control circuitry.) (Courtesy of *Digital Design*.)

independently servicing eight external devices. Here all I/O ports are synchronized with a single counter. One of the ports is the reference port with its address input directly compared to the output of the synchronizing counter. The other ports have their address input compared to the difference between the counter output and the number of bits separating these ports from the reference port in the direction of bit flow. Thus, all words stored in the closed-loop shift register are reference to one address system.

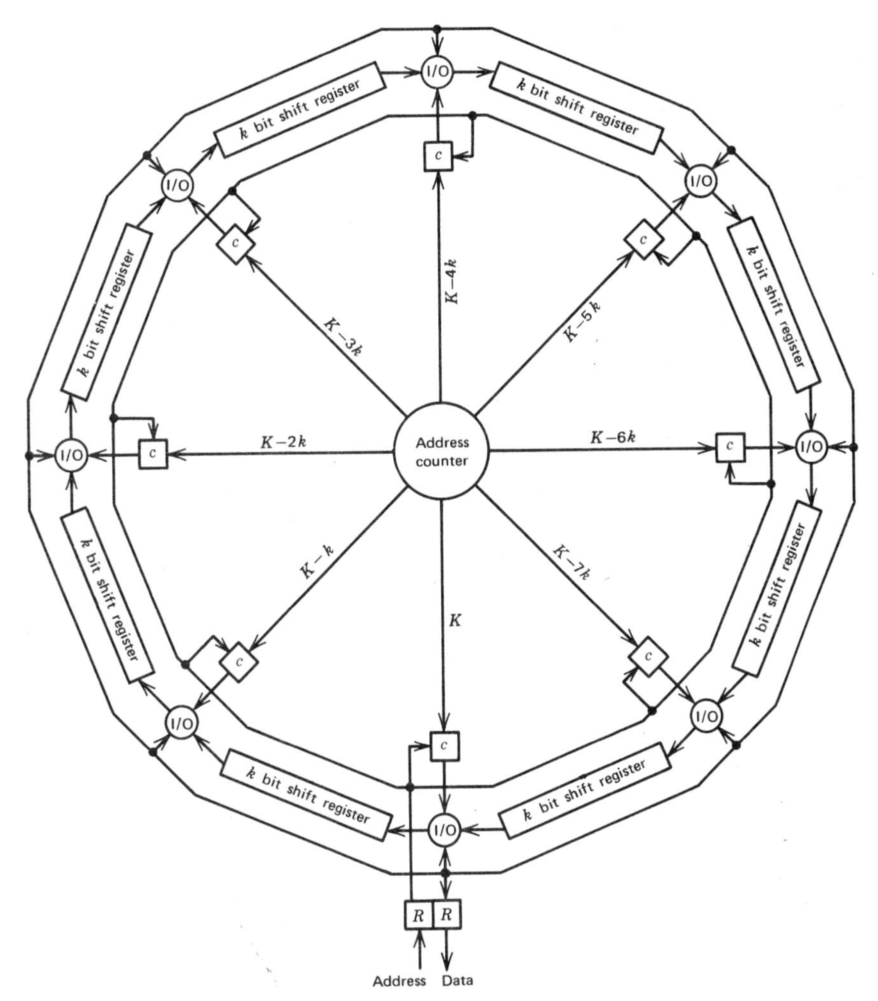

Figure 16.13. Cyclomemory with eight internal I/O ports merging into one external I/O channel. (Courtesy of *Digital Design*.)

High-Speed Transfer. The access speed of a multiport cyclomemory can be considerably increased if all I/O ports merge into one, as it is shown in Figure 16.13. In this mode the input address is compared to all port addresses and the port that reaches the input address first performs the data transfer.

The maximum access time of a cyclomemory is defined as follows:

$$AT_{\max} = \frac{1}{f_{\text{clock}}} \cdot \frac{\text{number of bits in loop}}{\text{number of ports}}$$

For example, a single-port 8K 16 bit cyclomemory of 20 MHz clock frequency has a maximum access time of

$$AT_{\max} = \frac{1}{20\text{ MHz}} \cdot \frac{8192}{8} = 51.2\ \mu\text{sec}$$

with the average being one-half that much or 25.6 μsec.

Applications

The cyclomemory finds wide use as a general-purpose peripheral storage device where in many applications with the aid of simple buffers it can operate at processor speeds. Specifically, the cyclomemory is ideal for CRT updating and refreshing, a function presently being implemented with disks and core memories. In this application the cyclomemory may be of single- or multiple-port configuration. In the latter case, the display can be easily modified by more than one source simultaneously.

The cyclomemory offers outstanding advantages over the fixed-head disk—namely, compactness, higher reliability, total solid-state configuration, less susceptibility to severe environments, flexible packaging, and lower power needs.

16.3 MAGNETIC MEMORIES

Similarly to the semiconductor memories, magnetic memories are also classified into random-accessed and sequentially accessed devices. The major types of magnetic memories are core, plated wire, tape, drum, and disk.

All approaches are based on the same principle, that of magnetizing ferrous material in one direction, or its opposite, with the two directions corresponding to the two values of a digital quantity. The above types of magnetic storage are described below.

Magnetic Core Memory

Magnetic cores are small toroidal ferrites that may be magnetized in either of the two directions along their circumference.

Magnetiziation is accomplished by means of a current that passes through the center opening of the toroid, and the direction of that current determines the direction of magnetization. Figure 16.14 illustrates the core magnetization characteristics and shows the relationship that exists between current direction and magnetization polarity.

Current I_f, the current through the core, magnetizes the core in a direction which is arbitrarily identified with logical state 1. A current in the opposite direction and of $-I_f$ amplitude will change the polarity of magnetization in the core bringing the magnetic flux from $+\phi$ to $-\phi$. The new magnetic state of the core is identified as logical state 0.

If the core is in the 0 state and a $-I_f$ current is applied to the wire that passes through it, the core polarity will remain unchanged. Application of a positive current greater than I_k will have an effect on the core polarity

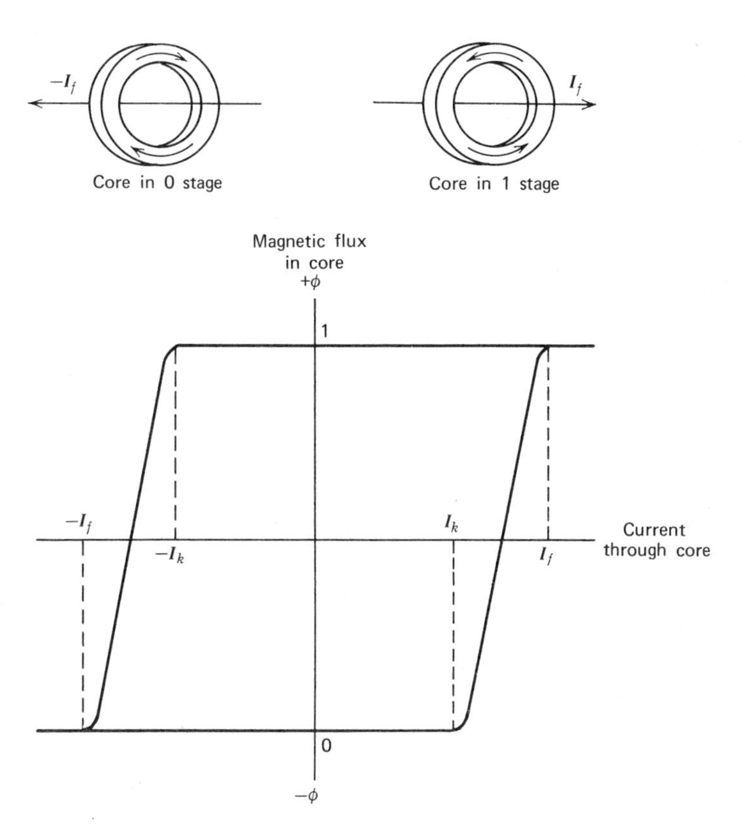

Figure 16.14. Magnetization, magnetization current, and logical states in a core where $I_k = 0.6I_f$.

tending to change it from $-\phi$ to $+\phi$. For a reliable change from 0 to 1 the positive current must be I_f or greater.

Currents I_f or $-I_f$ need not remain applied indefinitely but only for a short period of time which is a function of current amplitude and core characteristics. Core size is a parameter that directly affects current requirements and every effort is therefore made to reduce it. Standard cores are now down to 20 mils in diameter, while 20 years ago they ranged from 70 to 90 mils.

Core memories are arranged in X–Y planes, where each core is identified by stack X and Y coordinates. To minimize the number of wires that service a core memory, there are two wires that define the location of a core. If a core is set to 1 state, changing it to 0 is accomplished by simultaneously applying $-0.5\ I_f$ on each of the two X and Y wires.

To read the logical state of a core, current is passed through the center opening of the core through the X and Y wires and any change in magnetic flux is sensed. If the current enhances the already existing flux, no significant change will be sensed; but if it generates opposing flux, the polarity of the core will change and the change will be sensed. Figure 16.15 illustrates a magnetic core and the necessary sensing and driving circuits.

To read information present in a core, the appropriate X and Y drivers

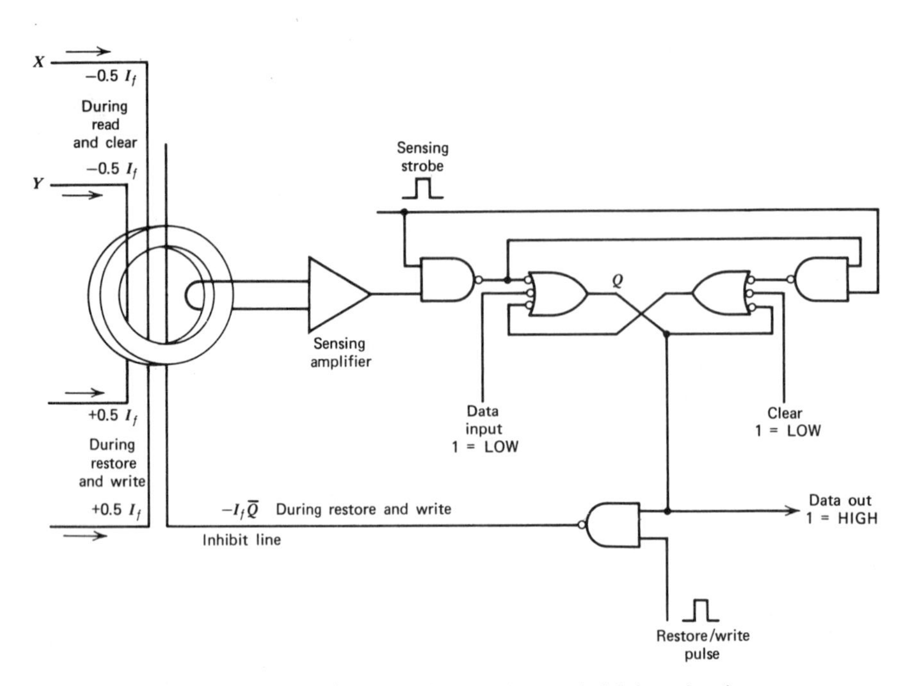

Figure 16.15. Magnetic core with sensing and driving circuits.

apply $-0.5\ I_f$, each forcing the core logical state to 0. If the core is already at 0, there will be no change in magnetic flux and the output of the sense amplifier during the sensing strobe will be LOW. If the core is at 1, forcing it to 0 will generate a significant flux difference which will be sensed by the sense amplifier.

During strobe the sense amplifier output will enter the storage element from where it will be available for further use. After this operation the core will be at 0, and the logical state in the storage element will be that of the core before the core was cleared. The readout was destructive and it is now necessary to restore the core to its original magnetic (logical) state.

This is accomplished by applying to the core a total current of I_f through the X and Y lines ($0.5\ I_f$ each) and another $-I_f$ if the readout content was 0. This additional $-I_f$ comes from an AND gate that ANDs the output of the storage element to a restore pulse. Thus, if the core was originally 1, no $-I_f$ is applied—that is, the reading process consists of two time segments, the read and the restore.

In the write mode, information to be entered in core is first stored in the storage element. To transfer it in core, the appropriate X and Y drivers apply $-0.5\ I_f$ each, forcing the core to logical 0. Following that, the restore/write pulse is enabled sending a current of $-I_f$ to the core should the bit to be written were 0. At the same time each of the X and Y wires provides a current of $+0.5\ I_f$ each.

Thus, if the bit to be stored in core is 0, no current will be applied to the core, resulting in no change of state. But, if the bit is a 1, I_f from the X–Y wires will be applied to the core switching its state from 0 to 1—that is, the writing process also consists of two time segments, the clear and the write. When cores are arranged to form X–Y planes, they all normally have a common sense and a common restore/write wire. In some designs a single wire serves both functions.

Plated-Wire Memory

The disadvantage of destructive readout, characteristic of core memories, is overcome by the plated-wire concept where information can be read from a magnetic simulated core without being altered during the reading process.

Plated-wire memory consists of copper wires plated with a ferromagnetic material. During the electroplating process the crystals of the plated material are strongly magnetized establishing an easy axis of magnetization aligned along the circumference of the wire. Thus, current flowing through the wire magnetizes the plated material either in the clockwise or the counterclockwise direction, as is illustrated in Figure 16.16.

The two directions of magnetization are arbitrarily associated with the

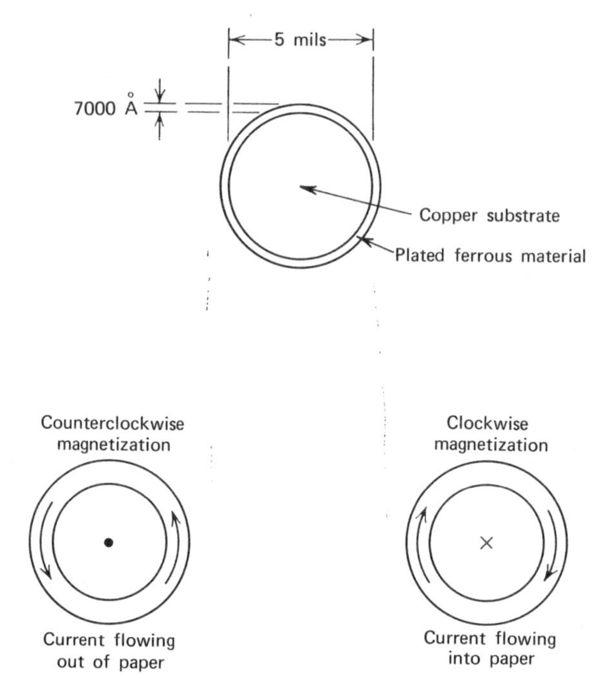

Figure 16.16. Cross-section of plated wire.

logical states 0 and 1, and the established direction of magnetization acts as the stored counterpart of a logical quantity.

Current through a conductor placed perpendicular to the longitudinal axis of the plated wire may affect the direction of magnetization and tend to shift that direction by 90°, as is illustrated in Figure 16.17. If the current is below a certain threshold, the magnetization change will be temporary and, upon removal of the current, the initial magnetization will be reestablished. However, this magnetization change which will occur during application and removal of the current will be sensed by an appropriate sense amplifier that will determine the stable magnetic state of the plated portion under the word-wire. Figure 16.18 shows the effect of the word-wire current onto the memory cell direction of magnetization.

Writing in a plated-wire cell is a joint effort of the word-wire and the plated wire itself. In this mode the word-wire supplies just enough current to set a threshold, so that application of current through the plated wire will set the magnetization in one of the stable states.

Neither the word-wire current nor the current through the plated wire is strong enough to permanently alter the cell magnetization alone. This way,

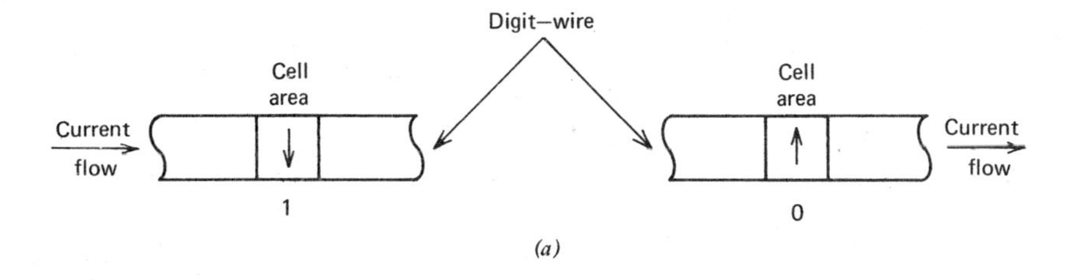

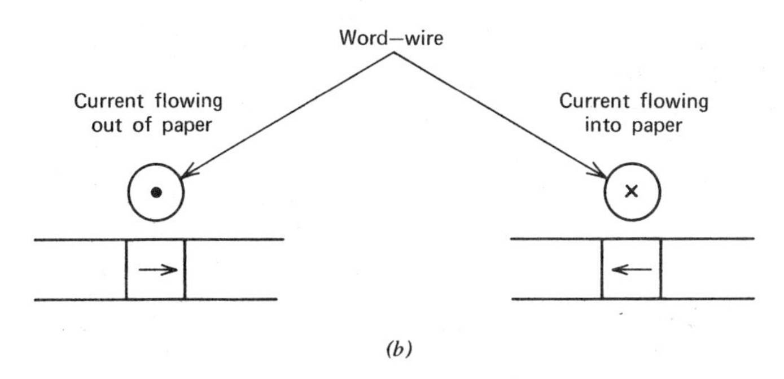

Figure 16.17. Side view of plated wire: (*a*) magnetization along the easy axis; (*b*) magnetization along the hard axis.

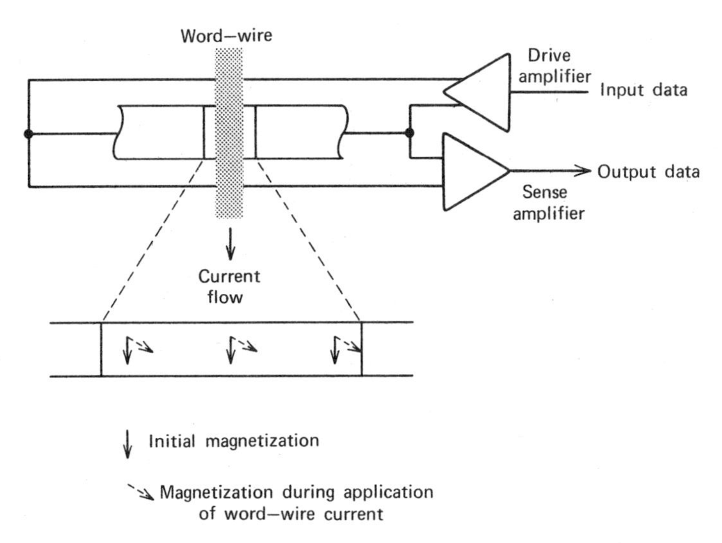

Figure 16.18. Plated-wire memory cell and its sense and drive circuitry.

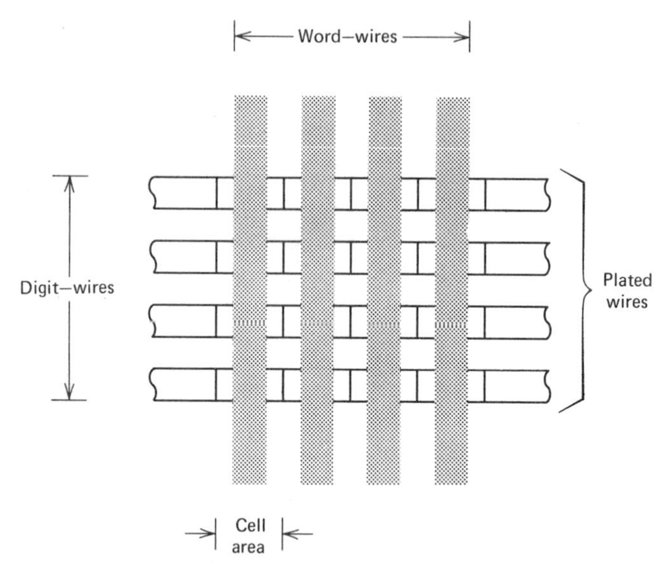

Figure 16.19. Simplified diagram of a 4 × 4 bit plated-wire arrangement.

only the coincidence of these two currents may enter new information into a memory cell. Figure 16.19 illustrates a simplified diagram of a 4 × 4 bit plated-wire memory arrangement.

Sequentially Accessed Magnetic Memories

The core and plated wire memories previously discussed have access times that are independent of the reference memory location. Magnetic principles, however, are similarly applied to memory systems that trade speed for higher density and less cost per stored bit.

These are the magnetic disks, tapes, and drums, and they all consist of a moving surface on which a ferromagnetic layer stores digital information. Reading and writing of stored data is accomplished by means of one or more magnetic heads, the same way music is recorded or played back on a commercial tape recorder. Magnetic disks, tapes, and drums are mainly used for bulk storage of digital data.

In addition to the semiconductor and magnetic memories are the mechanical memories, typical examples of which are the punched paper cards and paper tape.

16.4 COMPARISON OF MEMORIES

The ever-increasing need for bigger and faster memories will keep memory technology continually advancing, providing the digital designer with a wide

Type of memory	Advantages	Disadvantages
Semiconductor		
RAM	Fast access time (80–500 μsec)	High cost per bit ratio
ROM	Replaces logic and results in considerable hardware savings	Not easily altered
CAM	When properly used, surpasses all other approaches to direct content addressing	Very high cost per bit ratio
Cyclomemory	Replaces magnetic disk, offers higher speed	Presently more expensive than disks
Magnetic		
Core	Fast access time (300–900 μsec)	Destructive readout
Plated wire	Faster access time nondestructive readout low power	Higher cost
Disk	High density	High power
Drum	High density	High power
Tape	Very low cost per bits ratio	Slow speed
Mechanical		
Punched paper cards	Simplicity in altering data	Slower access
Punched paper tape	Ease in handling and storing	Requires more volume than other means of data storage

Figure 16.20. Comparison of commonly used digital memories.

variety of memories from which to choose. Figure 16.20 compares the advantages and disadvantages of the commonly used memory systems.

PROBLEMS

1. State the advantages of dynamic RAMs.
2. Using ROMs, design a θ to sin θ converter having the following characteristics:

$$\theta = 0\text{–}45^\circ$$
$$\Delta\theta = 0.1^\circ$$
$$\sin\theta = 10 \text{ bits}$$

Identify your design with currently available hardware. Show the ROM content for the first and last ten values.

3. Using currently available semiconductor hardware, design a memory system that will meet the following requirements:

   Word size = 16 bits
   Memory size = 512 words

   The four MSBs are content-addressable.

4. Determine the average access time of a 2048 bit cyclomemory operating at a rate of 20 MHz and having two internal I/O ports symmetrically spaced. The system has one external I/O port.

5. Design a magnetic core sensing and driving circuit that inhibits and senses using one wire.

6. Design a RAM system where information enters and exists at two independently variable rates, and a first-in first-out mode of operation is maintained. Consider a 256 word memory where each word has 8 bits.

APPENDIX A

General Algorithm for the Development of Detailed Algorithms for the Extraction of Integer Roots

The nth root of a number, regardless of the numerical system through which the number is expressed, can be determined by calculating 1 digit of the root at a time.

Let R be the nth root of number K where n is an integer; R can be written as

$$R = r_{im}r_{im-1}\ldots r_{i1}r_{f1}r_{f2}\ldots r_{fj} \quad \text{(A.1)}$$

where the integral part of R has m digits and the fractional has j. Equation (A.1) can also be written as follows:

$$R = (r_{im}\cdot r_{im-1}\ldots r_{i1}r_{f1}r_{f2}\ldots r_{fj})Q^{m-1} \quad \text{(A.2)}$$

where Q is the radix of the numerical system and r_{im} is different than 0. When R is raised to exponent n, (A.2) takes the following form:

$$R^n = (r_{im}\cdot r_{im-1}\ldots r_{i1}r_{f1}r_{f2}\ldots r_{fj})^n Q^{(m-1)n} \quad \text{(A.3)}$$

Let letter k represent the number of digits in the integral part of R^n; k is minimum when the expression in the parentheses is minimum. This is when r_{im} equals 1 and the remaining r digits are 0, as shown below:

$$\begin{aligned}(r_{im}\cdot r_{im-1}\ldots r_{i1}r_{f1}r_{f2}\ldots r_{fj})^n_{\min} &= (1.0\ldots 0)^n \\ &= 1\end{aligned}$$

(1 digit minimum) (A.4)

The minimum value of k is therefore the exponent of the radix multiplier Q plus 1. Hence,

$$(m - 1)\cdot n + 1 \leq k \tag{A.5}$$

Similarly, k is maximum when the expression in the parentheses of (A.3) is maximum. This is when all r digits equal $Q - 1$, which is the maximum digit value in numerical system Q.

$$\begin{aligned}(r_{im}\cdot r_{im-1}\ldots r_{i1}r_{f1}r_{f2}\ldots r_{fj})^n_{\max} &= (Q - 1\cdot Q - 1Q - 1\cdots Q - 1)^n \\ &= (10 - 0.00\ldots 1)^n \text{ (in the decimal system)}\end{aligned} \tag{A.6}$$

If the expression in the parentheses were only 10, the exponentiation would produce $n + 1$ digits.

$$\begin{aligned}10^n &= (1.0 \times Q)^n \\ &= (1.0)^n Q^n \\ &= Q^n \qquad (n + 1 \text{ integral digits})\end{aligned}$$

Since the expression in the parentheses is the closest lower approximation to 10^n and 10^n is the smallest number having n integral digits, n will be the number of integral digits in (A.6).

The maximum value of k is therefore the exponent of the radix multiplier Q, which $(m - 1)n$, (A.3), plus n. Hence,

$$k \leq (m - 1)n + n$$

or (A.7)

$$k \leq mn$$

Combining (A.5) and (A.7), we have

$$(m - 1)n + 1 \leq k \leq mn \tag{A.8}$$

It can be observed that when m increases by 1, the range of k increases by n. Also, for every unit of m there are n units in k. This indicates that for every integral digit of R, there correspond n digits in K, where $K = R^n$. Or, for every n digits of number K, there corresponds 1 digit in its nth root, $\sqrt[n]{K} = R$.

Let K be a number the nth root of which needs to be computed and let the desired root be

$$R = r_m r_{m-1}\ldots r_2 r_1 \tag{A.9}$$

For simplicity in the presentation of the derivation of the general root-finding algorithm, only the integral part of the root will be considered.

The weight of the r_m digit is Q^{m-1} and as a complete number it can be expressed as $r_m Q^{m-1}$ where Q^{m-1} indicates that there are $m - 1$ zeros between digit r_m and the fractional point. Root digit r_m must have a positive value that minimizes the difference

$$K - (r_m Q^{m-1})^n \tag{A.10}$$

Root digit r_{m-1} must have a positive value that minimizes the difference

$$K - (r_m r_{m-1} Q^{m-2})^n \tag{A.11}$$

Similarly, root digit r_1 must have a positive value that minimizes the difference

$$K - (r_m r_{m-1} \ldots r_1)^n \tag{A.12}$$

Equation (A.11) can be written

$$K - (r_m Q^{m-1})^n + (r_m Q^{m-1})^n - (r_m r_{m-1} Q^{m-2})^n$$

or as

$$[K - (r_m Q^{m-1})^n] - [(r_m r_{m-1} Q^{m-2})^n - (r_m Q^{m-1})^n]$$

or

$$[K - (r_m Q^{m-1})^n] - [(r_m Q^{m-1} + r_{m-1} Q^{m-2})^n - (r_m Q^{m-1})^n] \tag{A.13}$$

This is

[difference of preceding operation]
– [partial root + new digit)^n – (partial root)^n]

where each new digit must minimize the difference further.

If the minuend in the above subtraction is represented by M and the subtrahend by S, the expression for S will be

$$S_m(\text{for } r_m) = (r_m Q^{m-1})^n \qquad (M_m = K)$$
$$S_{m-1}(\text{for } r_{m-1}) = (r_m r_{m-1} Q^{m-2})^n - (r_m Q^{m-1})^n \qquad (M_{m-1} = M_m - S_m)$$
$$S_1(\text{for } r_1) = (r_m r_{m-1} \ldots r_1)^n - [(r_m r_{m-1} \ldots r_2) Q]^n \qquad (M_1 = M_2 - S_2)$$

The general expressions for S and M are

$$\begin{aligned} S_{m-(t+1)} &= [(r_m r_{m-1} \ldots r_{m-t} r_{m-(t+1)} Q^{m-(t+2)}]^n - [(r_m r_{m-1} \ldots r_{m-t}) Q^{m-(t+1)}]^n \\ &= [(r_m r_{m-1} \ldots r_{m-t}) Q^{m-(t+1)} + (r_{m-(t+1)} Q^{m-(t+2)})]^n \\ &\quad - [(r_m r_{m-1} \ldots r_{m-t}) Q^{m-(t+1)}]^n \end{aligned}$$
$$(M_{m-(t+1)} = M_{m-t} - S_{m-t})$$

The flow chart for the general algorithm is illustrated in Figure A.1.

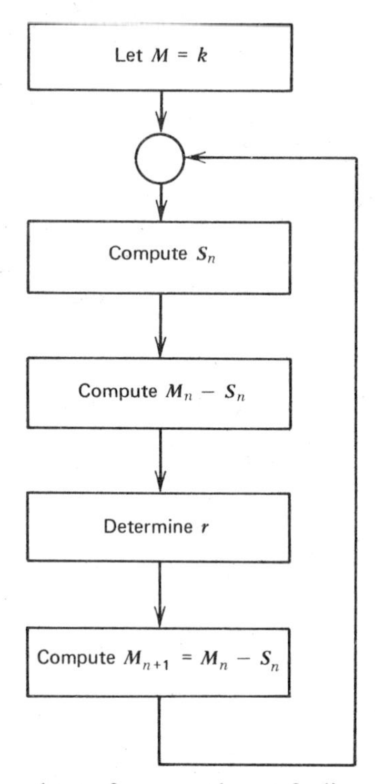

Figure A.1. Flow chart of a general root-finding algorithm. (*Note:* r is a nonnegative integer that minimizes the difference $M - S$.)

Normalizing the significance of the root digit currently being computed, expression S becomes

$$S_{m-(t+1)} = [(r_m r_{m-1} \ldots r_{m-t})Q + r_{m-(t+1)}]^n - [(r_m r_{m-1} \ldots r_{m-t})Q]^n$$

Letting $r_m r_{m-1} \ldots r_{m-t} = R$, standing for partial root, and plain r for the new digit, we have

$$S = [QR + r]^n - [QR]^n \tag{A.14}$$

Since in each iteration only n additional digits participate in the computation, the algorithm can be expressed as illustrated in Figure A.2.

To compute the expression of S for the square root and cube root algorithms n is substituted by 2 and 3, respectively, as shown below:

$$\begin{aligned} S \text{ (for square root)} &= (QR + r)^2 - (QR)^2 \\ &= (QR)^2 + 2QRr + r^2 - (QR)^2 \\ &= 2QRr + r^2 \\ &= (2QR + r)r \end{aligned} \tag{A.15}$$

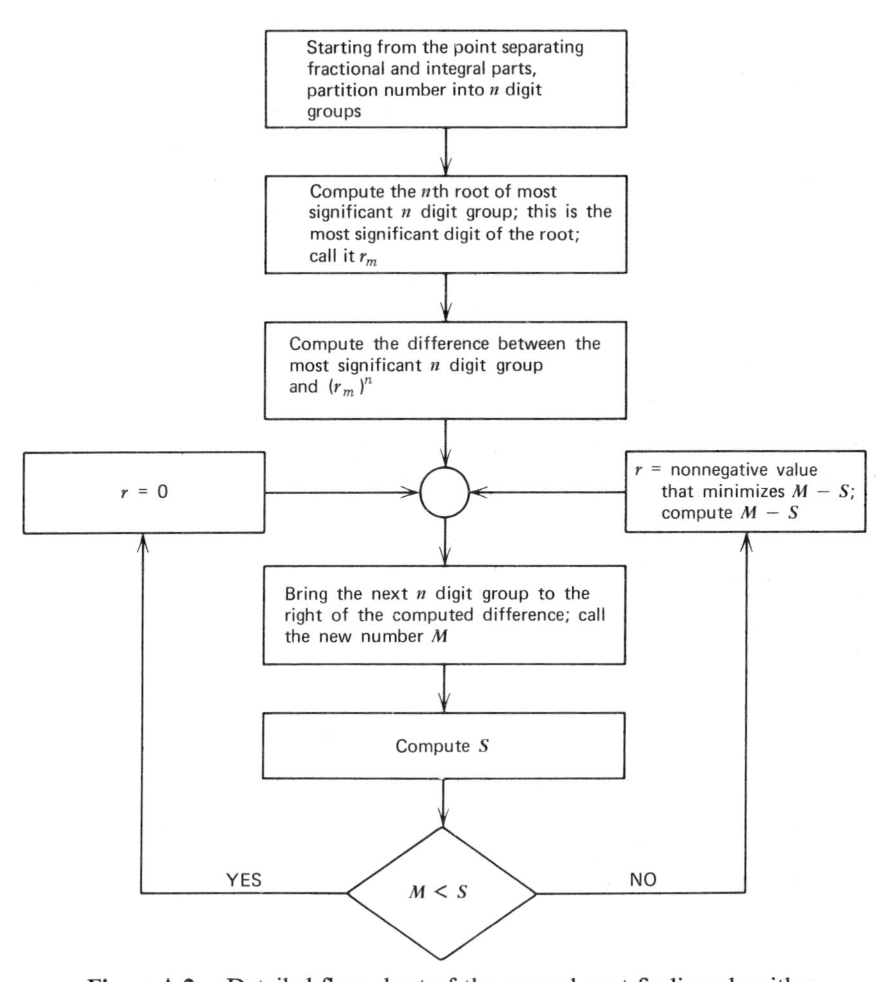

Figure A.2. Detailed flow chart of the general root-finding algorithm.

$$\begin{aligned} S \text{ (for cube root)} &= (QR + r)^3 - (QR)^3 \\ &= (QR)^3 + 3(QR)^2r + 3QRr^2 + r^3 - (QR)^3 \\ &= 3(QR)^2r + 3QRr^2 + r^3 \\ &= 3QRr(QR + r) + r^3 \\ &= [3QR(QR + r) + r^2]r \end{aligned} \tag{A.16}$$

To obtain the algorithm for the computation of roots in the decimal system term Q in expression S is substituted by 10. For the square root algorithm the expression for S becomes

$$S = (20R + r)r \tag{A.17}$$

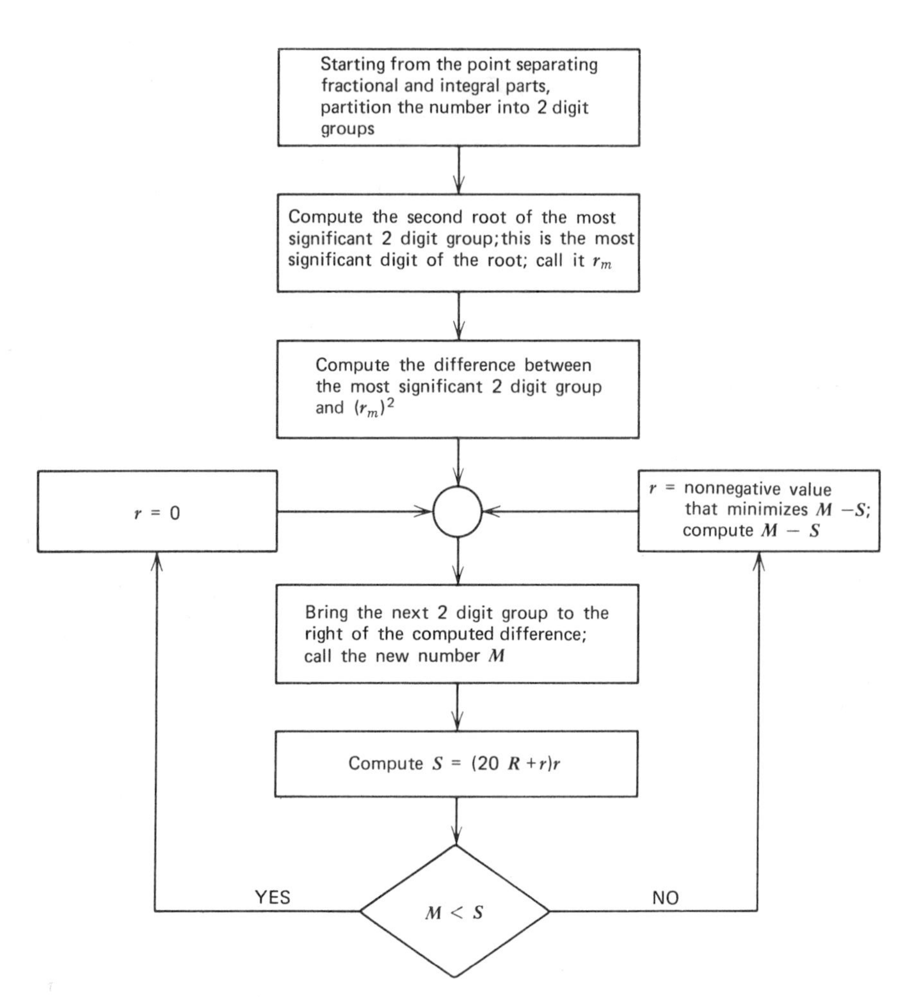

Figure A.3. Flow chart of a decimal square-root-finding algorithm.

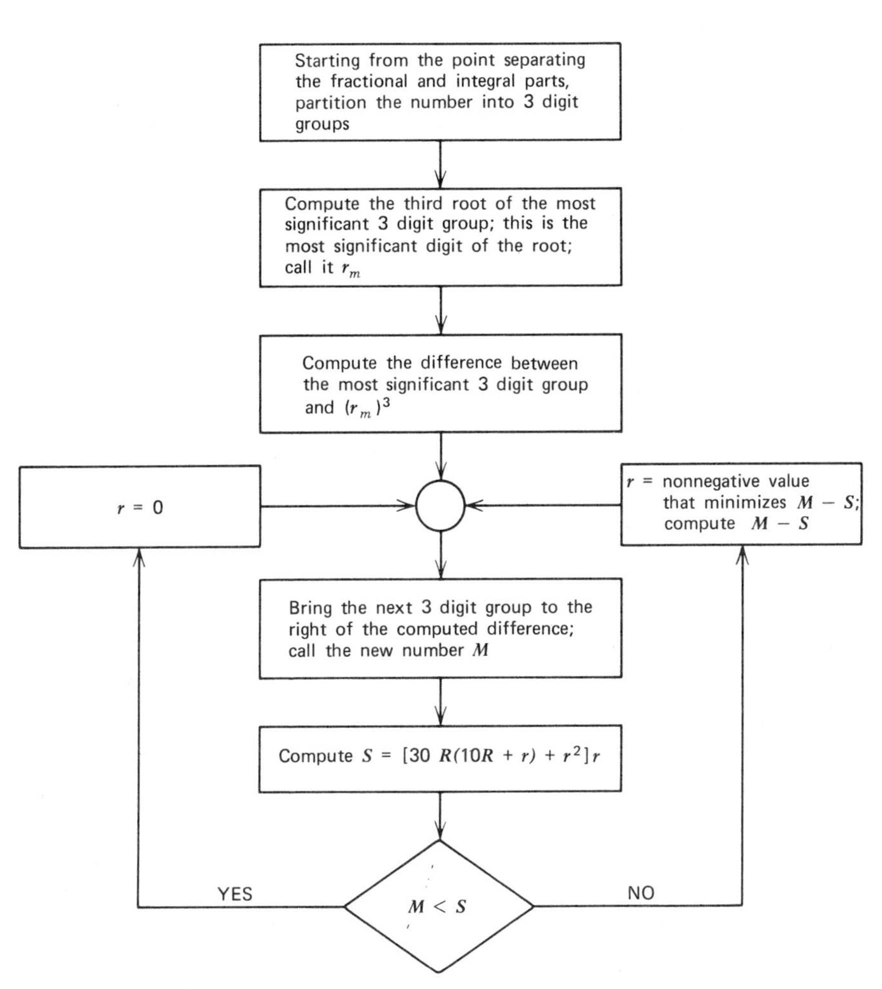

Figure A.4. Flow chart of a decimal cube-root-finding algorithm.

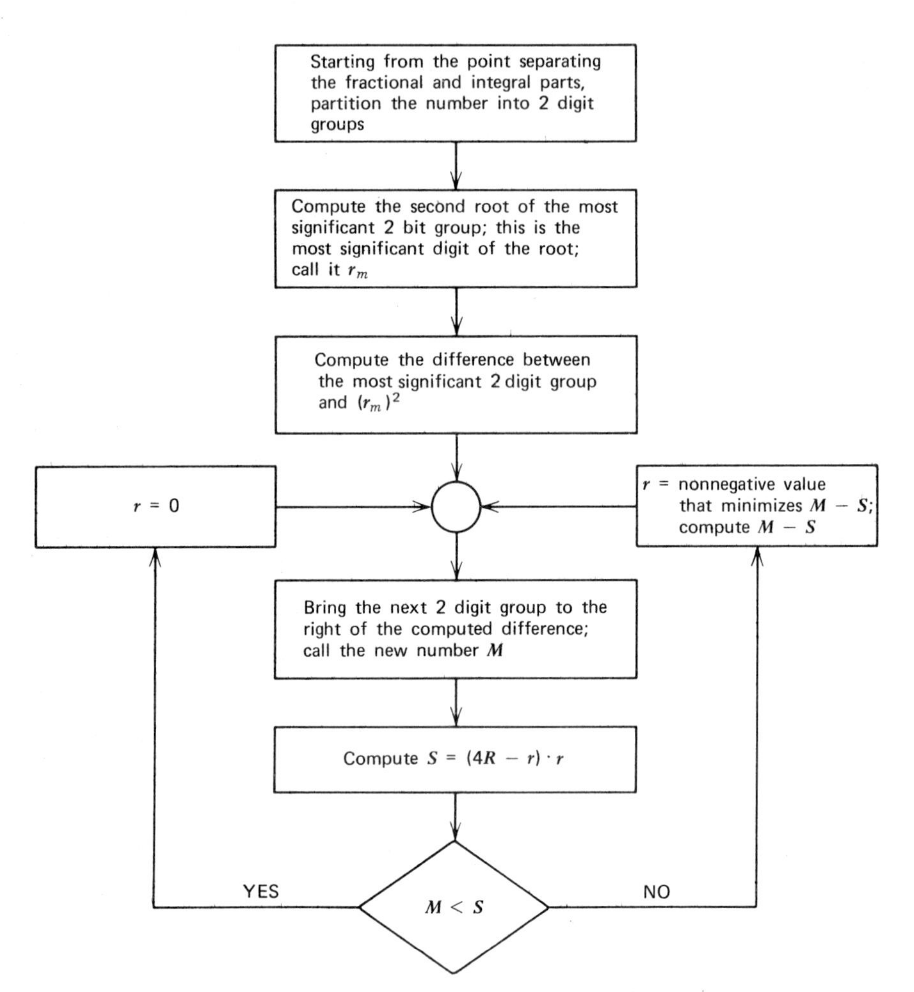

Figure A.5. Flow chart of a binary square-root-finding algorithm.

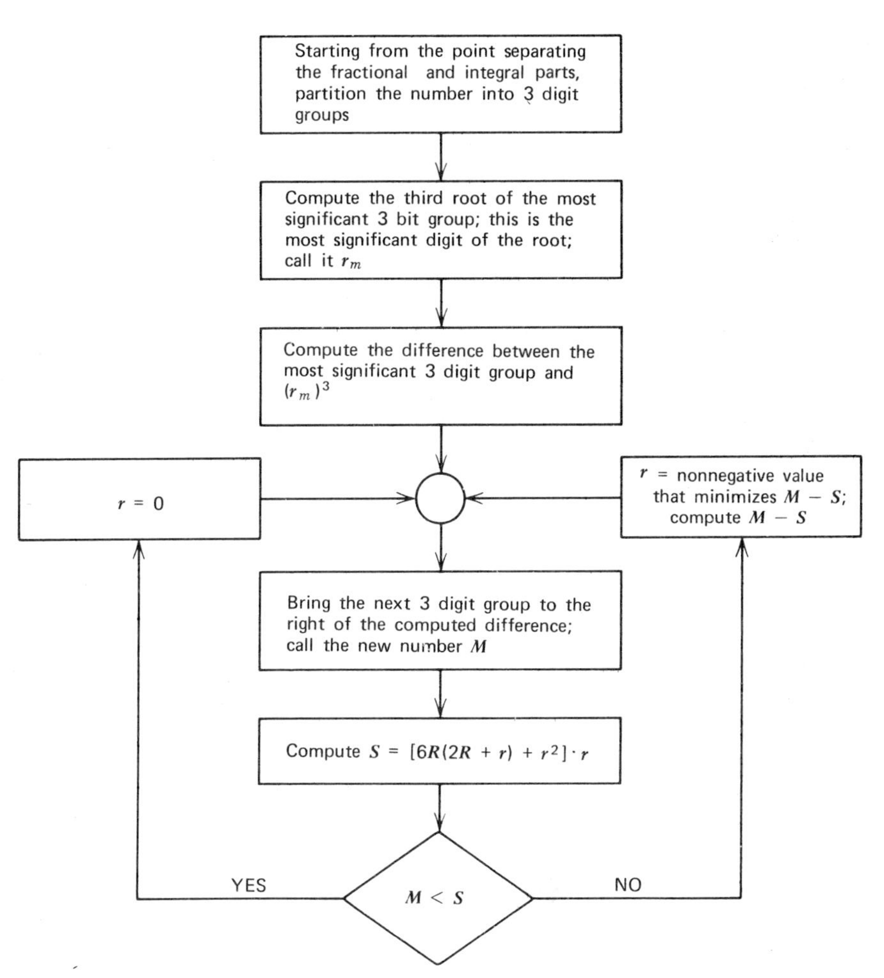

Figure A.6. Flow chart of a binary cube-root-finding algorithm.

Figure A.3 illustrates the flow chart of the decimal square root algorithm. For the cube root algorithm the expression for S becomes

$$S = [30R(10R + r) + r^2]r$$

Figure A.4 illustrates the flow chart of the decimal cube root algorithm.

Similarly, in the binary system for the computation of roots, term Q is substituted by 2. For the square root algorithm the expression for S becomes

$$S = (4R + r)r$$

Figure A.5 illustrates the flow chart of the binary square root algorithm. For the cube root algorithm the expression for S becomes

$$S = [6R(2R + r) + r^2]r$$

Figure A.6 illustrates the flow chart of the binary cube root algorithm.

APPENDIX B

Powers of 2

2^n	n	2^{-n}
1	0	1.0
2	1	0.5
4	2	0.25
8	3	0.125
16	4	0.062 5
32	5	0.031 25
64	6	0.015 625
129	7	0.007 812 5
256	8	0.003 906 25
512	9	0.001 953 125
1 024	10	0.000 976 562 5
2 048	11	0.000 488 281 25
4 096	12	0.000 244 140 625
8 192	13	0.000 122 070 312 5
16 384	14	0.000 061 035 156 25
32 768	15	0.000 030 517 578 125
65 536	16	0.000 015 258 789 062 5
131 072	17	0.000 007 629 394 531 25
262 144	18	0.000 003 814 697 265 625
524 288	19	0.000 001 907 348 632 812 5
1 048 576	20	0.000 000 953 674 316 406 25
2 097 152	21	0.000 000 476 837 153 203 125
4 194 304	22	0.000 000 238 418 579 101 562 5
8 388 608	23	0.000 000 119 209 289 550 781 25
16 777 216	24	0.000 000 059 604 644 775 390 625
33 554 432	25	0.000 000 029 802 322 387 695 312 5
67 108 864	26	0.000 000 014 901 161 193 817 656 25
134 217 728	27	0.000 000 007 450 580 596 923 828 125
268 435 456	28	0.000 000 003 725 290 298 461 914 062 5
536 870 912	29	0.000 000 001 862 645 149 230 957 031 25
1 073 741 824	30	0.000 000 000 931 322 574 615 478 515 625
2 147 483 648	31	0.000 000 000 465 661 287 307 739 257 812 5
4 294 967 296	32	0.000 000 000 232 830 613 653 869 628 906 25
8 589 934 592	33	0.000 000 000 116 415 321 826 934 814 453 125
17 179 869 184	34	0.000 000 000 058 207 660 913 467 407 226 562 5
34 359 738 368	35	0.000 000 000 029 103 830 456 733 703 613 281 25
68 719 476 736	36	0.000 000 000 014 551 915 228 366 851 806 640 625
137 438 953 472	27	0.000 000 000 007 275 957 614 183 425 903 320 312 5
274 877 906 944	38	0.000 000 000 003 637 978 807 091 712 951 660 156 25
549 755 813 888	39	0.000 000 000 001 818 898 403 545 856 475 830 078 125
1 099 511 627 776	40	0.000 000 000 000 909 494 701 772 928 237 915 039 062 5
2 199 023 255 552	41	0.000 000 000 000 454 747 350 886 464 118 957 519 531 25
4 398 046 511 104	42	0.000 000 000 000 227 373 675 443 232 059 478 759 765 625
8 796 093 022 208	43	0.000 000 000 000 113 686 837 721 616 029 739 379 882 812 5
17 592 186 044 416	44	0.000 000 000 000 056 843 418 860 808 014 869 698 941 406 25
35 184 372 038 832	45	0.000 000 000 000 028 421 709 431 404 007 434 844 970 703 125
70 368 744 177 664	46	0.000 000 000 000 014 210 854 715 202 003 717 422 485 351 562 5
140 737 488 355 328	47	0.000 000 000 000 007 105 427 357 601 001 858 711 242 675 781 25
281 474 976 710 656	48	0.000 000 000 000 003 552 713 678 800 500 929 355 621 337 890 625
562 949 953 421 312	49	0.000 000 000 000 001 776 356 839 400 250 464 677 810 668 945 312 5
1 125 899 906 843 624	50	0.000 000 000 000 000 888 178 419 700 125 232 338 905 334 472 656 25
2 251 799 813 685 248	51	0.000 000 000 000 000 444 089 209 850 062 616 169 452 667 236 328 125
4 503 599 627 370 496	52	0.000 000 000 000 000 222 044 804 925 031 308 084 726 333 618 164 062 5
9 007 199 254 740 992	53	0.000 000 000 000 000 111 022 302 462 515 654 042 363 166 809 082 031 25
18 014 398 509 481 984	54	0.000 000 000 000 000 055 511 151 231 257 827 021 181 583 404 541 015 625
36 028 797 018 963 968	55	0.000 000 000 000 000 027 755 575 615 628 913 510 590 791 702 270 507 812 5
72 057 594 037 927 936	56	0.000 000 000 000 000 013 877 787 807 814 456 755 295 395 851 135 253 906 25
144 115 188 075 855 872	57	0.000 000 000 000 000 006 938 893 903 907 228 377 647 697 925 567 626 953 125
288 230 376 151 711 744	58	0.000 000 000 000 000 003 469 446 951 953 614 188 823 848 962 783 813 476 562 5
576 460 752 303 423 488	59	0.000 000 000 000 000 001 734 723 475 976 807 094 411 924 481 391 906 738 281 25
1 152 921 504 606 846 976	60	0.000 000 000 000 000 000 867 361 737 988 403 547 205 962 240 695 953 369 140 625
2 305 843 009 213 693 952	61	0.000 000 000 000 000 000 433 680 868 994 201 773 602 981 120 347 976 684 570 312 5
4 611 686 018 427 387 904	62	0.000 000 000 000 000 000 216 840 434 497 100 886 801 490 560 173 988 342 285 156 25
9 223 372 036 854 775 808	63	0.000 000 000 000 000 000 108 420 217 248 550 443 400 745 280 086 994 171 142 578 125
18 446 744 073 709 551 616	64	0.000 000 000 000 000 000 054 210 108 624 275 221 700 372 640 043 497 085 571 289 062 5
36 893 488 147 419 103 232	65	0.000 000 000 000 000 000 027 105 054 312 137 610 850 186 320 021 748 542 785 644 531 25
73 786 976 294 838 206 464	66	0.000 000 000 000 000 000 013 552 527 156 068 805 425 093 160 010 874 271 392 822 265 625
147 573 952 589 676 412 928	67	0.000 000 000 000 000 000 006 776 263 578 034 402 712 546 580 005 437 135 696 411 132 812 5
295 147 905 179 352 825 856	68	0.000 000 000 000 000 000 003 388 131 789 017 201 356 273 290 002 718 567 848 205 566 406 25
590 295 810 358 705 651 712	69	0.000 000 000 000 000 000 001 694 065 894 508 600 678 136 645 001 359 283 924 102 783 203 125
1 180 591 620 717 411 303 422	70	0.000 000 000 000 000 000 000 847 032 947 254 300 339 068 322 500 679 641 962 051 391 601 562 5
2 361 183 241 434 822 606 848	71	0.000 000 000 000 000 000 000 423 516 473 627 150 169 534 161 250 339 820 981 025 695 800 781 25
4 722 366 482 869 645 213 696	72	0.000 000 000 000 000 000 000 211 758 236 813 575 084 767 080 625 169 910 490 512 847 900 390 625

APPENDIX C

Bibliography

CALDWELL, S. H. *Switching Circuits and Logical Design*. New York: John Wiley, 1959.

CHU, Y. *Digital Computer Design Fundamentals*. New York: McGraw-Hill, 1962.

CURTIS, A. H. *The Design of Switching Circuits*. New York: Van Nostrand, 1962.

HARRISON, M. A. *Introduction to Switching and Automata Theory*. New York: McGraw-Hill, 1965.

HOHN, F. E. *Applied Boolean Algebra*. New York: Macmillan, 1966.

KEISTER, W., A. E. RITCHIE, and S. H. WASHBURN. *The Design of Switching Circuits*. New York: Van Nostrand, 1951.

KRIEGER, M. *Basic Switching Circuit Theory*. New York: Macmillan, 1967.

LEWIS, P. N., and C. L. COATES. *Threshold Logic*. New York: Wiley, 1967.

MALEY, G. A., and J. EARLE. *The Logic Design of Transistor Digital Computers*. Englewood Cliffs, N.J.: Prentice-Hall, 1963.

MARCUS, M. P. *Switching Circuits for Engineers*. Englewood Cliffs, N.J.: Prentice-Hall, 1962.

MCCLUSKEY, E. J. *Introduction to Theory of Switching Circuits*. New York: McGraw-Hill, 1965.

MCCLUSKEY, E. J., and T. C BARTREE. *A Survey of Switching Circuit Theory*. New York: McGraw-Hill, 1962.

MILLER, R. E. *Switching Theory*, Vol. I: *Combinational Circuits*. New York: Wiley, 1965.

PHISTER, M. *Logical Design of Digital Computers*. New York: Wiley, 1958.

PRATHER, R. E. *Introduction to Switching Theory: A Mathematical Approach*. Boston: Allyn and Bacon, 1967.

SCOTT, N. R. *Analog and Digital Computer Technology*. New York: McGraw-Hill, 1960.

TORNG, H. C. *Introduction to the Logical Design of Switching Systems*. Reading, Mass.: Addison-Wesley, 1965.

WICKES, W. E. *Logic Design with Integrated Circuits*. New York: Wiley, 1968.

INDEX